Get *** CONNECTED to Packet Radio

by

Jim Grubbs, K9EI

PUBLISHING

P.O. Box 3042
Springfield, IL
62708

Get *** CONNECTED to Packet Radio

First Edition

Published by: QSKY Publishing
PO Box 3042
Springfield, Illinois
62708

Copyright 1986 by Jim Grubbs

ISBN 0-931387-22-1
Library of Congress 85-63490

Reach For The Stars

Many of us who were youngsters in the late fifties and early sixties grew up watching the space program develop. I have a Cub Scout scrapbook where I proudly displayed a clipping from a Honolulu newspaper explaining the first Sputnik. I remember vividly the sound of General Eisenhower's voice being relayed back to Earth by an early satellite.

I stood with my family searching the sky for the Echo satellite, a forerunner to today's communication centers in the heavens.

But most of all I remember sitting in my grade school classroom watching a black and white television as the United States launched its first man into space.

The goose-bumps never stopped. Every launch is still viewed with exhilaration.

Some years ago I made my first contact using the OSCAR satellite series. Just a short time ago I thrilled at hearing Owen Garriott, W5LFL, while sitting in my own radio shack listening to him transmit directly from the Space Shuttle.

As I wrote this book, amateur radio operators were preparing for another experiment from space, this one involving packet radio.

On Tuesday, January 28, 1986, my heart stopped for a moment. There were no words to describe what had happened. The exhilaration quickly turned to a feeling of profound loss. Seven of America's finest citizens lost their lives in trying to reach the stars.

A lot has been said in the last few years about interesting today's students in the sciences. At least one group has taken a very positive step in achieving that goal.

The Young Astronaut Council, launched at the White House by President Reagan on October 17, 1984 uses the excitement of both manned and unmanned missions of NASA's Space Program to provide the dramatic focus for the Young Astronaut Program. Students in elementary and junior high schools are encouraged to study mathematics, science and technology subjects. The Young Astronaut Program aims to instill a spirit of scientific adventure in our youth to prepare them for the challenges of the future.

It is my firm belief that if our space pioneers could deliver only one message to the youth of the world it would be to keep exploring. Most of us will never reach the stars, but we can go a long way in making it possible for our daughters and sons.

I encourage you to support the Young Astronaut Program. You and your children can receive more information by writing to:

> The Young Astronaut Council
> PO Box 65432
> Washington, DC 20036

Let us continue to build a living memorial to those who have given their lives in the exploration of space.

This book is dedicated to Gregory Jarvis, Christa McAuliffe, Ronald McNair, Ellison Onizuka, Judith Resnik, Francis Scobee and Michael Smith. You are with us always in our hearts and minds. Fly free.

> Jim Grubbs
> Springfield, Illinois
> March 1986

Pioneers

It was 1978. A strange sounding signal appeared on 20 meters. It was a beacon from VE7APU transmitted using packet radio.

December, 1980. Near San Francisco, Hank Magnuski, KA6M placed the first digital repeater on the air. He and other members of the Pacific Packet Radio Society would send packets through the OSCAR 10 Satellite.

Across the country in Washington, DC the Amateur Radio Research and Development Corporation (AMRAD) was promoting packet. Bill Moran, W4MIB placed a packet station on the air. It was now 1981. In October an article by Paul Rinaldo, W4RI and D. Borden proposed the possibility of a packet radio network. Paul was an early AMRAD participant and packeteer.

The first Computer Networking Conference was held. Shortly afterwards, Den Conner, KD2S, would move to Arizona and the Tucson Amateur Packet Radio group would be formed.

These are just some of the men and women who have brought packet radio to where it is today. Welcome to "Get *** CONNECTED to Packet Radio". You are about to learn more about this exciting marriage of radio and computer technology!

Table of Contents

Part One

Part Two

Part Three

Introduction

Do you enjoy being on the leading edge of technology? My guess is that you were one of the first in line to buy or build your own computer system. If you missed out on the initial days of S-100 bus systems built one card at a time, you probably got hooked a few years back when computers suddenly became one of the best buys around.

Regardless of your past radio and computing activities, you are now on the leading edge of technology that combines many aspects of the best of both worlds.

Just as building a computer from the ground up wasn't for everyone, the early days of packet radio weren't for everyone either. Beginning about the time of the 1985 Dayton Hamvention, packet radio took a turn toward becoming a very popular form of amateur communication. Just like computers, packet radio hardware has plummeted in price, skyrocketed in versatility and eliminated the need for you to build equipment unless you want to.

Whether you are brand new to packet communication, just trying to learn enough to decide what unit to buy and how to get going, or more experienced, already on the air but anxious to delve deeper, this is the book for you. The contents have been broken into two major sections. The first chapters address themselves to the basics of packet radio--what it is, what you need, where to find it and how to operate using it. The later chapters explore areas of packet radio of more interest to the advanced packeteer. Wherever you fall in the spectrum, you should find something of interest.

In the final pages you'll find a group of appendixes with useful information regardless of your level of involvement with packet radio.

As new terms are discussed, they are explained in the text. You may also wish to consult the glossary in Appendix A if you run across a term that is new to you.

Just a Peek

If you are reading this book, chances are good that you've already been exposed to packet radio--at least a little bit. Just in case you're starting from ground zero, let me tell you a little bit about the features you'll find in this exciting mode.

Packet radio is error-free transmission of data. Everything you send is checked for accuracy as is everything you receive. If even a single bit of data is incorrect, the information is discarded, but not lost.

Packet radio hardware and software keep track of each piece of information as it is received. If a piece of

information is missing, it is not
acknowledged, resulting in a
retransmission of the missing
information by the originating station.

There are times when conditions may
not allow communication to take place.
Rather than your screen filling with
garbage, it will remain blank.

RTTY enthusiasts generally would
consider 100 word-per-minute teletype to
be very fast by amateur standards.
Packet radio communicates at speeds
several times as large on HF and over
ten times faster on VHF.

But who can type 1200
words-per-minute? That's another one of
the advantages to packet radio.

Much of the time on a data channel
is spent in an idle or unused state.
You may be trying to figure out what to
say next for example. With packet
radio, your transmissions are held until
a certain size or you force a
transmission (usually by hitting the
return key). It takes only a second or
so for most packet transmissions. The
rest of the time your transmitter is
off.

Other stations on the same
frequency can use the idle time for
their own conversations! You'll be
learning about how a packet network
keeps track of all of this later. For
now, all you need to know is that
multiple conversations can take place on
a single frequency without affecting the
individual QSOs!

Packet radio also opens up the
direct transfer of computer programs
over the air in an efficient manner.

One of the features that makes packet radio very versatile is the ability of any station to serve as a digital repeater. Nothing special is required. This can allow you to send error-free messages over long paths even with a very low power radio!

Are you getting excited? I hope so, because packet radio is very exciting indeed. You're going to be exposed to a lot of material in the pages ahead, so get comfortable. I think you'll find it hard to put the book down once you start reading.

Before we "Get *** CONNECTED", it seems appropriate to go back a few years and find out how we got where we are today. Go North young people! (to borrow a phrase) All roads lead to VE land.

Chapter One
The Land of Aurora

While controversy began building in the United States over the possibility of a "no-code" license, Canadian amateurs quietly found themselves with a new class of amateur--the packeteer. When it was first introduced, the term packet communications seemed as foreign as a copy of CQ Ham Radio from Japan. I remember my total understanding of the new Canadian license was that it allowed only computer digital communication on short range frequencies. At the time, a "low-cost" computer system ran several thousand dollars or more. It required a wizard to make it work and took up the major portion of a room. These were the days when someone with 8000 bytes of memory had a "big" machine, disk drives were for the rich, and the "Kansas City Standard" ruled tape storage of computer programs.

In Canada the digital license required the applicant to pass a very stringent test on both radio and computer technology. There was no big rush of applicants and packet radio began very quietly.

It was 1978 when the DOC created the digital radio license. Under the

leadership of Doug Lockhart, the
Vancouver Amateur Digital Communications
Group collectively created the first
widely available terminal node
controller (TNC)--the hardware generally
used for effective packet communication.
It was a circuit board that you had to
"stuff" yourself--not exactly a
recommended first project. An early
form of packet protocol was also
developed by the group. As packet radio
grew, modifications were needed to allow
for greater flexibility.

In the Shadow of Kitt Peak

        Stateside amateur operators for the
most part stayed notably quiet about
packet communications until 1981. In
November of that year, Tucson Amateur
Packet Radio (TAPR) was formed as a
nonprofit organization. Tucson is rich
in high-tech industries. Engineers and
programmers, all enthusiastic hams,
banded together to tackle the task of
creating a TNC at a reasonable cost.
Not all of the work took place in
Arizona. TAPR quickly involved people
from all over the United States in
developing its TNC.
        I can't help but think about my
visit a few years ago to the Kitt Peak
National Observatory (KPNO) outside of
Tucson. It seems appropriate that the
TAPR organization sprang up in the
shadow of the Kitt Peak science
community. In fact, the second meeting
of TAPR took place inside the four meter
telescope at Kitt Peak! I imagine the
TAPR folks feel the same exhiliration as

the scientists at KPNO. Both groups
have done their part in providing me
personally with a great deal of
enjoyment as a result of their efforts.

The TAPR TNC was the first unit to
be made available as a complete kit with
software and extensive documentation
included.

The Amateur Radio Research and
Development Corporation (AMRAD) has been
a major contributor to packet radio
development as well. They became
involved early in 1981 with the
Vancouver packet system. AMRAD is
dedicated to advancing new technologies.
AMRAD and TAPR share many of the same
members, all dedicated to the
improvement of the packet radio art.

Keep in mind that at that time,
modern day digital communications were
not authorized in the United States nor
in most other countries. Even today
amateurs worldwide still rely on Baudot
(RTTY) for most communication of this
type. ASCII, the language of computers,
was approved for radio use only a few
years ago.

Fortunately the FCC listens with a
favorable ear when an individual or
group presents a well thought out plan
for implementing new technology in the
amateur service. We first gained the
use of ASCII code and later the use of
AMTOR code. After successfully proving
the value of packet transmissions under
special temporary authorization, it too
was approved for general use.

The history lesson is essentially
over, but it is to our benefit to
backtrack to those early days of packet

radio and take a closer look at the
mechanics of the mode. The details have
been refined, but the original concepts
remain largely intact. Packet
transmissions are very versatile and the
rules governing the specific make up of
packet continue to evolve.

Evolution

Let's start with something you may
be at least partially familiar with.
Many consider radio teletype (RTTY) to
be the first modern day form of digital
communication. RTTY became popular
during the 50s and 60s with the
availability of surplus military and
telephone company machines. They are
often large, bulky and very noisy.
Original teleprinters were based on the
Baudot code. With Baudot, five "data"
bits are available to encode
information.
Binary arithmetic reveals that with
five bits of data, thirty-two distinct
conditions can be represented. This is
barely enough to cover the alphabet. By
defining one of the codes as a shift
character, the machine can be made to
decode almost twice as many
combinations. That covers the alphabet,
numbers and the most common punctuation
marks.
As technology and needs grew
another code became the standard for
commercial use. Known as ASCII,
American Standard Code for Information
Interchange, it normally allows for
seven bits of data. Eight bit versions
are also common. The additional bits

16

over Baudot allow for both upper and lower case characters plus a wide range of punctuation. The eight bit versions even allow alternate characters to be defined.

One extra bit can also be used as a very crude form of error detection. The "parity" bit is either on or off depending on the make up of each character. Using even parity as an example, the number of bits set to one in each character is computed. If the number comes out even, the parity bit remains off. If the number is odd, the additon of a parity bit brings the total to an even number. Other forms of parity are possible but as we will see, these aren't needed for packet.

Stop and Go

In standard radio teletype each group of data bits is associated with a pair of extra signals.

Early teleprinters were mechanical with an inherent lack of close tolerance. Even though timing conditions were closely specified, gears would wear and clutches would slip causing errors.

By beginning each character with a start signal, the receiving machine knows exactly when to begin decoding a character. By adding one or more stop bits to each character, the mechanical elements of the teleprinter could come to rest before the next start signal was encountered. Actually amateur radio teletype standards call for a stop signal equivalent to approximately one and one-half bits.

Any code that uses the start/stop signal method of sending is considered asynchronous transmission. Any character can begin at any time, but the start of the character is signalled each time. Note that the start and stop bits do not convey any data. They are used only for synchronizing the sending and receiving stations. This takes up time that could be used for information exchange.

On VHF frequencies and above, virtually all teletype transmissions are made using audio frequency shift keying, or AFSK. This allows the use of any regular AM or FM transmitter and receiver for data transmission.

The on/off keying of the teletype machine is converted into tones, one representing the 0 or mark condition, the other representing a 1 or space condition. Amateur conventions now specify tones of 2125 Hertz and 2295 Hertz for a shift of 170 Hertz.

On the receiving end, it is necessary to translate the audio shift back into a voltage or current change. That's what terminal units or computer patches do for us.

The final consideration is speed. For many years, amateurs used so-called 60 word-per-minute machines. With the popularity of computerized teletype, 100 word per minute transmissions are common. Though higher speeds are possible, they are not in general use.

On the HF bands a slightly different form of transmission is used. Rather than using audio frequency shift

keying, the frequency of the transmitter
is shifted directly. On the receiving
end it becomes necessary to turn on the
BFO just as you would for morse or
single side band, to convert this
frequency shift keying (FSK) back into
audio tones. Other details of
transmission remain essentially the
same.

Identical technology has been used
for years for communicating data over
standard telephone lines. Demands for
higher speeds and greater reliability
have led to several alternatives to
traditional asynchronous data
transmission.

Getting In Step

With the widespread availability
and use of microcomputers for sending
and receiving information, many of the
limitations of mechanical predecessors
no longer exist. Recall that start and
stop bits are essentially used to allow
for "slop" in the working of gears and
other parts. A solid state
microcomputer has no such "sloppy"
parts.

It is still necessary to insure
that both the sending and receiving
stations start off together and remain
in step with each other. Several
methods can be used to accomplish this.

It is possible, if each station is
carefully timed to a standard source,
that information can be exchanged
without using start and stop bits.

Without a continuous standard for
both stations to lock onto, it is

possible to use a special locking sequence of data bits to start the transmission and then turn over the timing function to clocks in both the transmitting and receiving stations. These clocks must be reasonably accurate, but that is easy to achieve with today's technology.

Now if you are starting to argue and tell me that a start signal is still being used, you are of course correct. The difference is that rather than a start signal for each individual character, a somewhat longer start sequence proceeds each group of characters. We'll find that as many as 256 characters are included in the information field following a single start sequence!

Data transmission that requires precise synchronization between the sending and receiving stations is called synchronous communication. Remember, with asynchronous transmission, characters can begin at any time. With synchronous transmissions all characters must appear in a precise time slot. If this does not happen, garbage is printed on the receiving end.

Telephone engineers are paid to maximize utilization of the network. In their quest to do so many standards have been adapted. One important to our discussion of packet radio is the X.25 standard.

In their search for greater efficiency, engineers devised synchronous communications systems. The military also likes to use them because they are inherently more secure. The

X.25 standard is a packet form of
communication! Aha...there's the magic
word we have been searching for.

Searching for Perfect Packets

     A packet is a small bundle. For
data purposes, it is a small bundle of
information. The particulars of the
information are not important. Any
bundled information can be thought of as
a packet.
     Networking experts realized that
much of the time on a data circuit is
not actually used. Think about your own
use of a telephone based local bulletin
board. Even if you are a speedy typist,
there are still long periods of time
when you are sending nothing, or
searching for the next key to press. If
that time could be used for other
communications, the total amount of
information conveyed on any given
circuit could be increased.
     Coming from the other direction,
the bulletin board may indeed be capable
of sending to you at the full 300 or
1200 word per minute rate. Chances are
that it is actually capable of much
higher speeds. Again, if the
information could be bundled and sent to
you, the rest of the network could be
used to convey information to other
users.
     When you connect to services like
Compuserve via a telephone line, you
have tied into a packet network system!
Local telephone lines are relatively
inexpensive. Consequently, on a local
basis, dedicating a circuit to "talk" to

you at 300 or 1200 baud is relatively
cheap. It is expensive though to
connect you to the host computer located
hundreds or thousands of miles away in
this fashion.

If you are located in a medium
sized or larger city in the United
States, the chances are good that one or
more special data networks serve you
with a local telephone call. The
General Telephone Company's Telenet
service is one such example. Tymnet,
Datapac and others operate in a similar
fashion. There's a whole chapter
devoted to networking coming up later,
but here are the basics.

When you dial the local telephone
number, you are connected to a node of
the local network. Node is another
fancy term used by engineers. Webster
defines a node as a swelling or
thickened area. I like to think of it
as a crossroads where the pavement
widens to allow traffic to intersect
from two or more directions.

At the node, your asynchronous slow
speed ASCII transmission is converted
into a much higher speed form.
Additionally the information you send is
formed into packets for transmission
through the network. Other users are
having their information packetized as
well. Unless the network path you are
on is very congested, your message
arrives at the host computer without any
noticeable delay.

Coming back through the network the
same process occurs. The received
packets are buffered or held in a
special temporary memory and then

reconverted into the same asynchronous code you are using for output over the local telephone line to you.

If a radio path is substituted for the network we begin to approach the nature of packet radio!

Rather than hooking your computer up to a modem which in turn connects to the telephone line, your computer or terminal connects to a terminal node controller (TNC) for packet radio.

If you examine the terminology, things should start to fall into place. The TNC is your "node" in the radio network. The conversion from asynchronous to synchronous communications takes place in the TNC, just like it does at the node in a telephone network. You might also think of a TNC as a radio modem designed to accept data in one form through one port and convert it to another form through a second port. Most modern TNCs allow for full duplex operation, that is, transmission can take place in both directions at the same time, even though the "network" that is, the radio circuit itself is usually a simplex path. Buffering of inputs and outputs makes this possible. There are cases where the path itself is full duplex as well. The OSCAR 10 satellite is one such example.

If you can grasp these basics of packet communication you are far ahead of many others.

Later, when you've had a chance to digest the differences between synchronous and asynchronous transmissions, take a look at the chapter on packet protocol. In it we'll take a closer look at the contents of a typical bundle of information as it might appear in an amateur radio packet.

For now, making our first packet QSO is the major objective. To do that, we need a radio, a computer or ASCII terminal and a TNC.

Chapter Two
It's Not Terminal

To get involved with packet radio, you need three major pieces of equipment. You must of course have a radio transceiver of some kind. Since most packet activity currently takes place on 2 meters, a standard FM transceiver will work quite nicely. Everything from the simplest of hand held radios to complex, high powered transceivers are being used for packet communications. If you live in an area where packet activity is already well developed, you won't need much more than a watt or two. In many cases, more power would be wasted.

If you want to operate packet on HF, again, just about any unit will work. Power is perhaps more important since you will often be trying to connect under noisy conditions and changing band propagation to a station hundreds or thousands of miles from you. There's a special chapter devoted especially to HF packet operation later in this book.

Connections to the transceiver vary from unit to unit. Generally, audio-in, audio-out and the push to talk line need to be accessible. That's usually easy

to do through the microphone jack and
speaker connections, or alternately
through an auxillary plug if one is
available on your radio. Due to the
number of different possibilities it's
difficult to provide specific advice.
Generally the instruction manual that
comes with your TNC provides excellent
information for interfacing to almost
any radio. We'll explore some
additional considerations in the chapter
on making your first packet QSO.

Is a Computer a Terminal?

        The second item needed in the
packet radio station is a terminal. You
have a wide range of options available
to you. Most packet enthusiasts have
chosen to use inexpensive home computers
for this purpose. There are other
options.
        You need a device that is capable
of sending and receiving standard ASCII
characters at one of several standard
speeds. There are surplus mechanical
type ASCII terminals available at
reasonable prices these days. To the
novice they look like "teletype"
machines. They use the expanded ASCII
code, however, rather than the more
limited Baudot code common to amateur
radio teletype operations. If you
choose a mechanical terminal, make sure
that it is designed to operate using
RS-232 levels. If it isn't, and many
are not, you will have to convert the
input and output of the machine to
standard RS-232 levels. It's not a
particularly difficult task, but

generally more trouble than it's worth for the beginner.

Many of these machines operate at 110 baud. The slowest speed available on many newer TNCs is 300 baud. Be sure to check before purchasing to make sure that the terminal you are considering is compatible with your new packet radio hardware.

Incidentally, some enterprising individuals have constructed devices to allow old Baudot machines to be used instead of ASCII terminals. This requires some pretty sophisticated hardware and software combinations. The PK-1 packet unit from GLB does have an option to allow Baudot machines to be used. It can also be operated from a current loop rather than an RS-232 connection.

Another possibility is to use a dedicated electronic terminal. A terminal is strictly an input/output device. A simple terminal does nothing but display what comes in and send what's typed on a keyboard. Smart terminals generally employ a microprocessor to add additional features like buffering of input and output. Most terminals manufactured today are of the "smart" variety.

The option most usually chosen for the ham shack is the personal computer. A computer can be made to emulate or act like just about any kind of terminal you need. Simple software will turn most home computers into a "dumb" terminal. With the addition of a more sophisticated program, home computers can become very versatile terminals.

27

The primary requirement for using a
computer as a terminal in a packet
station is that the computer be able to
communicate over an RS-232 connection.
Most machines available allow for that
option.

The most popular machines in the
ham shack continue to be the Commodore
series. In recent times, however, other
machines have started to close the gap.
Let's take a look at a wide range of
machines, both new and old.

Commodore is a good place to start.
Whether you have a VIC-20, a new C-128
or the popular C-64, you have a good
candidate for use on packet radio.
There are two cautions to keep in mind.

First, the signal levels on the
Commodore user port are not standard
RS-232 levels. The needed signals are
there but not at the right levels. They
are at TTL levels rather than RS-232. A
simple interface overcomes the problem.
The VIC-1011A is one such device.
Others are available from Jameco
Electronics and other manufacturers.
You can build one yourself using a few
simple integrated circuits.

Recognizing the popularity of the
Commodore machines, some manufacturers
now provide a simple method for changing
their TNCs to accept TTL levels.
Kantronics makes one such unit. The new
MFJ TNC also will work without an
interface as will the AEA PK-64.

The second warning applies only to
owners of the VIC-20. The VIC-20
keyboard is scanned somewhat differently
than its newer brothers and sisters. As
a result, "control" characters can not

be sent in the usual manner. Normally
control characters are sent by holding
down the control key and then pressing
one of the alphabet keys. That doesn't
work on a VIC--at least not unless the
programmer has gone to some
extraordinary lengths in the software.
That's a real problem since most TNCs
are placed into command mode using a
control-C! Other control characters are
used for special functions when
communicating with the TNC. The problem
can be solved in software. The easiest
way out is to define the function keys
on a VIC-20 with control codes. For
example assigning a value of CHR$(3) to
the F1 function key will yield a
control-C whenever the F1 key is
pressed. If you are using a VIC-20,
make sure you can generate control
characters with the software you are
using!

Atari computers are another popular
machine in the ham shack. Unfortunately
machines like the 400 and 800 do not
have built in RS-232 communication
capabilities. An interface box that
often costs more than the computer did
is needed in order to use the machine as
a terminal.

Many TI-99s from Texas Instruments
found there way into homes a few years
ago. They suffer from the same problem
and the additional problem of
availability of the interface box. Some
individuals have successfully written
software to drive the joystick port as
an alternative.

Even the Timex Sinclair can be used
as a terminal, though hitting one key at

a time can be a problem with the standard keyboard. It too requires an external interface.

In the vast majority of remaining machines, from Apples to IBMs, the addition of an RS-232 or serial card is required. These cards typically run from about $100 to $200 dollars. If you already have an RS-232 card for your modem, you are in business. If you have a modem on a card that has its own built-in interfacing you are probably out of luck until you purchase a separate RS-232 card.

One interesting approach is to use one of the new portable electronic typewriters as a terminal. Bob Bruninga, WB4APR, reports success in using the Brother model EP44 as a portable terminal. Units like these are available for about $100 making them very attractive if you need a truly portable terminal.

So your terminal might be a large, noisy, mechanical device salvaged from the scrap heap for almost nothing, or it may be an ultra-new, mega-buck, state of the art personal computer. You'll find almost everything being used for packet radio.

If using a computer, you'll need some terminal software. It can be the same software you use for telephone data communication. Many features are available with some software packages. Split screen operation is possible, command sequences can be stored and recalled with a single keystroke, messages can be saved to a buffer for later viewing. The list goes on and on.

Later we'll take a look at one special package developed for packet radio.

    With a radio and terminal selected, the final and most crucial piece of equipment for packet radio is all that remains. The terminal node controller, or TNC, is the interface that connects the terminal to the radio. It's much more complex than standard teletype interfaces. It has a lot more to do! If you are getting ready to make a decision about buying a TNC, the next chapter is must reading. If you already have a TNC, go on ahead to the chapter on making your first packet contact.

Chapter Three
Selecting a TNC

It seems like only yesterday that
if you wanted to operate packet you had
very few choices about what hardware to
purchase. It was a bit like the early
days of the Ford Motor Company when
Henry decreed that the consumer could
have any color Ford he or she
desired--as long as it was black!
In the earliest days of packet a
bare circuit board was made available by
VADCG, the Vancouver, Canada based group
that helped get packet radio going. You
got nothing but a board. You had to
find the parts yourself, assemble the
unit, buy or build a modem and power
supply, calibrate the unit and scrounge
the software!
The TAPR organization saw a real
need to develop a complete TNC package,
which they did. For about $250 dollars,
packet experimenters could purchase a
complete TNC kit minus the cabinet to
put it in. The cabinet kit ran about
$70 more.
The TAPR TNC, now often referred to
as the TNC 1, became the defacto
standard for packet radio operation.
The programming for the TNC allowed for
both AX.25 and Vancouver protocols to be

used simply by commanding the TNC into the proper mode.

The TNC underwent extensive testing before release. Nevertheless, improvements in the software continued to be made. The use of an erasable/programmable read only memory chip made the upgrading of TNC software fairly easy.

At over $300 for a complete TNC with cabinet and the necessity for hours of construction, getting into packet radio was still not as easy as it might be. Still over two thousand units were sold around the world, and the packet radio community had a standard that proved packet radio was a viable mode at a reasonable price.

A New Breed

I'm a big believer in the image of the old time ham. He or she is an experimenter, not just a communicator. Even so, some of the most knowledgeable hams I know have great difficulty assembling modern day microchip based circuitry. Whether it's unsteady hands or poor eyesight, digital construction isn't for everyone.

The people at AEA saw the potential of packet radio both in the amateur community and for commercial applications. At the 1984 Dayton Hamvention they released the PKT-1 unit. The unit was manufactured through a licensing arrangement with TAPR and remained essentially a TAPR TNC 1. It comes packaged in a very professional looking cabinet typical of other AEA

designs. While the TAPR unit had been
designed with its own 110 volt AC power
supply, AEA modified the unit to operate
off of 12 volts DC, allowing for ease of
operation in a mobile or portable
environment. Otherwise the PKT-1 is a
commercial version of the TAPR TNC 1.
The PKT-1 immediately appealed to the
packet radio enthusiast unable to
assemble his or her own unit. With a
cost in the $500 range, however, the
price remained somewhat prohibitive for
many attracted to packet radio.

A breakthrough in TNC availability
has come. The number of options
increased drastically in 1985. Let's
take a look at each of them.

First There Were Three

The original TNC 1 from TAPR, which
is no longer available, the AEA PKT-1
and the Heathkit Packet unit model
HD-4040 are all essentially the same
design. The AEA unit has already been
mentioned. Besides the power supply,
the other variation in the PKT-1 is that
the parallel port has been made
optional. The port is not normally used
in day-to-day operation anyway.

The Heath company introduced the
HD-4040 under the same type of agreement
with TAPR that AEA had arranged. The
Heath unit also offers a commercial
looking package and the famous, almost
fool-proof, Heath construction manual.
While the PKT-1 is fully assembled, the
Heath version is in kit form. It's
priced at about $250 with the advantage
of the Heath guarantee and support.

34

To give you an idea of just how
closely these units resemble each other,
when TAPR discontinued selling the TNC
1, Heath offered to sell their cabinet
to TAPR owners still in need of one.

All of these units use the XR-2206
and XR-2211 chip set in their packet
modem. This modem serves the purpose
quite well on noise free VHF paths. It
can be configured for use on HF. The
TAPR TNC 1 type units do not allow for
an easy switch between the two modes.
However, an IC socket is available that
can be outfitted with interchangeable
headers to accomplish the change. A
header is simply a plastic form with IC
DIP pins on it that will allow you to
connect wires or compnents of your
choosing.

An external modem can be easily
connected if you wish to experiment.
Some hams are doing just exactly that,
exploring the use of high speed packet
transmissions on UHF frequencies.

VHF and UHF packet operations have
adopted the Bell 202 modem standard tone
frequencies. The tones are 1200 and
2200 Hertz. The 202 standard is almost
defunct in commercial applications.
That made finding used modems at a
reasonable price far easier. That was a
major concern in the early days since
the Vancouver TNC did not have a
built-in modem.

All of these units employ a nifty
device called a NOVRAM (non-volatile
random access memory.) Even with power
removed, a NOVRAM will retain the
information stored in it. Your callsign
and the numerous changeable parameters

associated with each TNC are stored
here, but can be changed easily by
keyboard command. NOVRAM's are
expensive and newer designs use a
somewhat different approach to parameter
storage. The TAPR units allow two
different sets of parameters to be
stored and recalled at the flick of a
switch, a handy option when using your
TNC with two different rigs or in a two
callsign family!

Doing Their Own Thing

Taking a somewhat different
approach to packet radio, GLB
electronics introduced the PK-1. In
it's lowest cost form, it runs about
$110. Many PK-1 units have found their
way into the packet radio revolution.
There are several options to choose from
including a cabinet. In late 1985 GLB
announced an enhanced version designed
for low power consumption and small size
making it ideal for portable operation.
Operationally the GLB unit has one
major drawback when compared to the TAPR
units as well as one major difference.
Because of the design of the PK-1,
it is not capable of both sending and
receiving packets and accepting input
from the computer at the same time. In
a two-way, real time QSO, situation this
can become somewhat of a problem. It is
possible however to write terminal
software for your computer that will
partially get around the problem.
GLB commands are structured quite
differently than TAPR commands. While
the same parameters can be set by the

user, someone familiar with TAPR's way
of doing things will find it necessary
to learn a new language when using the
GLB. Of course, if you've never learned
any command structure, that's a moot
point.

The original PK-1 offers no
parameter storage. When the unit loses
power it does not retain any operating
parameters.

The GLB PK-1 is particularly
popular for use in dedicated
digipeaters. Operating in this mode it
does an admirable job at a very
attractive price.

While the gang at TAPR worked away
on a new and improved TNC, activity in
the commercial field went full speed
ahead.

Kantronics released the Packet
Communicator. During its introduction
at the 1985 Dayton hamfest, Kantronics
personnel could be seen running around
the arena delivering initial shipments
to dealers.

Kantronics chose to use the basic
design of the TNC 1 but to modify it
somewhat both in hardware and software.
Initially the Packet Communicator
emulated the TNC 1 in almost all
functions. Noticeably missing was a
TRACE command and a highly modified
CALIBRATION mode. Clearly the Packet
Communicator was designed for the masses
rather than the serious experimenter.
After a few initial bugs were corrected
the Kantronics unit grew even more in
popularity.

Based on the response they received
from enthusiastic potential packet

operators, Kantronics moved to reduce the cost of the Packet Communicator to $219. The response has been tremendous.

The Kantronics unit deviates from the original TNC 1 type units by using a 7910 modem-on-a-chip, in place of the Exar chip series. By doing so, Kantronics made it easy to switch from VHF to HF standards. All it takes is a single command to do the job! Also recognizing that many TNCs were being hooked up to Commodore computers, a TTL option was included to eliminate the need for an RS-232 adapter.

In the fall of 1985, Kantronics made another significant improvement by offering an upgraded program that includes many of the commands made available in the TAPR TNC 2 described below. Early in 1986 the software was again improved, this time providing a full implementation of the AX.25 protocol including multi-connect. The latest upgrade includes a completely new instruction manual. It does require switching out a memory chip. Installation can be handled easily by almost anyone with moderate skill in using a screwdriver.

The Kantronics Packet Communicator has now evolved into the PC II. It includes the fully implemented AX.25 firmware and a new hardware design.

The TAPR booth was a hotbed of activity during the 1985 Dayton Hamvention. A prototype of the new, low cost TNC 2 was being demonstrated. In the summer of 1985 the TAPR unit became a reality. Priced at $185, the unit came as a kit only. It uses a different

processor than the original TNC, and has
a wider range of commands available.
While all the TNC 1 units offered
Vancouver protocol capability, the new
units dropped the feature. Vancouver
protocol has all but vanished in the
face of the packet stampede of the last
year or so. There are exceptions.
Australia for example has pockets of
Vancouver protocol activity.

The additional flexibility of the
TNC 2 resulted in many owners of TNC 1
units selling the old in favor of the
new. That's made more TNCs available on
the used market.

The new commands include several
that are particularly useful. Actually,
all of these things could be done in
your computer's terminal software.
Including them in the TNC software just
makes life that much easier.

A few of the more popular commands
include the ability to print a "time
stamp" on incoming packets. That way
everything that goes into your buffer is
automatically marked with the time it is
received automatically.

It's now easy to see what path is
being used between two stations. With
the TNC 1 software you had to know how
to use the TRACE command to decipher
that information.

A monitor program keeps track of
what stations have been on recently.
With the MHEARD command, you receive a
listing of stations heard recently and
the time they were heard. It's great
for checking band conditions or looking
to see if a buddy is around without
tying up the frequency. The TNC 2 is no

longer available directly from TAPR.
Several manufacturers have been licensed
to produce TNC 2 packages.
  What about all those TNC 1s out
there with old software? TAPR is
working on an update to implement many
of the features available on the TNC 2.
Additionally at least one third party
developer has a replacement chip
available that adds features not found
in the TNC 1. You'll find more on that
in the accessories chapter.

Magic in the Woods of Lynn

  Competition is an interesting
thing. When it works well, it brings
out the best in people and companies.
Competition in the packet radio arena
has certainly resulted in some excellent
values for amateur enthusiasts.
  Keep in mind that the folks at AEA
did a lot to foster interest in packet
radio. The PKT-1 was one of the first
full-function assembled TNCs offered.
It also reflected the quality of
workmanship that goes into every AEA
product. It is clearly a Cadillac from
a construction standpoint.
  With the competition building, AEA
was challenged to add additional
products that could compete and yet
maintain the high quality of the PKT-1.
The challenge has not only been met, but
exceeded with the PK-64.
  When rumors of the PK-64 first
began to circulate, it seemed a bargain
at the hinted price of $220. For
Commodore 64 and 128 owners, the PK-64
held promise for an alternative to the

Kantronics Packet Communicator.

When it was actually introduced, Commodore enthusiasts learned that not only were they getting a full-featured packet controller and software, they also had CW, RTTY, ASCII and AMTOR all in a single package!

With the optional HFM (high frequency modem), the PK-64 does the best job of performing on 20 meters of any TNC I have reviewed. The HFM runs about $100.

With so much in one package, the PK-64 could have been a very difficult package to operate. With software based on the popular MBA-TOR package, all of the modes have been successfully integrated into a smooth operating package.

Pages could be written about the PK-64 features. Suffice it to say they are numerous. You can even program an auto-response sequence to invite users connecting to your station to leave a message.

In years of reviewing amateur hardware and software, I have never seen a unit that offers as much dollar value as the PK-64 does.

The limitations in the PK-64 are few, but can be significant, depending on your needs. First, it is only available for the Commodore 64. It will run on the Commodore 128 when placed in the 64 mode. It is not compatible with any other computer.

If you are interested in experimenting with running your own packet bulletin board or wish to run terminal software of your own design,

the PK-64 is not for you. Because of
the integrated nature of the the PK-64
hardware and software with the C-64,
what you see is what you get. There is
no way to circumvent the supplied
program.

If that type of operation is a
requirement for you and you prefer AEA
products, the PK-80 is available. Just
as the PKT-1 is a commercially
implemented TAPR TNC 1, the PK-80 is
AEA's licensed implementation of the
TAPR TNC 2. You can expect all of the
TAPR features along with AEA's excellent
construction.

A Few Dollars Less

Some people like meat and potatoes.
I personally like steak and french
fries, but find myself eating a lot of
hamburgers and potato chips. Still, a
well done cheeseburger does taste good
and is a lot easier on the budget than
steak.

MFJ Enterprises have been offering
alternatives to hams for many years.
The range of their product line is
staggering, everything from CW filters
to computer patches. Now, there's even
a TNC!

MFJ introduced a TNC 2 clone packet
controller in early 1986. The MFJ-1270
set packeteers abuzz with a suggested
price of $129.95. At that price, MFJ is
offering an exceptional opportunity for
thousands of hams to give packet radio a
try without investing much money. MFJ
has successfully implemented TAPR
technology at a rock bottom price. If

cost is an important factor in your decision, the MFJ-1270 deserves special attention. MFJ equipment is adequately constructed. I've had several of their products operating for many years without failure.

When It Rains

If you have any doubt about how quickly packet radio is growing, just check the recent ads for packet controllers. Pick up a magazine from a year ago and see what you can find! The folks at GLB have arranged to offer the TAPR TNC 2 as a kit for $169.95. The TNC-2A joins the PK-1 and PK-1L (the low power portable version) in the GLB product line. Pac-Comm Packet Radio Systems has gone one step farther. Selling the TNC 2 design under the TNC-200 logo, Pac-Comm offers the option of two different assembled units all the way down to a bare board! Prices range from $219.95 for the CMOS version of the assembled TNC-200 down to $39.95 for the bare board. It is possible to get a full kit without cabinet using NMOS chips for $129.95.

Something Completely Different

One of the difficulties an active packet radio operator runs into is what to do when you want to use the computer for something other than packet radio. You could buy another computer system, or you could buy a dedicated packet radio terminal.

Packeterm is a complete terminal
and packet node controller. It's also a
licensed TNC 1 package with the addition
of the 7910 modem chip. All you do is
connect it to the radio via the
microphone and audio connections.

At $995, Packeterm is not for the
casual packet operator. It will be a
welcome addition in the shack of someone
who wants to operate packet radio on a
24 hour basis with a minimum of operator
concern and a maximum of operator
comfort. Just as dedicated AMTOR units
are popular with enthusiasts of that
mode, Packeterm is proving popular with
packet communicators. In many ways it
foretells the future. They are only
visions at the time of this writing, but
packet radios, that is, transceivers
with built in TNCs, are likely to become
a reality. The RS-232 connector may
become as common on a radio as the
microphone connector!

For IBM PC owners yet another
approach to packet radio is available.
The Hamilton Area Packet Network has
introduced a plug-in TNC for IBM and
compatible computers. It plugs directly
into one of the slots of the PC.

In addition to AX.25 protocol,
VADCG V1 and V2 protocol are available.
A bulletin board program and file
transfer program are offered by HAPN as
well.

The unit is available in several
configurations including just a bare
board for do-it-yourself enthusiasts.
It appears to be a very versatile TNC
indeed and should appeal to the growing
number of IBM owners.

With a wide range of TNCs now available packet enthusiasts can now pick the unit that best suits their needs. One factor affecting the cost of a TNC is the type of integrated circuits being used. Both CMOS and NMOS units are available. If power consumption is a primary concern for you make sure the unit you buy makes use of low power consumption integrated circuits.

Are We Having Fun Yet?

With the decision to purchase a TNC, you are about to embark on a journey to a fascinating form of communication. With all of your equipment assembled in one place, the instruction manuals near by, and this book balanced carefully on your knee, it's time to start having fun with packet radio!

Chapter Four
Who's On First?

Operating via packet radio is not a difficult science. The mode is new enough that many recent TNC owners have either never actually witnessed packet radio operation first hand, or they've only observed on the air operations for just a few minutes, perhaps at a friend's house.

While the instruction manuals for nearly all of the popular packet equipment contain a lot of worthwhile information, particularly on how to hook up the hardware, it's not often clear exactly what you need to do to begin communicating. If you are an old hand at the basics, skip ahead to the next chapter. This is "Get *** CONNECTED to Packet Radio" so I'll take a few pages to help the newcomer.

Perhaps by sharing my own first attempts at packet transmissions, I can put your mind at ease. I think you'll find I had many of the same questions you may have now.

The first TNC I used was the AEA PKT-1. It's a beautiful piece of equipment. As detailed in the chapter on selecting a TNC, it's a commercial version of the famous TAPR TNC 1 unit.

I had no trouble interfacing the packet
controller to my radio and computer, but
after that I was at almost a complete
loss. When I finally got my first "***
CONNECTED to" message, I felt like the
inventor of radio! So here goes a step
by step approach to making that first
connect message appear with ease.

It may seem old hat, but
please--read the manual for your TNC
before trying anything! Pay particular
attention to the general introduction,
the definitions of commands for your
unit and the interfacing instructions.
Only after doing this are you ready to
begin the journey to your first
error-free QSO.

You Gotta Hear 'Em First

The first two items for
consideration are the interfacing of the
TNC to your computer and your
transceiver. For now, let's assume you
will begin your packet activity using a
VHF or UHF FM transceiver.

Basically there are three must
connections between your radio and the
TNC. First, to receive packet
transmissions, you must connect audio
from the transceiver to the audio input
on the TNC. To begin, obtain this audio
from the speaker or earphone jack on the
radio.

If you live in an area with a lot
of packet activity and you already know
what frequency to monitor, after getting
the computer connected, you may wish to
skip ahead and do a little eavesdropping
before continuing. More on that in a

moment.

To transmit, two connections to the transceiver are necessary. The audio output of the TNC must be connected to the audio input of the transceiver. Normally, you will accomplish this through the microphone connector on the transceiver.

In order to key the transmitter, you must connect the push-to-talk line from the TNC to the radio. Here's the first place you can get into BIG trouble if you aren't careful!

Most TNCs are designed for use with transceivers that have a positive voltage push-to-talk line. An adjustment may be available to allow for operation with negative voltage PTT lines. Check your manual! If your transceiver keys a relatively high voltage you've got some additional interfacing to do. The best advice is to follow the instructions in the manual. If you have any doubts at all, either contact the manufacturer of the TNC directly, or take the advice of a fellow amateur who has successfully interfaced the same TNC with the same radio. While many installations will be the same, the exceptions can cause serious damage to your equipment. Don't take that warning lightly! AEA in particular has gone out of their way to provide listings of interfacing details for many radios. Remember though that their recommendations only apply to AEA products.

Make sure that in addition to the three connections I've mentioned, you provide a ground connection between the two units.

That wasn't so hard, was it? Now to connect the computer side.

Three Will Do It

While you could end up with a dozen or so connections between the TNC and your computer, you'll probably only need three to begin with.

Before discussing the hardware connections, let's talk about software for a moment.

If you are familiar with telephone based computer communication, using the computer for packet radio activities will be a breeze. Assuming for the moment that you have software designed to be used with a telephone modem, you are all set, software wise. If not, there are literally hundreds of communications programs ranging from free to hundreds of dollars in cost. Software is discussed in depth elsewhere. I'll go over some of the basics in a moment.

The next consideration in connecting the TNC to the computer requires that the computer have an RS-232 input/output port or equivalent. If you are already using a telephone modem, the chances are excellent that you have this connection or port as computer types call it. Typically it's a 25 pin DB-25 connector, just like the one on the TNC. There are, however, numerous exceptions.

Several manufacturers of telephone modems have eliminated the need for an RS-232 connection by building additional

hardware into their modems.
Unfortunately, while this works well for
telephone communication, it won't help
you get your TNC on the air. Two
specific examples include modems
available for the Apple II series of
computers and some available for various
Atari machines. With both of these
machines and others that do not come
equipped with an RS-232 port, an
additional card or interface box is
required to do RS-232 communication. In
some cases (Atari, Texas Instruments)
the interface box can cost several times
as much as the computer! For most other
computers an RS-232 card runs about $100
retail.

    Notably missing from the discussion
so far are the Commodore machines.
That's because their communication port
is somewhere in-between no port at all
and an RS-232 port. Depending on which
Commodore machine you own, most or all
of the RS-232 signals have been
implemented, but at TTL logic levels
rather than RS-232 standards. TTL
signals typically switch between zero
and five volts, while RS-232 signals
typically range between plus and minus
twelve volts. Keep in mind that these
are generalizations. The voltages for
your computer may vary considerably from
these values.

    Several companies, including
Commodore, sell RS-232 interfaces that
are made specifically to match the
Commodore signals. They are over priced
at about $50 retail, but I've been able
to find them on close out shelves as low
as $10.

Before you immediately assume you need such a device for your CBM computer, check the instructions for your TNC. Amateur manufacturers know that the Commodore machines (the C-64 in particular) are the single most popular machine in the ham shack. The Kantronics and MFJ units provide a simple means to allow direct connection of the TNC to Commodore computers without the need for an RS-232 interface. It's really not difficult since most of the signals running around in a TNC are at TTL level anyway and normally must be converted to RS-232!

Once assured that you have either an RS-232 connection or your TNC can be connected to the TTL signals on a Commodore machine, it's time to start wiring.

To get started, three wires are needed. First is ground, then a line for received data, and finally one more line for transmitted data. All of the TNC instruction books I've seen are pretty explicit about which lines are which. You may also need the reference guide for your particular computer to properly wire the other end of the cord. All of the TNC instruction books I've seen are pretty explicit about which lines are which. You may also need the reference guide for your particular computer to properly wire the other end of the cord.

That's it! You are wired and ready to try some packet communicating.

Before Connecting, Disconnect First

I would recommend that you disconnect the transmit audio and PTT lines between the radio and the TNC. Normally that just means disconnecting the microphone connector you wired from the radio end of the circuit.

After giving everything the "smoke" test, it's time to load the communication program into the computer. The specifics will vary from computer to computer and between programs. Once again, telephone modem users have the upper hand since they are likely to already be familiar with the software. If you haven't used such programs before, spend some time trying them out before attempting to use them for packet radio purposes.

With the communication software loaded and running, enter the terminal mode. With TNC 1 type controllers, a message will be sent from the TNC over the RS-232 cable to your computer at different speeds. Typically, you will be told to press the "*" key when you can read the message. This whole procedure is known as "autobaud".

TNC 2 units do not use this method. It is necessary for you to set the DIP switches on the TNC itself to match the speed of your computer.

Keep in mind throughout this process that you are setting the speed at which the computer and the TNC talk to each other. This does not affect the speed of transmission on the radio side of the circuit. On VHF, the radio port of the TNC should be set to its default

value of 1200 baud. For HF a speed of
300 baud is used.

    With TNC 1 units, once the key has
been pressed, the TNC adjusts
automatically to whatever speed your
computer is using. You know you are
successful when the copyright message
appears on your screen and you receive
the "CMD:" prompt on your screen. Note
that the prompt may be in either upper
or lower case.

    If you are having any trouble at
all at this point, here are a few areas
to investigate. Most TNCs are capable
of communicating at several different
standard speeds or baud rates. Most
computers are also capable of changing
communication speed. A limitation may
be placed on the speed by the software
you use with your computer.
Particularly for programs written in
BASIC language, speeds greater than 300
baud are not reliable.

    Besides speed, there are other
parameters associated with data
communications. They are: word size,
parity and number of stop bits. Your
TNC manual should tell you what to
select on your computer for proper
communication. Typically, 300 baud, 8
bit words with one stop bit and no
parity will work well, as will 300 baud,
7 data bits, even parity, one or two
stop bits.

    If adjustment of these parameters
does not correct the problem, first
check your wiring. If the problem
persists, other possible causes include:
trying to use a terminal program
designed to work with a "special" modem

such as described for the Atari and Apple computers or using software that checks for special start and stop signals that we have not yet implemented. In the latter case, it doesn't mean your terminal program won't work, it just means you've got more wiring and testing to do. More details on how to resolve problems follow in the next chapter.

You're In Command

With the "CMD:" (Command) prompt on the screen, you are now in control of the TNC. On TAPR-like TNCs, try typing: DISPLAY followed by a RETURN. A list of all of the TNC's settings will be returned to your screen. With some units HELP followed by RETURN will show a list of available commands.

If all is still going well, you are on your way to finding out just what your new toy can do. If you have been able to get the command prompt, but seem unable to send a command, you are probably having trouble with the transmit line between the computer and the TNC. Since you are receiving information, you should be OK with regard to speed and other parameters. Recheck your wiring. If it still doesn't work, enlist some experienced help!

As it comes from the factory, your TNC has a set of default parameters in it. The first one you will want to change is MYCALL. At the command prompt type: MYCALL KA9XYZ followed by a return. GLB owners should consult the

54

manual or the appendix in this book
since MYCALL is not used. Remember, we
are discussing the commands as they are
implemented on TAPR TNC 1 and 2 units.
Here's how it would look on your screen:

CMD:MYCALL KA9XYZ

TAPR TNC 1 units will respond with a WAS
message . The TNC 2 will answer with
MYCALL WAS NOCALL. If you want to
recheck the call at any time, just type
MYCALL followed by a return at the CMD:
prompt. There are a variety of other
commands you will want to make use of,
but for now our goal is the first "***
CONNECTED to" message. Make sure that
the "beacon mode" is disabled before
proceding. You can do this by typing
BEACON EVERY 0 at the command prompt.
     TNC 1 owners will want to insure
that new values are not lost. Typing
PERM at the next command prompt will get
the job done! TNC 2 remembers
everything so the PERM command is not
implemented with these units. The TNC
will now remember your call even when
the power is off.

Listen First

     If there is a lot of activity in
your area and you already know what
frequency is being used for packet, try
"reading the mail" for a while. To do
so, you must insure that the MONITOR is
on. Different TNCs implement the
monitor feature somewhat differently.
You want to set the monitor so that it
will print the maximum possible number

of packets. All packets do not contain text or data. Some packets only contain connect or disconnect information, acknowledgement data or repeat requests. One of the most common mistakes made by new packeteers is to assume that something is drastically wrong with their station because not every packet prints when the monitor is on. By implementing the maximum level of monitoring available, you may be able to see all the packets you hear. Particularly on older TNCs running early software, the monitor function may not allow all packets to be monitored. Some TNCs do include a TRACE command that will let you see what you hear, but it's a somewhat complex hexadecimal, bit-shifted dump that may be very confusing to a beginner.

You won't want to monitor everything for very long, but it's a good way to insure that the receive portion of your system is working OK.

If you aren't displaying packets at this point, several possibilities exist. The most likely one is that either you are not getting audio to the TNC from the receiver or the audio is too badly distorted for the information to be recovered. First check for the no-audio possibility. If your TNC has an LED marked DCD (data carrier detect) it should light when a packet signal is present. If it doesn't, an open circuit in the audio line is a good bet. I have several TNCs that will respond to "white noise" when the squelch is opened on my transceiver. Some newer units include true data carrier detect and will not respond unless the proper tones are heard.

If audio is reaching the TNC, try adjusting the volume control to different points while signals are being received. Contrary to the sage advice of "experts" it can make a big difference in some installations. Here's why. Packet transmissions occur on VHF and UHF right now at 1200 baud. For an audio frequency shift keyed signal, that's pretty high speed operation. Each transmitter and each receiver has different audio response characteristics. The two tones used for packet AFSK operation are one-thousand Hertz apart. This can lead to one of the tones being severely supressed in amplitude from the other, either by the transmitting station, or on the receiving end. Such distortion can introduce errors into the received information. Remember that packet is by nature an error-free medium, so if an error is detected, nothing is printed.

Except on an absolute noise-free path, it takes only one unwanted "spike" to render a packet useless. In my area, power line noise and perhaps more irratating, cable TV leakage, makes noise-free reception on weak signals nearly impossible, even using FM. I've found that even signals that sound strong often won't print. Careful adjustment of the volume control minimizes these problems.

. Many times the transmitting station will be found to be off frequency or be deviating either too much or not enough. Purists will tell you that the

audio going into a TNC should be taken
from the receiver after it has been
equalized but before going to the audio
amplifier section of the receiver. That
certainly is the best way to do it.
From the experience of several who have
gone to the trouble to do it that way, I
can only say that the results are often
disappointing. While it has helped
tremendously in some cases, no
difference has been noted in many
others.

Hopefully you are receiving some
packets at this point. If not, check
the items discussed. If most of the
signals in your area are weak, this may
not be a conclusive test. Go on to the
transmit tests.

Hello Packet People

With the microphone connector still
disconnected from the transceiver, try
sending this message at the command
prompt:

C KA9XYZ (followed by a return)

If your TNC has a PTT or XMT lamp it
should light for just a moment. It will
continue to do that at intervals of 10
seconds or so for several tries. Type a
"D" at the command prompt to stop it.
You will get a message saying: "Retry
count exceeded--Disconnected".

Even though you aren't actually
transmitting, if there is activity on
the air it may take longer than the
usual 10 seconds for the transmit
attempts to recycle.

I would suggest that you move off to a simplex frequency that is not heavily populated for the next test. Transmitting into a dummy load would be an even better idea. With the microphone connector hooked up to the transceiver, try the same message again, that is:

C KA9XYZ (return)

You should get the same results as before, but this time the transmitter should key. If it doesn't you've got push-to-talk problems and need to back up a few pages. Type a "D" for disconnect again and pull the microphone connection loose.

If the transmitter keys and releases OK, take a listen to the audio on another receiver if one is available. Otherwise have a buddy listen or momentarily skip on to the next step. The audio should sound good and clean, not too low in level nor distorted. There shouldn't be any hum. Packet data itself sounds like a buzz-saw so don't mistake that for hum!

Somebody somewhere is out there asking, "why not use the CALIBRATE command for the transmitter test and audio check?" Remember, we are on a quest for our first CONNECT message and want to minimize the learning necessary to start out with. CALIBRATE is implemented differently in different units, so if you need it, read the manual for your TNC.

Now Is The Time

   Are you ready for some hocus-pocus
that will make you a packet fiend
forever? Your time has arrived! With
the initial tests complete, it's time to
go back on the air and try for the first
QSO. That's easy if you have someone
waiting for you, particularly if you can
coordinate by land-line or on another
radio frequency at the same time.
   I remember my first packet
experiences. The PKT-1 got hooked up on
a weekday in January. It was about
minus ten degrees outside and everyone
was at work or trying to start their
car! It didn't help that only two
stations in my local radio range were
active at the time. There were one or
two others, but they didn't have
dedicated packet equipment and often
weren't on. Fortunately I knew the
callsign and frequency for one of the
stations. The following message at the
command prompt resulted in my first
packet connection:

C WA9KRL (return)

With a burst of packet activity my
computer terminal read:

*** CONNECTED to WA9KRL

It felt like my first contact via OSCAR
and almost as good as when I worked
Hawaii via OSCAR 7 for state number 50.
   After that, life was pretty
anti-climatic. Since the operator was
out of the room, I had no one to talk to

and no one to ask the hundreds of
questions I had. There was no book to
read then!
    With a touch of sadness I
disconnected from KRL with a simple "D"
command. It was over much faster than
it had begun.
    After doing a bit more reading in
the manual, I found out that I could
connect to myself. It was as simple as:

C K9EI V WA9KRL

In plain language, that says: I want to
connect to myself (K9EI) and I want to
do that "via" WA9KRL. That was my first
introduction to digi-peating, or using
another station's TNC to relay my
transmissions. In this case, my
transmissions were being relayed right
back to me!
    The packets really flew this time!
I tried typing something and hitting
return. My transmitter sprang into
action and a few seconds later my words
appeared again on the screen, repeated
through WA9KRL. I now knew that at
least something worked, some of the
time.
    Once you have connected to another
station, you are automatically placed
into the CONVERSATION mode. This can be
changed, but that's the normal setting.
While in the CONVERSATION mode,
everything you type is sent for
transmission by the TNC. In order to
issue any more commands you must exit
back to COMMAND mode by typing a
Control-C. You do this by holding down
the control key and the C key at the

same time. In some software programs,
you may also send a control-C by using
some other special key. VIC-20 owners
have to be careful. The keyboard is
scanned rather strangely in the VIC-20
and it takes a bit of software magic to
generate control codes with some
terminal programs. Some won't do it at
all!

With a final "D" for disconnect, my
transmissions to myself were terminated.

Keep in mind that all of these
instructions assume that you are using
one of the TAPR type TNC units. GLB
owners using the PK-1 will have a
somewhat different set of commands to
use. The appendix includes a partial
cross reference between TAPR and GLB
commands.

If you have the AEA PK-64 many of
the commands are similar. Many are not
commands at all but options selected
from a menu. Your PK-64 manual is your
best source for information on how to
change things. Since the vast majority
of packet activity is based on TAPR
units, I've used it here in the
examples.

On With The Show

You now have the basics for trying
out your new packet equipment, whether
someone else is available to help you or
not. I'd like to go just a bit farther
in this explanation and cover
multiple-repeats and how to figure out
who is around.

A strange thing happened once I
started sending packets that afternoon.

All of a sudden I started receiving packets from time to time where previously the channel had been quiet for several hours. The explanation involves the beacon capabilities of packet controllers and a special feature built into the TNC software.

Nearly all TNC units allow a beacon message to be pre-programmed. Check yours right now by typing BTEXT at the command prompt. What you will probably get is a commercial for the brand of TNC you are using. You can change the beacon message by typing something like this at the command prompt:

BTEXT THIS IS JIM IN SPRINGFIELD
ILLINOIS TESTING NEW TNC

followed of course by a return. Unfortunately if you have a TNC that uses NOVRAM for parameter storeage the beacon message takes up too much space to be saved, so you will have to type it in again if you power down your TNC.

In order to send your beacon, you have to do one of two things. You can either specify that your beacon be sent every so many seconds or minutes, or you can specify that your beacon only activate when other activity is first heard and then ceases for a certain period of time. In the first case Beacon Every is used. The latter case is implemented using Beacon After. Since the TNC likes to think in computer time units rather than seconds and minutes you have to compute the number of units (up to a maximum of 255) using the formula in your manual. With Beacon

Every your beacon will be transmitted
every time the timer counts down to the
amount you have set. It will wait its
turn, but will always beacon. With the
Beacon After option the beacon is only
transmitted when activity is heard. It
then waits the amount of time specified
after the channel goes quiet before it
sends a message.

In the early days of packet radio,
beacons proved to be quite useful. With
the increase in activity, beacons should
be used with a lot of discretion. In
populated areas, channels are very busy
and beacons just add to the congestion.
If you absolutely must beacon, set the
time interval for the maximum your TNC
allows.

The same goes for the CW ID option.
There was some question in the early
days whether a packet ID was legal or
not. Amateur convention is now to leave
the CW ID off. So, keep the beacons
infrequent or off and the CW ID
disabled!

What I found on my first day was
that several beacons both local and
distant had been activated by my
activity. By leaving the monitor
function on, I could see who else had a
station on the air at the time, giving
me more possibilities for communication.

With newer software, the MHEARD
command returns a list of stations heard
on the channel. It's a much more
effective way of checking on activity.

Repeating Yourself

One of the things that makes packet

radio so powerful is the ability of any packet station to serve as a digital repeater. As our final introduction to packet's most important commands, let's see how to create a path using several stations to cover a long path.

The basic "C" for connect message is used. For me to connect reliably to my brother about 70 miles to the north of me using my handie-talkie connected to my packet station, the message looks like this:

C WB9YJC V WA9KRL,K9CYW-1,K9CYW-2

As you can see, it takes three relays to get the job done. The first station is a local one with a fairly good sized antenna system and a bit of power. Normally, his antenna is pointed at a central Illinois wide area digipeater, K9CYW-1. The trustee for that station is K9CYW. He operates his own station at home and several digipeaters at good locations around the state. To differentiate between his several stations, packet protocol has an SSID or secondary station ID to be added to the callsign. His home station really is K9CYW-∅, but the zero is not normally displayed and need not be specified in a connect command.

The next station in the path is another digital repeater located closer to WB9YJC. It is identified as K9CYW-2.

My transmissions are relayed by each of these stations. The acknowledgements and other transmissions from YJC come back in the reverse order. It only takes several seconds for each sequence so the speed is fairly quick under good conditions.

Newer TNCs allow you to select a monitor option that will display the path being used. You soon learn to make a list of how to get to different stations. With the TNC 1 using older software, such options were not available. It is possible to get around this by using the TRACE command.

Hopefully, you've just completed your first packet radio QSO, even if it was with yourself! If not, check each step along the way, and if all else fails, call the experts who sold you your TNC. You have an advantage I didn't have. With the number of packeteers over 12,000 these days your chances of local help are excellent. When only 1000 or so of us were around, the usual response was "packet what?"

If you haven't been successful, read the next chapter. It contains information on some additional commands and some advanced hardware considerations.

Chapter Five
Beyond the First Connect

In the last chapter we took a look at a few simple packet commands necessary for establishing your first connection. In the vast majority of situations, the rest of the pre-programmed settings will suffice without modification. There are some instances where it will be necessary to make an adjustment to one or more additional settings. In other situations, modifying a preset value can improve your packet operation.

Count to Ten

Most modern transceivers are able to switch from receive to transmit very quickly. Some synthesized units require a few milliseconds to "lock" on the proper frequency.

If you are able to receive packets without any problem, but are having difficulty making or maintaining connection, one thing to try is increasing the TXDELAY. As with any command, at the CMD: prompt you type:

TXDELAY 6 (return)

for example. I've used many TNCs and
many different radios. With some radios
I've found that I must set the TXDELAY
fairly high in order to insure that my
packets make it through OK.

What does TXDELAY do? It simply
delays the actual data transmission for
a short period of time after the
transmitter has been keyed. If TXDELAY
is set excessively long, monitoring your
signal will reveal the sound of "flags"
being sent for a period of time before
packets are transmitted. In busy
networks this eats up the time available
for other stations to use. Most of the
complaints I've heard from other
operators about excessive TXDELAY are a
bit like some large city drivers who hit
the horn as soon as the traffic light
turns green. Still, the newcomer to
packet radio makes a better impression
if TXDELAY is set just long enough to
make things work properly for you.

Your own equipment is not the only
reason for increasing the transmit
delay. Some receivers are slow to
react, just as transmitters are. If the
station you are trying to connect to, or
relay through has such a unit, you may
experience difficulty. Though there may
be nothing wrong with your equipment,
you can often overcome the problem with
an adjustment to TXDELAY. There are
several stations in my area that work
unreliably without increasing my TNC's
TXDELAY.

Make The World Go Away

If you have no trouble connecting

68

to other stations, but they report being
unable to connect to you, check the
CONOK or connect OK item. Remember you
can check any parameter by typing just
the command in question at the CMD:
prompt. Normally CONOK will be on. If
you wish to leave your TNC active but do
not want to accept any connects,
changing CONOK to off will get the job
done. Why would you want to do that?
If you have only one computer and use it
for other activities, you might want to
turn CONOK off while you work on
something else. When you do that, your
TNC will respond to another station's
request with a "BUSY" message. It's a
way of saying that your station is on
the air and the path is good, but you
aren't in a position to accept messages
at the moment.

Rub Out

     One of the most common mistakes the
new packet operator makes is to have the
DELETE command improperly set. TAPR TNC
1 units normally have DELETE set on
which results in the delete character
being defined as a hexadecimal 7F. With
many machines (Commodore in particular)
the alternate choice of hexadecimal 08
is the proper choice. Make the change
by typing:

DELETE OFF

at the command prompt. If you own a TNC
2 or clone, hexadecimal 08 is now the
default value. Related to this setting
is BKONDEL or back space on delete.

Normal setting is on which is correct
for most video terminals. The OFF
option sends a backslash character which
may be preferred with mechanical
terminals.

Echo, echo, echo

    If you are seeing two of everything
you type, it's a good bet that the ECHO
command has been set ON. Your terminal
software may also be at fault. Normally
terminal programs allow for full duplex
or half duplex operation. In half
duplex, the computer software itself
will echo what you send from the
keyboard. At the same time the TNC is
echoing your input. Pick one or the
other, not both!

Waiting Forever

    Here's one that will get some of
the gang excited! One of the things
your TNC does is try its best to be
polite in its use of the network.
Occassionally things will go wrong and
two stations will both try and grab the
channel at the same time. That's called
a collision.
    Whenever you send a packet
requiring an acknowledgement, the FRACK
parameter controls how long the TNC
waits before trying the packet again.
When relays are involved, the effect of
the FRACK setting increases drastically.
The delay is equal to two times the
number of relays plus one, times the
value of FRACK. With a setting of 4,
the time delay increases to 28 seconds

70

if three relays are used!

On top of everything else an additional random delay is added to help avoid two units with the same settings continually knocking each other out! The value of DWAIT also comes into play.

DWAIT is defined as the number of time units the TNC will wait after last hearing data on the channel before it begins its own keyup sequence. Can you guess what happens if the station you are colliding with uses a lower value than you have? DWAIT needs to be agreed to on a local basis.

The setting of these two parameters can go a long way in cutting down on the number of collisions that occur on the packet radio channel. Keep in mind that while the actual transmission speed of packet communication on VHF is 1200 baud, anything that causes collisions and retries is detremental to the effective speed of data transfer. Even under ideal conditions the throughput is somewhat lower than 1200 baud.

Unfortunately, just like there are people who run more than the legal power limit, operate outside of their assigned sub-bands, and take great delight in bombarding repeaters with a kilowatt signal, there are those that decrease the FRACK and DWAIT settings to their minimums. If you remain a good guy and leave yours alone, when the offending station is on during a busy period, he or she will dominate the channel. You will always be waiting for the channel to be idle and it will never happen. Use FRACK with caution, but if necessary a decrease can be an effective tool that benefits everyone. Try and abide by the standard in your own area.

If you do a lot of message exchange over paths that are not very reliable, there are a few more areas you should look into. Adjustments may help not only you, but the rest of the network as well.

As defined for amateur packet purposes, packet data fields can be up to 256 characters in length. The more characters per packet, the longer it takes to transmit a single packet. Makes sense! The default value is 128. Under many circumstances that can be too high. I personally run with a length of 64. You change the parameter by using the PACLEN command. In my case typing:

PACLEN 64

at the CMD: prompt will do the job. Remember that the error checking computes the validity of a packet by doing a mathematical calculation based on the contents of the packet. Statistically, the longer the packet the greater the opportunity for error. If only a single bit comes out wrong, the packet is marked invalid.

When conditions are poor, packets may back-up as they stay unacknowledged. The TNC can keep track of up to 7 pending packets. Normally TNCs are shipped with a default setting of 4 on MAXFRAME. If the packets go unacknowledged and the operator keeps typing, the effective length of the transmission keeps getting longer and longer. That ties up the network, increases the probability of errors on marginal paths and generally causes a problem. So, on a noisy or marginal path set MAXFRAME to 1!

Finally, in this series of commands is RETRY. Your TNC will make up to 15 attempts to send a single packet. Default setting is 10. Once again, some individuals seem to delight in setting RETRY to zero. This results in an infinite number of retries! When helping someone work on a TNC on a seldom used frequency that may be OK. Otherwise, please, don't do it!

Give It A Permanent

Before going on, just a reminder to TNC 1 owners to send a PERM command to make all of your parameter changes permanent. Otherwise they will be lost when you turn off the power to the TNC. Again TNC 2 owners have the advantage of having everything saved automatically.

Hopefully if you left the last chapter still having problems, one of these commands has solved your difficulty. If not, chances are pretty good you have a "flow" problem. Often they can be corrected with a simple command change.

Go With The Flow

Recall that in the last chapter I suggested a simple three wire connection between the TNC and the computer. The reason is a simple one. The additional connections available assume that your terminal or computer will provide and

respond to some special signals. Not
all computers and terminals have these
signals. If a line that is suppose to
be high is found to be low, your TNC and
computer or terminal may not want to
communicate with each other. The
easiest way around the problem, for the
short run anyway, is simply not to use
the special lines.

There is a caution here in that
with some TNCs you may find it necessary
to place a jumper across two pins in
order to make the simple hook-up work.
Your TNC manual is the best source for
specifics. If none of the following
suggestions do the trick, suspect the
need for a jumper. If you can't figure
out where it needs to go, contact the
manufacturer!

Two Ways To Flow

Flow control can either be
implemented using software controls,
commonly called XON/XOFF or hardware
control usually called RTS/CTS
handshaking. Let's take a look at
XON/XOFF first.

Normally the default choice for
flow control is XON/XOFF. The choice is
controlled by toggling the XFLOW
parameter. Typing XFLOW OFF at the
command prompt enables hardware flow
control. With the TNC 2, hardware flow
is always on.

The standard in data communications
is for a control-S (XOFF) character to
be used to stop data flow. A control-Q
(XON) will allow information to flow.
You can use the XOFF and XON commands to

change these choices if you have a need to do so.

Some TNCs are full duplex devices. Others are not. Regardless, there are times when you may want to communicate to the TNC that data should be stopped. For example if you are in the middle of typing something and an incoming packet arrives, you don't really want it to interfere with you while you type. The FLOW parameter (normally set on) will allow output to the terminal to be halted until either a command line is completed or a packet is forced.

If you need to take your terminal off line for a short period of time, you may also send a control-S to the TNC just before doing so. When the terminal is back on-line, a control-Q will dump the contents of the TNC buffer to the terminal.

Hardware Is Sometimes Best

If your computer or terminal has RTS/CTS signals available, hardware control can be used rather than software control. When properly connected and implemented, the TNC will "assert" the CTS (clear to send) line. That tells the terminal or computer that the TNC will now accept data. Conversely the computer or terminal begins an attempt at transmission to the TNC by asserting the RTS (request to send) line.

You may also find reference to DSR/DTR handshaking. That's data set ready and data terminal ready. Your TNC is a packet modem and is considered a DSR device. The computer or terminal is a DTR device. If your TNC checks these signals, this may be where you need to place a jumper on the TNC end.

Why go to all this trouble? For everyday poking around on the keyboard, flow control is not a big problem. It becomes more important when you start trying to transmit large files or even programs. Often bulletin board operations fall into this category. Proper implementation of hardware flow control allows the TNC and terminal software to work smoothly at maximum transfer rates.

Unconnected and Unproto

One of the limitations in earlier packet protocol was the lack of ability to support "round table" discussions. Connection was limited to just two stations at a time. Recent changes are beginning to allow multiple connections. In the case of the PK-64, TNC 2 and the new Packet Communicator II, multiple connects are possible, but the conversations are still not round tables in the true sense of the word.

All is not lost however. Packet radio allows for unconnected transmissions to take place. No repeats are made automatically. If a packet gets lost or mangled, the operator sending it has to manually retype the message.

When all of the stations wishing to communicate are in the same area and all have good signals, that's not too much of a problem. Some have devised some pretty elaborate schemes to allow for

some error checking. Basically two
stations connect often using others in
the roundtable as digipeaters and then
leave the MCON command turned on. MCON
allows the monitor function to continue
to take place during connection. The
only problem is that you will see not
only the extra packets you want to see,
but also others. You can compensate for
this somewhat by carefully adjusting
your monitor or "buddy" lists. With TNC
1 hardware using old software, the MFROM
and MTO lists will accomplish the same
thing.
     For most newcomers, the simplest
way to start a roundtable is to do it in
unconnected fashion.
     When you are connected to another
station, your TNC automatically shifts
from command mode to CONVERS or
conversation mode. You can also
manually enter the conversation mode by
typing:

CONVERS (return)

at the CMD: prompt. After doing so,
everything you type will be sent out to
the world. It is addressed according to
how the UNPROTO parameter has been set
up.
     The default for UNPROTO is CQ.
It's a great way to generate a few
packets to let the world know you are on
when you have no idea what other
stations are on the frequency.
     In many areas, UNPROTO messages are
directed to a specific group, such as
the initials of the local club. In my
area, stations could direct all

77

unconnected packets to CIPRUS (Central
Illinois Packet Radio Users) or perhaps
SPI for Springfield. The monitor
function can be set up to scan for only
those packets addressed in such a
fashion.

Something that isn't always made
clear is how to use the UNPROTO command
to specify a path for beacon messages.
Particularly in the early days, everyone
was anxious to learn of new stations.
It wasn't unusual for stations to have
their beacon messages relayed through
several stations. For example, the
following command typed at the prompt
would allow all of my unconnected
packets (including beacons) to be
relayed first north to a Central
Illinois wide area repeater and then,
conditions permitting, on to a station
east in Indiana:

UNPROTO CQ V K9CYW-1,WB9QPG-1 (return)

The result would be messages addressed
to CQ but relayed through two
digipeaters. The CQ could have just as
easily been CIPRUS or any other
letter/number group, including a
callsign.

Beware TNC 1 Disconnects

One of the lessons I learned the
hard way is what happens when you are
connected to a distant station and the
path disappears. With the software used
in TAPR TNC 1 type units, if the
originating station times out and
disconnects, it is possible to leave the

other station still connected. This
makes the other station unavailable for
any other connections until manually
disconnected. It is possible if the
originating station can somehow get
reconnected, to then disconnect in the
normal fashion and all is well.

The first time it happened I think
both I and the other station thought it
was something I had done either
deliberately or through ignorance. That
wasn't the case as others have found
out.

The lesson was learned so well in
fact that TNC 2 type units have "watch
dog" type timers that can be
implemented. If no activity is detected
within a certain period, a packet is
sent out to check the connection. If
the path is no longer good, the TNC will
disconnect itself and return to the
unconnected mode. Incidentally, it is
also necessary on TNC 1 units to type a
control-C to return to command mode even
when a proper disconnect sequence does
take place. That's been corrected with
the TNC 2 units. They may optionally
return to command mode automatically.

The Beat Goes On

An entire book could be written on
just the types of commands available.
This is not intended as a complete
reference. Your TNC instruction manual
should do that for you. These are just
some of the commands so often both
intriguing and confusing to the
newcomer. We will look at the TRANS or
transparent command group in a chapter

all by itself. That's for advanced
users, though, and for now we want to
find out about the fascinating world of
packet on the low bands.

Chapter Six
Slower but Farther
HF versus VHF

Just about everything we've talked
about up until now assumes that you are
operating packet on VHF, probably two
meters. To review quickly, standard
operation on VHF and UHF frequencies
currently calls for stations to transmit
at 1200 baud using audio frequency shift
keying via an FM transceiver. The tones
used are the standard initially
developed by the Bell system for
landline based 1200 baud transmissions.
New technology has replaced the old Bell
202 standard for telephone use making
modems designed for the system readily
available on the surplus market. If
bandwidth was not a consideration, the
same standards could be used on HF
frequencies. The 202 standard calls for
tones of 1200 and 2200 Hertz. When
modulated with a 1200 baud signal, the
resultant bandwidth far exceeds the
permissible limit for digital
communications on HF frequencies below
28 megahertz.
     In order to allow packet
transmissions on 20 meters and other HF
frequencies, the amateur community has
standardized on using 200 Hertz shift

81

keying (Bell 103 standard). To further meet the requirements, transmission speed is reduced to 300 baud.

From a hardware standpoint, the difficulty in operating HF packet is dependent on the type of TNC you have. Original TNC 1 designs are optimized for VHF operation. Changing the baud rate is no problem. That can be accomplished with a simple command. Electronically speaking, changing the components to make the XR-2206/2211 chip set work with narrow shift is not difficult. Logistically it involves either changing components on the board or mounting components on "header" packages and plugging the right one in as required. Further, it's necessary to go through a calibration procedure to get the right tones.

Newer TNC designs include a command to switch between HF and VHF tones and shifts. That's gone a long way in promoting HF packet operations.

The major advantage to operating packet on 20 meters is the ability to communicate over much greater distances in a single hop. That's particularly helpful when the need arises to forward messages across the country or around the world.

Several problems surface on HF that can frustrate the packeteer. Keep in mind that noise is packet's natural enemy. Noise, fading and interference on HF are generally much greater than experienced on VHF. Remember, a single bit out of place will render a packet useless, requiring a repeat.

Another major problem is the proper

tuning of packet signals on HF. On VHF using FM AFSK (audio frequency shift keying) transmission techniques, everything is channelized. If you are tuned to the right frequency, you don't have to worry about tuning the tones themselves. They will automatically be correct.

HF operation is a different story. Rather than AFSK, frequency shift keying is used. Your receiver has to reinsert the "carrier". It does so based on where you have tuned the receiver. Even a small error in tuning will keep data from being received.

It's bad enough when you are attempting connection with another manned station. You can always tweak the tuning on your end to accomodate any differences in frequency. With a bulletin board system, you must tune to it, since it can not tune to you. Calibration accuracy is a must. Crystal controlled operation is better, particularly if you will be regularly participating in message forwarding using your station.

The same rules have applied to RTTY operations for years. Packet just adds another degree of difficulty to the task.

While it's possible to tune a packet signal "by ear" and get acceptable results, it is guaranteed to be a frustrating experience. Unlike RTTY signals where you usually have a rather lengthy transmission to tune to, packet radio is bursty in nature. A connect request for example takes only a second, not leaving much time for you to

tune. Additionally you can't check the
results of your tuning until the next
packet is sent. Remember, the entire
packet must be correctly received or it
is rejected. Tuning in mid-transmission
won't prove anything! A tuning
indicator is highly recommended.

An excellent example of an HF
tuning system is incorporated in the AEA
PK-64 when using the optional HFM
module. The tuning indicator is similar
to the system employed in the popular
CP-1 computer patch and is just as easy
to tune. I had almost immediate success
in copying HF packets using the PK-64
with HFM , where earlier attempts with
other TNCs had been frustrating.
Remember though, that for the most part,
the difficulty is not in the TNC itself,
but rather the lack of a good tuning
system. Several schemes have been
devised for adding such an indicator.
The October 1985 issue of the Packet
Status Register (PSR) shows several such
systems.

If you are looking for packet
activity, the best place to check for
starters is 14.103 megahertz.
Additional activity takes place on
10.147, 7.097 and 3.607 mhz. You can
place your transceiver in either
sideband (LSB or USB) or RTTY position.
As packet activity grows, favorite
frequencies will soon appear in all
bands.

What can you hear on HF packet?
While I have been working on this
chapter, I've had the monitor function
turned on with the TNC hooked up to an
R-600 general coverage receiver tuned to

14.103. In the course of several hours I've seen packets from numerous stateside stations from coast to coast, and I've copied DX signals from ZF1, YV5, SM7 and DL4. I get a good solid copy on the WØRLI system and I see beacons from many BBS locations across the land.

Getting on packet is certainly easier using VHF equipment and techniques. For the serious long distance packeteer, or someone interested in participating in long-haul traffic movement, HF packet is worth the effort. If you can, get your feet wet on VHF first, then tackle the low bands.

Chapter Seven
Organizations and Publications

Whenever I decide to tackle a new project there are two things I always try and do. Though it would probably amaze my high school English teachers, I have become a researcher at heart. I comb through all of the magazines and books I have searching for information.

Many times the subject of my interest is so new or obscure that I'm lucky to find anything. Once the first article or reference is found it usually leads to other sources of information. On a rather grand scale, that's how this book came into existence!

The other thing I do is to try and find other people who share my interest in whatever it is that has caught my fancy. Again, that's often not as easy as it sounds. Imagine trying to locate even one of less than a thousand individuals involved in packet radio just a few years ago.

The Best Of The Back

In the appendix to this book you'll find two lists that should help you go a long way in finding additional information about packet radio, both

from publications and individuals. I'd
like to take a few moments though to
make you aware of several special
sources of information that deserve your
attention.

Most of us depend on one or more of
the major amateur radio publications as
our main source of information about our
hobby. Each of the popular magazines
has devoted several articles or more to
packet radio in recent issues. QST in
particular seems to have a dedication to
publishing good, basic information about
this new form of amateur transmission.
Ham Radio has gone beyond the basics
with some very interesting pieces about
the software approach to packet (more on
that in a later chapter), comparing
packet to other transmission systems and
early implementations of packet radio
(Alohanet). CQ and 73 have published
some excellent articles aimed at
promoting packet radio.

Beyond the big four, you'll find a
wealth of information in several lesser
known publications. Of a commercial
nature, Computer Trader Magazine offers
a good source of information . The
recently renamed A-5 Magazine, now known
as Spec-Com devotes regular space to
information on packet. It still has a
loyal following of both slow scan and
fast scan television enthusiasts as
well.

In addition to the information you
find in QST articles, be sure and check
out the On-Line column. Topics vary
from month to month, but packet is
frequently mentioned. If you want a
more reliable source of packet

information from the league, a
subscription to Gateway is highly
recommended. It's a small, newsletter
format publication mailed 25 times a
year. Because it's distributed
frequently the information is often more
timely than found in magazines.
    Speaking of league publications,
you are in for a real surprise if you
haven't seen a recent edition of the
Radio Amateur's Handbook. Beginning
with the 1985 edition, packet radio is
given significant coverage. I
personally feel that this information
alone justifies the cost of a new
handbook! You'll find expanded coverage
of other subjects as well.

Three To Get Ready

    In addition to Gateway, three other
publications are available that every
active packet operator should at least
take a look at. The first is the Packet
Status Register or PSR published by the
Tucson Amateur Packet Radio people. In
1985 editorship of PSR was assumed by
Gwyn Reedy, W1BEL. The result has been
a quarterly publication that contains a
wealth of information on the current
state of the packet art. A subscription
to PSR is included in TAPR membership.
More on that in a moment.
    The only problem with PSR is that
because of its quarterly schedule, it
makes it a long time between issues.
The content is well worth waiting for.
Besides, it also lets you know what's
going on with TAPR from an
administrative point of view. Gwyn also

edits a monthly publication, Packet Radio Magazine, for the Florida Amateur Digital Communications Association (FADCA). Packet Radio Magazine is an outgrowth of the widely acclaimed FADCA Beacon Newsletter. PRM is unique in that it includes the same fine editorial content as Packet Status Register, and also incorporates the newsletters of a number of state groups. A growing number of clubs are adopting PRM as their official publication.

The Amateur Radio Research and Development Corporation, or AMRAD, is involved in many projects. For the past several years one of the main thrusts for AMRAD has been packet radio. The AMRAD Newsletter can be a good source for packet information. The monthly AMRAD Newsletter suffers from a somewhat irregular publication schedule, not unusual with any volunteer effort. In 1985 the Xerox 820 issue alone made the cost of membership in AMRAD worthwhile. I've picked up countless other bits and pieces of information here as well.

The final publication that deserves special mention is one written by Texas Instruments called Understanding Data Communications. It is a 272 page paperback book covering a wide range of communication topics. Included are chapters on synchronous modems and digital transmission, protocols and error control, packet networks and network design. What makes this book particularly attractive is that it can also be purchased from Radio Shack as their stock number 62-1389 at a price that makes it an outstanding value. You

may not find it in stock at your corner
Shack, but don't be shy about asking
them to order it for you. They should
have no trouble in getting it within a
week or two.

You've Got A Friend

     As you recall, I like to find other
people that share my interest in new
technology. There's nothing like having
someone give you a first hand
demonstration of how something works, or
sharing the "fix" to a problem they have
experienced. The best way to find
others like yourself is through a local
packet radio group. In some areas this
may be an outgrowth of an existing RTTY
organization. A list of contacts for
groups around the country is included in
the appendix. While many areas are
included, clubs come and go almost
daily. If you have no luck in
contacting one of the groups near you, a
national organization can provide a lot
of the information you may need as well
as putting you in touch with other
members near by. Two groups already
mentioned for their publications welcome
packet enthusiasts from around the
world.
     Tucson Amateur Packet Radio formed
as a club in 1981. It's largely through
their efforts that TNC equipment is
available today at attractive prices.
Development and research continue at an
amazing pace. TAPR will be happy to
send you some basic information on
packet radio and would be glad to have
you as a member. Write to them at the

address in the appendix.

AMRAD is involved not only in packet radio but also spread spectrum technology, and deaf TTY projects. They run several repeaters and bulletin boards. While amateurs in the vicinity of the Washington D.C. area may receive the greatest benefit from AMRAD membership, others are welcome.

There are additional groups located around the country that offer outstanding newsletters and other membership benefits. Your best bet is to try a few of the ones listed for your area. Ask around and then get active. Often groups sponser digipeaters just as voice repeaters have been sponsored by clubs for years.

The Electronic Connection

Things are moving so quickly these days that the mail seems to have become as slow as the Pony Express. Electronic mail on the other hand can be almost instantaneous. If you are interested in packet radio, chances are excellent that you already have a computer and may already be participating in telephone line based data communication. If that's the case there are several electronic outlets for information you should look into.

On the Compuserve Information Service, a special interest group (SIG) for hams run by Scott Loftesness, W3VS provides a wealth of late breaking news on all aspects of amateur radio. Scott himself is an active packet operator. You'll find many of the country's foremost packet experts active on HAMNET as the SIG is known. HAMNET features a monthly guest operator to help with your questions. Several recent guests on HAMNET of particular interest to packet operators have included Pete Eaton, WB9FLW, of TAPR fame and Howie Goldstein, N2WX, developer of packet software for the newer TNCs. Compuserve is a "pay as you go" service. It's easy to become hooked, so watch your time carefully.

No matter where you live, there are people who share your interest in packet radio. Use the appendix to get involved with others in your area and around the country.

Now That You're Connected

In the first seven chapters you've had a chance to learn quite a bit about packet radio. The roots of packet have been explored in a general sense, the needed equipment outlined and available TNCs reviewed.

You should now be comfortable with the basics of packet radio. Your touch on the keyboard is probably not as tentative as it was before. You know where to look for information and for others who share your interests.

Just like the song, "We've Only Just Begun" to explore this fascinating mode. Some of what follows in part two may be a bit beyond your interest for the moment. In time it will all make sense. While I've shied away from anything really deep, the subjects we are about to explore are definitely intermediate level or higher. Read them through and grasp what you can. You'll be far ahead of most newcomers to packet.

Remember too, the reference material at the end of the book. Packet has a very large set of terms that are brand new to most amateurs. The glossary has been compiled from many sources to help you pin down the meaning of the many new terms you are encountering. Use it often.

And now, for something completely different, let's get into the nitty gritty of forming packets. With an understanding of protocol, many of packet's features become much clearer.

Chapter Eight
Stuffing the Bits

A packet is a bundle of information. It is sent using synchronous data transmission methods. In order for the transmitting and receiving stations to remain in step, some method of periodically establishing synchronization is necessary.

A packet of information is a continuous stream of binary ones and zeros with no start and stop signals designating each character within the packet. However each packet does begin with a synchronizing signal or flag. It consists of an eight bit pattern that should never occur under other circumstances. It is represented in binary as 01111110. In the event five ones in a row are detected in outgoing data, a zero is inserted on the sending end and removed on the receiving end. That's called bit stuffing! This also insures a tone transistion after at most five bit times to provide bit synchoronization between the transmit and receive stations.

The rest of the information in a packet can be anything. You can even begin sending useful data at this point. Although you would be using synchronous communication, you would not be using packet protocol. Packet standards apply rules to the content of some of the information sent. These rules are translated into programming that implements the rules and allows the receiving station to do more than simply receive a stream of data. Protocol is the term used to describe such a system of rules. Remember that the international X.25 protocol sets standards for one form of telephone packet network. The amateur community has agreed on a set of rules known as AX.25 protocol, acknowledging its roots in telephone technology with the A standing for amateur much as it does in AMTOR (AMateur Teletype Over Radio).

Do You Have My Address?

Amateur standards dictate that an address field will follow the start flag of each packet. Amateur radio callsigns are used. The first callsign is that of the station for which the packet is intended. The second contains the callsign of the sending station.

Amateur protocol allows for up to eight stations to be designated as repeaters. Each call sign is allocated six spaces. On a short call, K9EI for example, the extra room is filled with ASCII space characters. To allow each station the flexibility of running more than one node with the same callsign, a secondary station identifier or SSID becomes the seventh character. K9EI-1 for example is a dedicated digipeater. K9EI-2 is a bulletin board. Up to 16

SSIDs can be assigned. The default value for the SSID is zero. So if I use only my callsign, it is actually transmitted in packet form as K9EI-∅.

It doesn't take all eight bits to designate the SSID; only four bits are required. The first bit of the SSID for the source call is set to 1 if no digipeaters follow, otherwise it is set to zero. The SSID itself follows in the next four bits. The next two bits have been reserved for local use by agreement. The eighth bit is used for "C" bits to identify link layer protocol.

If repeaters are specified the format remains similar. The last eight bit pattern of the last callsign group begins with a one again followed by the SSID. The reserved bits remain available. The final bit indicates whether the station identified has or has not repeated this packet.

The length of the address field can be as short as 14 characters if two stations are in direct communication with each other, or as long as 70 characters if all allowable repeater positions are specified. Remember, each callsign consists of six positions allowed for callsign and one more for an SSID. The minimum would be two stations at seven characters each, the maximum a total of ten calls at seven characters each.

Keeping Control

So far we are doing pretty well. The flag told the receiving station to

97

start doing its thing and the address field tells us where the packet is heading, where it came from and how to get it there. We've got one more thing to do before we try and send any information.

The control field consists of a single byte. It gives several useful pieces of information. The bit pattern indicates whether this particular packet frame is an information or supervisory frame. It also contains a frame number from zero to seven. As we will learn later, it is possible for packets to arrive out of order at the receiving station. We've got to mark them so we can put Humpty back together again!

The protocol identifier or PID is the first byte of the data field. It is used to identify the type of protocol being used at the next higher, or network, level.

With all of this behind us (actually it's the "front" of each frame) we can send some data in the information field. The current standards allow for up to 256 "octets" of data to be included. Conveniently an eight bit ASCII character is a single octet. We are not limited to sending ASCII however.

Our definition of a packet did not specify that its contents must be ASCII code. The truth is that the information in a packet can be in any form. Normally we use ASCII code, but the code could just as easily be Baudot, EBCDIC, binary program dumps or a code of our own choosing. If you're trying to think of possible applications for this,

computer program transfer over the air
is perhaps the most obvious. It is
entirely possible however to transmit
bit mapped graphics using packet
techniques and even digital audio with a
high enough transmission rate! That's a
bit beyond us for the moment, so let's
get back to the basics.

Check Your Packets

    One of the beautiful things about
packet transmission is that it is error
free communication! Simple parity
checking schemes allow some degree of
checking but are not foolproof,
especially on serial transmissions such
as packet. You may be familiar with the
much more reliable form of error
checking used in AMTOR transmissions.
Even that is not error free.
    Rather than a single parity bit
being computed, packet allows for a
frame check sequence field (FCS) which
contains a pattern of 16 bits that have
been computed by a complex formula
agreed to by international standards.
The transmitting station computes the
FCS value and places it in the outgoing
packet. The receiving station
recomputes the value. If the values
match, the packet is accepted and an
acknowledgement is sent back. If they
don't match no acknowledgement is sent.
    They say that all of life's a
circle and so it is with our packet. We
end where we began, with a final
01111110 flag.

Packet Personalities

With an understanding of the makeup
of each packet, let's turn to the
general types of packets that can be
sent.
The third element of each packet is
the control field. By paying close
attention to the control code and
applying some more rules of the road
through software in the TNC each packet
can be handled properly regardless of
its contents.
Several types of packets can be
sent, including: connect, information,
disconnect, unconnected and
acknowledgement packets.
To establish contact with another
station, a simple packet consisting of a
start flag, the callsign of the intended
station, your own callsign and a connect
request control code are all that's
required. No information field is
allowed.
If all goes well and the other
station is on and within range, a
connection will eventually be
established. That will require an
acknowledgment frame.
When the first connect request is
sent, the TNC starts a timer. If an
acknowledgment is not received to the
connect packet by the time the timer
runs out, the connect request is sent
again. TNC software can be set by the
user to adjust the number of times a
packet will be "retried" before giving
up. If an answer is not received the
request is eventually abandoned.

Information Please

Generally the purpose of connecting to another station is to send some kind of information. In the information transfer mode two types of packets can be observed.

The sending station will of course be transmitting information packets. Recall that each information packet will include a frame number in the control field. Up to seven frames of information can be tracked by using the number in the control field. The software will only allow up to seven frames to go unacknowledged. The number can be reduced by user definition but it can't be increased. Remember, there's only enough room in the control field to store the numbers zero through seven!

Upon reception, the FCS of the received frame is computed and checked against the received value. If they match, a "supervisory" frame will be sent that indicates acknowledgment of that particular frame. The frame number will be included.

If the information appears to be defective, that is, the FCS value does not match, the packet will be ignored and its contents thrown away.

Since the FCS is corrupted, the receiving station can't be sure the address field was correct.

Back on the originator's end the TNC is keeping track not only of what's been sent, but also what has been acknowledged. The lack of an acknowledgment after a defined period of time will result in the retransmission of the unacknowledged packet.

These events continue to take place
as long as information is being sent.
If the originating station ceases to
send information, the channel simply
remains idle, but the two stations are
still connected. Either station can be
the originator. In fact, in actual
operation both stations are frequently
sending information at nearly the same
time. While it's difficult for we
humans to keep up with such confusion,
computer chips manage to keep it all
straight.

Pulling The Plug

At some point the conversation or
information transfer will come to an
end. The users will want to disconnect
from each other in order to allow their
stations to become available for other
uses. A disconnect request may be sent
by either station, and acknowledged by
the other end.

Stations may also become
disconnected in other ways. If
information packets go unacknowledged
past the retry limit, the originating
station will automatically disconnect.
Particularly with older implementations
of packet protocol this could be a real
problem. From its own viewpoint the
station on the other end of the path is
still connected making it unavailable to
other stations.

New packet software employs a
fail-safe disconnect sequence. If
either station detects no activity from
the other station after a user-defined

period of time, a packet is transmitted
to "test" the connection. If
acknowledgment is received, all is left
alone. If there is no acknowledgement,
the station that sent the "test" packet
begins the disconnect sequence.

Communication Without Connection

While recent developments do allow
for multiple connections between packet
stations, some existing software will
only allow two stations to be connected
at one time. In some cases several
stations may wish to participate in a
round table discussion.
Unconnected transmissions are
allowed for. With TAPR-based units,
typing CONVERS at the CMD: prompt will
place you in the conversation mode.
Everything you type will be sent. The
packets sent will be unnumbered
information frames. No acknowledgment
will be received from the receiving
station or stations.
These are the same type of frames
used for beacon messages.

More To Come

The protocol just described is
officially Level 2, Link-Layer Protocol.
The International Organization for
Standardization devised a scheme to
describe any communication system. It
divides the system into seven levels.
Starting with the basics, level one
describes the actual kind of
transmission of data from one station to
another of bits of data, modulation

methods, data rates and so on. That's
the "physical" layer.

Link layer or level 2 arranges the
bits of data into frames and insures
their accuracy through the path. The
network layer, level 3, takes frames and
assembles them into packets.
Information on network routing is
included at this level.

Pausing here for a moment, today's
packet radio is difficult to pigeon-hole
into these distinct categories. As of
early 1986, only the link layer protocol
had been agreed upon. What we often
refer to as networking with the present
system is not really networking. We
have adapted the second level to allow
for repeating of packets. Ultimately, a
network layer protocol will be agreed
to, allowing more flexible packet
operation.

Even elements of the transport
layer, or level 4 can be seen in amateur
packet radio today. Officially, level 4
defines how messages are transmitted in
packets and how to reassemble packets in
the proper order on the receiving end.

The session layer, level 5, makes
the decision when a contact is needed to
exchange data. Store and forward
bulletin boards are an example of this
type of decision and are discussed in
their own chapter.

At level 6, presentation level
protocol actually displays the message
or files it in a standard fashion. For
example, teletext services available
both by telephone connection and through
transmission in the vertical blanking
interval of TV stations must use a

standard presentation level protocol so that all of the information sent by the service can be properly displayed by the customer. This often includes not only text but graphics as well. It could even define sounds so that music could be transmitted. One of the problems with the growth of teletext has been the lack of a single standard for transmission. We'd do well to learn from the mistakes of commercial data suppliers.

The final level, seven, is called the applications layer. It's here that the programs that interpret what you want to do actually make it happen.

Don't Be Confused

Entire documents have been devoted to discussions of only a single layer of protocol. If you wish to explore packet radio or other communications that deeply, refer to the bibliography. Several excellent documents are available.

Most amateur radio packeteers only need to know that there is a standard on which to model communication systems. You should now be able to nod in a knowing manner when the experts throw around terms like "presentation level protocol" and "transport layer".

Please keep in mind that I have described the current version of AX.25 link layer protocol. In packet radio's history other methods have been used. The first popular protocol was the Vancouver protocol. One of the major differences in the Vancouver system, is

that node addresses had to be assigned.
Callsigns were not used for addressing.
The protocol allowed for only 31 to 255
possible nodes in one network. This
could be a problem with today's
popularity of the mode.
    For now we do have a standard.
Keep in mind the need for additional
standards and don't forget the guys and
gals who got us where we are. Their
methods weren't wrong, just different.

Chapter Nine
Networking

Depending on your frame of reference, the word network conjures up many images. Amateur radio operators probably think of traffic and service networks or nets as we usually call them. A business person may well think of a people oriented network...who can I call that knows something or can help. Telephone engineers immediately think of the miles of wire and radio carriers that are all carefully intertwined to allow rapid connection of voice and data circuits.

All of these forms of networking share many of the same basics. There are just about as many approaches to networking as there are network designers. There is even a set of mathematical networking principles.

While packet radio networks may one day reach the complexity of the world's phone system, today we are at the very beginning of designing the system.

History is often a good teacher, so let's return to the days of Hiram Percy Maxim and his peers. Amateur radio began in a very haphazard, unorganized fashion. To be a ham, you only needed to build a transmitter and a receiver.

A license wasn't required.

Who Are You?

Early amateurs often used their initials or first name as an identifier. As long as there was only one JG in your communications range no confusion resulted.

As the technology improved and hams discovered that the "useless" ultra short wave frequencies (about 1.5 megacycles or 200 meters as it was called then) actually allowed communication over hundreds, even thousands of miles, a better identification system became necessary. At first just a number was added in front of the identifier. Call districts were born, though they have undergone some changes over the years. Finally by international agreement the prefix "W" was assigned to the United States. Suddenly "JG" was W9JG if he was lucky.

Most amateur communication was random at first. Transmitters were unstable and receivers were all but impossible to calibrate accurately. It made finding the same signal on the same frequency quite difficult.

Then came a challenge. With the state of the art improving, the better equipped stations across the nation and in fact around the world found that they could talk to each other "on schedule". A group of them set up a test to see just how rapidly a message could be relayed from one coast to the other. Radio networking on amateur frequencies had begun.

Early messages took minutes or even hours to relay. Tradeoffs were made. When possible, better equipped stations tried to make up time by relaying directly between themselves. When the path failed, traffic was routed through lesser stations where reliability was high, but the trade off was a delay in delivery of the message.

Out of these early efforts grew the National Traffic System. It still exists today, but rather than being a CW only net, it uses all modes of communication.

A Strong Chain

The concept is fairly straightforward. Generally a message from any particular town is first routed to a participant in a statewide network. If the message is for another point in the same state, it is relayed to a station as close as possible to the final destination using the statewide network.

Messages destined for more distant points are transmitted to the state's representative for a regional net. Depending on what part of the country you live in, the regional network serves a number of states. Normally only specially appointed stations operate in regional networks.

Traffic headed out of the region goes to yet another level. This is where messages make the big hop from region to region and across the country.

After working it's way up the chain of networks, messages then come back down in a reverse fashion and are finally delivered if all goes well!

Several key points can be gleaned from examining traditional amateur traffic networks. First, most amateurs will never have a need to participate in traffic handling beyond the state level. It's not a reflection on their abilities. Indeed, good traffic handlers are always needed for the upper levels, but they must be prepared to spend the time required. Generally they need to be appropriately equipped for their level of participation. Propagation has a way of being the worst when the traffic load is the heaviest and the messages are most urgent.

Secondly, you may have noted that participation in the upper levels of traffic handling is by appointment only. Again, it's not a matter of ability, but rather one of efficiency. At the upper levels, the traffic load is quite high. The members of the network have to work well together. The introduction of someone unfamiliar with proceedure can slow down traffic considerably. Duplication of effort becomes a problem. Messages get lost due to misunderstandings.

While the network just described is an entirely human one, the principles can be applied to networking of any kind. One step at a time, let's investigate the packet network and how it might be implemented. Some terms that may be new to you will be introduced, but keep the model just described in mind.

Weaving The Network

In computer networking terms, a
group of computers located in the same
close communications proximity is called
a Local Area Network or LAN. For
practical purposes you might consider
any station that you can connect to
directly on VHF or UHF frequencies using
reasonable power and an omnidirectional
antenna a member of your LAN. Each
station has a unique identification and,
generally speaking, any station can
speak to any other station within the
local area.

The model begins to have problems
when stations come on that have severe
power or antenna limitations. Not
everyone can put up a vertical antenna
at fifty feet.

What about a station located just
outside your LAN but isolated from all
other stations. With a bit more power
or a directional antenna that station
can effectively become part of your
local group, as long as the amplifier is
on and the beam stays aimed toward town.

The larger the number of stations
in the area the more likely it is that
you will run into the special
circumstances just described. Just like
FM, packet radio can make use of
repeaters. In fact, regular voice FM
repeaters can be used effectively to
relay packet traffic, but there is a
less costly way. Properly used a
traditional voice repeater can be better
than a digipeater.

Remember, any TNC has the
capability of acting as a digital

repeater, or digipeater. It can do so
in a manner that does not interrupt the
normal operation of the station.

In some communities that's exactly
what is done. One of the better
equipped stations becomes known as a
good relay for area communication.

If the funds and cooperation are
available, it is better to establish a
digital station solely for the purpose
of serving a particular locality. All
that's required is a transceiver, a TNC
and a single antenna. Although a
computer or terminal is needed to
initially set up the TNC, a dedicated
terminal is not required in such
operations.

Far And Wide

When a station is operated only as
a digital repeater and is located and
equipped to serve a fairly wide
geographical area (comparable to an FM
voice repeater for example) it serves as
the basis for a Wide Area Network or
WAN. With a system like this, a network
of low power VHF stations can
communicate with each other using only
one relay over a 20 to 50 mile radius or
more, depending on terrain and other
conditions.

The traditional definition of a WAN
is a collection of local area networks.
In many parts of the country the two
definitions tend to blur together.

I don't know about you, but all of
my local friends don't fall within that
radius. Apparently many others feel
that way too because methods have been

developed to allow Wide Area Networks to be linked together to allow almost infinite range. We'll find out why there is a practical limit in a moment.

If two or more WANs share the same frequency and their coverage areas overlap, it is possible to use the wide area repeaters to relay to a station in a distant LAN. You can see however how stations equipped to allow for such overlap could begin to interfere with each other.

The reality of radio networking in most areas of the country at this writing is that relay through multiple wide area repeaters is the only effective way to connect directly to a station in a distant LAN. During the evening and just before morning work hours, there are so many stations on in my area that while there is no problem with the path itself, the number of collisions and repeats required make it very frustrating when trying to get messages through. I have the advantage of being able to operate during the day time work hours and that's when I pass most of my traffic. Most are not so fortunate.

My area is without a dedicated digipeater. We have several well equipped stations that have the capability of repeating my signal into the nearest wide area repeater located nearly 70 miles away. The path is not reliable and stations are often down while their owners are on vacation, suffer power outages or for numerous other reasons. When several of the better equipped stations are on, their

signals are still heard by the wide area
repeater even though they may only be
communicating across town. That cuts
into the effectiveness of the WAN. The
effect is almost identical to the
problems encountered when FM repeaters
are located on the same frequency in
close proximity to each other, or when
band conditions are enhanced.

A Better Way

There are some excellent cures
available for these problems. They
follow the same techniques we learned
about in both telephone packet networks
and even the human national traffic
system.
Let's pretend for the moment that
Springfield, Illinois did have it's own
area repeater. With such a machine in
place, stations within a wide diameter
of Springfield would be able to
communicate with no problem using low
power and omnidirectional antennas, or a
simple beam aimed directly at the
repeater.
If a similar machine existed about
35 miles up the road it would be
possible to reliably connect to stations
within the local coverage of that
machine as well, provided that one of
two criteria were met. As noted, if the
distant machine shares the same
frequency and both machines can hear
each other, a relay can take place.
That's the method most used today and
the one that causes most of our
problems.

Put It On A Trunk

What if the local repeaters on both ends have a second port (computer jargon for input/output connection) connected to a transceiver on a different frequency, perhaps even a different band. With the proper addressing the relay between the two repeaters would take place on a trunk circuit established only for inter-repeater communication. Borrowing from telephone jargon again, a trunk is a communications path that is not directly accessible by the end user. When you place a telephone call across the country or even to a distant part of town your call is first routed via a line connection to the central office. You have direct access to this circuit everytime you pick up your phone.

At the central office the addressing you used (the telephone number) instructs the telephone equipment to connect you to a trunk that starts you on your way to your final destination. Depending on the distance involved your call will be routed through several levels of network before it finally gets connected to another line on the other end. You had no direct control of what path was used or the particular circuits used. The central office machinery made those choices based on its own programming and agreed to convention.

If longer distances are involved, toll switchers or long distance switching offices become involved. They work just like the central office

equipment, but are not concerned with
individual lines. They deal with trunk
circuits only. Their job is to connect
offices rather than individuals.

All it takes to implement this
technique in the amateur service is some
programming. The existing hardware is
flexible enough to handle the job. Even
as I type, amateurs are hard at work on
just this type of system. In some areas
of the country you may already have
access to limited trunking capabilities.

Another new term is fast becoming
commonplace. By now you know what TNC
means. New hardware, called a network
node controller, or NNC, is being
developed. For our purposes it can be
thought of as the equivalent of the
phone company's trunk switching
equipment. It is concerned with
connecting repeaters rather than
individuals.

Long Distance Carriers

By tying into high frequency
stations and satellite links, additional
levels of networking are possible. It
is also likely that in the future the
trunk operations will operate at eight,
sixteen or even greater multiples of the
1200 baud circuits currently available.
In fact, current TNCs have the ability
to communicate at 56,000 baud using
advanced modem techniques and special
software.

In the meantime, an answer to the
problem of long distance message
relaying has presented itself in the
form of store and forward operations.

The PACSAT program will utilize exactly
this technology and is described in its
own chapter.

Today, thanks to amateurs like Hank
Oredson, WØRLI, versatile radio bulletin
board systems have been devised that
will accept messages for distant
stations. The software contains a table
showing how to route mail to any
participating station. The main WØRLI
BBS operated by Hank on the east coast
is accessable on 20 meters (14.103 mhz)
allowing for worldwide message exchange.

On a predetermined schedule, store
and forward mailboxes attempt to connect
with distant stations and pass their
traffic. In my area, messages for
Illinois, Indiana, Michigan, Iowa,
Missouri and beyond are automatically
transferred between BBS stations in the
middle of the night.

The WØRLI software allows
connection of multiple ports to
facilitate moving traffic between two or
more frequencies. It's a nifty solution
to a real problem. Just like the
methods used by our forefathers, the
only sacrifice is the speed of delivery.
This is at least partially offset by the
automatic nature of the system. No
longer is it necessary for a human to
facilitate the relay.

This chapter is not intended as a
definitive work on networking.
Hopefully, it gives you an idea of the
way things work today and how they may
work in the near future. If you are
interested in some of the more esoteric
networking techniques, be sure and read
about the WØRLI system and the PACSAT

project, each in their own chapter.

Chapter Ten
The Xerox 820

"But I was so much older then, I'm
younger than that now," goes a song from
Bob Dylan. It's just another way of
saying, the more you learn, the less you
know.

As I began my quest for information
on packet radio, many terms new to me
started to surface. I thought I had
become at least semi-literate, packet
wise, after learning the meaning of TNC,
SSID and other selected collections of
alphabet soup. I was beginning to feel
pretty knowledgable until the day I got
hit with so many new terms. I didn't
know which way to turn.

It all happened when someone said,
"you know something about packet...where
can I get a Xerox 820 and a FAD board?"
If you already know the answer to those
questions, move on to the next chapter.
If not, and you are an experimenter by
nature, you may find this one of the
most interesting chapters in the book.
Regardless, the Xerox 820 is an on-going
part of packet radio history and
deserves at least a bit of your
attention.

Computers haven't always been
cheap. Just a few years ago a Commodore

64 or equivalent Atari machine cost
$500. A few years before that, you had
a large system if you could afford more
than 8K of memory! Most amateur
experimentation with microcomputers
began with S-100 based systems.
Generally, you had a box with a lot of
slots in it. Unless you were very rich,
most of the slots remained empty. The
system could be expanded as finances
allowed, with each addition occupying a
new slot in the box.

Many of these systems were based on
the 8080 microprocessor or its new and
improved cousin the Z-80. The first
time the chips appeared for hobbyist use
they ran $50 to $100 in single unit
quantities. A $10 bill will usually buy
you several high grade chips these days.

All On One Board

In 1980 a man named Jim Ferguson
designed a single board Z-80 computer.
Jim's hardware was originally marketed
by Digital Research of Texas. Xerox was
looking for something to base an
"inexpensive" computer on and purchased
the design. After some extensive
modifications to the packaging, Xerox
introduced the computer as the 820-I.
It is a multi-layer board which can make
finding some of the connections a bit
tedious. Incidentally, the original
circuit offered by Digital Research is
sometimes referred to as the "Big
Board"--a term you will hear used from
time to time among Xerox enthusiasts.

The reason for telling you about
all of this is that many of the Xerox

820-I boards have found their way into the hamshack. There's a very good reason. They come cheap!

Xerox went on to improve the 820 design, so much so that the new version is designated the 820-II. Because there are so many differences between the two, only the 820-I is described here.

We've come to think of computers in terms of complete packages. For the 820 enthusiast, that's not the case. Just having an 820 board won't get you much.

The Xerox 820 was chosen as the system to be used for developing network systems for packet radio. The decision was based largely on cost and availability.

Searching The Flea Markets

Before the 820 becomes a usable machine, you'll need an 820-I board. Prices very, but $50 is typical for an untested unit that has been pulled from service. You'll need a power supply. It can be homebrewed, but units are available that will do the job at a reasonable price.

You have to find a compatible keyboard. A video monitor is a must so that you can see what you're doing. You'll need one or more disk drives. Size and type can vary. The WØRLI system, for example, is designed to run on 8 inch drives. Others have had good success with 5 1/4 inch units. Finally, you get to find or construct all of the necessary cables to get it all hooked up. Some connectors are a bit hard to find, others are readily obtainable. As

you can see, being a Xerox 820
computerist is not for appliance
operators!

Since it is Z-80 based, the 820
lends itself best to running CP/M. Old
time computerists will recognize those
initials. It may be a new term for the
computer novice. CP/M--Control
Program/Microprocessor is an 8080 or
Z-80 disk operating system.

All of this is nice, but it still
doesn't get us operating on packet
radio. The rest of the circuitry needed
will depend on just what you want to do
with your 820.

What Will You Do?

If you are interested in using the
820 simply as a computer, such as
required by the WØRLI mail box software,
no additional major hardware changes are
needed. You can use standard TNCs to
implement the mail box.

If you are interested in making the
820 into a packet station of it's own,
as opposed to just using it as a
terminal, you've got your work cut out
for you.

Because packet techniques call for
a special type of data encoding,
additional hardware not found in most
computers is required. The magic
interface to accomplish this is called a
FAD--frame assembler/disassembler.
You'll also need a modem to accomplish
the radio interfacing.

If you like to experiment and make
something very useful from industry
cast-offs, the Xerox 820 may be your cup

of tea. I've included an appendix of
820 suppliers. I also highly recommend
that you invest in several publications.

Much of the information on the
Xerox 820 has appeared over a several
year span in the AMRAD Newsletter. The
August 1985 issue is a Xerox 820
compendium that's a must for the Xerox
experimenter. You'll find AMRAD's
address in the appendix.

The FADCA Beacon has on-going
coverage of the Xerox 820 machine.
You'll find many of the nitty-gritty
details of the FAD board and other 820
related topics in their excellent
coverage over the years.

Micro-Cornucopia Magazine, P.O.
Box 223, Bend, Oregon 97709 is a
favorite among single-board computer
enthusiasts. They've devoted a lot of
space to articles on the 820. It
deserves investigation.

The Xerox 820 isn't for everyone.
Seeing one in typical hamshack operation
is like a trip back in time. It reminds
me of my early days in radio. I can
still recall the first time I saw
someone working through the OSCAR
satellites of the early sixties. There
was an aura of science fiction to it
all. Even if you aren't that much of an
experimenter, you'll feel like a pro
when someone pipes up at the club
meeting and asks about big boards, FADS
and PADS.

Chapter Eleven
Store and Forward

There's an aspect of packet radio
that has its roots in RTTY operations.
In fact, while packet radio makes
bulletin board operation easier and of
course error-free, it has no magic power
over BBS operations.

In case you are new to radio
bulletin board operations, let's take a
moment to discuss the roots of bulletin
board systems and a typical setup.

Computer hobbyists have enjoyed
telephone line based message systems for
years. Using your computer and a
telephone modem you dial a number. On
the other end another computer running
special software answers. Generally you
are greeted with an opening message and
invited to log-in. Most systems require
a special log-in code and usually a
password. Often the sequence includes
your full name or an ID number assigned
by the system operator or SYSOP.

If your log-in is not recognized,
most systems open to the public will
prompt you for additional information to
help you become a member of the system.

Once you are logged on the system,
a menu will offer you a variety of
choices. The options vary, but usually

include a message exchange section,
bulletins and even a downloading section
for public domain computer programs and
text files too long to be included in a
single message.

Through a telephone based BBS you
can leave messages for other users.
Unless a private message feature has
been established and used, you can scan
messages to and from other users as
well. Often on-going discussions take
place on a variety of subjects,
depending on the board involved.

It is this type of operation that
led to the establishment of radio
bulletin board systems or RBBS. Some
systems concentrate on message storage
only and the term message storage
operation or MSO has become popular.
Most MSO or RBBS operations operate
using standard Baudot RTTY transmission
methods. There are hundreds of stations
operating systems, both on HF and VHF.
On the low bands 20 and 40 meters are
the favorites, while two meters carries
the bulk of the activity on VHF.

The first amateur MSO systems were
stand-alone computers. They were
expensive to build or buy. The HAL
Corporation led the way commercially
with MSO capabilities built into their
more sophisticated RTTY systems.

A few years ago when computers
became affordable for the home and
hamshack, amateur programmers created
MSO and RBBS software that allowed low
cost machines to be used for bulletin
board operation via radio. With the
widespread availability of terminal
units designed to work with personal

computers, RTTY operations flourished.

The natural language of computer communication is ASCII, the American Standard Code for Information Interchange. ASCII allows a wider range of characters to be sent, and generally takes place at higher speeds than RTTY. By using ASCII rather than traditional Baudot code it became easier and faster to exchange computer programs using amateur radio as the link.

With this background in mind, it's easy to see how RBBS and MSO operations became popular on packet radio. When the error-free nature of packet is added to the operation, message transfer and program exchange become extremely attractive.

I've pointed out elsewhere that packet transmission pays no attention to the contents of each packet. That is, you could send traditional RTTY data using packet radio. With the proper terminal software on each end, the result would still be an error-free path. For most purposes however, packet radio does make use of a slightly modified form of ASCII transmission.

While various systems are in use around the world as packet bulletin boards, few of them enjoy widespread availability. In many cases a landline based bulletin board software package is modified for use on packet. That's not too tough of a job for an experienced programmer. Chances are that if you have the knowledge to write a BBS system from the ground up, you'll have little trouble making the modifications necessary to allow it to operate

effectively on packet. Keep in mind
that menus and text should be kept to a
minimum when designing a system for
packet radio use. On the other hand, if
programming is not your forte, you'll
probably become quickly frustrated
trying to get the job done.

Showing The Way

At this writing one system has
become a defacto standard for BBS
operations on packet. Utilizing a
surplus Xerox 820 computer system, Hank
Oredson, WØRLI has designed an extremely
versatile software package with many
features. To give you some idea of how
Hank's system has grown and improved, in
late 1985 Hank released version 10.0 of
his program. Each version has included
major additions.
I won't go into the details of the
Xerox 820 computer here. It is covered
in a chapter of its own. Instead, let's
take a look at the WØRLI mailbox system
from a users standpoint.
Most users of a mailbox system are
interested in sending and receiving
mail. To get started a standard connect
message is sent. For example:

C WØRLI

You could of course connect using one or
more relays, but let's keep it simple
for now.
When the software at WØRLI sees the
connect message, it goes into operation.
A lot of things take place
automatically, but what you will see is

a welcome message. Hank's software keeps track of all users and their last access of the system. The opening message may include the time of your last access!

You'll receive a prompt requesting you to make a selection from the main menu. If it's your first time you won't know what your options are. However by typing an "X" a longer prompt with brief explanations will get you going.

You may want to read a message. If so the "R" command will get the job done. If you want to send a message the "S" option will get you started. The WØRLI system is loaded with features. To begin with though, you should master the send and receive commands since they will be your bread and butter. See the appendix for details.

Two of the more unique capabilities of Hank's system include gateway operation and store and forward capabilities.

Hitchin' A Ride

OSCAR satellite enthusiasts may already be familiar with the term gateway as it applies to the amateur satellite program. The concept of a gateway is that not everyone is equipped for specialized or long distance communication. By providing an easily accessible link to specialized communication, a gateway makes new modes available to a greater number of users.

In the case of an OSCAR gateway, a VHF repeater type of operation is generally used. Local amateurs transmit

to the gateway just as they would to
someone across town. On the gateway
end, the transmission is repeated using
equipment set up to operate on the OSCAR
input frequency. The receiver signal at
the gateway is retransmitted over the
local link to the user. All of the
things that generally make OSCAR
operation difficult become transparent
to the user. It's all taken care of by
the gateway equipment. It's a great way
to get people involved in new forms of
communication.

A packet gateway operation is
similar but is generally implemented for
a different purpose. Most gateway
operations on packet involve connecting
a VHF local link to an HF long-haul
link. This allows connection to
stations thousands of miles away using
nothing more powerful than a VHF
hand-held radio! Hardware wise, two
TNCs are required. One is connected to
a VHF transceiver and one to an HF
station. While the units could be
hardwired together, a better approach is
to use a computer with appropriate
software to establish the connections
and monitor the activity.

Borrowing From FIDO

When message systems exist in a
stand-alone configuration, all of the
users must be able to access the same
system to send and receive messages.
For example, it doesn't do any good for
me to leave a message for WØRLI (who is
located in Massachusetts) on the W9CD
MSO located in my area. Hank can't

access my MSO, so he'll never get the message. That's where store and forward operations come into play.

Using the WØRLI software, messages can be addressed to stations anywhere in the world, provided that certain conditions are met. Here's an example:

I want to send W3VS a message but my HF station is off the air so I can't access his station directly, but I know he checks the WØRLI system regularly. I would access the W9CD system, send my message and address it to W3VS @ WØRLI. The "@" is an indicator that the message needs to be forwarded. The second call sign is the "at" address. That is, send this message to W3VS and you will find him active "at" the WØRLI message operation.

The software at W9CD will search the forwarding file to find out if it has a way to get to WØRLI. If not the message is accepted but will just sit on the W9CD BBS until deleted by the SYSOP.

Let's assume that the forwarding file at W9CD shows a path to WØRLI by relaying on VHF to WB9FLW. WB9FLW is running WØRLI software and has both an HF and VHF port. The message will be relayed automatically on VHF to WB9FLW and then will be relayed again via a 20 meter link to the WØRLI board in Massachusetts. Scott, W3VS, could then reply to the message by addressing his response to: K9EI @ W9CD.

Forwarding takes place according to the instructions included in the software. In my area for example, VHF forwarding typically takes place in the middle of the night when other activity is low and the paths are generally the most reliable. Some systems try forwarding every hour. Still others will only attempt forwarding when the distant station has been heard recently.

The same type of system is being used for a land line based computer message system. It is called FIDO Net. Rather than using radio paths, FIDO uses standard telephone connections to forward messages. Obviously, long distance charges are incurred. Otherwise the forwarding scheme is very similar to that used by the WØRLI system.

I've included an appendix of WØRLI commands at the end of this book. It will serve as a good reference for using WØRLI systems. You'll probably find it helpful with other systems since many commands have similar meanings.

Different Drummers

While the Xerox 820 is a logical low cost choice for dedicated message operation, other personal computers have become equally attractive. Unfortunately little software has been developed for other machines. One notable exception is a package created by Bob Bruninga, WB4APR. Bob developed a system originally for the VIC-20. It used two Vancouver TNCs. Bob has since updated the software to work with TAPR style TNCs and the Commodore 64. With slight modification it can still be used on the VIC-20 and should be adaptable to other Commodore machines as well.

Bob has chosen to store all

messages in the computer's random access memory rather than on disk. He reasons that it makes little sense to spend $200 for a disk drive for a $100 computer to use in dedicated amateur service. To that end, the program is provided on tape. Once a day, the messages in memory are backed up to tape. Bob also uses a battery back up for the C-64 memory and reports little problem in continuous operation, even during power failures.

The limitation of course is that the system is not capable of storing either a large number of messages or files of great length. Bob has gone to great lengths to make his system compatible with WØRLI store and foward protocol. No modifications are necessary to accomplish this. The C-64 software also includes gateway capabilities.

If you are interested in obtaining software for packet BBS or MSO operation, both gentlemen have been kind enough to make their programs available at just enough to cover the cost of duplication.

To obtain the WB4APR software for the Commodore 64, send $5.00 to Bob Bruninga, 59 Southgate Avenue, Annapolis, Maryland 21401.

Perhaps the best way to obtain the WORLI software is to copy it from someone in your area who is already running it.

If you can't find a copy locally, TAPR is looking into a software distribution system similar to the AMSAT Software Exchange. To this end K7PYK in

Scottsdale has agreed to distribute the
SYSOP manual for the WØRLI sytem. It
continues to be written by Jon Pearce.
Drop a note to K7PYK to find out what
the current copying and postage charges
come to.

Dwayne Bruce, VE3FXI has offered to
make the source code for his WØRLI
compatible program available for the
cost of media and postage. Dwayne's
system is written in C language. He
prefers 5 1/4 inch disks operating under
OS-9. If that's Greek to you it won't
do you much good. The C language is
used by many professioal programmers and
not very common among amateur
computerists. You can reach Dwayne at:
29 Vanson Avenue, Nepean, Ontario,
Canada K2E 6A9.

If you are interested in running a
BBS on an IBM PC or one of its many
clones, Randy Ray, WA5SZL may be able to
help. In cooperation with the
Raleigh/Durham, North Carolina group,
they are making software available in
return for a diskette, mailer and return
postage. At this writing the system
does not support WØRLI forwarding but
they plan to include it in future
versions. You can write to: Randy Ray,
WA5SZL, 9401 Taurus Court, Raleigh,
North Carolina 27612.

Jeff Jacobson, WA7MBL has developed
a WØRLI type system for the IBM PC
written in Turbo Pascal. Jeff and
several other were testing the software
in early 1986. At this writing the
software was not yet available for
general release. Watch future issues of
PSR or check with Scott, W3VS on HAMNET

133

for the latest details.

Bulletin boards and message operations are only one aspect of packet communication. Such operations can be detremental to overall activity under some circumstances. Before you decide to put a BBS on the air, make sure that it will really be an asset in your area. If there is already a board in operation near you, little will be gained by adding another one.

Let's explore some of the alternatives and unique applications that have been dreamed up by packet visionaries.

Chapter Twelve
PAM-A Low Calorie MSO

With the proliferation of radio
bulletin boards on both packet and RTTY
frequencies, some frequencies have
become congested with endless requests
for help menus, message lists and other
files. There's nothing inherently wrong
with RBBS and MSO operations. The
problem occurs only when users don't
stop to think about the "big picture".
Since there are almost as many sets
of operating procedures as there are
message operations, users new to a
particular system will almost always
need instructions. Even at 1200 baud,
it takes a while to transmit a help
file. If we define a local message
system as being one that can be accessed
with no relays, that is, a direct
connection, a local user accessing such
a system gets the full benefit of the
relatively high speed of packet radio.
Except in areas of very high activity,
such systems don't cause most of the
problem.
Remember the excitement of being
able to key a distant repeater on two
meters when the band conditions were
good? When FM operation first became
popular and there were few repeaters,

135

that didn't cause many problems. As
activity grew, so did the number of
repeaters. With coordination efforts,
interference was minimized under normal
band conditions.

Unfortunately there are still those
stations that insist on running high
power to access a distant machine, often
keying several others when they do.
Just about any spring morning will
demonstrate what I'm talking about.

Packet radio is a bit different.
Certainly enhanced band conditions on
VHF allow additional range via packet
just as it does on voice. Even without
good conditions, the inherent
digipeating capabilities allow paths
hundreds of miles long to be created on
demand. All you need are stations at
key points along the path.

The problem is that once you start
a packet moving through such a path, it
becomes inaudible to you at some point.
You have no way of knowing what may be
happening several "hops" down the road.

The ever exploring amateur will by
nature want to try and access distant
stations. Again, there's nothing
particularly wrong with that until one
person's activities lead to a packet
frequency being dominated by one
individual for a great period of time.

Generally speaking the problems
created are entirely by accident. A
better educated packet community can
result in a better working system for
everyone.

For each relay included in a path,
it takes additional time for the same
amount of information to get through. I

frequently use a path that involves two
or three relays, depending on
conditions. During the day when most
amateurs are at work, the path works
well for me, and my traffic is delivered
rapidly. I'm able to get on and off
quickly leaving the frequency available
for others.

At night, when activity reaches a
peak, the same path becomes almost
useless. With several stations all
trying to make use of a path involving
one or more relays, collisions occur
frequently. Everyone suffers.

It's particularly frustrating when
some of the collisions are caused by
repeated attempts by other stations to
access a distant message operation for
no purpose other than to say, "I made
it".

I think the problem is aggravated
in my area by the lack of a full-time
local message system and the lack of a
wide area digipeater. In both cases the
nearest machines are about 80 miles
away.

There are some alternatives to MSO
and RBBS operations that lend themselves
well to packet radio.

Personal Packets

Since almost any standard computer
communication program can be used with a
packet controller, the features of the
more advanced software packages can be
put to good use. Many programs allow a
receive buffer to be turned on to
capture all incoming messages. By
simply turning the monitor function of

the TNC off, only messages sent by stations connected to yours will appear in the buffer. You can then read any traffic that has come in during your absence. The system is simple, and usually costs nothing extra.

There are a few drawbacks.

Unless the station connecting to you knows that you are saving connects to a buffer, he or she has no way of knowing that messages are welcome even when you are not available. There is no "welcome" message sent. Particularly with TNC 1 type units, there is no automatic disconnect feature. That is, if someone connects to your station and for whatever reason you do not receive a disconnect message, your TNC becomes "hung" and is unavailable to any other station. You have to manually disconnect once you find the condition. That can lead to bad feelings.

If for some reason your buffer should overflow, or the terminal program crash, everything is lost and you won't receive your messages.

Some terminal software allows for automatic periodic disk saves of the buffer. Still other amateurs have written simple software that automatically logs all connects directly to disk rather than just printing them on the screen.

The no "welcome" message and disconnect problems are still there.

New products such as the AEA PK-64 and TNC 2 work-alikes include an option to respond with a message of your choosing upon connection. The buffer is turned on so incoming traffic can be saved. A watchdog timer insures that your station is automatically disconnected after a specified amount of time of no activity. Things are starting to get better!

Somewhere between storing messages to a buffer and a full-fledged MSO lies the concept of a packet answering machine or PAM. Features vary just as they do with telephone answering machines. Generally, here's how one works. Special terminal software constantly checks for a connect message. When one is received an outgoing message is prepared and sent to the TNC. It may be as simple as "please leave your message and then disconnect" or a short menu may appear. I've seen PAMs implemented on Apple, Atari, Commodore and Xerox computer systems. I'm sure somebody somewhere has written one for just about every machine around.

Some of the best PAM programs I have seen include two major features. First they allow you to leave a message for the station operator. Your message is automatically stored. Generally the message is marked by the software with date and time or a message number so that it can be retrieved by the operator at a later time.

The second feature involves a limited outgoing message system. When you connect, if the station operator has left a message addressed to you, you will be prompted when you connect. Upon command, the message will be sent to you. No one else can retrieve the message and you can't read anyone else's message (at least not without illegally

logging in under their callsign). Such
a system allows for unattended
operation. You never have to worry
about trying to catch the other person
in the shack.

Some have implemented additional
features. For example, a "page the
operator" option might ring a bell or
flash a light to signal the operator
that someone has connected. If the
operator is not available the program
will revert to automatic operation
informing you to try again later.

Incidentally, sending a few
CONTROL-G characters may catch the
attention of the operator at any packet
station. That's the "bell" code. If
the receiving station has sound
capabilities, a tone will generally be
sounded by the computer terminal.

I've seen PAMs that allow for
automatic file or program transfer. The
more features they contain, the more
MSO-like they become.

A packet answering machine offers a
good alternative for stations not served
by a bulletin board, or who generally
only need to exchange traffic with a
limited number of other near-by
stations.

MSO and PAM operations aren't the
only things you'll find available via
packet radio. In my area, National
Weather Service data is available using
a HAL system. It operates much like the
standard HAL MSO but has been customized
to automatically update the forecast
files. It's a good example of a
specialized application. Perhaps you
can create a system that only allows

ARRL bulletins to be read on demand.
How about "propagation by packet"? Some
fairly simple additions to the popular
MINIMUF propagation program would allow
an on-line propagation service for your
fellow packeteers. Combine a PAM and
something like that and you are sure to
be a popular connect! Use your
imagination and I bet you can come up
with a unique application. It might
make a good club project. In any event,
make sure you are really adding
something and not just putting on
another also-ran.

Chapter Thirteen
PACSAT, SAREX 2, Meteors and More

Wayne Green, W2NSD, has a story he
likes to tell about what can be done
with today's pico-computers. Before
long, the story goes, you'll be able to
pull out your lap computer, type a
message to any similarly equipped
individual in the world, and have your
message delivered in just a short time.
It will all happen without using
telephone circuits or requiring any
special effort on your part. Everytime
I hear Wayne tell the story it gets a
little better. The last version I heard
involved having the pico translate
messages to any language based on a
pre-stored common series of phrases. It
might not be perfect grammar, but the
message would sure get through. The
system involves several techniques.
Packet radio is one of the major
ingredients. Pico computers are the
final interface to people sending and
receiving messages. Satellites are the
method to provide low cost, reliable
communication paths.
Not a single item in the chain
requires any new technology. It's all
here today.
Others have shared Wayne's thoughts

for sometime or parts of it anyway.
Within the Radio Amateur Satellite
Corporation or AMSAT as most of us know
it, a core group of enthusiasts have
been working for several years on
PACSAT--PACket SATellite. The concept
is pretty straightforward.

A satellite is placed into low
Earth orbit. Rather than the
traditional transponders common to
amateur and commercial satellites,
PACSAT operates on   specific
narrow-band channels. On board the
satellite is a large amount of random
access memory. Utilizing the error-free
techniques of packet transmission,
PACSAT continually listens for incoming
messages. The messages are acknowledged
and stored for retransmission.

Since the satellite is in a low
orbit, not much power is required for an
Earth station to be heard. Design of
the satellite is based on ground station
transmitter power in the 10 watt range
and an omni-directional antenna.

The omni-directional antenna is not
only convenient, but almost a necessity.
With the type of orbit planned, the
satellite will move overhead very
rapidly, making tracking difficult with
directional antennas. This also results
in a relatively short "window" of
visibility as the satellite moves along.

To optimize the usefulness of the
satellite, transmissions need to take
place at relatively high data speeds.
One of the major challenges in creating
a viable system is to design a
high-speed radio modem that can be built
inexpensively.

As the satellite continues its orbit, traffic for points located beneath its path are transmitted. The satellite will be in range of every point on Earth several times each day. If no major problems are encountered, any message should be delivered in 12 hours or less.

An additional advantage to a low orbit is that such launches are quite affordable now that the shuttle program is in full swing. Put somewhat crudely, a low-orbit amateur satellite has only to be "tossed" into space by the shuttle crew. Compared to the cost of traditional rocket launches or the booster rockets needed to achieve high orbits from the shuttle, PACSAT becomes very attractive.

The principles of PACSAT have already been demonstrated using UOSAT/OSCAR 11. The on board computer in the digital communications experiment (DCE) can be configured in many ways under ground control. A simplified PACSAT type operation has been tried successfully using this technique.

Where No Packet Has Gone Before

Perhaps even more interesting is another experiment proposed for a space shuttle flight. Officially known as SAREX 2 (Shuttle Amateur Radio Experiment 2) it involves placing a packet terminal node controller aboard the shuttle craft. Several problems are anticipated, such as overloading from the number of stations all trying to reach the shuttle at the same time. To minimize the problems, the hardware and software have been extensively tested on the ground. Project coordinators went so far as to invite packet enthusiasts to gang up on the test unit during evaluation and try to make it fail!

The Japanese are preparing to launch their own digital satellite. The JAS-1 satellite will be a packet radio only, store-and-forward system. Special modem adapters for most TNCs will be made available for operation on this satellite through a special arrangement between TAPR and JAMSAT, the builders of JAS-1.

AMSAT also has plans for a high-altitude (25,000 miles!) digipeater as part of a forthcoming Phase-3C satellite.

Meteor scatter propagation has been a favorite past-time of VHF enthusiasts for sometime. Commercial operations use random meteor scatter and some crude forms of data transmission to collect information from remote sites where the cost of a telephone line or high powered radio link is either not feasible or cost-effective. During an exceptionally good series of meteor showers in the fall of 1985, amateurs succeeded in establishing packet contacts over hundreds of miles using standard TNC equipment. The bursty nature of both meteors and packet radio work well together.

What else is in store for packet radio techniques? With enough speed, audio and video can be transmitted error-free using advanced TNC equipment. Digital audio is already a reality.

Making it occur in near "real-time" via packet will take a little longer. The same rules apply to video. With a much wider bandwidth, video will require even higher speeds to be effective. In the meantime, still pictures such as those transmitted via slow scan television are already a reality.

If anyone ever told you that packet radio was just another form of RTTY, they were looking through a very narrow window. While it's true that most of what is being done today with packet radio can be accomplished with some degree of success by other means, packet eliminates many of the restrictions of earlier forms of data transmission. It doesn't care what kind of information is being transmitted, it just delivers it--reliably and error free!

Chapter Fourteen
File Transfers

It's been mentioned several times
already, but it is good to note again
that the information contained in
packets can be anything. By now you
should realize that makes packet radio a
very powerful communication tool.
    Computer professionals and
hobbyists alike quickly developed means
to transfer program and text files using
standard telephone data communication
techniques. The error-free nature of
packet makes it a perfect candidate for
transferring files of any type.
    It's a good bet that you may go
through much of your packet radio
activities and never have occassion to
use the transparent mode available in
your TNC. This is the section for
advanced users so let's take a look at
the ins and outs of using this
additional feature.

I'm Looking Through You

    The normal default settings for
your TNC place it into conversation mode
when you connect to another station.
Alternately you can enter the
conversation mode by simply typing

CONVERS at the CMD: prompt. This assumes TAPR command structure of course.

For now, let's leave the default values alone and connect in the normal fashion. After raising the operator on the other end or connecting to an automatic file transfer system, it's time to figure out how to use the transparent mode.

Entering the mode is simple. First, type a control-C after you have connected. The CMD: prompt will appear. Then type TRANS and a carriage return. The TNC will now accept anything sent from your terminal or computer and pass it straight through to the communication channel without trying to interpret special characters. For example, in conversation mode the control-C is interpreted as a signal to the TNC that you wish to return to the command mode. That could be a disaster if you are transferring a program file that contains a control-C! The TNC would return to command mode and your file transfer would become corrupted.

That's simple enough, but if you can't send any special signals to the TNC, how do you ever get out of transparent mode when you are finished? It's time for some advanced commands.

Speaking Packet-Course II

The discussion that follows pertains only to TNC 1 like units. TNC 2 owners should skip ahead a couple of paragraphs.

On the TNC 1 the CMDTIME parameter

is a very important one. The value of CMDTIME determines how long you must wait after a file transfer to return to command mode. It works like this. Your file has been sent and you are all done sending data. After the last packet has been sent, the TNC starts running a timer. The timer has been preset to the value of CMDTIME. The software in the TNC demands that you must wait at least the value of CMDTIME before you try and exit the transparent mode.

After the appropriate time has elapsed, the TNC looks for a series of three control-C characters. They must all be sent in a row with no characters in between them. They must also all be sent not more than one period of CMDTIME apart. It may sound a bit confusing, but it's really not as tough as it seems. All you do is wait, then send 3 control-C characters and after a final wait of CMDTIME the TNC will return to command mode!

If for any reason something gets sent in between the control-C signals the TNC will send the sequence and remain in transparent mode. Don't panic if this happens. Just let the packet go and wait CMDTIME again and start over.

Things are quite a bit more simple for TNC 2 like units. By sending a hardware BREAK signal, the TNC will exit to the command mode from any data transfer mode.

You've just made your first transmission using transparent mode! There are a few more important things you should know about. You may panic if you aren't expecting certain things to happen.

First, when in transparent mode, the echoing of transmitted information is turned off. If you type anything, it will be sent, but it will not be visable on your screen unless you change your terminal program on the computer to allow local echoing of information. That one throws people the first time.

Secondly, the "force-packet" character (defined by SENDPAC and usually the return key) has no effect while in transparent mode. It's ASCII value is sent straight through just as any other character would be. So how does the TNC know when to send a packet? Another important parameter is PACTIME. In command mode the value of PACTIME can be adjusted to best suit your own needs. PACTIME works a bit like the BEACON parameter, that is, it allows for two options. PACTIME can be set so that packets are sent every so many seconds. Alternately packets can be sent only after no information is received by the TNC for a period of time. Either way, eventually all information will be packetized and sent. While we are here, let me point out that another command CPACTIME will allow the PACTIME value to have control during conversation mode as well as transparent mode. Normally CPACTIME is off.

Graduate Work

If you have written software that needs to be able to communicate all characters to a distant station, you can make good use of transparent mode.

Rather than having to manually enter the mode each time, you can set CONMODE (connect mode) to the TRANS (transparent) option. When you have selected this option your TNC is automatically placed into transparent mode when connecting to another station.

A final word of warning to TNC 1 users: don't set PACTIME or CMDTIME to zero! If you do, the only way out of the transparent mode will be to do a hardware reset, that is, kill the power and start over again. TNC 2 owners can escape by sending a BREAK as noted earlier.

Special Considerations

Since packet radio allows error-free transmission it is redundant to employ special transfer programs normally used on telephone paths that employ their own error checking. They should work, but why use a less reliable form of insuring the integrity of the data? Packet radio already takes care of that for us!

One area that can become important during file transfers is flow control. The two methods of flow control were covered in an early chapter of this book.

While very few of us can type much faster than 60 or 70 words per minute, computers are capable of sending information at much higher rates. On VHF a channel speed on the radio side of 1200 baud (about 1200 words per minute) is the usual case. Effective rates of transfer are slower, depending on how

busy the channel is. On the computer
side, the TNC can exchange data with the
computer quite quickly at speeds much
greater than 1200 baud if the computer
is able.

The TNC is only capable of
buffering so much information. If flow
control is not implemented correctly, it
is entirely possible to lose data on the
way to its destination. If that's the
type of problem you are experiencing,
take a look again at the information on
flow control in your TNC instruction
manual and in this book.

Why not give sending computer files
a try? Pick something relatively small
for starters. Figure out how you would
normally send a file over the telephone
line and give it a try over packet radio
using transparent mode. Most terminal
programs allow for transfer of ASCII
files, so you might want to give that a
try first before going on to binary
machine code files.

Packeteers experienced in
transferring computer programs by radio
suggest using hex files. They have also
found that the XMODEM program will not
work well due to packet timing
considerations.

On binary files, hardware flow
control becomes a must so that an XOFF
character doesn't accidentally get sent
and halt data flow.

Some users have experienced
hardware flow control problems with TNC
1 units due to a design problem with the
6551 chip.

Chapter Fifteen
Accessories

It's always fun to see what kind of additions and modifications become available when a new piece of amateur radio equipment is introduced. Amateurs are by nature curious and many are quite inventive. I don't know about you, but I have trouble leaving the covers on a piece of equipment for more than an hour or two. I want to see what it looks like and start thinking about ways to make it better.

Packet radio is new enough that the accessory market has not really had much time to take hold yet. Several individuals have already seen room for improvement. Even the major manufacturers are offering enhancement packages.

Keep in mind that many of these businesses are kitchen table-top affairs. Some will survive while others fall by the wayside. Some may even join the ranks of full-time operations. There are some mighty creative individuals out there so keep your eyes peeled on the classified and small display ads in the magazines. You may just find an accessory you can't live without!

This and That

One area that can definitely use
some improvement on most TNC units is
operation on HF frequencies. Anyone who
has tried to tune a packet signal by ear
can appreciate how useful a tuning
indicator can be.

John Langner, WB2OSZ found the
problem to be so frustrating that he
designed a digital tuning indicator.
The circuit is fully described in the
March 1983 issue of 73 Magazine. A
description of the unit is included in
the October 1985 Packet Status Register.
A limited supply of PC boards with
complete documentation are available
from John for $10.

The same issue of PSR describes a
tuning indicator for the TNC 1 and 2
designed by Dan Vester, KE7CZ. It's
pretty straightforward and should be
simple enough for even a novice to
construct. TAPR is now offering this
circuit in kit form for about $25.
Write to TAPR for details.

Earlier in the book I described the
requirements and difficulties in
operating HF packet. Particularly with
earlier TNC designs, switching to HF was
anything but easy. AEA has recently
introduced a unit that overcomes
virtually all of the problems associated
with switching between HF and VHF.

Known as the PM-1 Packet Modem, the
AEA unit connects to your existing TNC.
The audio connections and PTT lines from
your HF and VHF radios are connected to
the PM-1. The PM-1 in turn connects to
the audio and PTT connections on your

TNC.

By throwing a single switch on the PM-1 you can switch between HF and VHF. The PM-1 takes care of all the tone conversions. The same easy to use tuning indicator used on the PK-64 and CP-1 is located on the front panel. A squelch control is also available. Threshold settings are controlled automatically. The Packet Modem uses independent filtering of mark and space tones to improve the performance of the modem. That's particularly helpful when the signal is a bit on the noisy side or subject to interference as it is on HF.

The PM-1 is designed to match the PK-80 TNC in style. It should work with any TAPR compatible TNC. If you are interested in operating HF and find that your TNC doesn't allow you to do that easily, the PM-1 may be your answer.

If you are a do-it-yourself enthusiast, Hamtronics offers a packet radio modem kit. Their FM-5 UHF transmitter has also been used for high-speed trunking in packet networks. Drop them a line for their catalog.

Am I Connected?

One of the hits of the 1985 Louisville National ARRL Convention was the "connect siren" on the PK-64. Maxtec offers a device called the Kon-D-Kon for Konnect-Dis-Konnect for use with TNC 1 series packet radio controllers. It will provide both a beep tone and a lamp indication whenever someone connects to your station. It uses 5 volts DC and logic signals from the TNC. See the appendix for more information.

Connecting Some More

If you really have a hankering to be connected to more than one station at the same time you need a TNC that includes full implementation of AX.25 protocol version 2.0. Owners of original TAPR design TNC units with pre version 4.0 software don't have that capability.

Ronald Raikes, WA8DED offers an alternative that will allow for multiple connections. The software is in the public domain but is reserved for the non-profit use of individuals. Further information on obtaining a copy is available from Ron. The appendix includes his address.

One other interesting feature of Ron's software is how he handles the command mode. It is never necessary to change modes to give commands. If a line begins with an escape character that signals that a command follows. Otherwise data is assumed.

Even an auto-answer message can be implemented. The message is sent whenever someone connects. There are other differences. There is no transparent mode as such. Using a simple program, binary data can still be sent since the host mode is transparent. The decision on whether you'll like Ron's program or the original is strictly a matter of personal taste.

Wave Of The Future

One area of packet radio that is particularly attractive to enterprising amateurs is the terminal software market. Just about any telephone based terminal program will work fine for packet radio. With some of the more sophisticated packages their features can be specifically tailored with packet radio in mind. It would be a bit like using an overlay for a dBase application.

While several of the commercial suppliers are selling packet radio terminal software, most are pretty simple packages. There's little doubt that the theft of disk based programs has discouraged the commercial suppliers from continuing to support products in that manner.

There's much that can be done in terminal software to make operation on packet easier and more user oriented.

Apple Macintosh owners already have such a sophisticated software package available from Brincomm Technology.

Utilizing the windowing feature of the Macintosh split-screen operation is easily implemented. Incoming data is displayed on the top of the screen while the lower window shows what you have typed from the keyboard.

By allowing you to store connection paths and placing them in a menu, you no longer have to type long command lines. Commands are also available from a menu. It almost makes the instruction book obsolete.

MacPacket/TAPRterm allows up to 100

routes to be stored in the program.

Standard terminal features like save files to disk, send files from disk and print are included.

Kantronics Packet Communicator owners will be pleased to know that a version of MacPacket for their TNC is also available.

Radio Shack owners haven't been left out in the cold either. The Martin Company offers a packet terminal program that includes split screen operation, ten buffers, disk save, file transfer, printer output and a text editor. It will run on the TRS-80 Models I, III and IV. Cost is $20.

The popular Commodore 64 is being supported by the Texas Packet Radio Society. Version One of their TNC 64 package is $25 for non-members. All programs are tested in actual on the air use before being shipped.

Through some judicious programming, the TNC64 does a bit of magic and offers a 50,000 character capture buffer. It is menu driven, provides for automatic disk saves of incoming messages and the ability to load pre-programmed beacon text to the TNC. TNC64 is a blend of BASIC and machine language coding.

Version two which may be available by the time you read this will be comprised of four programs. Among other features it will allow for date and time tracking for first generation (TNC 1) TNCs plus keyboard macros. Many other features will be supported adding versatility to both TNC 1 and 2 hardware. Write to TPRS for further information and availability.

IBM PC owners will be happy to know
about the Pak-Comm terminal software
available from Kalt and Associates. It
includes many of the features mentioned
for other programs.

The list of features that this type
of program might include could go on for
pages. How about including a packet
answering machine as part of the
software? Ultimately it would be nice
to just tell the program who you want to
talk to and let it figure out how to get
there. To a great degree that is
possible. The program could even be
designed to try alternate paths if
connection is not possible through the
primary path. That's the type of
routing software that is being developed
for network node controllers.

So how about it? Are you handy at
programming? There could be a market
for your software. There are a lot of
different types of machines being used
in the ham shack. That means a lot of
opportunities for some good software.
Why not give it a try!

The FADCA software library has PAM
software available for most Z-80 based
computers. Drop them a line for more
details.

Packet Pictures

In an effort to provide additional
information on packet radio, Kantronics
has introduced a video tape presentation
hosted by Phil Anderson, WØXI. It's a
basic introduction to packet that should
be of interest particularly to club
program chairpersons. The tape is

available for sale to both clubs and
individuals. Contact Kantronics
directly for details.

Chapter Sixteen
The Software Approach

Everything we've talked about in this book is based on a dedicated hardware approach to packet radio techniques. It is assumed that you have a terminal node controller of some kind. A TNC is a complete microcomputer system in its own right. It has a processor, some stored program, random access memory and the means to communicate with other devices.

After thinking about that for a moment you may well have the same question I had when I was first exposed to packet radio. Why can't my personal computer be used to do the job rather than assembling or purchasing a separate unit to do the job. The answer is that it can be done, at least on some machines, but it's not as easy as you might think.

Ever since amateurs first began experimenting with packet radio they have used microcomputers to develop the necessary software to implement the needed protocol. There are several things about packet radio that make it difficult to implement using a single processor chip.

First of all to receive and send

packets you must be able to convert the data to and from the proper audio tones. That's pretty simply done with either a pair of modem chips like the XR-2206 and XR-2211 or with the "modem-on-a-chip" 7910 device. Hamtronics will even sell you a packet radio modem kit using the Exar chip combination as will the FADCA and VADCG organizations. See the appendix for more information.

Secondly, packet radio uses synchronous data communication rather than start bit/stop bit type asynchronous methods found in RTTY and regular telephone computer communication. The clock signal must be recovered from the data itself and then maintained very accurately.

The encoding of data in a packet signal is also different from what you may be used to for asynchronous modes like RTTY. In traditional systems, the levels or tones switch rapidly between mark and space levels with each bit of each character. Not so with packet! It's time for a new term, NRZI.

You've Got a Lot Of NRZI

Non Return to Zero-Inverted is what those strange letters stand for. In standard RTTY a mark tone is generally considered to be a zero state while a space tone represents a one. To send a one you must send the space tone. With NRZI it is the shifting from one tone to the other that represents a zero. No shift indicates a one. It is the shift, not the tone that communicates the necessary information.

Do you recall in the chapter on HF packet techniques that you were told that either lower or upper sideband both worked equally well? This unique encoding of packet data is the reason why. You don't care what the tones are, just whether they change or don't change.

One of the features that makes packet so appealing is the error checking that takes place. A rather complex mathematical formula is used to calculate the validity of all packets.

Individual pieces of data must be assembled into packets. The proper header information has to be added along with the synchronization information.

Ins and Outs

In addition to all of this, a packet controller has to communicate with you! It has to be ready to accept your input, whether it's data or commands. It has to tell you what it knows. That requires printing information to your screen or terminal.

Many of the operations have to occur at virtually the same time. Are you beginning to understand why a single 6510 processor such as you find in your Commodore, Apple or Atari might have difficulty in keeping up? Using the C-64 as an example, the 6510 has to do many other things associated with its own housekeeping. Those pretty pictures you can see on your C-64 screen don't come without a price. It takes processing time to keep those images in place. True, there are some special

chips that help, but the processor still
has to coordiante it all. Additionally,
clock speeds on many of the machines
mentioned are not as fast as some newer
designs. That aggravates the problem.
     Now that I've told you why
implementing packet radio on a home
computer isn't as simple as you might
hope, let's look at a few examples of
successful implementation.

It Ain't Trash

     Do you own a Radio Shack TRS-80
Model I or Model III? If so you are in
luck. Perhaps the best documented
implementation of amateur packet radio
is available for those machines. Robert
Richardson, W4UCH has developed a
software approach to packet radio for
the TRS-80 machines. If you are
interested in taking a quick look at how
it's accomplished, check the September
1984 issue of Ham Radio Magazine.
Through Richcraft Engineering you can
obtain "Synchronous Packet Radio Using
The Software Approach". It is published
in three volumes. Volume II contains
most of the information you are likely
to want. Disks are also available. See
the appendix for details.
     With over 5000 lines of source
code, the TRS-80 implementation of
packet radio makes it clear why using a
home computer for packet is not easy.

More Than a BBS

     The Xerox 820 computer is important
enough that it has earned its own

chapter in this book. If you haven't
read it yet and are interested in using
one for packet radio be sure and take a
look at chapter ten.

The 820 has been used with several
designs as a packet radio machine.
Several approachs require either
modification to the board, or the
addition of one or more auxilary boards,
called daughter boards.

The same AMRAD Newsletter mentioned
in the 820 chapter is one good bet for
general information on the several
methods amateurs are using. The FADCA
Beacon has published numerous articles
on using the 820 for packet radio.
You'll find full details on the FAD
board in past issues of The Beacon.

On The Commercial Side

By making use of the capabilities
of the home computer and suplementing
the hardware and software in it, the
cost of packet radio can be reduced.
One such approach in the commercial
category is the AEA PK-64. It is
designed to work only with the Commodore
64 computer. The 128 in 64 mode is also
supported.

Other than the chips required for
the packet modem, the main hardware
addition is an 8530 SCC chip. This is
the same specialized processor used in
several packet assembler/disassembler
boards. The addition of this
specialized chip takes a lot of the
pressure off of the home computer. For
a more complete look at the PK-64, be
sure and check out the chapter on

selecting a TNC. If you own the right
Commodore machine, the PK-64 deserves
your special consideration.

As time goes on the software
approach to packet radio may become more
common. New packet software is being
tested using IBM PC machines for
example. AEA has shown that with a
fairly simple addition the C-64 will do
a good job for packet transmission.
Maybe you will even be the one to make a
breakthrough in software development.
In the meantime you'll have a better
understanding of the complexity of
implementing packet software.

If you are serious about writing
packet software yourself, you would do
well to invest in the AX.25 link level
protocol document authored by Terry Fox.
It is available from the ARRL. You'll
find all of the protocol rules spelled
out. One look at the document is bound
to separate the curious from the serious
programmer!

Chapter Seventeen
Communicators Welcome

While this book has been written
for amateur radio operators, the concept
of packet radio is one that has a broad
appeal both for hobbyists and commercial
interests. Led by the experimentation
done by dedicated radio amateurs, packet
radio is just beginning to be used as a
viable technique for business
applications. Several amateur equipment
manufacturers have branched out into
commercial applications. Both the
cellular mobile phone industry and
pico-computer industry are utilizing
packet techniques to allow data
transmission from mobile phones and
truly portable computers.
      The implications for packet radio
in the computing hobby arena are
staggering. Rather than being tied to a
telephone line to exchange programs or
messages with fellow hobbyists, packet
radio offers the possibility for similar
operations without tying up the family
phone line.
      While the future holds promise of
wide-spread use of packet radio
techniques for computer hobbyists, for
now the only way to enjoy these benefits
is as a licensed amateur radio operator.

The Federal Communications Commission has considered the possibility of creating a public digital radio service. While many agree with the idea in prinicple, the biggest problem is finding room in the radio spectrum for the service. Commercial demands on radio frequencies continues to increase making it difficult to find the needed channels.

Now Appearing In Your Town

Amateur radio has undergone many changes over the past few years. Unfortunately, it seems to have grown away from its attractiveness to young people. That's truly unfortunate, both for the amateur radio hobby and our young people.

There are several movements within our society right now that are designed to re-create an interest in the sciences. The Young Astronaut program is going a long way in interesting youth at all levels in science careers. The visibility of astronauts that are licensed amateur radio operators has sparked a renewed interest in more than one young person in radio. The amateur radio community as a whole has sensitized itself to involving young people in our activities.

For now it is still necessary to take a radio theory examination, a morse code test, and show an understanding of federal radio regulations before you can obtain an amateur radio license. These requirements are subject to change at any time; indeed, the code requirement came very close to being eliminated a few years ago.

What has been done is to localize the testing program. Whether you aspire to an entry level license or the highest class available with the most privileges, you can usually take the test at a nearby testing center administered by volunteer amateur radio operators. This has gone a long way in making testing readily available in all parts of the country. The exams are often administered in a school or community center more familiar to the applicant than the sterility of a federal examination room in a distant city.

Once you obtain your first license, upgrades become effective as soon as you pass the required tests. There's no more waiting months for a new license to come from Washington before you get your new privileges.

Learning the necessary elements to pass an amateur radio exam is not very difficult. I was first licensed at age 11, my brother at age 13. If you are already a computer hobbyist, numerous self-instruction programs are available for a wide variety of computers to teach both theory and code. A special mention should be made of the AEA Morse University Package. The learning is so doggone much fun, you forget it's actually educational.

If you've ever even thought that amateur radio might be of interest, let packet radio be your reason for following through on your desires. Your best source for additional information

is the American Radio Relay League, 225
Main Street, Newington, Connecticut
06111. They can put you in touch with a
local radio club and usually find a free
or low-cost training class in your area.
    Computer hobbyists share many of
the same motivations as radio hobbyists.
Computer communicators are welcome!
Give it a try.

*** DISCONNECTED

In any good packet QSO, it's proper for the connecting station to take care of the housekeeping chores by issuing a disconnect request. It seems only a short time ago that I made my first connect and yet in packet radio terms it was indeed a long time ago.

I hope you have enjoyed Get *** CONNECTED to Packet Radio and have learned more about this fascinating mode of communication. Not since single side band voice communications has a new mode caused as much of a stir as packet radio has in the last few years. We are only at the beginning.

Welcome aboard, fellow packeteer! Connect with me anytime. K9EI is on 24 hours a day waiting to see your messages.

Before we disconnect completely, be sure and take a look at the appendix. It represents a massive amount of information on packet radio that has all been gathered in one place for your convenience. Enjoy!

Getting Around On HAMNET

Online since 1981, HAMNET is available to any Compuserve subscriber at no additional charge beyond the normal hourly connect fee.

To reach HAMNET, type GO HAMNET or alternately, GO HOM 11 at any Compuserve prompt.

The message and data library sections are divided into 10 sub-sections. They are:

0 The Roundtable
1 ARRL
2 News & W5YI Report
3 SWL / Satellite TV
4 Regulatory Affairs
5 AMSAT / OSCAR
6 Programming Library
7 Guest Operator
8 DX
9 Packet/RTTY/AMTOR
10 (reserved)

You'll find electronic editions of the GATEWAY newsletter and numerous other files of interest to packet operators in the data library. Typing DL9 at the Function prompt while in HAMNET will place you in the library for packet radio information.

A proper tour through HAMNET would take far more room than I can devote here. Contact the system operator, W3VS at the address in the appendix for further information.

Appendix A
Glossary

        Packet radio has its own special language.
Perhaps as you've read through the material a
particular abbreviation or term left you confused.
An effort has has been made to define each term as it
appears in the text. As an additional aid, here's a
glossary of terms often encountered when discussing
packet radio.

ACK--When information is received without any errors, the receiving station will ACKnowledge the transmission indicating that it has been received.

AFSK--Audio Frequency Shift Keying, is the method of transmission used to send packets on VHF frequencies. Data is first converted into tone frequencies by a modem.

ALJ-1000--Axial Lead Jig, a highly sophisticated solid-state device included with all TAPR TNC 2 kits to allow for the proper forming of component leads.

Aloha/Alohanet--One of the first packet radio networks, developed at the University of Hawaii. It implies a network where stations do not check the frequency first for activity before transmitting.

AMRAD--The Amateur Radio Research and Development Corporation. See text for complete information.

AMTOR--Amateur Teletype Over Radio, a sophisticated asynchronous form of communication that allows for a higher degree of error checking than RTTY, but not as great as packet.

ASCII--American Standard Code for Information Interchange, the type of code your computer speaks.

Asynchronous--In reference to data communication, asynchronous implies that each character is sent and received independently from other characters. Radio teletype is asynchronous since a character can begin at any time. See synchronous.

AX.25 (Level Two)--Amateur radio link-level packet protocol adopted in October 1984, by the ARRL. See X.25.

Baud--A unit of speed used for data communications. Typically it measures the rate of symbols (bits) per second.

bbRAM--Battery backed up RAM. Takes the place of NOVRAM in newer TNCs.

BBS--Bulletin Board System, see PBBS.

Beacon--In packet terms, beacons are unconnected messages sent on a periodic basis automatically by a TNC.

Bell 103, 202, 212--Standards for data communication hardware originated by the Bell System specifying speed, tones used and so on. Bell 202 tones are the standard for VHF packet transmission. Bell 103 tones are used on HF.

BER--Bit Error Rate, also BERT, bit error rate test, BLER, block error rate and BLERT, block error rate test, means of evaluating the reliability of a communication system.

Bit Stuffing--HDLC protocol does not allow for more than five one bits to be sent in a row, except during flag bytes. When this condition is detected the transmitting TNC adds a zero after the fifth bit. The extra zero is removed on the receiving end. This is necessary so that data will not be confused with the begin and end packet flag character. It also provides a method for the receive station to extract clock rate.

BOP--Bit Oriented Protocol, AX.25 is a bit oriented protocol. That is, it treats data on a bit by bit basis. Data can also be treated on a character oriented basis (COP).

Bridge--A node in a network that takes information from one frequency and transfers it to another frequency. See gateway, node and server node.

CTS--Clear To Send, one of the signals specified in RS-232 connections. See text for further information.

Collision--In packet radio a collision occurs when two station's transmissions are detected by a receiving station at the same time.

Connection--The condition that exists when two stations are linked together in an error-checking fashion.

CRC--Cyclic Redundancy Check, a mathematical calculation performed by the transmitting station and included in outgoing packets. On the receive end the CRC is recomputed and compared to that received. If they match the packet is considered valid.

CSMA--Carrier Sense Multiple Access, in packet radio use of CSMA helps to avoid collisions. The TNC checks for the presence of a carrier on the channel before transmitting.

DCE--Data Communication Equipment, in packet radio
your TNC is generally the DCE. In telephone
transmission, the modem is the DCE. It connects the
terminal to the communication channel. See DTE.

Digipeater--Digital Repeater, a packet station used
to relay packets from one station to another. All
packet stations are capable of serving as
digipeaters.

DRNET--A limited access telephone based message
exchange service used to tie together many separate
packet groups, hosted by the New Jersey Institute of
Technology.

DTE--Data Terminal Equipment, in most packet
stations, the home computer used as a terminal.
Alternately, a stand-alone terminal, either
electronic or mechanical.

FAD--Frame Assembler/Disassembler, the portion of a
TNC that is responsible for forming data into frames
of data for transmission including the header
information and synchronization flags.

FADCA--Florida Amateur Digital Communications
Assoc.   Publishers of Packet Radio Magazine.   See
Appendix D.

FCS--Frame Check Sequence, see CRC.

FIDONET--A telephone based bulletin board system
utilizing store and forward techniques.

Flag--In packet radio a special sequence that begins
and ends each packet or frame. In AX.25 protocol
01111110 is the flag used.

Frame--A block of data that includes not only the
information destined for the other station but
additional information necessary to get it there. A
packet can consist of multiple frames of information.

FSK--Frequency Shift Keying, the method of
transmission used on HF to send packets. See AFSK.

Gateway--A server node that translates information
from one network into the form used in another
network. Often gateway is used in the same sense as
bridge, though this is technically not correct.

HDLC--Bit oriented protocol developed by IBM. It is
part of the foundation of AX.25.

HDLC Chip--High Level Data Link Controller chip, a specialized microprocessor that performs several functions unique to packet communication. It helps to format data received from the computer into frames. The FCS is computed for transmission and checked on reception.

LAN--Local Area Network, in packet radio LAN refers to a group of amateurs all operating on the same frequency in the same geographical area. Often a common digipeater is included in the LAN arrangement. See WAN.

Modem--Short for modulator/de-modulator. Generally the device used to transmit and receive data over a communication channel.

NAK--Negative Acknowledge, the opposite of ACK, in some data transmission systems. If the information received is not considered to be valid the receiving station will send a NAK back to the originator. In AX.25 NAKs are not used; they are implied rather than actually being sent.

NAPLPS--North American Presentation Level Protocol Syntax, an example of level six protocol used for videotex transmission.

NNC--Network Node Controller, similar to a TNC, but generally used at the "trunking" level of a network. See TNC and Trunk.

Node--In packet terms, a node is an individual station. See server node.

NRZI--Non Return to Zero Inverted, a zero is represented by a shift from one state to another. A one is defined as no state change. RTTY operations on the other hand define a particular tone or level as one and another as zero. NRZI responds to changes, not the individual states.

NOVRAM--Non-volatile Random Access Memory, a special form of user changeable memory that does not lose its information when power is removed. Used in some TNC units.

Null Modem--Generally a cord used to connect two DCE or DTE devices directly. The received and transmitted data lines are switched on one end.

Octet--A group of eight bits.

Packet--A packet consists of one or more frames.

PACSAT--Packet Satellite, one of the programs
sponsored by AMSAT. It's a proposed store and
forward packet radio satellite.

PAD--Packet Assembler/Disassembler, access device to
a level three packet network.

PAM--Packet Answering Machine, a simple, personal
message system. See the text for more details.

Parity--A simple method of insuring the integrity of
transmitted data.

PBBS--Packet Bulletin Board System, a PBBS is
virtually identical to a telephone line BBS or a
radio-teletype RBBS operation except that it operates
on packet. The WØRLI system is an example of a PBBS.

PID--Protocol IDentifier, part of the information in
each frame of transmitted data. It identifies which
version of network protocol is being used so that the
station on the other end can adjust accordingly.

Protocol--A set of rules for communicating. This can
include the speed used, the number of bits, the mark
and space frequencies and much more. In packet
radio, AX.25 is the current form of protocol being
used. It defines many different aspects of the mode.

RBBS--Radio or Remote Bulletin Board, see PBBS.

SCC--Serial Communications Controller, a name used by
Zilog for its 8530 IC. This IC is capable of HDLC
operation and is used in some TNCs.

SDLC--Synchronous Data Link Control,   similar to
HDLC. SDLC was developed by IBM for their packet
type transmissions.

Server Node--A node that supplies some kind of
special services. See gateway, bridge and PBBS.

SSID--Secondary Station ID, in packet radio each
station can designate up to sixteen separate
operations. K9EI-1 for example might be a
digipeater, while K9EI-2 could be a PBBS operation.

Store And Forward--A system where messages are received, held until an appropriate time and then retransmitted to another station.

Synchronous--In data terms, a bit stream having a constant time interval between successive bits. The sending and receiving stations must be synchronized to a common clock signal. Packet radio is a synchronous form of data transmission.

TAPR--Tucson Amateur Packet Radio, see appendix on organizations for more details.

Throughput--While bit transmission on a packet channel takes place at a relatively high speed (1200 baud on VHF, 300 baud on HF) the actual effective speed of transmission can be much less. The actual effective transmission speed is referred to as throughput. Interference, multiple-repeats and other factors all will reduce throughput.

TNC--Terminal Node Controller, the device used to interface a radio channel to a terminal device.

Trunk--In networking terms, a trunk is a circuit that connects two networks together. Borrowing from telephone terminology, trunks are generally not directly accessable by the end user. The user makes a line or local connection. If the destination is outside of the local network a trunk circuit is selected and the transmission automatically routed over the trunk to the destination. In packet radio, trunks are controlled by NNC units.

VADCG--Vancouver Amateur Digital Communications Group, one of the grand-daddy packet radio organizations. See the organization appendix for details.

WAN--Wide Area Network, if a "super" digipeater is used to cover a wide geographical area, it can be thought of as the nucleus of a wide area network. Generally such systems work best only in areas where packet activity is low and the terrain allows stations over a very wide area to access the central digipeater. Also defined in the OSI network model as a collection of LANs.

X.25--A packet switching protocol designed by international agreement for packet switched networks. Many of the elements of X.25 inspired the amateur radio implementation of packet techniques, AX.25.

Are you a Commodore computer enthusiast? If so,
QSKY Publishing has several publications you may find
of particular interest.

"The Commodore Ham's Companion" is a 160 page
paperback book covering a wide range of topics of
interest to the Commodore ham. It's written for the
beginning to intermediate computerist.

"Command Post" is a 32 page reprint of K9EI's
pioneering column in Commander Magazine. While "The
Companion" is an extensive reference guide, the
Command Post reprint booklet is a "roll up your
sleeves and do it" approach to Commodore computing.
If you would like more information, just write for a
brochure.

"The Commodore Ham's Companion" has received
rave reviews, including, the January 1986 issue of
QST and Ham Radio Magazines, the February 1986 issue
of CQ and the September 15, 1985 issue of the W5YI
report.

Be sure and mention that you read about these
other QSKY publications in "Get *** CONNECTED to
Packet Radio"!

Appendix B
Bibliography

    Finding material related to packet radio can be
a frustrating experience. In the listing that
follows are articles that have been of help in
compiling this book.
    Numerous other articles have been published that
include information on packet technology. They are
either highly technical or very difficult to obtain.
They have not been included.
    This list is growing month by month. If you
have a favorite packet radio article that relates to
amateur radio that is not on the list drop us a note
so that it may be included in future editions.

Max Adams
Basic Amateur Radio Packet-CQ Magazine
November 1985 p. 13-20
An extensive introduction to packet radio. Very
good!

AMRAD Staff
XEROX 820 Compendium-AMRAD Newsletter
August 1985 p. 1-18
A must for the Xerox 820 owner.

ARRL Staff
Computer Network Conferences 1-4
Pioneer Papers on Packet Radio 1981-1985
Includes Gateway Newsletter through September 17,
1985.
Published 1985
Available from ARRL

ARRL Staff
The 1986 ARRL Handbook
Published 1986 p. 19-23 to 19-50
Expanded coverage of packet radio

Dave Borden
Applications of Packet Radio-QEX
January 1984 p. 6
Packet direction finding networks

Gary Cromack
Mobile Digital Communications-Mobile Radio Technology
May 1985 p. 62
A look at mobile packet operation

Perry Donham
QRX-Packet Ears-73 Magazine
December 1984 p. 7
Details on WØRLI BBS and ARRL's Gateway

Perry Donham
QRX-Packet Places-73 Magazine
May 1985 p. 8
Listing of popular packet frequencies

Perry Donham
QRX-Packet Prize-73 Magazine
June 1985 p. 8
Announcement of "The Golden Packet"

Cornell Drentea
Packet Radio and Local Area Networking-Ham Radio
December 1984 p. 38-49
Information on packet networking

Pete Eaton
Packet Radio-Micro Cornucopia
3 parts-June/August/October 1983
A complete description of packet radio

Terry Fox
AX.25 Amateur Packet-Radio
Link-Layer Protocol Version 2.0
October 1984
All the details of link layer protocol
Available from ARRL

John Gates
Packet Radio, Part I-Mobile Radio Technology
August 1985 p. 32-40
An overview of packet radio techniques

John Gates
Packet Radio, Part II-Mobile Radio Technology
September 1985 p. 46-
Using packet radio for business applications

Steve Goode
Modifying the FM-5 for 9600 Baud
Proceedings of the 4th ARRL Networking Conference
Modifying the Hamtronics UHF radio for higher speeds

Steve Goode
BER Performance of TAPR TNC-QEX
August 1983 p.   3
Evaluating the BER of the TAPR TNC

Jim Grubbs
The PKT-1, A Review-73 Magazine
June 1985 p. 66-67
Review of the PKT-1

Jim Grubbs
AEA PK-64 Pakratt Data Controller-73 Magazine
February 1986 p.   72
Review of the AEA PK-64

Jim Grubbs
The Kantronics Packet Communicator-73 Magazine
October 1985 p. 67-68
Review of the Kantronics TNC

Jim Grubbs
The Commodore Ham's Companion
August 1985 QSKY Publishing
Includes a chapter on packet radio

Ian Hodgson
An Introduction to Packet-Ham Radio
June 1979 p. 64-67
Packet radio in VE land

Stan Horzepa
What's The Difference-QST
February 1985 p. 64
Sorting out the different protocols

Stan Horzepa
Packet Radio for the Rest of Us-QST
October 1985 p. 64
A look at MacPacket/TAPRterm software

Stan Horzepa
Packet Radio for the Commodore-QST
February 1986 p. 70
Information on the TNC64 terminal program

Lyle Johnson
Join the Packet Revolution-73 Magazine
September 1983 p. 19-24
First of a "must read" series

Lyle Johnson
Join the Packet Revolution II-73 Magazine
October 1983 p. 20-31
Second in the series

Lyle Johnson
Join the Packet Revolution III-73 Magazine
January 1984 p. 36-44
Last in the series

John Langner
Automatic Frequency Tester-Ham Radio
December 1985 p. 41-52
Frequency and deviation adjustment by packet radio

H. Magnuski
National Standards for Amateur Packet Radio Networks
Conference Proceedings of the 8th West Coast Computer
Faire
Available from: WCCF, 245 Swett Road, Woodside,
California 94062

John Markoff
Bulletin Board In Space-Byte Magazine
May 1984 p. 88-94
A description of the PACSAT project

Lew McCoy
Kantronics Packet Communicator-CQ Magazine
November 1985 p. 56-59
Review of the Kantronics TNC

David McLanahan
A Packet Radio Primer-Ham Radio
December 1985 p. 30-39
An introduction to packet radio

Margaret Morrison et al
Amateur Packet Radio Part I-Ham Radio
July 1983 p. 14-18
Packet basics

Margaret Morrison et al
Amateur Packet Radio Part II-Ham Radio
August 1983 p. 18-29
Part two of the series

Harold Price
PACSAT-Personal Communications Magazine
September 1985 p. 17-18
Packet mailbox in a briefcase

Harold Price
What's All This Racket About Packet-QST
July 1985 p. 14-17
A quick overview of packet radio

Harold Price
A Closer Look at Packet Radio-QST
August 1985 p. 17-20
All about TNCs

Harold Price
The Digital Front-Satellite Journal
January/February 1985 p.6
Packet basics

Harold Price
The Digital Front-Satellite Journal
March/April 1985 p. 6-7
UoSAT packet information

Harold Price
The Digital Front-Satellite Journal
1985 Dayton Issue (April) p. 14
A timetable and specs for PACSAT

Harold Price
The Digital Front-Satellite Journal
May/June 1985 p. 8-9
Space digital communication

Robert Richardson
Synchronous Packet Radio:
The Software Approach
Published 1982, Richcraft Engineering

Robert Richardson
Packet Radio: The Software Approach-Ham Radio
September 1984 p. 63-66
Using a TRS-80 for packet radio

Paul Rinaldo
Making of an Amateur Packet Network-QST
October 1981 p. 28-30
Early Packet

Paul Rinaldo
ARRL Approves AX.25 Protocol-QST
December 1984 p. 35-36
An explanation of link layer protocol

Paul Rinaldo
Evolution of the Amateur Packet Radio Network
Conference Proceedings of the 8th West Coast Computer
Faire
1983

Steven K. Roberts
On Line and On The Air-Link Up
January 1986 p. 18-19
Layman's introduction to packet radio

Robert Rouleau
The Packet Radio Revolution-73 Magazine
December 1978 p. 192-193
An early discussion of packet

Doctor Martin Sweeting
UoSAT-B Experimental Spacecraft-Orbit
January/February 1984 p. 12-16
UoSAT packet experiment explained

Texas Instruments Learning Center
Understanding Data Communications
Available through Radio Shack as #62-1389
Overview of all data including packet

Bill Tynan
A New Tool for VHFers-QST
July 1985 p. 61
Using packet for weak signal work

Jeff Ward and Mark Wilson
TAPR, Heathkit & AEA TNC Review-QST
November 1985 p. 54-57
Review of TNC 1 type units

Appendix C
Frequencies

Packet activity on HF is relatively easy to find. By far, 20 meters is the most popular HF packet band. Stations are active on 80, 40 and 30 meters as well.

As activity showed a tremendous increase in late 1985, stations began to show up plus or minus about 5 kilohertz of the listed frequencies. On 20 meters steer clear of 14.100 to 14.103 to avoid interfering with propagation beacons. Here are the ones to check:

```
3.607
7.097
10.147
14.103 megahertz
```

The WØRLI bulletin board system and many others are operated on a fixed frequency of 14.103 to allow for exchange of messages. New boards are using 14.107 due to the heavy crowding on 14.103.

On VHF, frequencies are just beginning to become standard. For now most packeteers can be found on two meters. When in doubt, try 145.01 megahertz. You may also hear activity at 145.03, 145.05, 145.07 and 145.09 especially in densely populated areas.

Several exceptions exist. Some early packet operators standardized on one or more frequencies designated for general simplex work. Some used the old unofficial RTTY pair of 146.10/146.70. Still others got even more creative. Here are some of the exceptions to the 145.01 suggestion:

```
Atlanta 146.13/146.73
Chicago 144.95
Central Illinois Area 147.555
Colorado Springs 147.5
Dallas 147.57
Denver 145.70
Pacific Northwest 146.55
Southern California 145.36 and
146.745/146.145
Salt Lake City 146.10/146.70
Tucson/Phoenix 147.700/147.100
```

Appendix D
Organizations

     No matter what the subject, it's always nice to
have someone available who has some experience. You
can get a great deal of information by reading, but
nothing compares to first hand experience. No matter
where you are the chances are good that one or more
packet radio organizations have already been formed
in your area. If not, there are several excellent
national groups that will be happy to add you to
their membership.
     The following list is thought to be accurate at
press time, but we offer no guarantees. Since most
of these are strictly volunteer organizations, be
sure and enclose a self-addressed, stamped envelope
when requesting information.

Amateur Radio Research and Development Corporation
(AMRAD)
PO Drawer 6148
McLean, Virginia 22106-6148

Publishes the monthly AMRAD newsletter ($15) and
welcomes international members

Amateur Radio Satellite Corporation (AMSAT)
850 Silgo Avenue
Suite 601
Silver Spring, Maryland 20910

Publishes The Satellite Journal ($16) Good source for
PACSAT information and welcomes international members

Bavarian Amateur Packet Radio Group (BAPR)
c/o Thomas Kieselbach
Narzissenweg 10, 8031
Wesling, Federal Republic of Germany

British Amateur Radio Teleprinter Group (BARTG)
c/o Pat Beedle
Ffynnonlas, Salem
Llandeilo, Dyfed, Wales SA19 6EW

Publishes DATACOM newsletter

Central Illinois Packet Radio User Society (CIPRUS)
PO Box 1031
Bloomington, Illinois 61701

Central Iowa Technical Society (CITS)
c/o Richard Amundson
4621 SW 2nd Street
Des Moines, Iowa 50315

Cherryville Repeater Association
Box 308
Quakertown, New Jersey 08868

Chicago Amateur Packet Radio Association (CAPRA)
PO Box 8251
Rolling Meadows, Illinois 60008

Publishes the CAPRA Beacon

Cincinnati Amateur Packet Radio Experimenters Society
(CAPRES)
c/o John Schroer
984 Halesworth Drive
Forest Park, Ohio 45240

Eastern Packet Radio Of Michigan (EPROM)
c/o J. Nugent
307 Ross Drive
Monroe, Michigan 48161

Florida Amateur Digital Communications Association
(FADCA)
812 Childers Loop
Brandon, Florida 33511

Publishes Packet Radio Magazine and has previously
published the FADCA Beacon ($15)

Georgia Radio Amateur Packet Enthusiast Society
(GRAPES)
PO Box 1354
Conyers, Georgia 30207

Previously published the Grapevine, now part of
Packet Radio Magazine

Hamilton Amateur Packet Network (HAPN)
Box 4466 Station D
Hamilton, Ontario L8V 4S7

Los Angeles Area Packet Group (LAPG)
PO Box 6026
Mission Hills, California 91345

Melbourne Packet Radio Group (MPRG)
c/o David Furst
11 Church Street
Hawthorn, Victoria 3122 Australia

Mid-Atlantic Packet Radio Club
c/o Tom Clark
6388 Guilford Road
Clarksville, Maryland 21029

Minnesota Amateur Packet Radio (MAPR)
c/o Pat Snyder
565 Redwood Lane
New Brighton, Minnesota 55112

Mississippi Amateur Radio Digital Association
c/o Patrick Fagan
2412 East Birch Drive
Gulfport, Mississippi 39503

Mount Ascutney Amateur Packet Radio Association
c/o Carl Breuning
54 Myrtle Street
Newport, New Hampshire 03773

Mount Beacon Amateur Radio Club
PO Box 841
Wappingers Falls, New York 12590

New England Packet Radio Association (NEPRA)
PO Box 15
Bedford, Massachusetts 01730

Publishes PacketEar ($15)

Northwest Amateur Packet Radio Association (NAPRA)
c/o John Gates
750 Northstream Lane
Edmonds, Washington 98020

Publishes Zero Retries

Oahu Packet Enthusiasts Club (OPEC)
PO Box 1355
Pearl City, Hawaii 96782

Pacific Packet Radio Society
PO Box 51562
Palo Alto, California 94303

Packet of New York (PONY)
c/o Bill Schimoler
42-15 172 Street
Flushing, New York 11358

Radio Amateur Telecommunications Society (RATS)
RATS-North
c/o J. Gordon Beattie
206 North Vivyen Street
Bergenfield, New Jersey 07621

RATS-South
c/o Brian Riley
RD 2, Burnt House Road
Indian Mills, New Jersey 08088

Rochester Packet Group
c/o Fred Cupp
27 Crescent Road
Fairport, New York 14450

Rocky Mountain Packet Radio Association (RMPRA)
c/o Andy Freeborn
5222 Borrego Drive
Colorado Springs, Colorado 80918

San Diego Packet Group (SDPG)
c/o Mike Brock
10230 Mayer Circle
San Diego, California 92126

SOFTNET User Group (SUG)
Department of Electrical Engineering
Linkoping University
S-581 83 Linkoping
Sweden

Publishes Softnet News

Southern Amateur Packet Society (SAPS)
c/o Wayne Harrel
Route 1 Box 185
Sycamore, Georgia 31790

Southern California Digital Coordination Council
(SCDCC)
PO Box 6026
Mission Hills, California 91345

Saint Louis Area Packet Radio (SLAPR)
9926 Lewis and Clark
Saint Louis, Missouri 63136

Sydney Amateur Digital Communications Group (SADCG)
PO Box 231
French's Forrest
New South Wales 2086
Australia

Publishes The Australian Packeteer

Texas Packet Radio Society
PO Box 835873
Richardson, Texas 75083

Tucson Amateur Packet Radio Corporation (TAPR)
PO Box 22888
Tucson, Arizona 85734-2888

Publishes the Packet Status Register Quarterly ($12)
and welcomes international members

Utah Packet Radio Association (UPRA)
4382 Cherryview Drive
West Valley City, Utah 84120

Published UPRA Connect now part of Packet Radio
Magazine

Vancouver Amateur Digital Communications Group
(VADCG)
c/o Doug Lockhart
9531 Odlin Road
Richmond, British Columbia V6X 1E1

Publishes The Packet ($10) This is the one that began
it all!

Western Michigan Packet Radio Association (WMPRA)
c/o Len Todd
1819 Eloise Avenue
Muskegon, Michigan 49444

Wisconsin Amateur Packet Radio Association (WAPR)
PO Box 1215
Fond Du Lac, Wisconsin 54935

Appendix E
GLB/TAPR Command Cross Reference

While the largest number of TNC owners have a unit that uses TAPR commands or something very similar, the original GLB PK-1 units are also very popular. An entirely different hardware arrangement is used as well as a different program.

To help facilitate operation of the GLB PK-1 for the beginner, here is a very brief comparison of some of the most used commands. Your best source of information is the GLB instruction manual.

The GLB TNC works quite a bit differently from the TAPR type units. In order to begin, you first need to enter your call unless it was pre-programmed at the factory.

In order to make a connection you must enter the destination station and specify any repeaters being used. In TAPR command language that might look like this:

C WB9YJC V K9CYW-1

With the GLB each item must be specified individually. To set up the callsigns, here are the commands and their TAPR equivalent:

SC=MYCALL
SD=destination callsign (no direct TAPR equivalent)
SV=repeater callsign (no direct TAPR equivalent)

In addition to the callsigns you must enter the SSID separately. After typing SC you are prompted for your input. Next you enter the SSID.

For the destination callsign the procedure is similar, but you use the SD command.

Entering the repeat path is also similar but will continue prompting to allow up to eight repeaters and their SSID.

After doing all of this you can connect to another station.

You can do this by using the automatic command set.
You can also disconnect using the automatic command
set.

        AC=CONNECT
        AD=DISCONNECT

    You can also disconnect while in the chat mode
by pressing a control-C.
    For unconnected operation you must go through a
two step process. First you must enable the
unconnected mode. Normally it is disabled. The MU
command must be followed by an E for enable (D
disables the command). After doing that typing MS
does the equivalent of placing you in unconnected
CONVERS mode in TAPR language.

MUE & MS=CONVERS

    Beacon operations have similar commands to TAPR
language:

        BS=BTEXT
        BT=BEACON EVERY

    The beacon text is entered with the BS command
and the time interval is set with BT.
    Several other commands that may come in handy
include:

        SN=RETRY
        MX=TRANS
        SH=MAXFRAME
        SL=PACLEN

    Hopefully that will get you started. Operating
a GLB is just a little different than operating a
TAPR unit. Keep in mind that because of the hardware
used, the GLB can not handle data on both the radio
end and the computer end at the same time. Software
for your computer can be designed to buffer your
inputs to reduce this problem. Refer to the GLB
manual for details.
    Remember that all parameters are lost at power
off and no full-duplex operation is possible.

Appendix F
Xerox 820 Sources

If the Xerox 820 computer has caught your fancy, you're going to need some good sources of complete boards, parts, cables and information. Here is a list of companies that should be able to help you.
If you wish to contact the Xerox Surplus Outlet directly you may write to them at:

1301 Ridgeview Drive, MS 503
Lewisville, Texas 75067

If you live in the area and wish to visit in person, the address is:

4204 Lindberg Drive
Addison, Texas

Arlington Electronics
3636 Lee Highway
Arlington, Virginia 22207

BG Micro
PO Box 280298
Dallas, Texas 75228

Computer Shopper Magazine
407 South Washington Avenue
Titusville, Florida 32781
One of the best sources of both commercial and
individual 820 parts sellers

Emerald Microware
PO Box 6118
Aloha, Oregon

Ferguson Engineeing
PO Box 300085
Arlington, Texas 76010
Where the "Big Board" began!

Lolir Lectronics
13933 North Central Expressway
Suite 212
Dallas, Texas 75243

MicroCornucopia Magazine
PO Box 223
Bend, Oregon 97709
Lots of info on the 820 and an excellent series of
articles by Pete Eaton on packet.

Nuts and Volts Magazine
PO Box 1111
Placentia, California 92670

SW Computers and Electronics
3232 San Mateo, NE
Albuquerque, New Mexico 87110

SWP Microcomputer Products
2500 East Randol Mill Road
Arlington, Texas 76011

Wilcox Enterprises
PO Box 395
Nauvoo, Illinois 62354

Appendix G
Publication Addresses

Throughout the book I've referred to the many publications that have included information on packet radio. If you are interested in subscription information or details on how to obtain reprints, here's a list of addresses for your use:

AMRAD Newsletter
See listing in Appendix D for address

Computer Trader Magazine
1704 Sam Drive
Birmingham, Alabama 35235

CQ Magazine
76 North Broadway
Hicksville, New York 11801

Gateway (see QST)

Ham Radio Magazine
Greenville, New Hampshire 03048

Link Up Magazine
143 Old Marlton Pike
Medford, New Jersey 08055

Micro Cornucopia
See listing in Appendix F

Mobile Radio Technology
5951 South Middlefield Road
Littleton, Colorado 80123

Packet Radio Magazine
812 Childers Loop
Brandon, Florida 33511

Personal Communications Technology
4005 Williamsburg Court
Fairfax, Virginia 22032

Packet Status Register
See TAPR listing in Appendix D

Personal Communications Technology
4005 Williamsburg Court
Fairfax, Virginia 22032

QEX (See QST)

QSKY Publishing
PO Box 3042
Springfield, Illinois 62708

QST
American Radio Relay League
225 Main Street
Newington, Connecticut 06111

Satellite Journal and Orbit
See listing for AMSAT in Appendix D

Spec-Com
PO Box H
Lowden, Iowa 52255

73 Magazine
WGE Center
Peterborough, New Hampshire 03458

Electronic Forums

American People Link
3215 North Frontage Road, Suite 1505
Arlington Heights, Illinois 60004

Compuserve Information Services
PO Box 20212
Columbus Ohio 43220

HAMNET
Scott Loftesness, W3VS
16440 Rustling Oak Court
Morgan Hill, California 95037

Appendix H
Packet Suppliers

The list of manufacturers supplying packet radio
related items has taken a major upswing in the first
months of 1986. To help you locate all that's
available for packet radio, here's a quick reference
of all the suppliers we knew about at press time. A
brief description of the packet products offered
follows each listing. Enjoy!

Advanced Electronic Applications
PO Box C2160
2006 196th SW
Lynnwood, Washington 98036-0918

AEA offers the PKT-1, PK-80 and PK-64 TNCs. They
also offer the PM-1 packet modem and HFM add on for
the PK-64

Bill Ashby and Son
Box 332
Pluckemin, New Jersey 07978

Offers a Vancouver TNC style board only or complete
kit

Brincomm Technology
2980 Wayward Drive
Marietta, Georgia 30066 33535

Software supplier of the MacPacket terminal programs
for the TAPR and Kantronics TNCs

GLB Electronics
151 Commerce Parkway
Buffalo, New York 14224

Makers of the PK-1, PK-1L and TNC-2A TNCs

Hamilton Area Packet Network
Box 4466 Station D
Hamilton, Ontario L8V 4S7

Offers a plug-in TNC for the IBM PC as well as
additional software

Hamtronics
65 Moul Road
Hilton, New York 14468-9535

Hamtronics sells several kits for the experimenter.
Included is a packet radio modem and a special fast
switching 220 mhz power amplifier for packet
networking

Heathkit
Benton Harbor, Michigan 49022

Heath's version of the TAPR TNC 1 is offered as the
HD-4040

Jameco Electronics
1355 Shoreway Road
Belmont, California 94002

Offers an RS-232 adapter for Commodore computers

Kalt and Associates
2440 East Tudor Road
Suite #138
Anchorage, Alaska 99507

Suppliers of the Pak-Comm IBM terminal software

Kantronics
1202 East 23rd Street
Lawrence, Kansas 66044

Kantronics offers the Packet Communicator (both original and model II)

The Martin Company
PO Box 982
Marysville, Washington 98270

Terminal software for the TRS-80 computers

Maxtec
3721 Spring Valley Number 111
Addison, Texas 75244

Connect alarm and lamp indicator

MFJ Enterprises
921 Louisville Road
Starkville, Mississippi 39759

The MFJ-1270 is a TAPR TNC 2 unit

Pac-Comm Packet Radio Systems, Incorporated
404 West Kennedy Blvd.
Suite 620
Tampa, Florida 33609

The TNC-200 is offered in many configurations from a bare board to a fully tested unit. It's a faithful TNC 2 clone. Also offers MacPacket software and an IBM PC TNC card

Packeterm
PO Box 835
Amherst, New Hampshire 03031

The Packeterm unit is a totally integrated packet
radio system with TNC 1 based technology

Ronald Raikes, WA8DED
9211 Pico Vista Road
Downey California 90240

TNC 1 Firmware-write for details

Richcraft Engineering
#1 Wahmeda Industrial Park
Chautauqua, New York 14722

Sells TRS-80 software that makes the Model I or III a
packet TNC. Also offers a three volume set of books
on the software approach to packet radio.

Tucson Amateur Packet Radio
See Appendix D for address

TAPR is the designer of both the TNC 1 and TNC 2
units. Manufacturing of the units has been turned
over to the suppliers listed above who pay a royalty
on each unit to TAPR. Some parts, manuals and other
miscellaneous items may be available directly from
TAPR. Write for details.

Appendix I
WØRLI BBS Commands

     The popularity of the WØRLI Packet Bulletin
Board System is apparent in most areas of the
country. Through Hank's efforts, many of us have
access to not only a fine message system, but to
gateway operations and automatic message forwarding
as well.
     The newcomer to packet radio when first
encountering a WØRLI PBBS often has to take a lot of
channel time to download the help files. We are glad
to include with Hank's permission a summary of the
commands used on his PBBS. Some examples should help
to further clarify how to operate on the WØRLI
system.
     Other boards may very well use a similar command
structure so you may wish to reference this guide for
them as well.

You connect to a WØRLI system in the same
fashion you connect to any other station. When you
finally get connected you will be greeted with a
short message and prompted for an input. The
following commands will help you navigate around the
board.

B-Bye, log out of the Mailbox

G-Activate the Gateway (Note: Not all systems
will have a gateway active. Remember a gateway
links you from one band to another)

H-Displays the help text. That's what you enter
when all else fails!

I-Displays information on the system.
Typically, where it is located, what kind of
equipment is being used, frequencies and so on.

N-Add your name to the user log.

J-Displays the calls of stations recently heard
or connected so you can see who has been on.

T-Talk to the system operator. If the SYSOP is
near by and sees or hears your request he or she
can answer directly from the keyboard. If the
operator is not available the system will tell
you that.

X-Switches between the short and long forms of
the menu. Please--use the short form as soon as
you feel comfortable with it!

Here are the meat and potatoes commands:

L-List messages

This command has many forms and can be confusing
to the beginner. Just typing L results in all
the messages entered since you last logged in
being displayed. That takes time, so take a
look at the options that can be added:

A-ARRL Bulletins only

B-General interest Bulletins

L-List the last N messages (more on this
one follows)

M-List messages addressed only to you

206

T-List National Traffic System messages only

>-List messages TO call given only

<-List messages FROM call given only

@-List messages @ call given only (see forwarding information for explanation of @ symbol)

Those options can be a bit confusing so here are some examples:

Typing LL 10 will list the last 10 messages only.

Typing LM will list only the messages addressed to you.

Typing L@ WB9FLW will list all messages on the system routed to WB9FLW regardless of who they are addressed to.

Sending and reading messages are somewhat simpler.

R-Read a message
RM-Read all messages addressed to you.

The command R 526 would allow you to read message number 526. You learned the number when you listed the messages!

S-Send a message

SB-Send a bulletin.

SP-Send a private message (keep in mind that anyone will be able to see the message as it is transmitted. It is only private in the sense that only the party it's addressed to can retrieve it.)

ST-Send traffic for the NTS

All Send messages should be followed by the callsign of the station it is to be sent to. For example:

SP WB9YJC sends a private message to WB9YJC

For forwarding the sequence "S call1 @ call2" is used, where call1 is the station you want the message delivered to and call2 is the BBS where he or she is active.

Once you have become an old hand at using the PBBS you may want to upload or download files. Remember that the WØRLI system expects standard CP/M filenames: 8 character maximum name, 3 character maximum extension. For example:

DOC.TNC HFCALLS.HRD CALLS.HRD

> D-Download a file from the mailbox followed by the name of the file you want.

> U-Upload a file to the mailbox followed by the name of the file. To terminate the sending of your file you must send a control-Z!

If it is implemented the Gateway feature may allow you to access the BBS on VHF and retransmit your packets on HF. You enter the gateway by first using the G command. After the Gateway menu appears, the following commands apply:

> C-Attempt a connection. This takes the same form it would as if you were using your own TNC. For example:

> C W3VS via W9XYZ,N7ML

> J-Display calls of stations recently heard or connected.

> U-Call CQ. Essentially you are in the unconnected mode just as you might be from your own TNC. Whatever you type will be transmitted through the gateway.

> R-Returns you to the main mailbox menu and disconnects gateway operation.

That's it! PBBS operation can be a lot of fun. Keep in mind that requesting large files to be sent during busy activity periods can hurt everyone. Use discretion in your BBS activities.

AV INSTRUCTION
TECHNOLOGY, MEDIA, AND METHODS

AV INSTRUCTION
TECHNOLOGY, MEDIA, AND METHODS

Fifth Edition

James W. Brown   Professor of Instructional Technology
San Jose State University

Richard B. Lewis   Emeritus Professor of Instructional Technology
San Jose State University

Fred F. Harcleroad   Professor of Higher Education
University of Arizona, Tucson

McGraw-Hill   New York St. Louis San Francisco Auckland
Book Company   Düsseldorf Johannesburg Kuala Lumpur London
Mexico Montreal New Delhi Panama Paris
São Paulo Singapore Sydney Tokyo Toronto

AV INSTRUCTION
Technology, Media, and Methods

1 2 3 4 5 6 7 8 9 0 DODO 7 8 3 2 1 0 9 8 7 6

Also available in Spanish: Brown, W., B. Lewis, y Harcleroad, F., INSTRUCCIÓN AUDIOVISUAL: TECHNOLOGÍA, MEDIOS, Y MÉTODOS, Editorial Trillas, Mexico, D. F., 1975. (En español de la cuarta edición publicada in inglés por 1973, McGraw-Hill, Inc.)

Library of Congress Cataloging in Publication Data

Brown, James Wilson, date
  AV instruction—technology, media, and methods.

  First-2d ed. published under title: AV instruction—materials and methods; 3d ed. published in 1969 under title: AV instruction—media and methods.
  Bibliography: p.
  Includes index.
  1. Audio-visual education. I. Lewis, Richard Byrd, date, joint author. II. Harcleroad, Fred F., joint author. III. Title.
LB1043.B75   1977   371.33   76-3407
ISBN 0-07-008165-4

This book was set in Helvetica by York Graphic Services, Inc. The editors were Stephen D. Dragin, Janis M. Yates, and David Dunham; the cover was designed by Jo Jones; the production supervisor was Thomas J. LoPinto; Margot and Donald Wolf were consultants for page layouts and prepared them. New drawings were done by J & R Services, Inc.

R. R. Donnelley & Sons Company was printer and binder.

CONTENTS

PREFACE

This fifth edition of *AV Instruction* presents an overview of media used for instruction and communication. As were previous editions, this book is designed for use as a text for pre-service and in-service teachers, media specialists, and librarians, and for curriculum and course developers from preschool education through college and continuing education. The book may also be used for independent study. Furthermore, the principles, practices, and resources discussed are relevant for the training programs for business, industry, government, and other organizations.

The examples in the book are selected from many subject fields (see the Subject Field Reference Guide) and from various levels of education. In the case of programs or fields not treated specifically, readers should have no difficulty in interpreting and applying the basic concepts presented.

Numerous changes have been made with the intent of improving this book over previous editions. First, the text itself has been condensed and is considerably shorter than that of the fourth edition. However, neither the content nor the number of illustrations has been reduced. Illustrations are now placed within the text proper so that they carry much of the content. Picture stories have been included, as before, and they do not duplicate the text. In making these changes, we have responded to the recommendations of students who used previous editions and to the suggestions of instructors and other professionals who volunteered or were invited to make recommendations.

In terms of learning, optimum results in any instructional program are attained by using various learning activities and appropriate media selected and arranged in interrelationships by a systematic procedure. The cluster of four opening chapters of the book specifies both the rationale and means for such instructional development. Following chapters deal in depth with individual media: their selection from commercial or other sources outside schools, their production by teachers, students, media technicians, or media professionals, and their use to achieve learning objectives. A final chapter projects the future of education and of educational media and technology.

Four of the six Reference Sections, which constitute a considerable portion of the book, include technical information and suggestions for operating equipment and improving facilities for instruction with media. The last two Reference Sections provide selective lists of sources of information and references—both in print and in audiovisual form.

We make special mention of the use of color to achieve greater visual harmony and emphasis in the section of illustrations printed in four colors. In changing the second color of the text from magenta to green in this section we are applying principles discussed in Chapter 5, "Displaying and Some Fundamentals of Visual Communications." Fortunately, printing technology permits us to do this at no additional cost.

This book is one of a group of other revised and newly released McGraw-Hill publications on media: *AV Instructional Technology Manual for Independent Study,* containing more than sixty student exercises; *Instructor's Manual,* accompanying this textbook; and a booklet, *Transparency Masters,* which is also correlated with it.

While we take responsibility for the contents of this book, we wish to acknowledge the generous and extensive assistance of many individuals in many organizations: the editorial and production staffs of our publishers; the many publishers, producers, and distributors of printed and audiovisual materials; the manufacturers of equipment and instructional supplies; our friends and colleagues within and outside the profession of education. All these people have given freely of their advice, assistance, and supportive friendship. Our gratitude goes also to our students, who over many years have been our continual teachers.

We regret that we cannot name everyone who has helped us, even though each deserves mention, but here are some who gave us very special assistance. Ollie Bissmeyer, Jack Blake, Barry Brown, Winifred Brown, Jaclyn Caselli, Stephen Dragin, David Dunham, Ralph Ferguson, Roy Frye, Robert Gerletti, Moyne Harcleroad, Kenneth V. Hill, Richard Hilts, Harry Johnson, Kenneth Jones, Jerrold Kemp, Cleo Kosters, Henry McCarty, Gerald LaMarsh, Rex Lee, Lawrence Liden, Thomas LoPinto, Hardy Pelham, Elinor Richardson, Walter Robson, James Russell, Bert Seal, Maxine Sitts, Laird Stiegler, Richard Szumski, Mary Lou White, John Witherspoon, Ken Winslow, Margot and Donald Wolf, Judy Yarborough and Janis Yates.

A special note of appreciation is given to Mary Auvil, who executed several new drawings and displays, and to Mary Jane Lewis, for her valued editorial assistance. Finally, to Professor Phil C. Lange, Teachers College, Columbia University, we express gratitude and admiration for his insightful criticisms and suggestions about how to improve the book.

James W. Brown
Richard B. Lewis
Fred F. Harcleroad

SUBJECT FIELD REFERENCE GUIDE

Page listings in the following Subject Field Reference Guide are to remind users of this book of the ways media resources and techniques may be applied in different curriculum areas and instructional levels.

Examples selected for the list are both general (i.e., applicable to all types and levels of instruction) and specific to particular subjects. Many other examples not included here may be found throughout the text.

MATHEMATICS

ELEMENTARY: 6–11, 31, 37, 84, 87, 88, 92, 104–108, 118–119, 144, 147, 154, 170, 199, 229, 293, 303, 333

SECONDARY: 6–11, 31, 37, 48, 54, 57, 87, 88, 92, 104–108, 118–119, 129, 144, 147, 199, 293, 303

POSTSECONDARY: 6–11, 22, 31, 37, 48, 54, 57, 87, 88, 105–108, 118–120, 144, 229, 293, 303

MUSIC AND PERFORMING ARTS: 6–11, 25, 26, 32–33, 35, 39, 46, 51–52, 53–54, 56, 57, 69, 80–81, 103, 147, 161, 162, 174–175, 183, 188, 195, 200, 211, 226, 241–243, 244, 258, 275

NATURAL SCIENCES

ELEMENTARY: 6–11, 20, 31, 40–46, 48–49, 51–52, 59, 104–112, 114, 127–142, 180–181, 189, 199, 207–208, 210, 270–274, 276, 283–285, 328–329, 342

SECONDARY—BIOLOGY AND HEALTH: 6–11, 22, 24, 26, 27, 31, 37, 40–46, 48–49, 51–52, 53–54, 56, 58–59, 71, 88, 103, 108, 112, 114, 134–135, 153, 158, 165, 180–181, 183, 184, 189, 199, 207–208, 210, 226, 230, 232, 240, 263, 270–274, 275, 276, 283–285, 289, 328–329

SECONDARY—ENVIRONMENTAL SCIENCES: 6–11, 22, 24, 27, 31, 33, 35, 37, 40–46, 48–49, 51–52, 53, 56, 59, 85, 95, 97, 108, 114, 117, 137–138, 141–142, 144, 153, 155, 180–181, 189, 199, 207–208, 226, 232, 240, 259, 270–274, 328–329

SECONDARY—CHEMISTRY, PHYSICS, AND EARTH SCIENCES: 6–11, 24, 31, 33, 35, 37, 40–49, 51–52, 53–54, 58–59, 66, 90, 97, 112, 117, 129, 132–136, 141–142, 144, 180–181, 189, 199, 207, 232, 239, 282, 283, 289, 328–329, 345

POSTSECONDARY: 6–11, 22, 24, 27, 33, 37, 40–49, 51–55, 56, 57, 59, 66, 108, 117, 180, 183, 199, 207–208, 230, 232, 240, 357

PHYSICAL EDUCATION: 6–11, 56, 60–79, 85–87, 98–101, 103, 147, 189, 211, 232, 240, 258

SOCIAL SCIENCES

ELEMENTARY: 3, 6–11, 17, 19–20, 26, 32, 40–46, 53–54, 58–59, 65, 68, 85, 86–88, 92, 97, 104–107, 109, 125, 165, 170–171, 178, 199, 200, 210–215, 225–226, 258–259, 294, 297–298, 312–330, 338

SECONDARY—GOVERNMENT AND CONTEMPORARY AFFAIRS: 6–11, 17, 19–20, 25, 26, 33, 40–46, 53–54, 55–58, 66, 70, 86–88, 97, 104–107, 109–112, 116–125, 147, 165, 171, 180, 198, 199, 200, 203, 210–215, 225–226, 244–245, 259, 270, 293–294, 312–330

SECONDARY—GEOGRAPHY AND ECONOMICS: 4, 6–11, 17, 23, 30, 33, 40–46, 56, 58, 65, 66, 86–88, 105–107, 116–125, 165, 170–171, 178, 183, 189, 192, 198–199, 200, 203, 208, 210–215, 259, 270, 297–298, 312–330

SECONDARY—HISTORY: 6–11, 25, 33, 45, 46, 47, 54, 56, 57, 58, 65, 66, 69, 70, 71, 86–88, 92, 97, 109, 116–125, 154, 165, 178, 180, 184, 188, 198–199, 200, 210–215, 225–226, 244–245, 270, 314, 315–330, 345

POSTSECONDARY: 6–11, 22, 33, 45, 46, 47, 54, 56, 57, 58, 71, 105–107, 109, 116–125, 154, 165, 200, 210–215, 225–226, 244–245, 294, 321–330

AV INSTRUCTION
TECHNOLOGY, MEDIA, AND METHODS

1 MEDIA AND THE SYSTEMATIC APPROACH TO TEACHING AND LEARNING

chapter purposes

- ⊗ To suggest the importance of media in the processes of teaching and learning.
- ⊗ To emphasize the importance of the student as the central concern in all instructional planning.
- ⊗ To introduce the model of instructional planning used in organizing this textbook.
- ⊗ To reassure you that the authors of this book are interested in helping you to help your students learn, a goal toward which all efforts in using media and the equipment of instructional technology should be directed.

Westinghouse Learning Corporation

This is a book about media and the many activities that you can plan for your students because of the great variety of instructional resources available.

What are instructional media, and why a book about them? Why are there courses in the use of media and in instructional technology? Answers to these questions may be found in one idea: *The resources for learning that you and your students use can influence the effectiveness of your instructional program.* Creative uses of a variety of media will increase the probability that your students will learn more, retain better what they learn, and improve their performance of the skills they are expected to develop. But, of course, just the use of media in instructional activities will in no way guarantee results. Much more is involved.

That you may achieve positive results from your use of media, we will review briefly the process of planning for instruction and point out when and how you and your students may consider media, select them, and decide how they will contribute to learning.

After an introduction to planning for the use of media, chapters explore the selection and production of media and general principles of media utilization; the roles of media in learning centers, and the implications for media utilization as a result of student groupings, are reviewed. Because of the trend toward education programs outside formal classrooms, a discussion of the community as a learning center concludes the first group of chapters.

The extensive variety of contemporary instructional resources is discussed: audio, visual, audiovisual, and real—where to find them, their special advantages, their most favorable applications in learning activities, and their limitations. Further, assistance is provided for you to develop basic skills that will enable you to use various types of equipment as well as the materials to be used with them.

Finally, ways are suggested for you to develop skills and techniques of your own for producing various types of instructional materials, and also for your students to produce interesting materials and learn as they create.

the terminology of media

Many people have written about the "systems approach" to instruction and instructional planning. Many experimental programs have been conducted in a majority of the fields of curricula, each following a "systems approach," or systematic planning. In nearly every case, media are regarded as central elements in the approach to systematic instruction. Here the goal of the authors is to aid teachers to

improve instruction and stimulate learning by increasing the effectiveness of their work, from developing objectives to evaluating results.

3M Company

In discussions of media, another phrase that represents a stage in the evolution of knowledge about teaching is "instructional technology." Not many years ago, media were called "audiovisual aids." These were physical things—tools for instruction. No doubt the first "aids" were sticks used by teachers to scratch the ground; in more recent times slates, chalkboards, audio materials, and, even more recently, television and computers have appeared where students are learning. These many technological resources for instruction have taken their place among the products of another technology—printing the various materials on paper. For one unacquainted with the term "instructional technology," this definition may be useful:

. . . instructional technology goes beyond any particular medium or device. In this sense, instructional technology is more than the sum of its parts. It is a systematic way of designing, carrying out, and evaluating the total process of learning and teaching in terms of specific objectives, based upon re-

search in human learning and communication, and employing a combination of human and non-human resources to bring about more effective instruction. . . .[1]

This definition implies that, for improvement of instruction, systematic planning and the wise and skillful use of the products of technology are basic prerequisites.

Another concept that is constantly in the minds of successful teachers and instructional planners is the adaptation of technology to protect, to free, and to enlighten humankind. Thus the use of teaching resources in education has a dual purpose: to improve learning and teaching and to permit teachers and students to interact as human beings in a climate where people control their environment for their own best purposes. That is, when we plan to use media—books or films, projectors or graphics, cassettes or television receivers—students with their goals must motivate and guide our efforts.

"Systems" are presented in numerous models, each representing an attempt to find an effective way to describe planning for instruction. We, too, have devised a model for a systematic approach to instruction. We have attempted to keep it simple, and, as nearly as possible, to present it in terms familiar to you. Our model is designed to help you consider the full range of media you may use in your teaching—whether you are using resources and procedures prescribed by a course of study or devising your own program of instruction.

In respect to all such terminology, whether you find yourself involved with "systems planning," "instructional technology," "audiovisual instruction," or "educational telecommunications," you will not be out of tune with the times if you are working systematically to improve your teaching techniques.

Finally, the countless number of decisions that must be made about media selection is markedly simplified when a systematic planning process is used for the development of the entire instructional experience.

---

[1] *To Improve Learning,* A Report to the President and the Congress of the United States by the Committee on Instructional Technology (USGPO 40-7105), Washington, 1970, p. 5.

MEDIA IN USE BY PLAN

Inclusion of media in the processes of instruction requires carefully thought purposes as well as judiciously selected kinds of resources appropriate to the subject, the students, and the environment. Successful learning by students in various groupings may depend upon the availability of the right media to implement instruction. Pictures: top, *Eastman Kodak Company;* bottom, *Sony Corporation of America.*

systematic instructional planning: a model

For as long as schools have existed, and wherever they have existed, four basic questions have had to be answered about them:

⊗ What should schools accomplish? What goals should schools help learners achieve?

⊗ What learning activities should students undertake in order to reach those goals? In what modes should those activities be conducted?

⊗ In what physical environments should those activities be undertaken? What resources—personnel, facilities, materials, and equipment—should be employed?

⊗ What evidence should be gathered, and through what means, to aid in judging the extent to which learners actually reach goals? As a result of studying that evidence, how may the system be improved and better results ensured the next time around?

The accompanying chart, called "Systematic Approach of Instructional Technology," plots the ele-

Rondal Partridge

ments of instructional planning represented in these four questions. Note particularly that our approach focuses upon *students*—their needs, their capabilities, their special interests and motivations, and their styles of learning, which, together, are basic to all decisions about *individualizing* learning activities. Remember too that this approach recognizes that students exert, or should exert, a continuing influence upon decisions at each separate stage of instructional planning.

Therefore, analysis of the students is essential before design for instruction is initiated. This is a familiar process, but essentially it includes such items as: What do they *need to know*? What do they know *now*? What are their problems in learning, as indicated by records? Are they, or can they be, ready for the proposed instruction? After the analysis, the development of the plan for instruction can be started. In examining the accompanying chart, note the four broad areas for the convenient grouping of the steps in the process of systematic development of instruction:

A. What goals are to be achieved?
B. How, and under what circumstances, will students seek to achieve these goals?
C. What resources are required for necessary learning experiences?
D. How well were goals accomplished? What needs to be changed?

Within these groups are specific steps; taken point by point, and in preferred order of occurrence, there are seven things to be done:

1. Define (or accept) objectives and select content.
2. Select appropriate learning experiences, and seek to individualize them.
3. Select one or more appropriate teaching-learning modes to carry out the learning experiences.
4. Assign personnel roles.
5. Select appropriate materials and equipment.
6. Choose physical facilities for the learning experiences.
7. Evaluate results and recommend future improvements.

After you have studied the relationships of the elements of the chart, proceed to the discussion of what is involved in each step of the process.

the systematic approach
of instructional technology

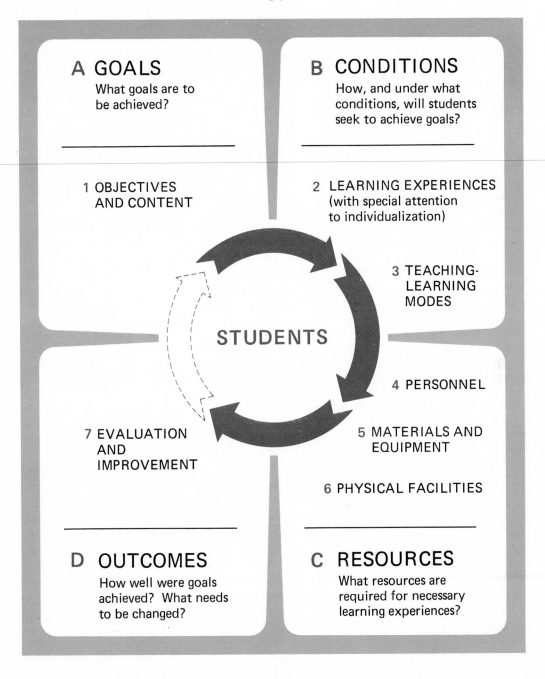

A GOALS
What goals are to
be achieved?

B CONDITIONS
How, and under what
conditions, will students
seek to achieve goals?

1 OBJECTIVES
AND CONTENT

2 LEARNING EXPERIENCES
(with special attention
to individualization)

3 TEACHING-
LEARNING
MODES

STUDENTS

4 PERSONNEL

7 EVALUATION
AND
IMPROVEMENT

5 MATERIALS AND
EQUIPMENT

6 PHYSICAL FACILITIES

D OUTCOMES
How well were goals
achieved? What needs
to be changed?

C RESOURCES
What resources are
required for necessary
learning experiences?

goals to be achieved (A): objectives and content (1)

To determine what is to be done in instruction, a process is required to develop (or accept) a set of instructional objectives and to select subject matter content with which to achieve them.

Frequently, the *objectives of instruction* in any particular facet of the curriculum will have been prescribed by an officially approved course of study or a curriculum syllabus.

It is hoped, however, that teachers will have participated extensively in the development of the program and will, therefore, be inclined to achieve the objectives prescribed.

A careful study of the categories classified by Bloom and others[2] is helpful in any consideration of instructional objectives. To these authors, all such objectives may be placed under one or another of three main headings: (1) knowledge and information, or *cognitive;* (2) attitudes and appreciations, or *affective;* and (3) skills and performance, or *psychomotor;* as follows:

⊗ KNOWLEDGE AND INFORMATION To assure that students acquire essential knowledge and information, or cognitive background, is an obvious (and perhaps oversubscribed) objective of education. Along with knowledge of facts must come *understanding* of principles, concepts, trends, and generalizations. Students are expected to grasp and to reorganize separate and sometimes seemingly unrelated learning experiences into meaningful higher-level abstractions and to use them, in turn, in successively more complex learning tasks.

⊗ ATTITUDES AND APPRECIATIONS Objectives

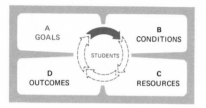

relating to changes in attitudes and appreciations are difficult to define. They are even more difficult to measure and to evaluate. When and how does a student prove that he "appreciates," for example, or that he has developed a revised and desirable attitude toward some condition or person? Is respect for law a desirable attitude? How do you "teach" it? How is it "learned"? How do you prove that the student has actually achieved a change in attitude? But, despite such uncertainties, teachers continue to seek to develop in their students attitudes reflecting sensitivity to others, tolerance and broadmindedness, respect for truth and for rights of others, and a desire to participate responsibly as members of society. What you do, what your students do, and the instructional resources and procedures you use in arranging appropriate learning activities will determine to a degree the extent to which any of these goals is achieved.

⊗ SKILLS AND PERFORMANCE Objectives in the psychomotor, or skills, category relate to activities required of young people serving in formal roles as students (studying, using a library, handling data, doing critical thinking) as well as in carrying out activities of life employment or fulfilling obligations of personal living. If an individual must talk, how *well* must he or she talk? Is one able to communicate and exchange ideas efficiently? How does one measure competence in speech? In writing? In driving a car? In using a slide rule or a pocket calculator? The skills students will need to develop must be recognized, described, and accommodated in our teaching plans.

Answers to several important questions will make your objectives clear.[3]

⊗ What will your students be *doing* when they are demonstrating the proficiency you describe in your objectives?

⊗ Under what *conditions* will these behaviors occur?

⊗ In each case, what is to be the level of *acceptable* performance?

---

[2]Benjamin S. Bloom (and others), *Taxonomy of Educational Objectives.* (See Reference Section 6.)

[3]Adapted from Robert F. Mager, *Preparing Instructional Objectives* (2d ed.), Fearon Publishers, Belmont, Calif., 1975.

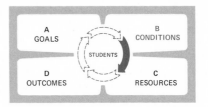

To be clear and specific in describing your objectives, avoid using terms that may be misinterpreted. Some of the great offenders are "to know," "to understand," "to *really* understand," "to appreciate," "to *fully* appreciate," "to grasp the significance of," "to enjoy," "to believe," or "to have faith in." You will obtain better results by using terms with clear-cut meanings, such as "to write," "to recite," "to identify," "to name," "to differentiate," "to classify," "to contrast," "to solve," or "to compare."

Selecting content which leads to learning experiences that achieve planned objectives is another priority of the systematic approach to instruction. As with objectives, decisions relative to subject content may have been determined by state, county, or district school officials, or by the local department chairman. But even if those decisions have been made, you will still have opportunities to expand the content prescribed.

Although the *outline of required content* may appear in the course of study, together with a list of recommended or required readings (often in textbooks), you may expect to have the right to range far beyond it. The likely result will be that the actual content that you and your students settle upon will be found in films, filmstrips, field trips, transparencies, recordings, television programs, and other media—as well as in books—all of which provide learning experiences to further the accomplishment of prescribed objectives.

conditions (B):
learning experiences (2)

The second step of the systematic approach to planning for instruction requires that, from among an almost unlimited number of alternative learning experiences, you choose those offering best promise to achieve for your students the outcomes you seek. If you support *active* learning—the idea that learning results from *doing*—you will want to involve your students in many different kinds of activities which, together, constitute *learning experiences.* You may ask them to read, recite, or listen, or involve them in discussions and debates; they may

even want to make speeches. You may recommend or require that they write, report, or develop interpretations of what they have read, heard, or observed. They may construct, experiment, collect, exhibit, or display. They may play games simulating real-life experience. They may visualize, create mentally, or invent. They may dramatize or act out; often they will draw, graph, or paint. Sometimes, too, they may tell picture stories through photographs which they themselves have taken. They may interpret, criticize, judge, or evaluate. They may sing, perform on musical instruments, or watch or listen to the performances of others. Or they may work on real problems and real things—sometimes for pay as well as for academic credit.

Early in the process of systematically selecting and developing learning experiences, you will need to learn about your students as individuals. Characteristics particularly important for you to know about are intelligence, reading ability, socioeconomic background, emotional maturity, past experiences, study habits, special interests, and levels of motivation in different subject areas.

With respect to uses of educational media, still other information is important: What are the learning styles of the students in your classes? Do they not vary considerably? What types of learning procedures and activities promote the most rapid and successful learning by students who have deeply ingrained patterns for their own learning? Are there in the group students of high verbal ability who will have little difficulty in following, or who might even prefer, completely oral or written, perhaps quite abstract presentations? Are there others whose inadequate command of verbal communication skills requires extensive use of visual experiences as a basis for improving their capacity to read and to communicate through effective speech? Selecting the best learning experiences for a student may make the difference between success or failure.

To individualize instruction is a most important goal!

The accompanying chart, entitled "Experiences Leading to Learning," itemizes these and other activities through which students may be expected to learn. Can you think of others that should be added?

experiences leading to learning

Thinking

Discussing, Conferring, Speaking, Reporting

Reading (Words, Pictures, Symbols)

Writing, Editing, Scripting

Listening

Interviewing

Outlining, Taking Notes

Constructing

Creating

Drawing, Painting, Lettering

Photographing

Displaying, Exhibiting

Graphing, Charting, Mapping

Demonstrating, Showing

Experimenting, Researching

Problem Solving

Collecting

Observing, Watching

Traveling

Exchanging

Audio Recording

Video Recording

Dramatizing

Singing, Dancing

Imagining, Visualizing

Organizing, Summarizing

Computing, Programming

Judging, Evaluating

Working

conditions (B): teaching-learning modes (3)

The third step of the systematic approach to planning for instruction calls for selecting or arranging teaching-learning modes that will accomplish the purposes of the learning experiences you have in mind. Since the type of objective may indicate the most advantageous type of mode that can be chosen, the display, lower right, suggests the implied relationship between objectives and teaching-learning modes. Note the boxes X, Y, Z: There are circumstances when presentations to large or very large groups will be both efficient and desirable. A very large group might be a television audience, distributed over a wide but local area, or it might be nationwide. A large group might be in a lecture hall, from 100 to several thousand people, or it might be in a single school where 300 students may watch a program in ten classrooms. Instruction in the latter case might be face-to-face, lecturer and students; or the program might be mediated—that is, in motion pictures only, or with lecturer plus motion pictures. The exact number of students required to formulate each of the above groups will vary, of course, depending upon the circumstances, the purposes of such classifications, and the preferences of those who compose them.

A similar analysis may be made of each of the other two types of modes suggested, small or medium groups, or a student working alone with resources. With the alternatives of various objectives and these alternative modes, plus the types of experiences just outlined to the left, the problem of making wise decisions is evident. Will learning, for example, be as effective in a *large* group as in *small* or *medium* groups for the objectives of instruction specified? Or will learning be best achieved in *independent study*?

In some cases, you may be tempted to choose one or another of several alternative instructional modes simply to provide your students with a new experience. However, your choices should be made on the basis of the appropriate mode for the immediate instructional experience. If you judge a film sufficiently valuable to be viewed (but not necessarily *immediately* analyzed and discussed) by each of 100 or more students in three sections of a history course, for example, for which you think alternatives of timing or grouping for that experience will have little effect upon end results, you might better arrange a single large-group showing rather than a time-consuming series of individual or small-group viewings. Or, if you have adequate learning resource center facilities, you may simply ask the students as individuals (or in small groups, if they prefer) to see the film at their convenience in available viewing areas.

Again, if your purpose is to provide oral instructions to clarify what your students should do in completing their individual study assignments, you might also see this as an appropriate application of the large-group mode, but only if the questions likely to be raised during such sessions would not be too numerous for you to handle fully and well.

While the reading of an assigned book or back issue of a magazine preserved as a microfiche transparency is obviously appropriate as an independent study activity, the activities of discussing and debating issues, points of view, or facts contained in it would have to be a group experience.

In the ABC and XYZ arrangement below, the crisscrossing lines suggest that different objectives may be achieved by different groupings; the problem is to seek the optimum arrangement in terms of ultimate learning, time required, costs involved, and satisfactions for the participants.

Whatever the modes and processes used, many

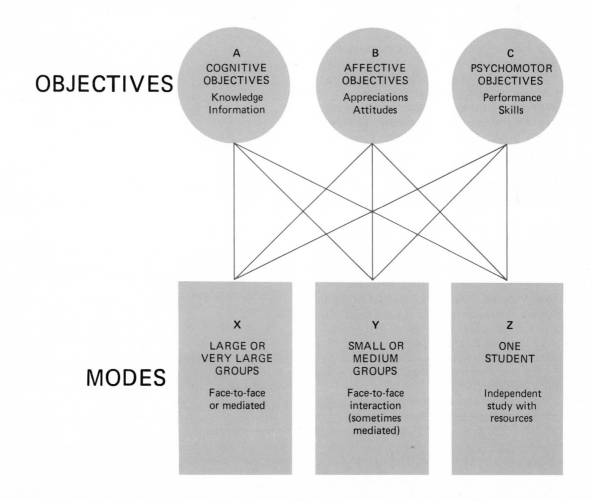

kinds of learning can and do take place at the same time: mastery of skills may engender changes of attitudes; when knowledge is being acquired, favorable or unfavorable attitudes toward the subject or toward learning may be developed simultaneously. For each situation, there are resources that can facilitate learning, and in each situation, both students and teachers have specific roles.

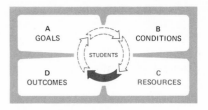

resources (C): personnel roles (4)

Systematic planning requires that responsibilities and activities be specified for all personnel who are to be involved in teaching and learning: Not only the roles of teachers and students are to be specified, but also those of the support personnel and the paraprofessionals.

As a teacher, one of your roles is to be a transmitter of information. In that role you will talk (communicate). To ensure that your communications are meaningful to your students, you should use media—such as chalkboard, slides, filmstrips, and bulletin board—that will most appropriately aid your presentations.

Perhaps the most important role you may perform is structuring the learning environment and learning experiences. In this role you should draw upon your professional insights and skills, including your sensitivity to specific interests, needs, and abilities of your students, and you should be alert to the many options you and your students will have in choosing media resources and experiences.

Another important teacher role is that of *discussion leader*. With your students you will be involved in exchanges of ideas, information, and points of view. To lend depth and meaning to the discussions, you should use tape recordings, films, television programs, resource visitors, books, displays, and other relevant media.

You may also be expected to assume the role of *educational coach* or tutor. When skills are to be learned, for example, you may guide and supervise student practice. Sometimes you will arrange to have students observe, correct, and perhaps tutor

other students. You may use various means to record behavior for later study and comparison.

You may be asked to assume another role as *a member of a teaching team.* In this role you will work with other teachers and with support personnel who contribute to instruction.

Student roles in the teaching-learning processes will also be varied. Students will assist in setting goals and in defining the bases of measuring achievements. They may be asked to assist with planning and managing learning activities such as selecting media or planning experiences—previewing and showing materials, arranging bulletin boards, tutoring, arranging field trips, or inviting resource visitors to the school.

resources (C): materials and equipment (5)

The systematic approach to the selection of materials and equipment involves attention to the following variables:

⊗ The variety of intellectual capacities and divergent backgrounds of the learners in your class

⊗ The number and variety of learning objectives to be achieved

⊗ The suitability to your purposes of particular media types or specific media production activities

⊗ The alternative learning experiences for achieving objectives

⊗ Materials and equipment available to you

⊗ Physical facilities

It is appropriate to remind yourself again that *the resources you use and the ways you use them should be determined by what you want your stu-*

dents to know, how you want them to behave, and what levels of accomplishment you wish them to achieve.

As an exercise, relate materials in the "Educational Media for Learning" list (right) with activities mentioned in the earlier chart, "Experiences Leading to Learning." Which materials seem to be most promising as resources to provide the different kinds of learning experiences named? What reasons can you give for your choices? Do the materials you select seem to have particular relevance to the kinds of experiences named? Why? Are there other materials that should be added to the list?

In selecting learning experiences to meet instructional objectives, the availability of either materials or equipment may influence your decisions. In any case, it seems wise always to seek the simplest mode of communication that will meet program requirements, and, fortunately, there are many alternatives from which to choose. Many types of equipment are listed below. How many of them are you competent to use?

equipment for learning

Record Players, Tape Recorders, Radios
Slide and Filmstrip Projectors and Viewers
Overhead Projectors
Motion-picture Projectors and Viewers
Television Receivers
Video-tape Recorders, Players, Viewers
Teaching Machines
Computer Terminals and Print and Image Reproducers
Electronic Laboratories: Audio/Video/Access and Interaction Devices
Telephones with or without Other Media Accessories
Microimage Systems—Microfilm, Microcard, Microfiche
Copying Equipment and Duplicators
Cameras—Still and Motion

educational media for learning

Textbooks
Supplementary Books
Reference Books, Encyclopedias
Magazines, Newspapers
Documents, Clippings
Duplicated Materials
Programmed Materials (Self-instruction)
Motion-picture Films
Television Programs
Radio Programs
Recordings (Tape and Disk)
Flat Pictures
Drawings and Paintings
Slides and Transparencies
Filmstrips
Microfilms, Microcards
Stereographs
Maps, Globes
Graphs, Charts, Diagrams
Posters
Cartoons
Puppets
Models, Mockups
Collections, Specimens
Flannel-board Materials
Magnetic-board Materials
Chalkboard Materials
Construction Materials
Drawing Materials
Display Materials
Multi-media Kits

resources (C): physical facilities (6)

A casual visitor is usually amazed by the apparent smoothness with which learning activities are carried on in classrooms, media centers, laboratories, or large-group teaching facilities. Students and teachers work together, use materials, operate equipment, discuss, read, listen, and engage in innumerable activities—often with only minimum confusion and delay.

Such smooth-running activities do not just happen. They reflect application of careful planning, teamwork, organization, financial support, professional insight, and a *physical environment* that lends itself to a wide variety of learning activities. The learning environment affects the performance of students and teachers in many different ways. Fixed spaces and immovable furniture and walls, for example, often freeze teaching methods and curriculum goals in undesirable, unsatisfying ways. On the other hand, functionally varied, modifiable, well-equipped spaces will free thinking and activity and encourage the highly prized flexibility that much of modern teaching requires.

Ideally, of course, the physical facilities you finally select will offer a congenial and functional environment for learning activities. However, you may often be required to accept compromises between the facilities you would like to have and those you can get. It is the peak of frustration for a teacher to plan an otherwise effective instructional sequence, only to discover that a particular classroom cannot be darkened for projection or that the only available sound projector is in the shop for repairs. Still, it is a hopeful trend that more and more school districts do have educational plants that are suitably designed, equipped, and arranged to accommodate the systematic approach to instruction.

The accompanying chart, lower left, entitled "Facilities for Learning," lists the physical facilities frequently available in the schools. Should others be added?

When you enter any instructional area, such as the resource center below, make it a habit to take a mental inventory of facilities that you might expect to see and observe whether or not they have been provided. Are the resources, furniture, and traffic patterns arranged to facilitate the achievement of learning functions for which the area is assigned? Do you readily identify changes you might recommend that would improve circumstances for inde-

facilities for learning

Lecture Halls

Classrooms
  Divisible,
  Undivided

Independent Study Areas

Discussion Rooms

Laboratories

Shops

Theaters

Studios

Libraries

Resource Centers

Electronic Learning Centers

Playing Fields

Community Resources

Home Study Centers

pendent study or provide opportunities for small groups to work together on projects?

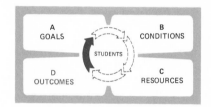

outcomes (D): evaluation and improvement of the plan (7)

Evaluation procedures are used continually in the systematic approach to education. As each element of instruction is completed, evaluation is needed to reveal the extent to which students have achieved previously stated objectives and are ready for the next steps. When the entire program is completed, a final evaluation of both student success and that of the program will be necessary. Evaluations of student performance provide essential information about (1) faults in the instructional plan—content, procedures, materials, and other elements; (2) the readiness of students to undertake instruction; and (3) the adequacy of objectives or of the instruments used to measure student achievement.

In preparing evaluation procedures, look for opportunities to use nonverbal audiovisual instructional materials and techniques. This practice will (1) reduce dependence upon reading ability as a sole means of comprehending test stimuli; (2) permit the simultaneous presentation of parts of questions without necessity for step-by-step, piece-by-piece buildups or descriptions on which strictly verbal questions are usually based; (3) enable students to visualize relationships among various parts of test data; (4) present fairly realistic pictorial or graphic representations of things, events, or situations, prompting students to see relationships between problems in school and problems in life; (5) provide variety in testing procedures to improve student attitude toward test taking; (6) cause some students to realize they are better able to demonstrate their ability in nonverbal than in verbal tests; and (7) in some instances, permit the measuring of aspects of objectives which cannot be measured by strictly verbal means.

Organizing and administering a systematically planned and technologically oriented approach to instruction is expected to bring to attention needs for program adjustments and improvements such as these:

⊗ Clarification of objectives by improved wording

⊗ Better ways to obtain, interpret, and use data about student capabilities and needs

⊗ Selection of new content or change in emphasis on content elements previously used

⊗ Changes in methods of teaching the course

⊗ New designs for learning experiences

⊗ Better use of materials, equipment, or facilities

⊗ Reassignment of responsibilities to individuals associated with the plan, or bringing in new persons to assist

⊗ Change in the timing of the plan, varying the amount of time devoted to some phases; restructuring the order of events

⊗ Refinement of instruments and techniques of evaluation and use of results to produce a more effective future plan

putting it all together

On the opposite page, another chart is presented to integrate the ideas in this chapter. On the left side of the chart is the "Analysis and Decisions" section, representing the initial planning stage for instruction. At this stage, most of the effort is headwork and paperwork, or work with 4- by 6-inch cards in an interchangeable arrangement of the ideas in the developing instructional plan. The early decisions at this phase of planning markedly facilitate the later work with details. In the first phase, you see the ABC/XYZ factors listed; these were discussed on page 9. To these are added some of the controlling factors that you may discuss with your colleagues in terms of how they influence your instructional decisions. Personnel can mean teachers and students, and sometimes other personnel: aides, technicians, media specialists, and resource visitors. Media, equipment, and facilities have been discussed and will be treated extensively in the rest of the book. Although costs are a less familiar subject for many teachers, they warrant both thoughtful discussion and continual consideration.

Policies are interesting in scope: they may be requirements by some official board, or may reflect the attitudes of administrators or peer groups; they may often be controlling factors in instruction.

At the right, in the implementation phase are the areas of actual development and production of every detail of the instructional program, the evaluation of its effectiveness as it evolves, and again when it is completed. The repetitive list under "Program Development" is indicative of the scope and great intensity of the work that must be done. In each case, problems that arise must be solved, actual facilities and equipment must be arranged for, and the responsibilities of all personnel involved in the instructional process must be spelled out in clear details. If materials are located, they must be ordered; or if they must be made, production must be arranged. Cost problems must be solved, or some alternatives found. And evaluation activities and tests must be prepared or designed.

Conspicuous in the implementation phase of the development process are the trial run(s) and the feedback process that lead to possible redesign of problem areas in the program. It is at this point that learner verification activities are relevant and necessary. As units or modules are completed—even in preliminary or rough fashion—they are given to students to try; student reactions lead to acceptance or modification of the instructional plan or materials, or both. This practical testing-reworking-testing in the development process is quite separate from a complete course validation, which can be a costly, exacting, and highly technical process. There is often complete justification for course validation, but the process of student verification should be a commonplace and routine procedure. Even a few, well-selected, typical students may serve as a reliable guide to the effectiveness of an instructional plan and instructional materials. Use their assistance.

When the program is put into routine use, the degree of student success, the amount of student satisfaction, and the efficiency of instruction are used as indicators of needs for possible revision of the instructional program.

The arrows representing feedback of information imply the same cyclical process presented in the first chart in this chapter, "The Systematic Approach in Instructional Technology" on page 5.

With this frame of reference, subsequent chapters will deal with the very practical problems of deciding

Education Service Center, Region 17, Lubbock (Texas)

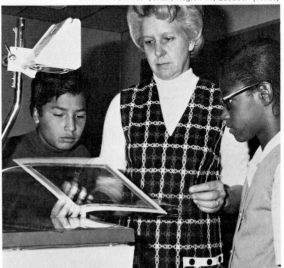

TOWARD APPROPRIATE CHANGES IN STUDENT BEHAVIOR					
ANALYSIS AND DECISIONS			IMPLEMENTATION		
Analysis of target population (students)					
GOALS AND OBJECTIVES	PROMISING MODES	CONTROLLING FACTORS	PROGRAM DEVELOPMENT	TRIAL RUN(S)	ROUTINE USE
A Knowledge B Attitudes C Skills Content Evaluation: Test activities Test items	X Large and very large groups Y Small and medium groups Z Independent study Methods and procedures	Personnel Media Equipment Facilities Policies Costs	Methods and procedures Media: Selection or Development Personnel assignments Facilities Equipment Policies Cost analysis Evaluation/ tests	Evaluation Feedback analysis Redevelopment (all or parts) EVALUATION AND FEEDBACK	Continuing evaluation Revisions

when media can contribute fundamentally to the instructional process, what media to use, and how to make that use effective. Circumstances of media use are discussed, as well as how media can be used advantageously with large and small groups, and for independent study. Emphasis will also be placed upon the use of media in individualizing instruction.

The media we discuss have a variety of characteristics: they are produced to assist in achieving many different purposes; they are of different levels of difficulty and complexity; some pay special respect to the student and to individual processes of learning, others do not. It will be your responsibility to choose from among alternative media, and from among many separate items represented in each type of medium. To gain maximum benefit for a particular student or student group, you will often need to adapt these media to your own patterns of use.

summary

The systematic approach to planning instruction discussed in this chapter provides a useful framework in which to consider applications of educational media resources and tools to processes of teaching and learning. Many models of the systematic approach as applied to education have been developed. While the one supplied here is in many ways similar to others, ours emphasizes how media relate to (1) instructional objectives and content, (2) types of learning experiences, (3) modes or groupings for instruction, (4) personnel roles, (5) materials and equipment, (6) physical facilities, and (7) evaluation of results and improvement of the instructional plan.

In this model, the student is central. *Planning* for teaching, as well as teaching itself, should be *student-centered.*

Stress is placed upon the fact that following an orderly system of planning facilitates the selection of appropriate resources to meet specified objectives.

Chapter 2 discusses modularization and individualization of instruction and includes a number of case examples of systematic planning applied in practice. Also discussed are classroom and school learning centers and ways they serve in planned instruction with media. This topic is continued in Chapter 3, The Community as a Learning Center.

Principles of media utilization are presented in Chapter 4, Choosing, Using, and Producing Media, which serves as an introduction to later chapters which deal with media types that are commonly used in education and with others that have been recently introduced.

Throughout this book, the authors try to assure that your reading time will be well invested. We hope the suggestions and recommendations offered will help you to become a challenging and creative teacher.

2
MEDIA AND INDIVIDUALIZED LEARNING

chapter purposes

- ⊗ To describe ways media are used to facilitate individualized learning in large-group, medium- and small-group, and independent study modes.
- ⊗ To compare characteristics of several widely used individualized and independent study systems and the patterns of media utilization in each.
- ⊗ To present standards and criteria for effective classroom and comprehensive school learning centers.
- ⊗ To introduce the concept of learning modules and to explain how they are designed and used to facilitate individualized learning.

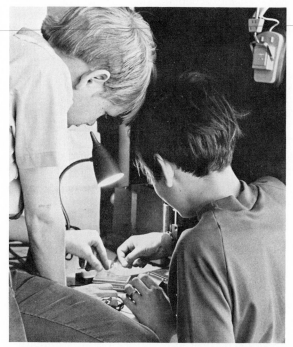

San Diego City Schools

Educational media of all types play increasingly important roles in enabling students to reap benefits from individualized learning. It is fortunate that the potentialities of modern technology may be combined with educational planning to provide resources needed for this purpose. The desirable result of this effort should be a viable system involving purposes, processes, people, materials, machines, facilities, and environments leading to "the best for each"—the cornerstone of a democratic society.

individualization and media

There is continuing emphasis in schools upon ways and means of individualizing instruction and learning. The goal of this emphasis is to provide all students with the experiences and resources they need to work to the best of their abilities and at their own rates. The first step toward achieving this goal is basic to systematic instructional planning and management: accepting students as the central, all-important guides to processes and resources used to help them reach planned objectives.

If you plan to offer individualized, independent, and modular forms of teaching and learning for your students, you will be concerned with these questions:

⊗ Are they for all students? If not, what are the characteristics of students for whom they are suited?

⊗ When is a student ready for individualized instruction? How is readiness determined?

⊗ What principles should guide the design and management of media resources for individualized learning?

aids to individualization

While much instruction seems likely to be continued in medium-sized groups (typically twenty-five to thirty-five students), the trend at all levels is toward increased use of independent study procedures and techniques. Opportunities exist to individualize learning with groups of nearly any size, but to do this requires systematic planning.

The discussion presented in Chapter 1 emphasized this point. There we recommended that teachers exploit the somewhat unique advantages of teaching and learning that occur in each of several group sizes—small, medium, and large—as well as when students work alone. The chart below depicts relationships of *general* and *individual* learning experiences derived from large/extended group instruction such as auditorium film-based teaching sessions or broadcast, small/medium groups, and situations that involve students working alone or in pairs, or being tutored. Notice that in the chart individual experiencing occurs in all three modes, but in different amounts, and, of course, in different ways. Stress here is upon the degree of individualization that is possible in the different modes: individualization in terms of such matters as content, presentation techniques, and rate of delivery.

Students may achieve individualized learning by being permitted options. Some students may prefer to attend large group presentations, and to select a number of independent study activities to meet the learning goals required. Others, by choice, may prefer to work almost totally in independent study activities, supplemented by an occasional lecture or small-group session. What the individually programmed student learns from the selected sessions may be what is needed at a particular point in the study sequence, and in the optimum modes.

INDIVIDUALIZATION AND INSTRUCTIONAL GROUPING

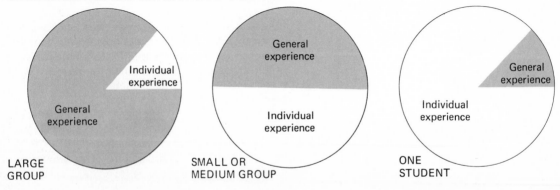

LARGE GROUP — General experience / Individual experience

SMALL OR MEDIUM GROUP — General experience / Individual experience

ONE STUDENT — Individual experience / General experience

Whether this requirement is satisfied through either large-group or independent experience is not considered important; it is important only that learning goals be achieved.

This textbook is directed especially toward assisting with all possible processes of individualizing learning—through appropriate uses of media resources. Each chapter describes numerous ways to use media—for large-group, medium-group, or small-group instruction or for independent study. We emphasize, with each, how media resources may be used to individualize learning as well as to make it interesting and useful.

San Diego County (California) Schools

The media we discuss have a variety of characteristics: they are produced to assist in achieving many different purposes; they are of different levels of difficulty and complexity; some pay special respect to the student and to individual processes of learning; others do not. It will be your responsibility to choose from among alternative media, and from among many separate items represented in each type of medium, the resources that match your needs and those of your students.

It is essential, of course, that the media you and your students choose be in accord with objectives. But these media need not be the same for all. Various students often have different ways, as well as preferences, for meeting objectives. This is why you should plan to use a variety of media in providing options for dealing with these differences. As a result, you will seek an effective match between the media chosen and the learning styles, experience backgrounds, previous accomplishments, learning skills, and preferences of the students who use them.

Many options are available to you to individualize instruction and learning through media and media utilization practices. For example, you may be stimulated to:

⊗ Search for media resources that represent a variety of structures—some formalized such as workbooks, programmed materials, textbooks, expository films, some with little formal structure such as current newspaper articles, recreational reading materials, or historical records. In addition, provide a variety of reference materials that facilitate study and discovery such as bibliographies, encyclopedias, and almanacs.

⊗ Prepare students to expect to engage in an extensive number of different learning activities or projects involving media; encourage capable students to move quickly through assignments and on to others; help those who progress more slowly to maintain quality performance without overemphasizing quantity of production as the goal.

⊗ Determine in advance individual differences among group members with respect to objectives and criterion performance standards; then, if necessary, provide preliminary instruction, often with self-instructional media and procedures, to bring all to the minimum proficiency level before beginning study.

⊗ Assign or offer options of different content, or treatment of content, for students of differing interests, goals, or preparation.

⊗ Similarly, assign different options of *media* in which subject content is presented for study to suit individual learning styles or proficiencies.

⊗ Also, accommodate to differences in beginning proficiencies through use of differentiated assignments or learning options, many suited to independent use and involving different types of media.

⊗ Organize assignments—through contracts, for example—that call for varying amounts of work to be done, varying levels of work difficulty, and varying rewards or assessments upon completion.

Perhaps, through these and other efforts to individualize instruction, you will achieve the ideal described by one student, who said: "Here, everything is geared to me."

INDIVIDUALIZATION IN ACTION

There are opportunities in every classroom to individualize instruction through varied learning activities. Nearly all such options will require media; with some, equipment will also be needed. Study the pictures on this page, and then list the special skills that teachers must have to organize instruction in these ways. *Jo Mackey sixth-grade center, Clark County Schools, Las Vegas (Nevada).*

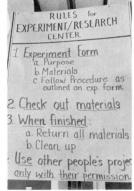

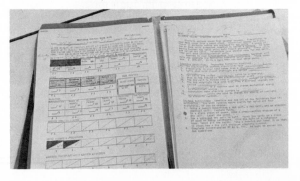

modularization of instruction

Interest in individualization of learning has led to increased use of modular forms of teaching and of media resources required for them. With this system, pretesting is used to permit students to determine their own entering levels of competence as compared with stated performance criteria and to skip entirely or to choose only some of the instructional units or minicourses recommended to achieve mastery. Alternatively, students may be given complete programs of self-instruction to follow until they believe they are ready to take required proficiency tests.

The essentials of the processes of modular instruction, including their interrelationships with media, are traced in the "Flow Chart of a Modular Plan," below. Note particularly that, at the start, cumulative records are consulted and a pretest is given to determine a student's readiness for the module. Those ready to proceed with the assignments then receive a syllabus outlining procedures, standards (criteria by which the student's efforts and products will be assessed), activity details and

directions, and other information describing how instruction will be managed.

The first module assignment requires students to carry out one of three different reading assignments, each designed to accommodate to needs of students with particular backgrounds, reading skills, and interests. Some choices may be made by the teacher; others by students. A large-group lecture-demonstration, backed with suitable media resources, is then attended by all. The syllabus indicates the varying assignments and expectations for students in the separate tracks. Following the syllabus, students in one track proceed with independent reading; those in other tracks see a film. Further on, all students, as a group, see and discuss a film. Depending on track assignments, syllabus directions then call for students to engage in one or more small-group activities. Finally, at the end of the module, a test is administered—either taken as a self-test or given by the teacher. Those who perform at the criterion level are judged to have passed; those who do not are required to review and complete additional learning exercises before retaking the entire test or taking subtests which cover only objectives and content in which they performed below standard.

FLOW CHART OF A MODULAR PLAN

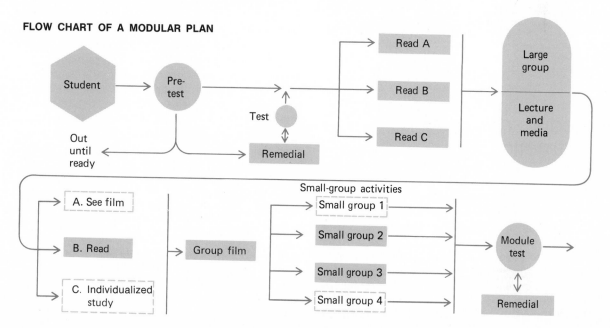

San Diego City Schools

ENVIRONMENTAL BIOLOGY	BIOLOGY IN THE CONTEMPORARY WORLD
1. The Ecosystem: Order or Disorder? 2. The Ecosystem: Energy and Materials Flow 3. Cycles in Nature 4. Interactions 5. Biomes 6. Succession	1. Man and Drugs 2. The Problem of Population 3. The Biological Basis of Race 4. Disease and Technology 5. Cancer
ANIMAL STRUCTURE AND FUNCTION	PLANT STRUCTURE AND FUNCTION
1. Complementarity of Structure and Function 2. Digestion 3. Circulation 4. Gas Exchange 5. Excretion 6. Biomechanics 7. Reproduction 8. Nerve Transmission 9. Hormonal Regulation	1. Leaves 2. Roots and Stems 3. Seeds and Germination 4. Water Relations 5. Plant Reproduction 6. Light in Plant Growth and Development 7. Hormones in Plant Growth and Development
	Minicourses in Biology

W.B. Saunders Publishing Co.

media and instructional management systems

The several instructional systems described in this section are functional variations of ways to manage tasks involved in individualizing learning and teaching. As mentioned earlier, the purpose of each system is to adapt goals, media resources, subject content, instructional methods, physical environments, equipment, and instructional personnel to the unique as well as general requirements and interests of particular students. Again, the config-uration of each is essentially the systematic approach to teaching and learning presented in Chapter 1.

The systems described briefly here are: (1) PSI: Personalized System of Instruction; (2) CMI: Computer Managed Instruction; (3) IPI: Individually Prescribed Instruction; (4) IGE: Individually Guided Education; (5) LAP: Learning Activity Packages; (6) The Contract Plan; (7) ATI: Audio-Tutorial Instruction; and (8) HEP: The Hawaii English Program. We suggest that, in reading these vignettes, you give special attention to the elements and identify those that appear to be compatible with your concept of ways in which teaching is best conducted and learning is most effectively achieved.

PSI: personalized system of instruction

The Personalized System of Instruction (PSI), or the Keller Plan, is used in an estimated 2,000 colleges and universities throughout the United States. Although the system may be implemented differently, it is basically distinguished by the following characteristics. It is:

⊗ Self-paced. Students proceed through assignments as rapidly as they desire and are able to do them.

⊗ Criterion-referenced and mastery-oriented. Students work to meet clearly stated performance criteria. Grades earned usually reflect the degree to which criteria are met by each individual. Students repeat study-test cycles as necessary, regardless of time required, until they perform at mastery levels.

⊗ Tutorial. Instructors and advanced students often help beginners or those with special problems. Printed study guides offer study suggestions, identify useful media resources, suggest worthwhile projects and activities, and present sample test items to be used in preparing for examinations.

⊗ Eclectic and varied. The printed study guides may also suggest that students voluntarily attend special enrichment discussions or lecture meetings at which films or other audiovisual media may be shown or guest lecturers or demonstrations presented. A variety of other similar activities and resources may be introduced in the course.

CMI: computer-managed instruction

Schools of the near future are expected to make increased use of computers in managing instruction as well as in performing some tasks which heretofore have consumed valuable hours of teacher time. Rapid access to up-to-date computerized information about student backgrounds, interests, achievements, and needs will help teachers as well as students to decide what to emphasize, what to review or repeat, and what to omit in their studies.

Project PLAN (Program for Learning in Accordance with Needs), which is an example of computer managed instruction, was first developed by the American Institutes for Research, Palo Alto, California, under project financing by the Westinghouse Learning Corporation, AIR, and several cooperating school districts. PLAN is sometimes supported by the computer, but the computer itself does no teach-

Westinghouse Learning Corporation

ing, and the student never works directly with it. To achieve the aims of the program, students work together with the teacher to develop clear objectives relevant to their individual needs. Stored in the computer is full information about each student —his or her special aptitudes, patterns of learning, interests, and background of previous experience.

Each day, as students complete individualized work assignments, data on their performance are fed to the computer. Available from the computer, also on a daily basis, is a comprehensive file of teaching-learning units from which appropriate student assignments can be made. Selection of each assignment is influenced by previous performance of the student, both in assignments and on periodic tests and other evaluative measures. If students

Westinghouse Learning Corporation

demonstrate that they have already achieved some objective, they may proceed to another assignment. Choice of curriculum materials for the program is made largely from existing items which have been brought together as modules (manageable segments called Teaching-Learning Units, or TLUs). In the computer, the modules are indexed in several ways, according to what students are expected to learn from them and for what type of student and situation each is recommended.

IPI: individually prescribed instruction

IPI, or Individually Prescribed Instruction, was first developed by the Learning Research and Development Center of the University of Pittsburgh and Research for Better Schools, Inc., Philadelphia. One characteristic of IPI that distinguishes it from ordinary forms of instruction is that it requires the entire curriculum of each school subject area to be broken down into relatively small instruction units. With mathematics, for example, some 430 specific instructional objectives have been identified and grouped into approximately a hundred units, each with its own specific objectives and criterion measures of success and accomplishment, and all to be completed in predetermined order.

The results of pretesting at the beginning of the school year reveal the competencies and levels of competence of each student and suggest special needs in selected areas. Upon this basis, an appropriate study plan containing possible unit sequences is designed. Each individual is assigned only the particular units that he or she is judged to need; there are some options. Work on each unit is usually performed as independent study and requires about an hour of the student's time. Assistance from the instructor is available to the student whenever needed.

IPI procedures involve a sequence of activities that call for uses of many different kinds of educational media, all of which are treated in later chapters of this book.

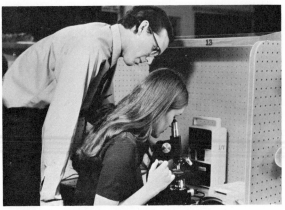

San Jose State University

San Jose State University

San Jose State University

When a student completes a unit, the teacher assesses achievement, usually by means of a written unit test, and assigns another unit appropriate, at that point, in the total study sequence. This assignment-study-complete-test process, which is usually repeated daily, may occupy up to half of a student's schedule.

Although IPI began as a nonautomated program, it has been adapted to the computer-managed instruction form known as the Individually Prescribed Instruction Management and Information System (IPI/MIS). With the automated system, data are collected regarding each student's background and progress, prescriptions of recommended assignments are made, and student learning difficulties are analyzed.

IGE: individually guided education (UNIPACs)

The Charles F. Kettering Foundation, through its Institute for Development of Educational Activities, Inc. (I/D/E/A), and the Wisconsin Research and Development Center for Cognitive Learning are known for their work in developing an experimental teaching/learning system that has come to be known as IGE, or Individually Guided Education.

Under IGE procedures, teams of three to six persons (one of whom—a teacher—is designated as the unit leader) work with groups of approximately 150 students. The usual application of the plan provides for: (1) nongraded classes—i.e., students of various academic levels work together, some as tutors; (2) team teaching; (3) differentiated staffing —with team members (teachers, apprentice teachers, clerical aides, and other paraprofessionals) performing various specialized roles; (4) individualized learning activities and pacing of progress through assignments; and (5) as much attention as possible to individual differences among students, especially with regard to learning styles, aptitudes, and needs and preferences in choices of learning assignments and media employed with them.

The several unit leaders of a single school, together with the principal, compose an instructional improvement committee whose functions are to test, evaluate, and seek ways to improve the system, including the media resources and services provided.

LAP: learning activity package

Learning Activity Packages (LAPs), first developed at the Nova Schools (Florida), continue to receive widespread attention and application.

LAPs developed by individual teachers are often exchanged. They usually consist of sets of learning assignments organized in sequence to achieve specific behavioral learning objectives. Typically, they offer alternatives as to how, where, what, and when to carry out assignments. They utilize a wide range of media resources. Students proceed at their own pace through assignments relevant to their purposes and levels of achievement. Opportunities are

Aurora (Illinois) Public Schools

provided for self-assessments as well as teacher assessments of accomplishments at various points in the process. LAP packages provide all of the information a student needs to locate and use various learning resources required for the completion of assigned tasks. The character of LAP assignments can be inferred, to some extent, from the following two excerpts (both addressed to students) that have been drawn from an English LAP package:

Exercise 1.
This tape provides an exercise for you in matching poetry and music. For each set of three brief musical excerpts, select the one that best matches the mood, style, sensual appeal, or structure of the poem supplied with it. Then write a brief statement telling why you chose it over the others.

Exercise 2.
Listen on this tape to the introductions to speeches of three politicians. As you listen, record on the form provided: (1) the main topic of each speech, (2) the point of view each speaker wishes you to accept, and (3) two probable arguments each will present during the remainder of his speech. Stop and start the tape player according to directions you will hear on the tape.

When you have completed this exercise, continue listening to the rest of the tape. Then turn in your response sheet to be scored by the teacher.

the contract plan

The contract plan, which is really not new but is experiencing renewed use today, provides breadth, organization, opportunities for individualization, and control of independent study. In organizing contracts, the teacher takes into account the range of interests and abilities of students, the several different objectives set up for the study of a particular unit, the resources available to students, and the time allowed to complete the work.

A written syllabus, or contract plan, provided for the students usually spells out (1) the overall objectives of the study, (2) bases for grading or evaluating student products and efforts to achieve those objectives, (3) details of a full range of activities to be conducted by students as parts of their contracts

San Diego County (California) Schools

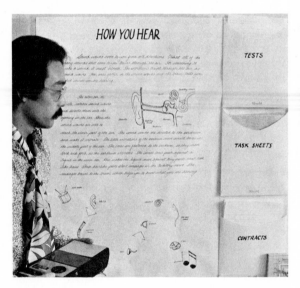

—some required (and therefore to be completed by all), some optional (to be completed as extra or above basic requirements of the course and therefore to serve as incentives to maximum achievements). Details of activity assignments are so clearly stated that students are expected to work on them independently—at their own speed and, at least in some cases, in whatever locations they choose. Student choices of assignments relate to their individual interests and needs. Accomplish-

ments are evaluated on the basis of the quality, amount, and depth of work completed. Often, an overall rating received for the total contract is derived not only from assessments of the quality of work for each assignment but also from contributions made to the class (including participation in discussions), and one or more examinations.

Typical assigned activities of the contract plan include (1) reading—in textbooks, reference books, fiction—and completing assignments; (2) listening —to tape or disk recordings; (3) making out-of-school visits to study community sites; (4) completing multi-media learning-kit activities; (5) viewing —films, filmstrips, or television programs; (6) developing oral or written reports; (7) conducting surveys and interviews; (8) completing creative projects —original stories, amateur films or slide-tape sets, construction projects, or art work.

ATI: audiotutorial instruction

The basic media resource that integrates audio-tutorial learning assignments is the audio-cassette tape which contains the instructor's remarks, directions, and interpretations of learning assignments and study resources. Such audio recordings may be purchased commercially, but they are more often developed locally to accompany readings, 8mm films, filmstrips, charts, realia, and other study materials.

A quite sophisticated version of audiotutorial instruction has been developed and refined by Dr. Samuel Postlethwait of Purdue University. There, botany students check in on their own sched-

ules for laboratory assignments, on an independent basis, moving from station to station within converted laboratory spaces equipped for independent study. The instructor-taped voice tutors students through independent learning activities, which in-

volve such varied assignments as reading from the text, performing experiments, collecting and analyzing data, using the microscope, studying motion pictures, examining specimens, charts, diagrams,

Michigan State University

or photographs, or listening to brief lectures and discussions. Records are maintained of student participation in and completion of laboratory study and assignments.

Students following the plan at Purdue University also attend integrated sections in which they use various media and quiz each other about subjects studied. As part of the course, each student must complete two miniature research projects—one structured and one unstructured.

Other student instructional groupings, often involving the professor in charge, range from small buzz sessions to seminars and large lectures, lecture-demonstrations, and film showings. Qualified lecturers play significant roles in motivating students to learn. Finally, this system provides time for the instructors to give personal attention to students with problems.

HEP: Hawaii English program
(multimode; multimedia)

Because of several unique characteristics, the Hawaii English Program may be considered to be the most comprehensive, complete, and systematic curriculum yet developed in the United States. Reasons for this status include the fact that HEP originated as a decision of a State school system to concentrate massive effort on a curriculum project considered to be of highest priority—the English program. The outcome is a totally developed program from grades K through 6 based upon tested learning theory and teaching techniques; further, the program includes all elements required in a system: goals and objectives, specified outcomes, student and teacher roles, a vast bank of materials, a specified learning environment, and a diverse system of evaluation and record-keeping.

Another important characteristic of the HEP is that it can be replicated with reliable results, for example, it is being used in Guam, Samoa, and Trust Territories, and by Project ALOHA, centered in Santa Clara County, California, from which the program is available to mainland schools.

The HEP is designed for grades K through 12. The K through 6 program was started in 1966 and completed in 1971. The curriculum for grades 7 through 12 is scheduled for completion in 1977. Design of the HEP permits its installation in a school system vertically (by one grade level per year) or horizontally (by grades K through 3 in several schools each year, while others are added as feasible). From 1975 on, the HEP was included in the list of exemplary programs about which information is being disseminated by the U.S.O.E.-sponsored State Facilitator programs.

Honors received by the HEP include: "Innovative Project" and "Educational Pacesetter" (1972); one of twenty most successful development projects in the nation (American Institutes of Research, 1974); selection for national recognition by the Right to Read Effort (USOE, 1975) as one of twelve ". . . outstanding validated reading programs in the United States." In 1975 The National Council of Teachers of English facilitated inclusion of the HEP in the ERIC System.

The photographs to the right in the picture story on HEP were supplied by The ALOHA Project.

SEEING IS
BELIEVING

There is no substitute for a visit to an HEP class, where the processes, the student involvement, the teacher leadership all can be seen as a continually changing drama. These pictures show only a few of the things to look for when you visit HEP in Hawaii or at an ALOHA project. The checklist below will help you observe that HEP provides advantageous learning conditions, such as the following:

A total instructional system in language skills, literature, and the nature of language for grades K through 6 (K through 12 by 1977).

Total individualization of learning in terms of pacing and alternatives in modes and patterns.

A multi-media, multi-mode content presentation, as opposed to a single text, to accommodate different learning styles of students.

A great variety of materials developed by teachers and curriculum specialists, or selected by them and acquired from commercial sources. Audiovisual equipment including typewriters, audio-card readers and cassette players, phonographs, loop film projectors, and other resources as required.

Numerous opportunities for student choices, self-direction, self-instruction, and self-evaluation.

Emphasis on inquiry and discovery and such activity-centered learning experiences as games, simulations, creative drama, improvisations, art activities, and writing.

Materials, equipment, and a management system with a structure based upon goals and precise learning objectives.

Materials used to facilitate peer-tutoring as an integral part of the system with benefits to both learner-pupil and tutor-pupil.

A structured management system and objective record-keeping procedures for effective evaluation in qualitative as well as quantitative terms. Change in teacher role from a single source of knowledge and direction to the role of catalyst, consultant, diagnostician, guide, evaluator, and example/model.

A shift to effective early education in language arts and a decreasing reliance upon deferred remedial instruction.

An environment that promotes a dynamic, ever-changing pattern of activities, where individual responsibility and cooperative effort appear on every hand.

During independent study, or with help when necessary, students enjoy literature and improve their reading

Teacher and teacher aide consult with and assist students in the informal classroom

Each day, the English period starts with a meeting in which students report their plans

Peer-tutoring is a key mode for learning and is designed into the HEP

Below are displays of only a portion of the materials for the HEP at Project ALOHA

comprehensive school media centers

Ideas about what a school media center ought to be have evolved with the concept of individualized learning. No longer is just a library or an audiovisual center adequate. Educational media centers are now regarded as laboratories—places where students pursue a full variety of learning experiences, often independently.

Instructional media resources recommended for and available within a single school will vary according to the range of grade levels taught, financial support provided, needs and abilities of the students, curricular emphases, and the number and qualifications of teachers.

National standards for educational media and equipment have been developed jointly by the Association for Educational Communications and Technology and the American Association of School Librarians. These guidelines are contained in

Media Programs: District and School, available from either organization.

Visit your own school media center, or a similar center in some other school, and determine the scope and quantities of resources provided. Inquire too about media items and services that may be obtained from outside sources. Also, review policies and procedures used in the center to govern services. Compare your findings with the recommendations and guidelines of *Media Programs: District and School,* mentioned above.

school media services to teachers

As a teacher, you may expect that most of the following resources and services will be available to you through your school media center:

⊗ Catalogs and lists of all types of instructional media directly from the school center as well as from district, regional, or other school sources. Besides the simple bibliographical data, additional helpful information may include a description of each item, the grade levels for which it is recommended, and levels of difficulty.

San Diego County (California) Schools

Orange (California) Unified School District

⊗ Assistance in arranging details for field trips, and advice about procedures for making use of other community resources.

⊗ Experts to produce simple audiovisual materials for your classes, and instruction for both you and your students to teach you how to produce them.

school media services to students

The several typical resources and services that the media center offers teachers suggest those that should also be provided for students. They may include:

⊗ Library materials in printed form: books, pamphlets, encyclopedias, reference books, recreational reading materials, and a variety of supplementary study materials for students of differing abilities and interests.

⊗ Many types of audiovisual equipment designed for individual student use or for use by groups of students. The possible range of equipment resources is extensive and may include 8mm and 16mm projectors or viewers; audio devices for disc recordings; audio cassettes or audio tapes; programmed instruction books or machines; filmstrip viewers; games; manipulative devices for studies in mathematics, sciences, or other subjects; microscopes; video tape recorder-players; machines for duplicating printed materials, slide projectors, slide cartridges, and accessories. No matter how many items are listed here, you will think of more that may, or should, be available to your students.

⊗ Catalogs of materials available commercially or for loan from other than school sources for examination.

⊗ Professional consultation services by a specialist trained in media of all kinds, who knows how teachers use or have used the materials on hand and can suggest materials to meet your needs and the needs of your students.

⊗ Equipment and materials that you may take to your classroom when you need them. These are maintained in ready condition by the resource center.

⊗ A place where you may plan and prepare your lessons, near the materials you may wish to examine or review, with access to a typewriter.

⊗ Technical assistance in placing orders with outside sources for films, books, filmstrips, and other resources needed in your classes; assistance in setting up room libraries, grouping special collections, and establishing learning centers that support activities in which your students engage.

⊗ Paraprofessional or teacher aide personnel to assist in collecting materials required for your classes.

San Diego County (California) Schools

⊗ Proper places for students to work as they undertake a variety of learning activities. These facilities may include comfortable lounge areas for reading, for study, or for enjoyment, and carrels where one student may work independently or where several students may work together. The carrels may be "wet"—that is, they may be equipped with power sources to run audio or visual equipment or an electric typewriter. They may also have lines from central audio, television, or data sources, or installed equipment such as cassette players or television monitors. Other carrels may be "dry"—that is, they may have no equipment provided in them, consisting of only a semiprivate space for independent study, with good light and a comfortable chair. To encourage group work, there should also be areas where students may confer without disturbing others.

⊗ Facilities in which students may produce media. At a minimum there may be a recording room in which students record audio tapes or simulate broadcasts. With some additional features, a simple television studio may be provided, with an economical, small-format video tape recorder, tele-

ALOHA Project (San Jose, California)

vision camera(s), and essential accessories. Provision of materials and equipment for the production of transparencies is not an unusual service for students, and the photographic equipment necessary to make slides or simple 8mm films is fast becoming standard in the school media center.

⊗ Places for quiet reflection, writing, or project planning. Such areas may contain a light box for viewing slides, audio devices for listening through earphones to musical or nonmusical recordings associated with lessons, and a drawing table for sketching or making plans.

⊗ In some cases, a workshop with tools for woodworking or modelmaking or for simple construction to be used in photography, puppet productions, or other projects.

It is apparent that the resources and accommodations listed here emphasize *activities,* both quiet and energetic, that are directed toward enabling students to learn. Your role as a teacher is to encourage students to discover suitable projects and activities that move them to dynamic processes of creation and communication. In this program, the school media center becomes a principal work center and its professional and technical staff a significant source of support and enrichment necessary to achieve this goal.

Portland Community College

media resources from outside the school

Although media services within the school are of top priority, no less important in many ways are the additional media services provided by school district centers and various other regional and state educational agencies. Typical of the functions and services supplied by such units are (1) making short-term loans of motion pictures, multi-media kits, models, displays, and similar items; (2) offering expert supervisory assistance in improving media utilization practices and in organizing or reorganizing single-school media programs; (3) gathering data to reinforce requests for increased support of media programs; (4) producing or reproducing, as a service to teachers, certain media (including transparencies, slides, filmstrips, original or dubbed tape recordings, printed booklets or syllabi, and radio and television programs); (5) organizing workshops and in-service training programs to help teachers develop or improve their media production or utilization skills; and (6) conducting large-scale programs for the selection, appraisal, and evaluation of media resources to be purchased for county, district, or single-school collections.

Similar services are sometimes supplied cooperatively by regional administrative units. A cooperative effort in establishing regional or area centers helps school districts reduce costs and at the same

Montgomery County (Md.) Schools

time expand the quantity and quality of media resources available to them.

Many states support central agencies which provide specialized instructional resource services. Some supply book, library, and film library services —at times through a self-supporting program of media rentals.

To discover what resources are available in your own district center and how to obtain them is an interesting first responsibility, and you may depend upon your school media center staff to help.

the community as a learning center

The community in which a school exists constitutes a valuable educational laboratory, filled with resources to strengthen and enrich learning.

Thus, it is recommended that you become as familiar as you can with the resources that surround your school—the special facilities, expert persons, government institutions, industries, and businesses; the geographic-geologic-political-cultural facets available for selective use in your instructional program.

The school media center staff usually take the lead in locating and organizing such data, but you and your students may be asked to assist. The following chapter deals in detail with the community as a learning center and suggests ways to utilize community resources to the greatest possible advantage.

3M Company

classroom learning centers

Learning centers, like the comprehensive school media centers just discussed, also provide places and materials in which students perform independent learning activities of many types. Typically, they are arranged as self-contained stations or carrels in classrooms, in the media center itself, or wherever appropriate space is available. As previously stated, these facilities may be "wet," i.e., equipped with projectors, playbacks, or other devices that require electricity, and sometimes with line inputs for audio or television programs or computers. Or they may be "dry," i.e., without such facilities and offering merely a study or work space in which nonpowered items—printed materials, realia, scales, measuring instruments, and similar items—are placed.

types of learning centers

There are several types of learning centers: (1) those of a modular nature that focus on some particular segment of unit study; (2) those that provide materials or exercises for general enrichment or drill in a special subject such as math, reading, or social studies; or (3) centers which contain collections of

San Diego County (California) Schools

TYPES OF LEARNING CENTERS

The diagram to the right (opposite page) portrays several different types of learning centers that may sometimes be found in a single classroom. Those shown are for: (1) diagnosis, (2) review of prerequisite skills, (3) introducing concepts through model-

miscellaneous materials on subjects of general interest. In the latter, students are encouraged to participate in free-time activities of exploration and creativity. In the first type, learning tasks are laid out clearly and accompanied by written directions for completing them. A pretest is often used as a measure of initial competence. With all types, most, if not all, of the materials needed to complete the tasks are provided, or clear, written information is given as to where and how to find them.

Typical tasks performed by students in learning centers will require media resources of many types, as mentioned in the next sections. Not all modular assignments and tasks require students to remain in one place—or even in the school. The activities they are to complete may take them to the media center, to other parts of the building, to a community facility (a museum, for example), to their own homes, or wherever they may best fulfill their objectives.

commercially produced learning center modules

Commercially produced learning center modules play prominent roles in individualized learning and independent study. Such modules are expected to have been formulated after extensive, thorough testing and to have been influenced in their development by the performance of students upon whom they were tested. Concepts and processes of *learner verification,* which are basic to assessments of all teaching-learning resources, including modules, will be discussed in more detail in Chapter 4.

ing, (4) developing key concepts, (5) concept enrichment, (6) practice and reinforcement, (7) skill drill, (8) application through games, and (9) criterion-referenced evaluation. Other kinds of learning centers and the media related to them are also pictured. With each, full directions are provided (in written or printed form) to guide students in carrying out assignments and in checking their achievement of objectives at satisfactory performance levels. Diagram to right, courtesy of EDL/McGraw-Hill; photos and diagram below, courtesy of Dr. Leonard Espinosa, San Jose State University.

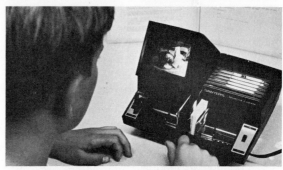

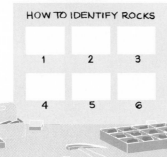

producing or adapting your own teaching modules

Locally produced learning modules of the types you and your students may prepare, often with the assistance of school media specialists and district professional curriculum personnel, ought to meet the same standards as those obtained from commercial sources. But the need for developing such modules may be derived principally from teacher insights and experiences gained largely from working closely with students who use them. However, even in such cases, reuses of materials would be expected to produce cues for module improvements.

Several suggestions can be inferred from the experiences of teachers who have produced and used their own modules:

⊗ Produce modules of different, but functional, types. Remember that many will be simple contracts, on paper, that will not require use of stations.

⊗ Be clear as to the objectives of the module (what it is expected to teach); state the objectives in accompanying job cards or direction sheets.

⊗ Involve students in designing, constructing, writing, and testing modules.

⊗ Pretest modules with one or two students before using them with others; revise or refine, as needed; obtain continuous feedback with further use.

⊗ Don't attempt to teach too much in a single module. Don't make the steps toward module goals too difficult or too easy, as suggested by the drawings below.

⊗ Establish reasonable performance standards, state them clearly, and adhere to them.

⊗ Provide suitable means for students to evaluate their performance as often, and at such points, as necessary.

⊗ Inject a little fun and humor into the assignments whenever possible and appropriate.

⊗ Make modules success-oriented; make it possible for students to succeed (not fail) with reasonable effort.

⊗ Offer alternatives for various types of break or rest periods during use of the module.

⊗ State deadlines, if appropriate, for end of the assignment and for checking of interim products.

⊗ Identify, at least in a general way, resources needed for the assignment and where they may be obtained. At the same time, do not be so specific in this that all stimulus is removed for students to be inventive in their search for appropriate media to complete their projects and assignments.

⊗ Don't try to invent everything. Draw upon good ideas you learn from others, or adapt other modules with which you are acquainted to your particular requirements.

⊗ Be sure to provide for suitable replacements of lost parts of module kits and for inserting better parts, when called for, as you gain experience with them.

⊗ When appropriate, protect module materials by laminating or covering them with plastic spray.

MATCH CONCEPTUAL DIFFICULTIES WITH STUDENT ABILITY

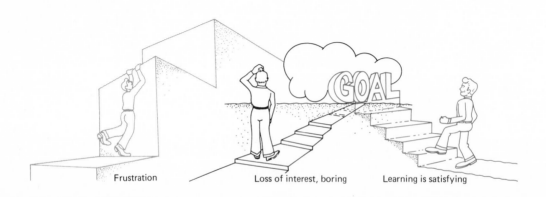

Frustration   Loss of interest, boring   Learning is satisfying

individualized learning options

Many different options are available as learning assignments and activities to be used in preparing modular materials and learning centers. Experienced teachers often stress the functional benefits of varying assignments that may be developed through the following activities:

⊗ Reading—textbooks, magazines, newspapers, reference books, nonfiction books, encyclopedias, almanacs, pamphlets, atlases, anthologies, dictionaries

⊗ Listening—prerecorded audio tapes, phonograph records, live lectures, or music

⊗ Viewing—8mm or 16mm films, video tapes, sound and silent filmstrips, flat pictures (sometimes accompanied by audio tapes), charts, displays, bulletin boards, slides and transparencies, models (some of them of the "working" type), television programs, specimens, real things, microscopic specimens or slides, microforms, and through telescopes or binoculars for stargazing or bird watching

⊗ Visiting—museums, institutes, factories, other schools or universities, businesses, planetariums, zoos, newspaper offices

⊗ Writing—analyses, reports, letters, scripts, self- or teacher-prepared tests, newsletters, questionnaires

⊗ Interviewing—other students, or teachers, or individuals in the community either face to face or by telephone

⊗ Discussing and debating—in small, medium, or large groups

⊗ Experimenting and researching—gathering data in search of solutions to problems; ordering the data; drawing conclusions

⊗ Constructing—devices, instruments, models

⊗ Illustrating—drawing, producing schematics

⊗ Photographing—flat pictures, transparencies, 8mm or 16mm films, video tape recordings

⊗ Gaming and simulating—inventing or using already-invented games; modeling; simulating reality

⊗ Collecting, classifying, displaying—realia, specimens, stamps

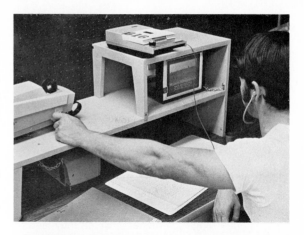

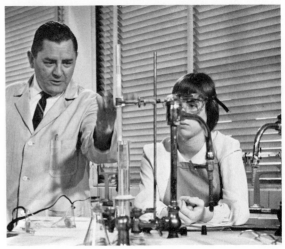

CHEM Study, University of California

⊗ Reporting—orally, in writing, in visual or audio-visual forms

⊗ Demonstrating—processes and operations, elements

⊗ Interning, or working—as on-the-job experience, sometimes for pay

⊗ Operating—developing skill, practicing

Are you able to identify other types of learning activities to be added to this list? For example, where would you classify the experience of working interactively with a computer in learning a body of information?

summary

Educational media of the types discussed throughout this book make special contributions to the individualization of learning—in small, medium, and large groups, and in independent study modes. Current advances in the systematic design, production, and use of media resources and related devices have served to extend benefits of instructional technology to all educational levels. This approach to teaching and learning promotes functional correlation of educational purposes and goals with processes, personnel, media, machines, facilities, and learning environments which may lead to providing "the best for each."

Eight forms or systems of instructional management, each involving extensive use of educational media, were described: (1) PSI, Personalized System of Instruction, (2) CMI, Computer-Managed Instruction, (3) IPI, Individually-Prescribed Instruction, (4) IGE, Individually-Guided Education, (5) LAP, Learning Activity Packages, (6) Contract Plan, (7) ATI, Audio-Tutorial Instruction, and (8) HEP, Hawaii English Program.

Resources and services provided to teachers and students through comprehensive school media centers, together with other important media services and resources from outside the school, in the surrounding community, and elsewhere, were also described as being basic to individualized learning, as were several different types of classroom learning centers. The selection, local production, and use of modular learning packages are crucial processes in achieving successful results with individualized instruction.

3

THE COMMUNITY AS A LEARNING CENTER

chapter purposes

- ⊗ To explore the possibilities of using the local community as a learning center.
- ⊗ To identify community resources that can provide learning activities for students at all levels.
- ⊗ To describe how to conduct community studies that can identify community resources which include places, organizations, and people.
- ⊗ To describe procedures for planning and conducting field trips into the community.
- ⊗ To encourage the use of resource persons from the community to enrich the school program.
- ⊗ To describe typical instructional programs provided by such institutions as museums and zoos.
- ⊗ To note the use of community resources by alternative schools.

Dayton (Ohio) Public Schools

The previous chapter dealt with the concept of learning centers as a means of providing appropriate places for student educational activities. Here the discussion includes the entire community as a learning center. What is the community? It is the neighborhood, the town, the city, the county, township, or state—whatever the area in which students will live, make homes, and work or create.

This chapter is especially important today, since interpersonal, intergroup, and interoccupational relationships seem to be increasingly complex and fraught with conflicts. The gap between the world as experienced by the student within the school and the real world outside widens continually. Later, the student faces difficult but necessary learning problems simply because the outside world is so different from the school environment.

However, schools that make opportunities for interaction between the community and the instructional program prepare students to bridge that gap. The more that students of all ages in all schools develop a feeling that the community life is part of their life and that they belong in the community, the better prepared they will be to solve their adjustment problems as members of an adult community wherever they go in the outside world.

Many of the alternative schools, which are discussed later in this chapter, have succeeded, especially those in the cities, in overcoming the gap that exists between real life and traditional school life. Since the community is filled with natural resources from which to learn, one function of every school is to design experiences that will take students into the community to search for objective, complete, and accurate information to be integrated with their basic studies.

Exploring the community is hazardous, just as is any adventure. Our plan to use the community as a learning center should include the objective to bring the students back alive—alive in social awareness and in good health, and alive in interest and intellectual stimulation.

Whatever activities students may pursue in community studies, work experience, or in learning interpersonal relationships, media for communication can and usually do play an important role. In this book, therefore, much attention is given to how media can be used to expand the experiences of students as they work in the community as a learning center.

Lawrence Hall of Science

making a community resources inventory

To use the community as a learning center, teachers need extensive information about the agencies, people, places of interest, and services that are accessible and interested in serving the school. Frequently, however, teachers have only limited knowledge of the community in which their students live, grow up, and eventually go to work. Sometimes they do not even live there and know very little about local projects and facilities. Often, too, while the school media center may have quite complete resources for studying distant places, it may have failed to develop resources and files of materials relating to the local environment.

Such conditions should challenge teachers to rectify the imbalances they find. Without question, nearly every community is rich in situations and activities to study and enjoy. Teachers, as individuals, need to know about them and to pay particular attention to those areas that students regard as basic and of most interest. Here are some ideas about how to acquire the needed information.

organizing a resources survey

Ideally it should be the responsibility of the central school district office to prepare and maintain a community resources inventory. The use of cumulative files of data for the inventory, and the regular screening of the files, should keep the inventory comprehensive and up to date. But it may also be that, in the absence of such central office assistance, you and your students must develop your own inventory.

The end product of a community resources survey is usually a handbook which may be prepared in loose-leaf form that can be easily augmented or updated. Information provided in it may include such items as

 A single alphabetical list, or several separate alphabetical lists, of agencies, companies, and organizations willing to cooperate in providing educational resource services to schools: government agencies, municipal service units (fire stations, police stations, post offices), museums, transportation

⊗ Could these same benefits have been obtained in some better way? If so, what?

⊗ Were there any unexpected problems? What caused the problems? Inadequately briefed guides? Students? Poor planning? Unexpected trip conditions that could have been avoided? What should be done next time to eliminate such problems?

⊗ Should the trip be recommended to other classes studying similar topics? If not, why not?

You and your students should also take time to evaluate yourselves as takers of field trips to see whether your planning needs to be improved. One teacher and his class decided that their biggest mistake had been to give too little attention to needs for food and rest. It was their first all-day trip, and they found that they needed longer rest periods and less pop and hot dogs. Another group felt it had wasted the time going and coming, and so they prepared a list of things to look for and do en route the next time.

The authors have asked teachers to note their various experiences in taking field trips. The three following excerpts from their answers indicate what can happen with and without careful preparation:

I once took a group of students to a museum. We just wandered around from place to place. We had set up no specific things to find out and I am sorry to say we didn't find out very much.

Instructions about where to meet, and the parents who were to bring cars, were not checked and double-checked. So we started late. The whole trip was a mess.

I took my classes to downtown Chicago to make a study of unemployment. Before the trip, we used motion pictures, still pictures, and recordings to get ready for the trip. Members of the class were interested during the trip because they were already familiar with what they were seeing.

Inevitably, then, the success and benefits of the time, effort, and expense of a field trip will depend upon the thoroughness and skill with which the event is planned in every detail. Consider, for example, student and teacher activities that would precede a trip to Independence Hall!

INDEPENDENCE HALL: A NATIONAL TREASURE

Independence Hall, with its Liberty Bell in a location nearby, is typical of many state and national historic monuments now restored for the enjoyment and enlightenment of the people.

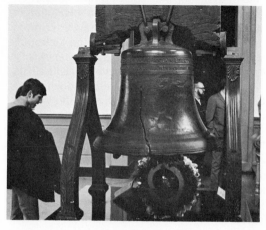

going where the action is

An increasing number of institutions that provide exhibits of the arts, sciences, industries, and history now include many displays that permit active participation by the visitor. Typical of outstanding institutions that reconstruct the past are The Farmer's Museum in Cooperstown, New York; The Provincial Museum of British Columbia, in Victoria, B.C.; Greenfield Village, Michigan; and Williamsburg, Virginia. A walk through one of these places is a way to enjoy history, to relate to persons living in the past, and to have a simulated contact with another time.

Science museums are represented by a wide variety of types: extensive collections of dioramas of animals in their native habitats; dynamic exhibits of people or mannequins working in ancient mines; institutions with working computers, telescopes, or moving devices exhibiting physical principles; and areas where visitors may work or be taught. It is the latter types of exhibits that visitors find most interesting. Here, then, is some information about places where the action is, where the trend is toward active learning experiences for the serious student or the casual visitor.

museums

The United States, Canada, and Mexico have many of the world's best museums.

Some of the most widely known museums in the United States are The Smithsonian Institution in Washington, D.C.; the American Museum of Natural History, the Museum of Modern Art, the Metropolitan Museum of Art, and the Brooklyn Museum, all in New York City; Chicago's Art Institute, Natural History Museum, and Museum of Science and Industry; the Los Angeles County Museum; and Fort Worth's Museum of Science and Industry. Mexico City has one of the finest archeological museums in the world, with associated museums in Guadalajara, Colima, and other regional centers. In Canada, from Quebec west to British Columbia, provincial cities have museums, library-museum centers, and zoos of high quality which attract thousands of visitors.

SMITHSONIAN INSTITUTION
Among its many divisions, the National Air and Space Museum attracts thousands every year to see the record of United States achievements in the air . . .

All these institutions offer programs which furnish educational assistance to the regions they serve. The Fort Worth Museum, for example, provides informative programs in all natural sciences, anthropology, history, medical science, health, and archeology as well as in such specialized areas as meteoritics, early firearms, and Texas history. It also conducts a field school in Mayan archeology and sponsors annual expeditions to the Yucatán peninsula. As part of its Children's Museum program, it operates a 1,500-acre nature center and refuge on an island near the city. A planetarium projector and theater provide a center for studies in astronomy, astronautics, and some earth sciences. In its "Life Begins" section, sculptured models show the steps of human conception and birth. Special exhibits are constantly being developed, adapted, and changed. A television program for children reaches about 500,000 persons once each week. Information about the different exhibits is provided by recorded materials broadcast through wireless loops and heard by listeners using small, rented wireless headset receivers.

An extensive program of museum classes is also offered at Fort Worth. There, preschool programs are available for children three years before they enter school. Nearly 60,000 Fort Worth students attend organized school study tours each year; in addition, the museum conducts its own educational courses for some 6,000 completely voluntary stu-

. . . and in space.

July 4, 1976, the National Air and Space Museum opened its new building to the public, another great museum where the action is!

dents per year. Hundreds of other museums offer programs somewhat comparable to those of the Fort Worth Museum.

In the United States today there are nearly four hundred children's museums or departments in regular museums which cater to children's interests. The junior museum has developed in many communities, offering such specialties as guided nature walks, displays and loans of animals, field trips, and other science activities.

Some specialized children's museums (the Boston Children's Museum, for example) also provide media kits containing all the materials, equipment, supplies, and instructions students and teachers need to develop their own classroom exhibits. In many cities and counties, the museum-mobile in a large van takes special exhibits right to the schools.

The Anacostia Neighborhood Museum in Washington, D.C., and MUSE, Brooklyn's Bedford-Lincoln Neighborhood Museum, represent some of the most promising museum developments for the seventies. The Anacostia Neighborhood Museum, a branch of the Smithsonian Institution, is housed in an old, previously abandoned theater which has now been renovated. MUSE, an outgrowth of the Brooklyn Children's Museum, is housed in a renovated automobile showroom and a pool hall. Both museums cater to an inner-city clientele and rely on neighborhood advisory groups to assist with planning programs. Children using these museums

know that their ideas and wishes will continue to be reflected in presentations and exhibits. Much of the work involved in the museums and in their programs is done by these same student users. Members of the Youth Advisory Council at Anacostia, for example, designed and silk-screened the museum greeting cards, worked on a film documentary, and planned an exhibit on the history of jazz. One of Anacostia's most famous new exhibits grew out of youth council suggestions that the rat problem be illustrated. "The Rat: Man's Invited Affliction," a very effective ecological exhibit and program series, resulted.

Another example of an outstanding exhibit combined materials on Afro-American history with highlights of the life of Frederick Douglass. But beyond such innovative exhibits, these two museums thrive on emphasizing the personal, neighborhood nature of their programs and continue to encourage direct, active user participation in their work. Objects are to be touched, felt, smelled, or created. MUSE, for example, offers daily "Walk-in Workshops" in art, writing, geology, anatomy, and other fields, requires no formal enrollment, and is open to any young persons wishing to attend. Open twelve hours a day and offering users everything from painting, stage work, and dancing to karate and moon watching, MUSE sets a new pattern for museums that may deserve to be copied, more than it has been, by the institution called "school."

other places for experiences

Some institutions have a variety of options for active participation by students and casual visitors, offering both formal and informal instruction. One such institution is the Lawrence Hall of Science in Berkeley, California. This institution, privately endowed and privately operated, on the University of California campus, provides numerous opportunities for students to enjoy firsthand experiences with science and scientists. Lawrence Hall provides classes for students with exceptional ability, taught after school hours by visiting scientists or scientists from the university. Exhibits in Lawrence Hall range from well-displayed photographs and graphic designs presenting up-to-date scientific information or new science education materials to a wide variety of special devices that can be manipulated by the students as they learn principles of mathematics, how computers work, what lasers are, and similar subjects that capture student enthusiasm and interest. Film showings and demonstration programs are also available on a schedule. Students from elementary and secondary schools from miles away are brought regularly to Lawrence Hall; others come on their own or with their families to enjoy the inviting environment for active learning.

The Lawrence Hall of Science also offers workshops, courses, and resource exploration experiences for teachers in the region. Science teachers, for example, may attend seminars and regular classes in methods of teaching science subjects, or they may expand their knowledge of current developments in science fields.

In various college and university centers of the United States, research programs in science education as well as workshops and seminars and institutes similarly serve to maintain the skills and interest of teachers.

In your community, state, and region, look for similar programs offered by centers of learning, such as the Lawrence Hall of Science.

The following picture story illustrates the extent and variety of services offered by this outstanding institution.

SOMETHING FOR ALMOST EVERYONE

Provisions for casual visitors as well as for serious children, youth, and adults are made in a variety of modes: small and large group presentations and individual activities. And the programs and participation displays serve different interests: those of teachers of science and other subjects, high-ability secondary school students, and drop-in visitors of all ages who are merely interested. All of the photographs are supplied by the Lawrence Hall of Science; the name of the photographer is below each picture.

Large groups hear lectures illustrated by the giant periodic table
(Jackman)

Small groups of secondary students write their own computer programs
(Charles Frizzell)

Young students attend a special after-school class in microbiology
(Jack Fishleder)

Pond biology is studied near the main facility
(Charles Frizzell)

The quest for information is motivated in the laboratory and in the display areas
(Above, Jack Fishleder; below, Charles Frizzell)

Each summer, teachers work on laboratory experiments learning to use Project Physics Materials
(John Quick)

neighborhood and school museums

In response to the needs of children and youth, some major museums and local communities have developed an innovation in children's museums. The trend in these institutions is toward creation of activity centers and displays where students are given complete freedom to touch and play with a variety of devices and exhibits directed to learning through the senses. Large museums, such as the Museum of Science and Industry in Chicago, have had such great success with exhibits that invite visitor participation that many smaller institutions have been established for the sole purpose of involving visitors in activities and experimentation with the exhibits. A far cry from the ''look but don't touch'' characteristic of some museums in the past, children's museums and neighborhood centers emphasize learning by direct experience.

The neighborhood museum in Anacostia, Washington, D.C., for example, sponsored by the Smithsonian Institution, involves young people and adults in selecting, arranging, and maintaining exhibits.

Anacostia Neighborhood Museum of the Smithsonian Institution

Anacostia Neighborhood Museum of the Smithsonian Institution

Community participation in the neighborhood museum project is expanded when the public visits the exhibits and recognizes not only the content of the displays, but the contributions made by the young people in this constructive enterprise.

Anacostia Neighborhood Museum of the Smithsonian Institution

In this metropolitan area center young people are given the opportunity to participate in fundamental and practical ways in the development, care, and operation of facilities. A sense of responsibility and pride are products of the experience.

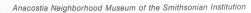
Anacostia Neighborhood Museum of the Smithsonian Institution

The Dayton, Ohio, Public School Museum provides children with many vivid sensory experiences—hearing, seeing, smelling, feeling, acting out roles, or testing and trying. Do similar opportunities exist in your community? Are there agencies that might develop centers to provide experiences such as these shown below?

Dayton (Ohio) Public Schools

Dayton (Ohio) Public Schools

Dayton (Ohio) Public Schools

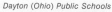

zoos and aquariums

Zoological gardens are also to be found in major cities of the United States and Canada. Some of the most widely known in the United States are those in Cincinnati, Washington, D.C., New York City, Chicago, Detroit, St. Louis, San Diego, and San Francisco. Zoos the country over attract millions of visitors each year. The total yearly attendance at Chicago's Brookfield Zoo recently amounted to approximately 2 million persons, with more than 1 million of them as free admissions, mostly students. With such huge numbers of visitors, careful planning is vital. The curator of education at Brookfield Zoo asked the authors to "caution the teachers to check with the institution *each year* before they visit, as institutional policies change with the times. . . ."

Zoos, and museums too, depend heavily on volunteers known as *docents,* and almost all such institutions have programs to recruit and train them. At Como Zoo in St. Paul, Minnesota, any interested person over fifteen years of age can volunteer for such service. Many high school students might be encouraged to investigate opportunities to serve as docents. Docents receive brief training, and they conduct zoo study tours, give animal or slide shows, help exhibit animals in the Children's Farm Zoo, or help in other ways wherever the need arises. Other zoos, such as the Houston Zoo, limit docents to those who are eighteen years of age or older. Whatever the rules, and wherever the zoo, volunteers may help a great deal in the educational programs.

The Kansas City Zoo provides an educational program that is somewhat typical of such programs elsewhere. Classes are offered which supplement existing elementary and secondary school curricula. Emphasis is placed on the study of animals and their relationships with their environment and human beings. Zoo study classes and independent research study can be arranged for senior high and college students. Special classes are also available in such subjects as endangered animals, camouflage, and zoological realms, and there are courses dealing with birds, waterfowl, African animals, and primates. Art or photography students are also assisted in learning skill in animal photography.

An outstanding example of an educational program presented by an aquarium is that conducted by the Vancouver, British Columbia, Public Aquarium, which provides guided study tours of general interest. In addition, a special biology laboratory for study of seashore ecology presents sessions for up to thirty eleventh-grade students as a group. In these sessions, trained docents provide the instruction for which study guides and locally produced, taped audiotutorial exercises are available. Just ahead of the students' visit, the teachers spend an evening at the aquarium to perform the same laboratory exercises that their students will perform the following day.[2]

These examples suggest that museums, zoos, aquariums, and similar institutions provide a variety of learning experiences which supplement those offered by the school. Often they provide students with subject matter that would otherwise be unobtainable. Investigate the resources that may be immediately available for your classes. Learn what artifacts may be borrowed, what tours might be arranged, what relatively inexpensive or free publications may be acquired, and whether radio or television programs publicize these community centers of exciting information. Community resources are not lacking; encourage your students to seek, to explore, and to report. Here, for example, students are making a study of koala bear behavior in the San Diego, California, Zoo.

The San Diego Zoo, widely known for its diversified collection and for open exhibits, serves different groups with experiences appropriate to their needs and interests. Below, medical and veterinary medicine students, from the University of California at San Diego, intern for a time at the Zoo hospital.

San Diego Zoo

And, some students, a bit younger, get very well acquainted with some citizens of distant lands.

San Diego Zoo

San Diego Zoo

[2] Sharon Proctor, "Biology Laboratory at the Vancouver Public Museum," *Drum and Croaker,* vol. 11, pp. 14–15, September, 1970.

people as instructional resources

Some of the people who live in the school community may also constitute a useful instructional resource. Exceptionally well-informed people are almost always interested in helping the schools and will agree to do so if asked. Are you studying current events, for example? Travelers who have returned from the Middle East, Europe, Africa, or Asia and who live nearby may agree to give their personal observations of conditions in the areas visited. Are you studying the history of your country or your state? A few early settlers who are still alive may be able to describe their pioneering years. Resource persons of any age, and from the most improbable places, can provide valuable firsthand information of a kind that may be available nowhere else.

finding resource people

Names of individuals who may be identified as prospective experts may come from a wide variety of sources. The father of one of your students may, for example, work in a steel mill and be quite capable of offering interesting and completely accurate descriptions of the processes of manufacturing steel.

The mother of one of your students may be a member of The Urban League and would be pleased to offer an interesting description of the purposes of that organization and the way it functions. Still another student's father, who works in a greenhouse, may provide similarly expert advice about plants that grow best in your region. A city attorney, the city manager, and a judge will be able to furnish full accounts of the functions of their offices.

The questionnaire mentioned earlier in the description of procedures for conducting community surveys suggested how to develop inventories of resource people.

Also directories of industrial, business, and government organizations, or information provided by the public library, a college or university public relations office, or newspaper articles about interesting people will furnish good prospects, as will the contacts you make in taking field trips.

Media specialists, always on the alert for useful programs, may recommend as a guest an author-illustrator, such as Leo Politi, who cartoons as he tells stories in a Santa Barbara, California, school. Mr. Politi was made available through his publisher, Scribner's.

using resource persons

Resource persons may be invited to come to school to make a presentation to your students, or one or more students may arrange to visit them at their home or place of work. But whether they come to school or are interviewed outside, similar procedures apply.

⊗ ENGAGE THE EXPERT By telephone or by letter, the students or the teacher communicates with the resource person, describing the class needs and extending an invitation to assist the school. Arrangements are made in detail. A definite time is set for the visit to the school or for the out-of-school interview. Oral invitations are always confirmed in writing. If the guest requires special equipment or facilities (projectors, charts, maps, darkened rooms) for the presentation, make a list of them and be sure to provide them, as well as necessary assistance with logistical problems such as parking, carrying materials, projectionist services, and packing up at the end of the program.

⊗ CLARIFY THE PURPOSES OF THE VISIT Make sure that the visitor understands the purposes to be served by his visit, and the class size, ages of students, grade level, and useful facts about the background of the class for understanding the subject. In writing, provide exact directions for reaching the school and the classroom; be clear about the time limits of the meeting, and, if not inappropriate, send along a list of questions students have asked about the topic.

⊗ PREPARE THE CLASS In addition to formulating questions for the visitor, class members prepare themselves for the visit. They decide who will meet the speaker and obtain data to be used in the introduction. They determine who will take notes and who will make a final expression of thanks. They may also make arrangements for return transportation if needed. Also, they may schedule and be ready to obtain, operate, and return any audiovisual or other equipment needed by the guest.

⊗ ARRANGE THE FOLLOW-UP After the visit of the resource guest, the first concern of the class is to send a note of appreciation, acknowledging *specifically* the contributions made to them and their studies. What the students do, however, to use the information will to a great degree determine the value of the time, energy, and thought devoted to the activity. An article could be written about the event for the school or community paper. Featuring the guest, the topic, and the visit, this article could engage the talents of a committee and would also please the guest. Sometimes a picture of the guest with the students—taken by a member of the class— will be welcome. Often the students will write personal essays or summaries of the points they have learned; these might be bound informally and sent to the guest with the letter of thanks. By further research into the topic presented, students may expand their background in the subject and prepare reports, displays, or presentations. Occasionally, artists or writers who visit classes stimulate some students to creative effort, which can be encouraged by teacher and fellow students alike. And often the contributions of a guest, such as a scientist, may suggest other specialists in the same or related fields who might be invited to extend the class experience.

Finally, of course, some written record should be made of the visitor's contributions. It is a good idea to place in one file all materials pertaining to the visitor—including correspondence, records of calls, and clippings—and to retain the file in a special place in the school media center. It may be consulted on other occasions by those interested in extending an invitation for a repeat visit.

There are many, many options when the opportunities to bring resource visitors to a school are analyzed. A typical option is illustrated below. A dentist who specializes in work with children gives an illustrated lecture to the Ward School, Tempe, Arizona. He has brought handouts and posters from the dental association, and the Media Specialist has provided models, charts, and additional handouts. In preparation for the visit, students have prepared a scroll theater that reinforces ideas they have learned previously.

Tempe (Arizona) Elementary Schools

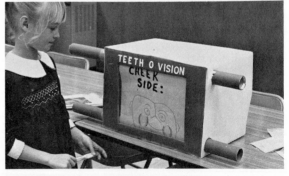

Tempe (Arizona) Elementary Schools

In another school, the law enforcement officer has granted permission for high school students to video tape his presentation on safe driving in order that it may be used again.

The visitors, below, are sharing the stage in a psychology class at the University of Nevada, Las Vegas. From long experience as an entertainer, the guest discusses with the students animal behavior and training, a topic of value in the course.

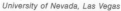
University of Nevada, Las Vegas

interviewing resource persons

Interviewing a resource person outside the classroom requires just as careful advance planning as an invitation for a class visit. An interview is primarily a question-and-answer session, and students who are to conduct interviews often need some coaching and practice in advance on procedures and courtesies involved. Questions prepared ahead of time by a class or a committee are helpful. Pictures may be taken or publications or other items collected for use back at school. Permission may also be obtained to copy one-of-a-kind items.

Student interviewers must realize that their primary responsibility is to obtain information. They should show interest, listen intently, and express appreciation.

Interviews are sometimes conducted by telephone and amplified to permit an entire class to hear. This type of telephone interview has many advantages and can also have high motivational value to a well-prepared group of students. The facilities are not difficult to arrange, either through the local telephone company or through the use of a telephone amplifier that can be obtained from local electronic suppliers for only a few dollars. Remember too that many resource persons willing to be interviewed over a telephone circuit may also be willing to be recorded, a procedure that can be arranged with your telephone company to meet requirements of the Federal Communications Commission for a beep-tone indicator of recording.

students conduct community studies

With the community accessible as a learning center, a logical development from the use of field trips is to give students opportunities to experience a much more sophisticated use of the community: to learn how the community functions, what careers are typical of the locality, and how, as students, they may relate school studies to real world activities nearby. In making community studies of extended scope and significance, students may apply many of the academic skills which they have developed in their schoolwork. Such community studies give practical opportunities for individual students and groups of students to expand their creative talents, their use of the media of communication, and their developing abilities to work with adults, to their mutual advantage and respect.

One example of a successful high school program in community studies has led to actual part-time student community service. Evanston Township High School, Evanston, Illinois, offers a course titled "Social Studies: Volunteer Opportunities." Each student in the class works three to five hours a week, during school time, in one of the many agencies of Evanston: ranging from day-care centers to the police department. Seminars are held once a week; research projects reported at the end of each semester are intended to aid the agencies involved.

The pilot "4-1" program (four days each week in school; one day in the community) of the John Dewey High School, Brooklyn, New York, is another example of the extensive and exciting use of community resources. The marine biology program of this school utilizes surrounding beach areas and the Coney Island Aquarium. Students also study anthropology in the Museum of Natural History, art in the Brooklyn Museum, and science related to a number of medical specialties in the Brooklyn Downstate Medical Center. Special interests of students are further accommodated as in the case of one student who participates as a free-lance reporter for a local newspaper, while others work in commercial and industrial enterprises.

A college program that reaches into the realities of public community concern is represented at the University of Wisconsin–Green Bay by student-selected topics similar to the following:

Our Lake: Its Quality (Physical and Biological)
Our Water Supply: Its Present and Its Future
The Impact of Snowmobiles (or Trail Bikes) on the Rural Environment
How to Make the Vacant Lot a Busy, Useful Place

Topics and areas of study, as reflected by the programs just referred to, suggest student awareness of the need for environmental control, regional analysis, urban analysis, modernization processes, population dynamics, community growth and development, human adaptability, and human identity and values. Activities growing out of these programs help to develop a way of life that nurtures active participation in community affairs.

Permitting students to pursue subjects in these ways will be a challenge to any teacher, regardless of grade level or subject interest. As for the students, there is ample evidence that they are ready and eager to engage in real studies of the real world with which they are familiar through mass media and daily living. Opportunities frequently arise to encourage them to express choices in such matters, to settle on priorities, to help them go after the details, to put their studies in order, and to communicate their findings to each other and to their communities.

In real life studies within a community, opportunities to use the skills of communication are brought into play: writing, speaking, recording, taking pictures, making graphics, preparing visualized reports and presentations to school and community groups, and developing interpersonal relations.

Mastery of audiovisual and publication techniques is motivated by the nature of the investigations: collection of data; discussions with informed persons; preparation of organized information; selection and design of films, slide sets, video tapes, exhibits, and displays; and preparation of materials for news media in the form of reports and press releases—all integrating past training and experience in a real world environment. Such opportunities are available in every community.

alternative schools and the community

Alternative schools are included in this discussion of the community as a learning center because many of them include extensive student experiences in the local community as one aspect that differentiates them from the conventional public schools. Some of them are called *storefront schools,* thereby indicating their location close to business areas. Generally, however, the intent of the alternative schools has been directed toward some type of alternative to existing schools that seemed to sponsors to be failing to serve student needs. Some alternative schools have been directing their efforts toward assisting drop-outs to find attractive and meaningful activities in a nonacademic atmosphere, while at the same time providing opportunities for them to acquire learning and develop as individuals. Others have been directed toward serving small groups of students, often from one social stratum, with instruction characterized by novel procedures and considerable freedom of choice on the part of the students. And some schools have been designed and sponsored to provide a strongly traditional program of instruction, in the belief that the freedom of choices and learning activities permitted in many public schools has been too free, that learning was inadequately structured, and that standards and demands upon student efforts were too low.

Readers of this book should consider the current status of alternative schools and alternate programs. Some may be firmly established. Some, early supported by foundations or by federal funds, may now be part of school system programs. Some may be privately sponsored for small groups of students. Some may be extended broadly into a large community concept, even using open university and television college methods and neighborhood meeting places for participating students. And, in all of them, the importance of the community and of media in the instructional process will vary within the programs according to teachers, the philosophy of the school, available resources, and the point of view of the sponsors and the clients of the schools.

A number of the schools-without-walls especially represent a type of alternate school that has utilized the community extensively as a learning center. Philadelphia's Parkway program utilizes the entire city as a campus and as a basis for the curriculum. Parkway has no school buildings, for classes are held in city and state facilities, hospitals, businesses, and educational institutions, as well as in private homes, churches, and offices. In its report on alternative schools, *Matters of Choice,*[3] the Ford Foundation stated: "Each unit of students has a meeting place and ten full-time accredited teachers plus some university interns. This staff teaches about half of the courses—mostly basic curriculum such as reading, math, social studies, and science—and supervises weekly tutorials. Community volunteers with special skills offer on-site programs and internships in academic, commercial, and vocational subjects. A Parkway student may study science at a nature center . . . law and justice at City Hall, library skills at the Philadelphia public library, office procedures at a bank, mechanics at a local auto body shop."

Other examples of alternative programs that use the community as a learning center are On Target High School in Berkeley, California, Metro in Chicago, Illinois, and City-as-School in New York.

Among the advantages suggested by those working in community-involved alternative schools are: natural bringing together of students of many backgrounds and abilities; small groups of students involved at one time with able teachers and needed community talent; opportunities to expand experiences in both career fields and in practical applications of basic academic skills; and provision of learning resources beyond those traditionally available.

It is possible that many of the alternative schools will disappear, while others with sufficient experience will have stabilized and will have continued to meet the needs of the particular student groups to which they have given principal attention. Success, at least in the past, seems to be determined by the skills, attitudes, and dedication of the teaching staff. And the role of the community in the alternative school programs is peculiar to each situation.

---

[3] *Matters of Choice: A Ford Foundation Report on Alternative Schools,* Ford Foundation, New York, 1974. p. 7

generating community resource materials

Several types of successful community activities will be reported here, especially to indicate the solid groundwork that can be laid by community-school planning and effort, all to the end that students will have free and adequate access to the resources they want to study.

⊗ WARREN (MICHIGAN) COMMUNITY RE-SOURCES WORKSHOP The Warren, Michigan, community resources workshop involved more than fifty teachers in a cooperative effort with twelve local industries and the Chamber of Commerce to learn about the socioeconomic environment in which they worked. An outgrowth was a handbook for a community action involvement program, with information on eighty local places recommended for field visits. Data were organized under the following headings: (1) commercial, (2) communications, (3) culture and history, (4) finance and banking, (5) food preparation, (6) industry, (7) government, (8) recreation, (9) transportation, and (10) utilities.

Participants communicated with more than a thousand potential resources, which included community leaders, citizens with special talents and skills, and hundreds of potential field trip locations from which the final eighty were chosen. Thus many community contacts were made and community resources explored that were of interest for possible future use.

⊗ NEW JERSEY'S SCHOOL-INDUSTRY SCIENCE PROGRAM The New Jersey School-Industry Science Project represents a statewide program of cooperation between education and industry initiated to keep up-to-date scientific information flowing to schools from industry. For this project, local and regional committees facilitate contacts between industrial specialists and teachers. To activate this program, an extensive bibliography of available instructional materials was prepared and many field trips were conducted by teachers and students. Now industrial specialists in relatively large numbers go directly to school classrooms or participate in regional forums for teachers and students. The Bell Telephone Company's production, "The Sci-

ence of Semiconductors," was prepared for high school physics classes as a part of this program.

⊗ SAN DIEGO COMMUNITY EDUCATIONAL RE-SOURCES PROJECT The San Diego County Schools initiated an extensive program to "reduce the time lag between the development of new knowledge and its availability to the teacher and learner . . . and to identify and organize the community resources . . . to bring about their most effective and efficient utilization in the instructional program."

Community leaders and cooperating educators in several task-force groups established (1) a common area library directory for the more than seventy special libraries in the San Diego region; (2) a special botanical garden containing rare live-plant materials in an organized ecological situation for students to visit and study; (3) a program for making new information available about space, using sound filmstrips, educational films, information pamphlets, and picture sets; (4) a carefully selected set of large study pictures to be studied by groups making field trips to the local zoo, together with a filmstrip to show in classes preparing for these trips; and (5) sets of biology slides from a local United States Navy hospital for study in biology classes, plus tissue cultures, x-ray films, and twelve experiments using radioactive isotopes. Other materials were also developed to aid studies of nuclear energy, plasma physics, cryogenics, human communication, oceanographic research, modern medicine, and computers.

Through this project, the specialized talent of individuals in local industries and agencies was marshaled into an organized effort to provide up-to-date instructional materials. These materials were carefully developed, pretested, and evaluated for accuracy. In many cases new information was made available to students in a variety of media within less than a year of the time of its discovery. In addition, in-service seminars were developed to explain details of the materials and to keep teachers abreast of modern knowledge. This program, it should be noted, was an early and successful attempt to create multi-media kits for modular instruction, and included print, audio recordings, pictures, and motion pictures.

COMMUNITY EDUCATIONAL RESOURCES

That many communities in the United States have the resources to develop materials locally is demonstrated by the San Diego Community Resources Project. In fact, numerous large city school and regional systems use the San Diego program as a guide and emulate it. The resulting programs are as varied as the talents and activities of the local community.

San Diego County (California) Schools

. . . "Living and Working in Inner Space" is based upon real activities of the industrial, business, and scientific community within a county . . .

. . . the drug problem . . .

. . . and airline transportation

summary

People and places in communities offer a wide range of resources for learning. Organized, districtwide surveys by students and teachers can locate and catalog such resources and thus develop a list of valuable activities.

In many school districts, the professional media personnel will assist in preparing up-to-date, well-organized lists of community resources.

A review of local community activities under such headings as Commerce, Industry, Finance, Transportation, Utilities, Culture and History, Government, Recreation, Health, and Welfare can produce numerous community resources usable in the school program.

The programs and services of local museums, zoos, aquariums, and related institutions provide rich resources for student activities. Industrial and business organizations, as well as government agencies, also provide services that can be adapted for instructional purposes. Utilization of these community enterprises is emphatically justified.

Individuals in the community constitute a resource that deserves thorough exploration. Often, the most economical and effective procedure with resource persons is to bring them into the school. Presentations by these individuals can be supplemented by audiovisual media.

Community agencies and activities can often best be studied on location, in which case field trips may be desirable. In each situation the investigation of community resources requires thorough planning. Sometimes, using audio recorders and cameras, students and teachers can interview resource persons at their locations—with great success if the event is well planned. Such planning is not the job of the teacher alone. Students, too, must play a role in it—and thus assume their full share of responsibility for making the experience meaningful, profitable, and pleasant.

Many alternative schools have taken unusual advantage of the values of student contact with the adult community. Some of them have designed their regular instructional program to include part-time volunteer work in businesses, industries, government offices, libraries, museums, zoos, and other public agencies. In other schools, community research projects are a formal part of regular student activity. Student contacts with the community are designed to bridge the gap between the more or less isolated experiences in schools and the experiences in the outside world of adult life.

One responsibility of all schools is to utilize community resources for instruction, both within the school and in the field. To achieve that end, there are almost endless opportunities for interesting and productive activities in the community as a learning center.

4 CHOOSING, USING, AND PRODUCING MEDIA

chapter purposes

⊗ To indicate the need for an extensive variety of resources in any instructional program that proposes to serve individual interests and diverse needs of students.

⊗ To highlight the many uses of educational media resources in systematized teaching and learning.

⊗ To show that formulas for using media may be helpful as guides in teaching, but that there are limitations and hazards in applying them.

⊗ To present criteria and guidelines to be used in carrying out media selection programs.

⊗ To highlight characteristics and values of imitative, adaptive, and creative media design and production activities.

Borough of North York, Toronto (Canada), Schools

Perhaps one of the most valuable contributions of instructional resources to your activities as a teacher is the stimulation they give to your creative abilities. Many teachers find that seeking and finding resources that produce results when students use them is a distinct pleasure among the multitude of teaching details. Media activate students, and as they learn actively, you can enjoy the evidence of their progress. Your inventiveness, your resourcefulness, and your enthusiasm have direct and visible results. In this chapter you will find guidance in many variations of procedure for selecting, using, and producing media. Whatever the resources you and your students select, the need is for a creative approach in using them.

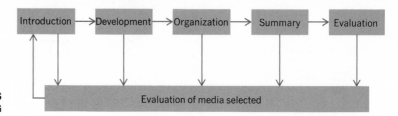

Introduction → Development → Organization → Summary → Evaluation

Evaluation of media selected

**MEDIA SELECTION IN PHASES
OF TEACHING**

media uses in teaching and learning

The urgency you and your students feel for access to media resources relates closely to the setting and viability of the instructional program. In a traditional school environment, for example, where long-range lesson planning is required and followed to its structured ends, media resources may be preselected and delivered on fixed schedules. With that kind of closed curriculum, unanticipated access to a variety of other media will not be particularly urgent.

But if you and your students are encouraged to approach instruction with freedom to build on your own insights and priorities as they emerge, your media needs will be considerably different. In a

Orange (California) Unified School District

Educational Facilities Center, Chicago

more open classroom that is responsive to various changes of student interests and that encourages individualization of learning, you will surely require ready access to many different types of media of varying difficulty levels and dealing with a multiplicity of subjects.

But no matter how you organize a learning environment and develop instruction, you will want to have students experience these types of learning activities: (1) introduction to or overview of what is to be studied and accomplished, (2) development of the study, (3) organization of the findings, (4) summary and perhaps presentation of the findings, and (5) evaluation of the total effort—to ascertain the degree to which objectives are reached and the extent to which there is student readiness to proceed to new tasks. Each of these five types of learning activities involves the selection, use, and evaluation of media resources and media-related activities. To assist you in developing those most appropriate for particular purposes, consider the following suggestions.

media in introduction

The start of a unit or a topical study is the time for you and your students to examine or develop objectives. It is also the time to clarify, through an overview conducted with and participated in by the entire class, what will (or may) be studied and accomplished. This, too, is the time to motivate students to participate in required and optional learning activities that may be undertaken. Pretests may be used to measure initial student competencies. Results will often suggest kinds of study assignments that will be needed later.

The introductory phase of teaching and learning is also the time for a systematic review of options for uses of media or media-related activities that will be undertaken throughout the period of study. It will be appropriate, at this time, to invite school media center personnel to work closely with you and your students in bringing together a variety of materials to be found in collections in your school, in the district media center, and possibly elsewhere. It may also be appropriate to prepare exhibits of some of these items and to provide time in which students are encouraged to scan, audition, or review them before selecting some to be studied later in detail. Numbers of such items may be placed in classroom learning centers or in storage racks for convenient access during later study. At this time, too, arrange-

San Diego County (California) Schools

ments could be made for the guided use of multimedia packages, books, recordings, video tapes, films, and other resources in both the classroom area and the school media center. Production facilities and supplies in the center could be pointed out and optional production activities identified. Similar explanations could be made on the possible uses of community resources (museums, specialized facilities, experts, equipment).

Some contributions of media during the introduction phase are suggested by the following:

⊗ Flat picture displays in game form can be used to explore what students already know or have questions about.

⊗ A 16mm film or 2- by 2-inch slide showing can provide a challenging overview and raise questions about the subject to be studied. Reuse of the item at the conclusion of the unit could provide a useful means of comparing prestudy and poststudy understandings and attitudes.

⊗ An audio tape or disc recording can dramatize one or more problems related to the study. Perhaps it may be stopped part way through, and students can hypothesize solutions or developments appropriate to the situation.

Pistor Senior High School, Tucson (Arizona)

media
in development

During the development phase of study, students will locate, examine, assess, and use or reject information in many different forms. In some classes, this phase may be carried out chiefly through independent study assignments. In others, it is likely to involve teacher-student planning, committee assignments, and project work. Problems studied will sometimes be modularized (that is, broken into small related segments) to permit study by individuals as well as by groups. There will almost certainly be extensive teacher-student discussion and decision making about activities to be undertaken and about the formation of teams and committees.

Students will be encouraged during this development phase to seek and use a variety of media relevant to their studies: pictures, maps, globes, models, real things, games, encyclopedias, pamphlets, magazines, films and slides, recordings—anything useful for the many activities that lie ahead.

Most of the investigative activities undertaken during the development phase will be conducted as independent, individualized learning. Classroom and media center learning stations will be used. Each will be self-contained, and each will provide a means of measuring the abilities of students to satisfy the criterion performance standards set for the particular activity.

Sources of media required for the development phase will again range from those in the school to those in district media centers and the community. Often, students will consult source lists in a search for free and inexpensive materials. This extended treasure hunt can be highly productive of student interest and lead to the development of questions suggesting new directions for exploration. The broader and deeper the search during this phase, the more challenging and productive may be the next phase.

San Diego County (California) Schools

media in organization

The third, or organizational, phase of study is the time to assemble ideas and information derived from previous research and study activities in readiness for presenting them in understandable form to others of the group. Student effort at this stage, then, is to bring together the results of individual investigations to produce a coordinated, integrated whole.

Student production of media may thus be expected to occupy whatever time is necessary. Principles of visual, audio, and verbal communication will be practiced and perfected. Students will begin to prepare written reports illustrated by graphs, tables, still pictures, and drawings; some of these products will be bound for "publication," while others will be transformed for presentation through audiovisual forms such as bulletin board displays, three-dimensional displays, or sound or silent sets.

San Diego County (California) Schools

Some students may produce working models, transparency sets, relief maps, sand table displays, puppet plays, demonstrations, or dramatic presentations—a simulated radio broadcast, for example. You or some of your students may also select films or filmstrips to be used in reviewing the findings of the unit for the entire group. As effective means of concluding the study, summary panel discussions may be planned.

media in summarizing

The summarizing phase of instruction and learning provides opportunities for students to communicate results of work completed. Sometimes it may be marked only by a day or two of discussion about what has been learned and why it is important. More commonly, it will be used as a major presentation period during which every member of the group participates by using products referred to in the preceding discussion of the organization phase.

The bringing together of resources at this time emphasizes the totality and the interrelation of student experiences. Such presentations also offer the reinforcing experience most students need to understand and appreciate the significance of what they have learned.

During the summarizing phase you may also bring in new resources—a new film, for example, to facilitate the transition to the evaluation phase, which is discussed next.

Hans J. Noecker

media in evaluation

The evaluation phase of study is the time for students to take stock, to see what they have learned, or to understand how they have changed as a result of study.

Comparisons of results of posttests with those of pretests will suggest the degree to which objectives have been achieved and whether or not further work—some of it corrective—will be required. At this time, too, student opinionnaires about preceding study phases may be returned and analyzed.

In testing student achievement at the end of a unit of study, you may find it particularly helpful to use visual or audio materials as bases for questions. Such materials may be displayed, projected, played,

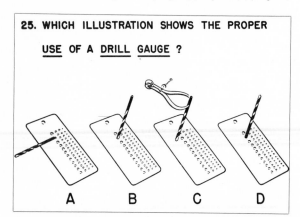

Jeppesen PRIVATE PILOT COURSE

ATTITUDE INSTRUMENT FLYING SECTION

Fig. E

Fig. F

10. In Fig. E the airplane is turning to the _____.

11. In Fig. E, the altitude is constant but the airspeed is decreasing. To maintain the same rate of turn and a constant altitude, _____ must be increased.

12. If power is not added, the airspeed will continue to decrease and altitude will start to _____.

13. In Fig. F the airplane is _____ and turning to the _____.

14. In Fig. F the rate of turn is proportional to the _____ of _____.

Jeppesen Company (Denver)

McGraw-Hill Films

or passed from student to student. Sometimes, extensive arrangements of materials may be placed on tables, with items numbered in sequence. Students may move about from station to station, answering questions which appear on cards displayed at each.

Also, during the evaluating phase, students will be expected to appraise media and media services they have used. Aspects commented upon will relate to communication value, technical quality, and cost effectiveness. The intent, in every case, will be to judge whether or not items used were worth the time, energy, and skill invested in them or whether alternative media resources should be used when repeating the study.

media utilization procedures

Discussions of ways to use educational media usually stress the need to be consistent with instructional objectives—with what you want students to learn from the experiences that media provide. Faced with translating this recommendation into *action,* teachers sometimes search for a simple formula or set of guidelines. To bring that search into focus, we identify such a basic formula—but we follow it immediately with a case example that suggests useful and imaginative variations as well as a few of the pitfalls of using it slavishly.

a basic utilization plan

One basic plan for utilizing educational media has sometimes been described as requiring the instructor to "prepare, present, and follow up." This simple formula involves five steps, as follows:

1. PREPARE YOURSELF Actually preview the film, for example, or listen to the recording; sort through and examine pictures comprising a picture set; or

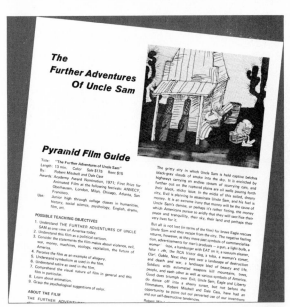

The Further Adventures Of Uncle Sam

Pyramid Film Guide

Title: "The Further Adventures of Uncle Sam"
Length: 13 min.   Color   Sale $175   Rent $15
By: Robert Mitchell and Dale Case
Awards: Academy Award Nomination, 1971; First Prize for Animated Film at the following festivals: ANNECY, Oberhausen, London, Milan, Chicago, Atlanta, San Francisco.
Use: Junior high through college classes in humanities, history, social science, psychology, English, drama, film, art.

POSSIBLE TEACHING OBJECTIVES
1. Understand THE FURTHER ADVENTURES OF UNCLE SAM as one view of America today.
2. Understand this film as a political cartoon.
3. Consider the statements the film makes about violence, evil, war, money, machines, ecology, capitalism, the future of America.
4. Perceive the film as an example of allegory.
5. Understand symbolism as used in the film.
6. Understand satire as used in the film.
7. Comprehend the visual nature of film in general and this film in particular.
8. Learn about animation.
9. Grasp the psychological suggestions of color.

ABOUT THE FILM
THE FURTHER ADVENTURES

The gritty city in which Uncle Sam is held captive belches black-gray clouds of smoke into the sky. It is encircled by highways carrying an endless stream of scurrying cars, and further out on the ruptured plains are oil wells pouring forth their black, sticky issue. In the midst of this soiled, dreary city, Evil is planning to assassinate Uncle Sam, and his fuel is money. It is an extreme irony that money will be the cause of Uncle Sam's demise, or perhaps it's rather fitting, the money which Americans pursue so avidly that they will sacrifice their peace and tranquility, their sky, their land and perhaps their very lives for it.

But all is not lost (in terms of the film) for brave Eagle rescues Uncle Sam and they escape from the city. The negative feeling returns, however, as they move past symbols of commercialization, advertisements for men's products — a gun, a light bulb, a woman's shoe, a hamburger with EAT on it, a vacuum cleaner, Clark Gable. Next they pass over a landscape of destruction and death and war, a landscape bled of beauty and life. Soldiers with automated weapons kill mountains, trees, people, and each other as well at various symbols of America. Good does triumph over Evil; Uncle Sam, Eagle and Liberty do dance off into a cheery sunset, but not before the filmmakers, Robert Mitchell and Dale Case, have had an opportunity to point out our perverted use of our inventions and our self-destructive tendencies.

Robert Mit...

Pyramid Films

survey a field trip you plan to take. Study available guides or note files about the item; take more notes during preview or on advance visits to the community site. Develop a *plan* to use the item which describes how you will introduce it, what you will do and ask your students to do during and after using it, and how you will tie the experience to the flow of activities and thus make it useful and relevant.

2. PREPARE THE ENVIRONMENT Arrange necessary materials and equipment required for proper viewing or hearing. See that equipment is reserved, on hand, and properly set up, loaded (or threaded), and ready to go when the time comes to use it. Check room ventilation and temperature; check transportation and safety matters connected with any field trip you may plan to take.

3. PREPARE THE CLASS Introduce the item; make clear why it is being used at the particular time it is; briefly describe what it covers; stress what is important to be learned from it. Tell students what they will be expected to do after using the item. Should they expect a test? Should they be prepared to discuss points raised by it? Should they simply sit back and enjoy it, after which they may, if they wish, discuss its content or ideas?

Indiana University

4. USE THE ITEM Show the film, for example, properly. Be sure that images are in proper focus and that they are projected above the heads of viewers; be sure, too, that sound volume and tone are properly adjusted so all may hear, understand, and enjoy the message. End the showing professionally— turning off the lamp as final images fade (to avoid blinding and distracting screen glares), and turning

down sound (to avoid noisy popping and clicks) after the film is finished.

5. FOLLOW UP After use, invite and answer (or discuss) questions about the film, filmstrip, recording, or field trip used. Review the experience; perhaps give a test. Supervise student performance or demonstration of skills expected to be learned from the experience. Write thank-you letters to those who cooperated with the field trip. Assess the value of the experience; pass along comments (via files maintained in the school media center) to help others who may use the item at some later date.

San Diego County (California) Schools

This simple, almost classic five-step procedure is sometimes recommended as the basic guide to film utilization. It is true that, under some circumstances, inexperienced teachers can vastly improve classroom results by following it. But teachers who wish to use materials in more creative ways, and who do so through applying the more detailed and systematic planning procedures outlined in Chapter 1, will vary the plan—especially in its steps 1, 3, and

5—in an effort to increase the effectiveness of film use, to provide opportunities to individualize instruction, and to permit variations of patterns of student interaction and grouping. The case example which follows illustrates how one calculated departure from the traditional routine of film utilization produced superior results.

case example of film utilization

A secondary school teacher had seen and liked the widely used 16mm sound film *Why Man Creates.* Created and produced by Saul Bass & Associates for the Kaiser Aluminum and Chemical Corporation, the film is applicable to many subjects and includes many concepts about creativity—both as a historical phenomenon and as a personal characteristic. It is entertaining and stimulating to both youthful and adult viewers—depending upon how it is used. Structured as several discrete but related subunits and nearly a half hour long, it is packed with captivating images that portray large concepts and ideas.

In this case, the teacher first undertook to follow the traditional pattern for film use, as just described. He prepared his class by having the students study the notion of creativity, creative people, and their impact upon world history, He then used a class session in which to discuss creativity and what to look for when seeing the film.

But when the film was shown, both he and the students were disappointed. The experience fell flat; the film produced little impact; discussion was dismally stilted and halting.

What went wrong? The teacher had followed an accepted procedure, but the result was unsatisfactory. He could not, and did not, blame his students; they had cooperated. What, then, should he have done differently? The answer was not immediately apparent, but his reasoning suggested that the film was too full of ideas, too full of complex concepts and of novel visual impressions to evoke orderly and clear student responses. Try again!

On the second showing, he decided to run only three of the several film segments before stopping to discuss and review. This time, students reacted spontaneously, started discussions, and indicated a strong desire to express opinions about what they

Taken from Why Man Creates, *created and produced by Saul Bass & Associates for Kaiser Aluminum and Chemical Corporation.*

From Why Man Creates, *created and produced by Saul Bass & Associates for Kaiser Aluminum and Chemical Corporation.*

had just seen. Comments suggested that they had seen much more during the second showing than they had in the first. Believing that the procedures followed in showing the film were important factors in the changed results, the instructor resolved to use still another approach with the film when showing it to another class.

The following term, the instructor had that opportunity. He observed his usual media planning procedures, bringing the new class to a point of readiness to see the film. Again, this preparation required time during which students discussed, read widely to develop questions and to obtain information about creativity and creative persons, and searched magazines and newspapers for items about creativity in the realms of national and international politics, business, local government, sciences, and the arts. A bulletin board exhibit was developed around the theme "Little-known Creative Men and Women of Modern Times." When the question "Why *does* man create?" emerged, the film was brought to class. This time the teacher showed only the first two subunits, then immediately allowed time for group reflection, reaction, and dis-

cussion. After an intervening day, he followed with other film subsections—through what is popularly known as the "ping-pong ball sequence." After one more day, he reran that sequence and continued on to the ending. Several students requested a repeat showing of the sequence "A Digression," which involved the imaginary conversation of two snails. A spirited discussion of creativity immediately erupted. Students examined evidence of creativity that might be observed in class, in school, throughout the community, and elsewhere in the country. In deciding to *act* upon their new interest in creativity, one student chose to write a haiku poem, another to design a "people carrier" based on ideas put forth during the pre-showing discussion. Several others volunteered to form a work group to plan improved school waste-disposal procedures. Extensive conversations developed among students about how creativity is thwarted or encouraged. All this in direct contrast to the routine results of his earlier use of the five-step procedure!

Were these differences in results chiefly attributable to changes in the way the instructor used the film? He thought they were. To class differences? He thought they were not. What do you think?

From Why Man Creates, *created and produced by Saul Bass & Associates for Kaiser Aluminum and Chemical Corporation.*

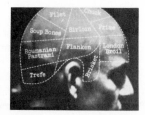

guidelines for media selection and use

Although much still remains to be learned about how to use media resources to best advantage in teaching and learning, research and accumulated experience provide valuable guidance for the process.

As mentioned earlier, the five-step procedure for media utilization is basically a reminder of the considerations involved. Obviously, more inventive approaches are required if one expects to capture and hold student interest and to increase student involvement in learning.

The generalized principles that follow suggest several ways such inventiveness may be achieved with all kinds of media: whether print, audio, audiovisual, or real; whether preserved on tape, paper, plastic, or film; whether presented in machines or manipulated by hand.

⊗ Recognize that no *one* medium, procedure, or student experience is necessarily best for learning a particular subject, for acquiring a particular skill, or for developing a specified desirable attitude or level of appreciation. By their nature, some subjects appear to be better suited for presentation by one medium rather than another. A typical example of optimum fit between one kind of medium and the subject studied is the use of audio recordings in learning foreign languages. History, it is widely believed, can be brought to life and students can be motivated to learn the facts and concepts of the subject by the use of a variety of media—books, motion pictures, still pictures, and maps and globes. Yet it may be that not all students need such variety. Instead, some may prefer to learn history through reading historical novels or textbooks or through listening to recordings or participating in simulations.

⊗ Be sure your uses of media are consistent with your objectives. If your chief purpose in using a motion-picture film with a history class is to develop a favorable attitude on the part of your students toward further study and exploration of the period involved, for example, think twice before invariably following a film showing with a fact test. At a high moment of student interest, such testing may kill an otherwise favorable attitude and create aversive

generalized principles of media selection and use

⊗ No one medium is best for all purposes.

⊗ Media uses should be consistent with objectives.

⊗ Users must familiarize themselves with media content.

⊗ Media must be appropriate for the mode of instruction.

⊗ Media must fit student capabilities and learning styles.

⊗ Media are neither good nor bad simply because they are either concrete or abstract.

⊗ Media should be chosen objectively rather than on the basis of personal preference or bias.

⊗ Physical conditions surrounding uses of media affect significantly the results obtained.

reactions to the topic—and even to the study of history.

⊗ Recognize that, to adapt materials to specific program purposes, you must know them thoroughly—their content, how they may be used to best advantage, the levels of difficulty in relation to competencies of your students—and conditions of availability—that is, when and for how long you may have them. As you examine media, also consider their suitability for use in the instructional mode you have in mind: large-group instruction, small-group activities providing for interaction, or independent study. It now seems obvious that, for independent study, the trend is toward simple cassette audio equipment, for example, rather than reel-to-reel machines; yet the latter type is sometimes thought to be more convenient for making master tapes, for editing, or for large-group auditorium presentations. A motion picture selected for use with a large group in a lecture hall will probably need to be in 16mm size; but, for a small class or for independent study assignments, an 8mm format may be preferable. Large (overhead) transparencies are considered to be especially suitable for large-group presentations; but they are also effective with small discussion classes. And, of course, they may be used by students in presenting visualized reports to classes of almost any size.

⊗ Be aware that student experiences, preferences, individual interests and capabilities, and learning styles are likely to influence results of using media. Students who read well and who enjoy reading, for example, may be expected to benefit more from reading books than those who do not. But to attract and retain the interest of slower readers and to help them understand and profit from what they read, you may need to introduce, along with print items, various correlated audiovisual media.

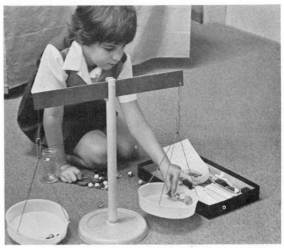

San Diego County (California) Schools

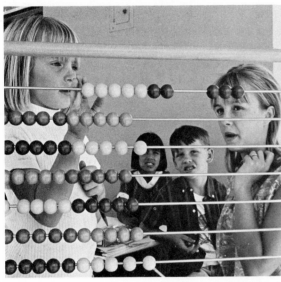

San Diego County (California) Schools

⊗ Be aware, too, that resources and learning experiences are not necessarily good or bad simply because they are presented through concrete or abstract means. It is more nearly accurate to consider media resources as having special (but not necessarily unique) advantages for certain teaching purposes. Usually, one or more of several different items of the same class that deal with the same subject may be used to accomplish a particular purpose. In such cases, the worth of items selected may depend more upon how they are used than upon any built-in advantages or disadvantages related to concreteness or abstractness.

⊗ Don't let your own preferences for particular

media stand in the way of providing learning experiences your students need. Teachers sometimes say they won't use a particular film again because they have tired of it. If you have ever said this, perhaps you ought to develop a new point of view about the film in question. Remember that, in all probability, your students will not have seen it and that, in any case, it may be the very best means of providing the experience you have in mind. In such cases, your reward for seeing the film for perhaps the tenth time or more should stem from what your students learn. But it is also possible that your own boredom with some media item may signal your need to examine it critically to be sure it is still up-to-date and suitable for uses you have in mind. It may also suggest your need for a new approach in using it—as an independent study assignment, for example, rather than as a group activity.

⊗ Finally, be aware that some otherwise excellent media resources may become second-rate in their instructional effect when they are used in inadequate or inappropriate environments. Improper acoustics, uncontrolled or poor lighting, overheated rooms, stale air, noise pollution, and similar distractions are hazards to be avoided, preferably by means that you control. Thus in each instructional effort, as you select media, be concerned with the conditions under which they will be used.

University of Oregon

selecting media for purchase

Anyone who has attempted to read the steady stream of brochures, catalogs, and other advertising releases of educational media publishers and producers ought to appreciate the seriousness and scope of the problem of choosing from among them the relatively few items that can be purchased by a single school or school district. While there are numerous guides, source lists, and books to aid in accomplishing this task, the usual practice is to conduct formal, in-hand reviews of most items before issuing purchase orders.

As a professional educator, you may be asked to assist in assessing and evaluating media for:

⊗ Single-school purchases for classroom use or for placement in the school media center.

⊗ Systemwide purchases of items intended for specified types of schools, especially opening day collections for new schools.

⊗ Systemwide lists of approved educational media resources from which schools choose items suited to their special requirements.

⊗ Systemwide collections (a 16mm film library, for example) from which items may be borrowed by teachers and students in all schools.

In addition to advising on purchases of media resources on a one-time basis, media selection committees make recommendations for or against adoptions of materials (workbooks, textbooks, and maps, for example) to be purchased throughout a specific period of time—at least a year, probably longer.

In most school districts it is usual to apply a procedure for materials selection that provides for preliminary screening of potentially valuable instructional resources. This screening eliminates materials not immediately applicable to school needs. Materials considered worthy of further consideration are then brought in for final evaluation. It is at this stage that you are most likely to be involved in the selection process. Typically, you will be provided with a list of objectives for the selection of media and with criteria to apply in reviewing them. Sometimes you will also be asked to critique or revise such lists as part of the selection process.

media selection criteria

Commercially produced, ready-made instructional materials of the types you will be appraising have in common the fact that they are in *finished* form when they reach the school in which they are to be used. A possible important advantage of such materials is that, because of the large anticipated sales, considerable time and money may have been invested in testing and improving them before placing them on the market. But one of their possible disadvantages (which only adds to your selection responsibilities) is that the content, organization, form, and treatment of commercially produced materials have been *preselected* to satisfy needs in all parts of the country—not those of just one school, one district, or one region. Thus, as a member of a selection team, you will be expected to recommend, from among the numerous media resources of a general orientation, those coming closest to meeting local requirements, at the same time, of course, meeting other stated standards of excellence.

Criteria such as the following are often used to guide media selection:

Northern Illinois University

⊗ CONTENT Does the item deal with significant curricular content? Is it up-to-date? Is it accurate? Is it pitched at proper levels of difficulty and sophistication for students with whom it will be used?

⊗ PURPOSES For what significantly important instructional purposes may the item be used?

⊗ APPROPRIATENESS Is the medium used suited to the message it seeks to communicate? If the topic is essentially one that requires portrayal of motion, for example, does the medium show motion? Or if color is essential to the message, is the item in color?

⊗ COST Is the item likely to be worth what it costs, as measured by educational results derived from its use? Would other less expensive items, in other media formats, for example, be better choices?

⊗ TECHNICAL QUALITY Is the item technically satisfactory in photography (color, exposure, angles, focal lengths of lenses used), editing (cuts, dissolves, continuity), and sound?

⊗ CIRCUMSTANCES OF USE Will the item function effectively in circumstances and environments in which you are likely to use it? Is it suitable for large groups, small groups, individual study? If it must be projected, for example, will images be sufficiently large and bright for all to see?

⊗ LEARNER VERIFICATION Is evidence supplied that the producer of the item has improved it through systematic trial and revision before offering it to purchasers? Are such data available? Are characteristics of the trial groups sufficiently similar to those with whom it is likely to be used in your situation?

⊗ VALIDATION Are reliable data available and supplied which prove that students do learn accurately and efficiently through use of the item? Again, are the characteristics of the trial groups sufficiently similar to those of the students with whom the item will be used in your situation?

If you plan to use students to help with the selection of media, how could you have them assist in specifying criteria? Is it possible to have them arrive at a workable list with minimum direction from you? How would they benefit from the exercise?

Two processes—learner verification and validation—are of special concern to professional teachers who choose and use educational media.

San Diego County (California) Schools

Are wrong impressions created by outdated pictures?

By continually collecting learner verification data and critically assessing validation statements, you will be ensured of improved quality of future media productions. The professional media staffs of your school and district media centers are important links in the process. The pioneering, and continuing, work of the Educational Products Information Exchange (EPIE) in this aspect of education is discussed in more detail in Chapter 16.

useful media selection guides

School media centers usually contain catalogs of selected media producers as well as some that describe loan collections of the school district media center, universities and colleges of the area, and the state department of education.

More general sources of media information may also be available to you from school or district media centers. Those mentioned below are typical of the offerings. Bibliographic data related to each of them will be found in Reference Section 5.

Learner verification—when applied to media—identifies a procedure through which producers are able to revise and refine their products by testing them with small groups of learners *before* releasing them. *Media validation,* on the other hand, generally refers to the process of determining, *after* release, that students do learn from the media to the degree and in the manner projected. Individuals responsible for recommending media items for purchase should study statements of producers about both types of product analysis.

As you and your students make daily use of media, you may find continuing learner verification to be useful. The films you borrow from your district media center, for example, may be accompanied by user evaluation forms which invite ratings of: up-to-dateness, importance of subject, organization, sound quality, photographic quality, and overall value. Discussions with your students may bring out significant facts about the quality and suitability of other media, a textbook or multi-media kit, for example. Do they contain factual errors, some, perhaps, the result of changes that have occurred since publication? Is some of the writing unclear?

general media selection aids

Several general media selection aids will provide assistance in locating specific source lists or general information about the selection process:

Educational Media Yearbook (Bowker)

Guide to Reference Books for School Media Centers (Libraries Unlimited)

Index to Instructional Media Catalogs (Bowker)

Media Reviews and Review Sources (University of Maryland)

Selecting Media for Learning (Association for Educational Communications and Technology)

print media selection aids

A few of the many selection aids pertaining to print media are:

Children's Books in Print (Bowker)

Children's Catalog (Wilson)

Elementary School Library Collection (Bro-Dart)

Junior High School Library Catalog (Wilson)

Magazines for Libraries (Bowker)

Opening Day Collection (Choice)

Paperbound Books in Print (Bowker)

Picture Books for Children (American Library Association)

Recommended Paperbacks for Elementary Schools (Book Mail Service)

Resources for Learning: A Core Media Collection for Elementary Schools (also available for nonprint media; Bowker)

Senior High School Library Catalog (Wilson)

nonprint media selection aids

The following suggest the scope of publications useful in assisting in the selection of a variety of nonprint educational media:

Catalog of Audiovisual Materials (National Audiovisual Center)

Core Media Collection for Secondary Schools (Bowker)

Educators Guide to Free Films (Educators Progress Service)

EFLA Evaluations (EFLA)

Feature Films on 8mm and 16mm (Bowker)

Harrison Tape Guide (Weiss)

International Index to Multi-Media Information (Audio-Visual Associates)

Media Review Digest (Pierian)

National Center for Audio Tapes (catalog)

NICEM Index titles (various; National Information Center for Educational Media)

local media production

There may be times when you prefer to produce materials for your own presentations as well as times when such production activities will be especially beneficial as learning experiences for your students. In materials that follow, we discuss the benefits of teacher production and student production of media resources. Techniques to use in producing specific types of media resources are presented in later chapters.

3 M Company

teacher-produced materials

As a teacher, you consider producing instructional media in one or both of two general categories. First are materials to use in your own presentations and in leading class discussions or for testing or carrying out other teacher-directed activities; second are those you wish to put directly into the hands of your students for their study and use.

In the first category, simple, projected materials are common, and overhead transparencies typical. In addition, teacher-produced 2- by 2-inch slide sets, audio cassettes with edited recorded material, flat picture sets, posters, graphics, and displays are common.

Materials you may produce for student use, the second category above, provide needed media resources on local topics. Though basic and quite general resources may be available (a film on the phenomenon of ecological balance, for example), locally oriented or specialized resources on the subject may be missing. In such cases, producing them yourself or with the help of students may be the best, perhaps the only, way to get them.

The current widespread recognition of such needs for locally produced media resources has encouraged the development of production centers in individual schools and school district or regional media centers as well as in colleges and universities. Such a center in your own institution may offer a number of professional and technical services which you may find reason to use.

A side benefit to be realized in planning and producing your own materials, in directing others to do it for you, or in assisting your students to prepare them is that in doing these things you will nearly always clarify your own objectives and ideas about how best to organize your program. When making a picture or a drawing or producing a graphic illustration for any kind of presentation, for example, you must be sure about what you want to communicate. Self-discipline is involved in deciding exactly what you are to present and how and when you will use the media resource involved. Many teachers attribute substantial improvement in their professional ability to the experience of applying themselves in this way to the development of visual resources or other instructional materials.

In some states and provinces, district and regional agencies provide mobile production centers for the schools. Staffed with trained personnel and equipped with adequate production equipment to make slides, transparencies, audio or video tapes, or other media, these units may be driven to your school and remain there for one or more days while expert personnel produce the materials you order.

student-produced materials

Students involved with you in media production may obtain many benefits from their experiences. They may discover, for example, that there are standards of quality that must be met, and that following specifications and meeting deadlines are important. They may also encounter new ideas and develop previously unrecognized creative talents which they can enjoy as hobbies, or perhaps for a vocation, for the rest of their lives. Talents of students for making or providing displays, independent study resources, tapes, transparencies, slides, single-concept motion pictures, or countless other media resources should not be underestimated.

Beyond the production services students may provide to assist you in obtaining needed classroom resources, as just described, there are numerous opportunities for them to produce items they themselves will use to accompany their own class activities. In such cases, it is important that several points be kept in mind:

⊗ Remember that the end product is often not

San Jose State University

Charles Beseler Company

nearly as important as the processes and experiences used to reach the objective. Capitalize on the skills learned, and watch for opportunities to suggest further use of the practical production techniques that students have mastered.

⊗ As students invest their time and effort, be sure that they continue to improve and that they feel rewarded and stimulated by their progress. Your aim should not be to exploit them or to cause them to work beyond their ability, but rather to give them opportunities to be productive and to enjoy production activities.

⊗ Be sure that, before any project is undertaken, students produce plans on paper. Help them learn how to lay out projects and, before actually commencing production work, to specify details of steps to be undertaken.

⊗ Help students to learn to meet occasional failure without frustration and to persevere and try again, recognizing that occasional failure is normal in learning something new. Ensure, too, that they understand that some projects are likely to require several trials, and sometimes even accidental success, before they are ultimately completed.

⊗ Seek ways to correlate student media productions with activities of your classes or of other classes in the school.

levels of productive effort

As you proceed through the following chapters of this book you will encounter numerous examples and suggested techniques related to the processes of *creating* educational media.   We start with simple, everyday techniques with which all teachers need to be proficient if they are to be able to help their own students develop such skills themselves: drawing letters, designing bulletin boards, arranging displays, using chalkboards.   These *fundamental* skills are also involved when producing other materials: slide sets, large transparencies, an audio or video script, a puppet play, a motion picture sequence.

To ensure optimum benefit from the various media production activities, remember that there are essentially three levels of productive effort, each related to creative originality.   These are identified as: (1) imitative, (2) adaptive, and (3) creative invention. These three overlapping levels suggest infinite opportunities for you to provide individualized learning projects for students of different experience, talents, and interests.

The chart below emphasizes that the essential ingredient of creative invention is freedom of individuals to apply their own ideas as they seek solutions to problems associated with media production.

imitative media production

Much of the world's work requires following specific directions to carry out tasks that have been laid out

Creative Playthings

and tested by others.   Production of media resources through such means may be described as *imitative* because the producer uses the models or directions of others.   Imitative production may be illustrated by cutting out letters according to a pattern, for example.   Copying pictures or maps by tracing projected images, and constructing a flannelboard or producing a terrarium or puppet theater according to printed instructions are all forms of imitative production.

To be of value to students, such imitative assignments should meet important criteria.   The student should (1) use appropriate skills and knowledge, (2) maintain agreed-upon performance standards, (3) produce accurate and neat work, (4) meet sched-

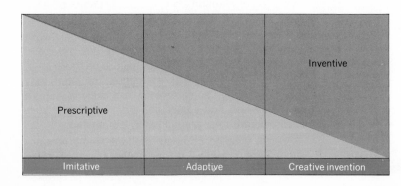

Prescriptive

Inventive

Imitative   Adaptive   Creative invention

**LEVELS OF
MEDIA PRODUCTION**

From Wyoming Science Fair, *Jaime Brun, Photographer*

creative invention

Creative invention involves the ability to define and solve problems in original ways, without extensive dependence upon written directions or the experience of others. The production of materials that are original in concept and treatment represents a form of creative invention. A class- or student-produced sound-slide presentation, an 8mm film with synchronized sound, a simulated radio program, and a play written, produced, and recorded for presentation on a television system are all examples of production in which a high degree of creative invention could be realized. Encourage students to be more creative, inventive, and original than they are. Recognize and encourage originality in student work; help students to appraise the worth of their productions, using value standards as guides. Finally, be sure they recognize the value of their creative efforts whether or not it is discovered, later, that someone else had already done something similar. (Such is the character of inventiveness!)

ules, (5) complete projects, and (6) develop improved capacity for self-direction and self-evaluation.

But even with imitative production activities, students should be encouraged to exercise some ingenuity and to find improved or simplified ways to do the work while still meeting standards.

adaptive media production

Adaptive media production requires the creation of new forms or new ways of using media resources that already exist. Here there are no set directions to follow. A prototype or pictures of some prototype may exist and be useful as a guide; but the ultimate production requires that the individual involved work out numerous problems personally. Adaptive production calls for individual judgment, self-direction, and initiative. Quite often, some forms of inventiveness will also be needed if one is to make progress or to complete the project undertaken.

One example of an adaptive project might be the preparation of a transparency through the revision of a set of paper masters that, in their present form, lack some essential for the class in which they are to be used.

Orange Coast Community College (California)

summary

The processes of choosing, using, and producing media require participation of students, teachers, and professional specialists of several types.

The general media utilization procedures presented here have emphasized the need for teachers to provide an essential interface between materials produced for use in all parts of the country and requirements of particular students in a particular school.

The several phases of teaching and learning in which media resources are applied—for different purposes and emphases—have been identified as (1) introduction, (2) development, (3) organization, (4) summary, and (5) evaluation.

Teachers are involved, with the assistance of students, in recommending media for purchase by their schools or school systems. Criteria used to guide the process are related to (1) content, (2) purposes, (3) appropriateness, (4) cost, (5) technical quality, (6) circumstances of use, (7) evidence of learner verification, and (8) evidence of validation of usefulness and quality.

Special contributions of learning to be derived from local media production processes have also been discussed. Media production projects, particularly those at the level of creative invention, frequently help students to learn the things they need to know, or need to be able to do, but which they have previously failed to learn through other means.

Emphasis is given to the fact that throughout the remainder of this book examples will be given to illustrate the many ways that media resources may be selected from among those produced commercially, produced locally by students, teachers, or professional media specialists, or obtained from other sources within the typical community.

5

DISPLAYING AND SOME FUNDAMENTALS OF VISUAL COMMUNICATION

chapter purposes

⊗ To indicate basic skills helpful in the preparation of a variety of visual resources; to assist you in acquiring sufficient competence with those skills to guide and encourage your students to learn to use them too.

⊗ To suggest ways to improve your own use of some of the simple and readily accessible resources for instruction: bulletin boards, cloth boards, chalkboards, and exhibits.

⊗ To suggest a number of standards for producing effective and attractive visuals that will: (1) help you select teaching resources; (2) help you evaluate your students' work and your own work with visuals; (3) encourage you to use more visuals, knowing they can motivate students to learn; and (4) suggest display activities that will involve your students in creative uses of visual media in communication.

Charles Mayer Studios, Inc.

For purposes of instruction, the simplest, most direct means of communication is usually desirable. Readily available in most schools are bulletin boards, chalkboards, and magnetic boards, all of which are both economical and easy to use. Each permits opportunities for creative visual communication that excite the interest of students and can involve them in the development of displays which add life and color to learning activities.

In this chapter some of the desirable skills in displaying are discussed, with principles presented that you can use and that your students can use as you encourage them to communicate with pictures, bulletin boards, exhibits, and displays. You will find that students will need minimal motivation to work with displaying activities, and their work will give you insights into their perceptions and their talents.

The basic principles discussed in this chapter are relevant to all types of visual communication. In addition to those mentioned above, we can add others such as graphics, transparencies for overhead projection, slides, and motion picture and television visuals. Thus, you may wish to refer to this chapter often as you plan and produce examples of the many media for communication.

bulletin boards and teaching displays

Bulletin boards and their extended forms—teaching displays and exhibits—are among the least expensive instructional resources. Both teachers and students plan and prepare bulletin boards for a variety of purposes, but the more that students participate in creating displays, the better and more effective their school experiences are likely to be.

instructional functions of bulletin boards and teaching displays

Bulletin boards and teaching displays serve various important functions:

⊗ TO FACILITATE CLASS STUDY OF SINGLE-COPY MATERIALS When only one copy of a useful resource is on hand, the bulletin board provides a means of making it available for study by groups. Especially valuable materials may be placed behind clear plastic sheets or protected by lamination.

⊗ TO STIMULATE STUDENT INTEREST The jackets of books, for example, displayed on a bulletin board may encourage reading. Posters may also encourage visits to museums or other centers of community resources.

⊗ TO SAVE TIME Bulletin boards and teaching displays allow students to examine or to study materials which cannot be discussed or studied during class because of lack of time.

⊗ TO ENCOURAGE STUDENT PARTICIPATION Problems presented in bulletin-board or teaching displays may be studied and discussed by student groups as a part of class activity.

⊗ TO PROVIDE A REVIEW A class may be divided into small groups to prepare bulletin boards or displays that review main ideas studied or that summarize information related to specific instructional objectives.

⊗ TO HELP STUDENTS LEARN TO COMMUNICATE THEIR IDEAS VISUALLY In thinking through the objectives of a bulletin board or a display, planning the content, and developing and arranging the display, students use and extend the communication skills and abilities usually found among them.

⊗ TO VISUALIZE PORTIONS OF A TEST A well-planned visual display, showing things to be identified, related, or contrasted, may cause student responses that help in the evaluation of their knowledge. And, on such a basis, evaluation of student progress may be more accurate than when developed strictly as a verbal test.

⊗ TO PROVIDE A MEDIUM FOR INDIVIDUAL OR GROUP REPORTS When students present oral reports, they can prepare displays to illustrate discussions. They can show photographs, printed materials, handmade charts, graphics, and even selected real materials according to a plan that helps to organize and makes specific the content of the reports as well as adding interest and action to the presentation.

⊗ TO MAKE THE CLASSROOM DYNAMIC, RELEVANT, AND ATTRACTIVE The importance of student participation in the creation of the bulletin boards and teaching displays cannot be overestimated.

Beverly Hills (California) High School

planning
bulletin boards
and teaching displays

Achievement of instructional objectives through activities involving bulletin boards and teaching displays will be aided by the following procedures:

⊗ Decide early on a theme or key idea to be expressed by the bulletin board or teaching display. Seek a new, fresh approach to content. Begin early to manipulate ideas for a printed title or main heading to carry the theme idea. Find a wording to catch the attention of viewers. Think of putting the title in an eye-catching location; considering the expected size and location of the audience, plan for lettering and arrangement that can be seen at the normal viewing distance.

⊗ Start early to determine exactly what the display is to communicate. Have the students consider the need for a display and suggest what service it might perform. A general teaching purpose may, for example, be to provide the students with an activity project that will lead to an organized and visualized presentation in review of a unit of work, such as "Efforts to Achieve Ecological Balance in Our Community." Specific student goals may include the following: to give at least eight students active responsibility for planning and making a bulletin board display; to provide opportunity for development of headings, captions, and other written materials that require clearly executed lettering; to provide for student-made photographs of local situations, institutions, and persons relative to the purposes of the display, and so on.

⊗ Plan the display well ahead of time—on paper. Start by making some *thumbnail sketches,* small drawings of possible arrangements for the display. Typical shapes of layouts are offered in later illustrations. Have students put their sketches on the tackboard; select those considered most promising for development. Make several preliminary plans. Let imaginations run free. Make notes on fresh ideas as they are presented. As possible arrangements and visualizations are suggested, ask students to make additional sketches.

⊗ As the final planning stage for the display approaches, continue working with the sketches and with available photographs, printed materials, or such real things or constructed forms as may have been obtained. Using the sketches as a guide, try arrangements of the materials and drafts of headings and captions (even in rough, schematic forms) to see how they fit and whether they will contribute to the main purpose of the bulletin board or display. Soon, the need to add some materials and to eliminate others will become apparent, and the display will begin to take final form.

⊗ Keep in mind the persons who will see the bulletin board or teaching display—*their* interests, special characteristics, habits; in other words, *identify your audience* and *identify with your audience.* Think of their past experiences and of the language they will understand.

Beverly Hills (California) Unified Schools

⊗ As aspects of the construction task become clear during planning, assign responsibilities to students. They will need guidance as they work together and strive for an end product that has coherence and viewer impact.

⊗ Consider too where the bulletin board or teaching display could best be exhibited to achieve instructional purposes. Placing it in a classroom is not always necessary; it might best be seen in a corridor, in a special display cabinet, or in some other location. If the display is for teaching purposes, however, remember that busy corridors and lunchrooms will not permit attentive viewing and careful study. Such busy places might be good locations for displaying split-second-impact posters; but they are unsuitable for displays that require leisurely examination and study.

⊗ Because students will soon discover that development of an effective display can become something of a treasure hunt, they will be greatly helped if there is a file of display materials maintained ready for use, either in the classroom or in the school instructional materials center, where they may look for pictures and other visual materials, as well as real things, that may contribute to the presentation. Ideas for headings, captions, and arrangements will be suggested by the display.

⊗ Add motion to displays. Exhibits that move are used in stores for advertising. The small motors, usually operated by flashlight batteries, can be used to introduce eye-catching action in student-made displays. Students can watch for moving displays in stores and store windows and ask the store to save the motor for the school poster file. These motors are relatively inexpensive and can be pur-

chased through display advertising companies if not otherwise available.

⊗ Think of other attention-getting devices. String, colored yarn, cardboard arrows, cards with felt-pen lettering on them, colored pins, and strips of colored paper or tape to make lines are among the many graphic materials useful for catching the eye of the viewer. Use lift cards, pushbuttons, strings to be pulled, items to be touched or handled, and any number of other techniques to invite viewers to react to or to study the display. Give the viewers choices: ask them to make decisions; challenge them to avoid making mistakes in responding.

⊗ Use color. Tastefully used, color can contribute materially to the attractiveness of the display and to the viewer's enjoyment of it. Color can make important content stand out.

⊗ Incorporate audiovisual devices. Many slide projectors have an automatic slide-changing feature; have students develop photographic or hand-drawn slides or titles. Arrange the projector at one side or out of sight, but design the screen area as part of the display. A small rear-projection screen can be made of thin paper such as vellum, a front-projection screen of a piece of matte white paper.

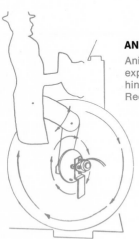

ANIMATED DISPLAYS

Animation is produced by a small, inexpensive electric motor mounted behind this exhibit. Hankscraft Motors, Reedsburg (Wisconsin)

Atlanta (Georgia) Public Schools

chalkboards

Most competent teachers do make special efforts to use the chalkboard effectively. For lesson elements that develop from simple to complex ideas, for example, they build their explanations on the chalkboard, point by point. They add symbols, charts, maps, or outlines at just the right time to emphasize the content involved.

During class planning or group discussions, other good chalkboard techniques are useful. Students may write random topical headings on the board for analysis and reorganization into an orderly and clarified structure. The class follows the discussion, topic to topic, as ideas are listed; the visual communication directs the attention of the whole class toward the purpose of the discussion.

Before a class session, a teacher or students may use the chalkboard for displaying material relevant to the day's work. To cover the material until it is needed, they may pull down maps or projection screens, hang a curtain on a wire in front of the board, use sheets of drawing paper (attached with masking tape), or employ other convenient procedures. Then, at the appropriate time, the material may be uncovered. Illustrations for speeches, diagrams, test questions, assignments—any kind of information that is enhanced by a visual presentation—may be handled in this way.

Students often do practice work directly on the board. Some solve arithmetic problems while others work at desks. Chalkboard work permits immediate feedback information on both accuracy of the work and knowledge of procedures. And, because of the attention given them while they are on their feet actually performing, many students are stimulated to optimum performance by doing chalkboard work.

When the chalkboard is used for student or teacher presentations, a few techniques should be remembered. Keep writing brief and to the point. Write a comment, or make a portion of a sketch, and then turn back to the class to ask a question. Keep the class on the alert as you work, and keep yourself alert too! As students use the chalkboard, call their attention to the communication technique they are learning.

Peg boards often facilitate the presentation of displays by simplifying the use of three-dimensional objects, and also flat pictures, as shown above. A variety of brackets is available for hanging objects or for making shelves.

A slide projector can be used as an effective spotlight on a display; colored cellophane or plastic in 2- by 2-inch slide mounts can tint the light, if appropriate. Be sure to test the effect of colored light that falls on colors in the display. In fact, in early planning, tests of various colors for backgrounds and for lettering should be made with the colors of light that might be used. If your school has a slide-sound synchronizer, both audio and visual material may be used. (See Chapters 8 and 10 for more information on this subject.) If sound alone will add materially to the effectiveness of your display, investigate the possibilities of endless-loop cassettes on which messages may be repeated at intervals. The headphone distribution box used for in-class listening might also be used with some displays to permit viewers to listen to your message without disturbing others in the area.

By using various types of easels, room dividers, corrugated paper, portable chalkboards and bulletin boards, and other devices, you can easily expand bulletin board space. In Reference Section 4 are numerous practical and economical suggestions for providing new bulletin board areas.

chalkboard drawing aids

Three simple techniques for improving your chalkboard drawings are templates, pounce drawings, and projected drawings.

⊗ TEMPLATES Templates, which are prepared forms of cardboard, plywood, or other materials, are useful timesavers when certain shapes are to be repeated on the chalkboard.

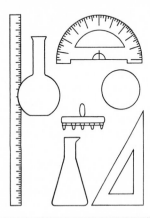

⊗ POUNCE DRAWINGS For the pounce technique, original or copied drawings are transferred to tough, lightweight paper or to window-shade material. A leather hole spacer pushed along the lines in the drawing makes a perforated line in the paper or window shade. If a leather hole spacer is not available, use an ice pick, a large pin, or a nail to punch the holes. To transfer the dotted line to the chalk-

board, pat the pattern with a heavily chalk-dusted eraser. Draw finished lines by following the dots.

⊗ PROJECTOR DRAWINGS With an opaque projector, original drawings and book or magazine illustrations can be projected in enlarged form on the chalkboard. Only lines that are essential for communication need to be traced. When preparing drawings on the chalkboard in this way, stand in front of the beam of light occasionally in order to check the partly finished work.

Instructional figures and pictures from slides and transparencies can also be transferred to the chalkboard by the same technique. Map slides, for ex-

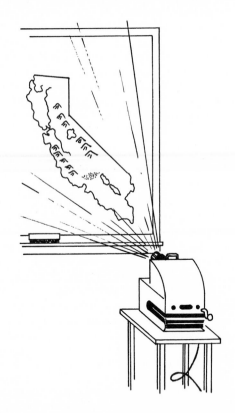

ample, can be easily projected to any desired scale and drawn accordingly. Further, it is easy to modify such transferred drawings by omitting irrelevant information or by adding shaded or lined areas for emphasis. Partially completed drawings can be projected and completed later during instruction.

Effective use of the chalkboard does not come by accident. Some planning is called for. Both you and your students have to think before you write or draw—to consider what you want to show and how best to arrange it. Rough out complicated sketches in advance—perhaps lightly chalking in shapes to speed up later drawing in broad, firm lines. (You might also write yourself a few notes about what you want to do when the time comes to do it.)

Always walk to the back of the room and ask yourself whether or not what is on the board is legible, interesting, and helpful to student under-standing and motivation. Encourage students to evaluate *their* work by the same procedure. As you walk along school corridors, look in at the class-room chalkboards and appraise the work of other students and teachers. Do you see definitive draw-ings: a meaningful cartoon, sketch, or graph; well-presented, orderly outlines? How do you judge the effectiveness of the work you see? Can you identify the criteria that you are applying as you criticize what you see?

Alliance Wall Corp.

chalkboard lettering

Many potentially fine chalkboard presentations are spoiled because the lettering is inadequate. The most frequent fault is that letters are made too small and weak to be seen from the back of the room.

For most writing on the chalkboard, judgments are made by observation and by checking with stu-dents, especially those who sit farthest from the board. Note that legibility depends upon style of letters (usually bold block letters carry best), the contrast of the lettering with the color and condition

of the board, the lighting that illuminates the board, the spacing of letters within words, the space be-tween words and between lines, and the accuracy of letter forms. In the lower grades, teachers choose lettering styles that are most familiar to the children. As a general rule, for comfortable reading from the back of a 32-foot-long room, lettering at least 2 to $2\frac{1}{2}$ inches high is required; the lines forming letters should be nearly $\frac{1}{4}$ inch wide.

If the above rule is extended to apply to rooms or lecture halls where the farthest student is 60 or more feet from the chalkboard, you can estimate the size and amount of lettering and the scale of draw-ings your students will be able to see. The im-portance of legibility is one of the reasons for the great popularity—even necessity—for overhead pro-jectors instead of chalkboards in large rooms used for instruction.

To overcome the natural tendency to write either uphill or downhill on a chalkboard, place a starting dot on one side of the board and another dot at the same level on the far edge of the board. Use the first dot as a starting point and aim your line of writing toward the farther dot. Other aids to at-tractive, legible chalkboard work include light guide-lines made with chalk and a meter stick or a long board used as a guide.

A word about the care of chalkboards: Since there are many different kinds of chalkboard sur-faces, and since each may require special treat-ment, check with your custodial department for instructions.

magnetic chalkboards

For variety in presentations and for opportunities to use imagination in developing materials for chalkboards, consider *magnetic chalkboards.* These are simply chalkboards with a backing made of steel, so that magnets will cling to the board. With the magnetic chalkboard, the value and convenience of a standard chalkboard are preserved, but many additional display techniques are possible. Here is a list to stimulate teaching with them:

⊗ Lettering, sketches, or photographs may be mounted on cardboard, to the back of which magnets may be temporarily attached with masking tape or permanently attached with contact or other strong cement. The materials can then be arranged or moved about on the surface of the chalkboard as desired; chalk drawings or lettering can be added as appropriate between the magnet-supported visual display materials.

⊗ There is a variety of commercially available materials for use on magnetic chalkboards, including magnetized rubber strips—plain or in colors, and in different sizes; clips with magnets attached to support charts or other sheet materials; magnetic letter sets, arrows, and other symbols; and magnets themselves in a wide variety of sizes, shapes, and strengths.

⊗ For special instructional purposes, lines can be applied to the surface of magnetic chalkboards by using permanent paint or temporary adhesive tapes. Symbols, balsa-wood models, or signs may be used to represent real objects or people.

⊗ Some magnetic chalkboards are white or some other light color, and therefore they may serve conveniently as projection screens. Though the board may be used conventionally to show films or slides, inventive teachers have also used it as a base on which to project such images as circuit diagrams, genetic probability charts, weather maps, and process outlines. Then, using magnet-backed cards, they have added symbols, lettering, or real things, and they have placed these items on the projected images. When appropriate, they also have added chalk lines and lettering. Thus, an orderly and sequential display can be built during a presentation.

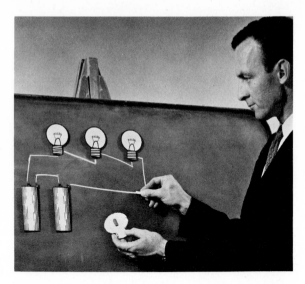

If a permanent or portable magnetic chalkboard is not available, one can be constructed with minimum cost and skill. One, illustrated in this chapter, is made from an oil-drip pan, used on the floor of garages under automobiles, which can be purchased from an auto supply store. After cleaning the surface, spray or brush-paint the metal surface with chalkboard paint, available from paint stores. Apply at least two coats and allow time for drying and very light sanding or rubbing with fine steel wool between coats. To cure the surface before writing on it, be sure the paint is filled with chalk dust. Pat the chalk on with an eraser, then brush it smooth. Failure to cure the surface will cause writing to become permanent in the pores of the paint.

Resourceful teachers have also made a temporary magnetic chalkboard by using the smooth, painted back of a steel filing cabinet or storage locker. Some teachers have extended the versatility of magnetic boards by mounting a flannel or Velcro board on the back of their portable magnetic chalkboard to permit use of additional visual display techniques.

Recognize, of course, that small, portable chalkboards such as those described here will not be useful in very large classrooms or lecture halls; but for classes of from five to thirty students and for television presentations, they will be entirely practical.

cloth boards

Another visual display device used in teaching, really an old-timer that continues to be popular, is the cloth board. Often, such boards are called by the name of materials used in their construction, such as flannel boards, felt boards, or Hook-N'-Loop boards. Flannel—always inexpensive and easily available—has been extensively used to make cloth boards because pieces of flannel stick together when gentle pressure is applied. Coarse sandpaper also sticks to flannel, and so will fuzzy yarns, some types of flocking, and other soft, rough-surfaced materials. Thus, by attaching small pieces of flannel or sandpaper to pictures or to other things attached to lightweight cardboard, these objects can be held on a piece of flannel stretched over and attached to a sheet of particle board or a similar material. With this simple device, a variety of presentation techniques is possible; and, as a discussion evolves, the visualization can be displayed step-by-step.

In primary and preschool instruction, for example, teachers use cloth boards to display the visuals that accompany a storytelling session. With adult groups, in education and in the worlds of business, industry, and government, cloth boards are used to develop such diverse visualizations as organization plans, traffic patterns, economic and mathematical theories, grammatical structures, and historical events. As in other uses of visuals, the creative imagination of the teacher often determines the usefulness and effectiveness of the technique chosen for instruction.

kinds of cloth boards

As just mentioned, flannel is a convenient material for inexpensive cloth boards and can be obtained in a number of colors. Felt, with a soft nap, is also very satisfactory; though usually more expensive than flannel, felt is durable and is available in rich colors. One disadvantage, however, of both flannel and felt boards is that, though the materials will adhere to each other and support cardboard or other light substances, the attachment is often minimal. Thus sometimes, as a display is manipulated, one or more items will fall off or slip down the face

of the board. As with the magnetic chalkboard, there is an improved—a better—cloth board. A sturdy nylon material called Hook-N'-Loop is made in a variety of colors and dimensions and in two types of surfaces. The surface of Hook-N'-Loop yard goods material is covered with a very fine, fuzzylike surface made of tiny nylon loops; the companion material, used for attaching things to the surface of the board, usually comes in rolls or strips of tapelike cloth which has a surface of coarse, hooklike texture. When the hook material is pressed on the looplike surface of the loop cloth, the two surfaces stick firmly together with strong holding power. A very small patch of hook material, attached firmly to a solid object—a book, a can of paint, or a three-dimensional display—will support a surprising weight. And objects backed with the hook material and displayed on these Hook-N'-Loop boards will not slip or change positions until they are firmly pulled away.

So reliable are these nylon Hook-N'-Loop materials that they are now used widely—for example, to replace zippers or buttons on clothing; to line walls of space capsules, so that astronauts may attach small objects to wall surfaces and thus prevent them

Charles Mayer Studios, Inc.

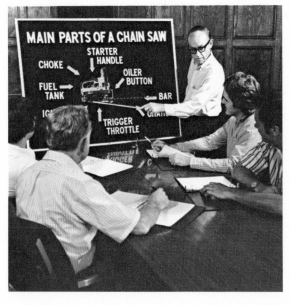

from floating about the cabin; and to attach covers to the headrests of airplane seats.

using cloth boards

A few reminders may help you avoid errors that sometimes occur when inexperienced people make and use cloth boards:

⊗ Respect legibility. As with chalkboards, cloth boards must be viewed by all class participants, and the materials used must be clearly recognizable. Writing should be readable from the viewpoint of every student.

⊗ Keep visuals simple. A few strong symbols or key words are better than a complex or wordy display. And careful use of color can enhance presentations.

⊗ Preserve visuals if they are to be used in future lessons; use labeled envelopes for convenient filing and to keep the materials clean and fresh in appearance.

⊗ For very small groups, such as children in a reading circle, use a small cloth board held on the lap, with prepared materials conveniently at hand on a nearby chair or table. With larger groups of up to thirty students, place boards, perhaps 4 by 5 feet or larger, on an easel. Display materials should be of proportionate size.

Charles Mayer Studios, Inc.

Here are a few suggestions that may encourage you to consider when and how a cloth board can be of help in your class or seminar room:

⊗ Use cloth boards to promote interaction among students and between students and materials displayed. Drill exercises, for example, can be presented in sequence as students respond either orally or by going to the board and matching question with answer, symbol with word, or symbol with symbol according to lesson requirements.

⊗ Reinforce learning by combining visual with verbal responses. Encourage students to observe and participate mentally as problems are set up on the board. The possibilities for applying cloth-board techniques in mathematics, in reading and spelling, in subject identification or classification, and in other topics will become apparent as you experiment.

⊗ Animate stories or historical events on the cloth board. This activity captures attention, helps students retain ideas presented, and clarifies interrelationships discussed. Persons, background scenery, and important physical elements can be incorporated in these visualizations.

⊗ Don't miss opportunities to involve students in making their own visuals for presentations, from simple "show and tell" experiences to more elaborate research reports which require statistical or other graphic data.

⊗ Capitalize upon clothboard potentialities by posting titles or backgrounds for titles for films or

Instructo Corporation

slide stories or for television productions. Even simple animation techniques can be used.

General directions for making cloth boards are provided in the following step-by-step instructions.

making cloth boards

To make a cloth board, start with a sheet of Celotex building board about 24 by 30 inches and a piece of cotton or felt about 28 by 34 inches.

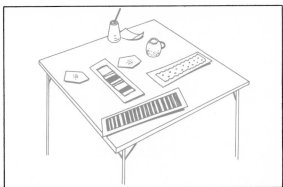

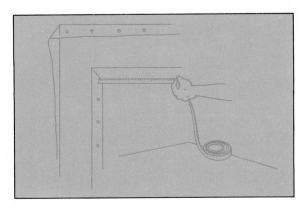

material, and mounted if necessary, it will stick to the flannel board. The display can be manipulated by students as they practice communicating with words and visuals.

Fold the cloth over one edge of the board and tack it. Stretch the cloth lightly; fold over the opposite edge and tack. Follow the same procedure for the two remaining edges. For a smooth job, use wide masking tape to hold down the edges and loose ends of the cloth.

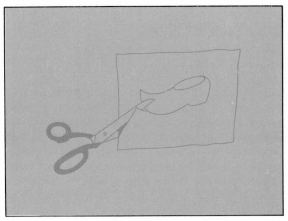

making cloth-board display materials

Many materials such as felt, flannel, corduroy, and sandpaper will stick to flannel boards with no special processing. Other materials—cardboard for example—should be mounted on one of these materials that stick easily. To be sure of a successful display, test the holding power of any backing you plan to use. Bright-colored felt and similar materials can be cut into many shapes, symbols, and designs. Invite students to use magazine pictures or sketches as outlines for drawing desired patterns on the material. When the pattern is cut out of the

art fundamentals: layout, sketching, lettering, color

Whatever display projects you and your students may undertake—bulletin boards, exhibits, teaching displays—they need not be great works of art to be effective. But they should be attractive and colorful; they should communicate their messages well. To these ends, this chapter includes the following discussion on planning, designing, layouts, lettering, and sketching. Color, and how to use it effectively, are also discussed.

As you progress through this section, be aware that the techniques and skills presented are relevant not only to the large-scale displays just discussed. They apply also to designing transparencies for overhead projection, for titles, and for illustrations for slides, filmstrips, and motion pictures, as well as for television visuals.

planning the layout

A plan for a visual presentation may be prepared on paper in rough form, with details suggested rather than filled in. This plan—or *layout*—provides a convenient way to judge qualities of the project as proposed before attempting any time-consuming or costly construction. Further, improvement of the layout can result from a try-and-test process, which will clarify ideas and simplify both the content and the means of presentation. The ultimate success of *any* display project can be attributed largely to the thoroughness of layout and pretest activities.

Practically every advertising display, animated motion picture, slide series, exhibit, or poster starts with one or more layouts. Layouts may be planned life size, or they may be developed as small-scale, *thumbnail* sketches of proposed final products. If the students prefer to work in life-size scale, use an area of floor space or a large table which approximates the shape and size of the display area. When the display is finished, transfer it to a wall or other display surface.

In analyzing the layout to determine whether the final display will be successful, a number of fund-

amental *characteristics* should be studied, each of which may contribute some element of appeal:

⊗ BALANCE Of the two main types of balance—*formal* and *informal* (both illustrated in accompanying sketches)—informal balance is usually regarded as the more interesting. Casual, informal arrangements for bulletin boards, chalkboard displays, and exhibits usually appeal to students. Sketches for planning bulletin boards permit evaluation of design effectiveness before any extensive finish work is undertaken. The value of rough sketches is evident

in the two examples. Note how balance (formal and informal) can be tried and how contrast, placement of component elements, and the usefulness of a cartoon sketch are tested.

Balance and interest in visuals can satisfy a viewer's aesthetic sense, for, as in nature, balance is sought to give boundaries and stability to things.

⊗ SHAPE A configuration pattern is usually found in effective displays. It may be established by the directionals that are developed to guide the viewer to see details in proper sequence. Whether the shape is subtle or obvious, it must be present; it should be clearly evident in the original layout. Here are three basic layout shapes (I, T, and Z) and their development in thumbnail sketches.

A POSTER EVOLVES

In this column are illustrated the steps that lead to a finished poster. Decisions to be made: determine appropriate balance, arrange content, choose lettering, and select color. Ultimately, of course, communication is the goal of a display, and the response of the viewer is a test of display effectiveness.

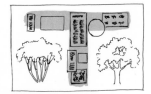

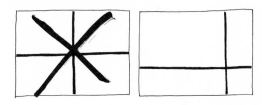

These configuration patterns exemplify well-organized layouts. Arrangement of photographs, or emphasis through panels, color, or textured materials, may establish basic directional lines. By using these and similar shapes as a starting point, you will find it easy to develop effective displays.

⊗ EMPHASIS Through the proper use of lettering, one or more dominant colors, and/or directionals (all illustrated in the drawings), a well-designed layout will emphasize a central idea.

⊗ CONTRAST If the parts of displays are to communicate, they must be noticed. A skillfully arranged exhibit which contains contrasting areas of light and dark will catch the eye of viewers. One way to achieve such contrast is to use dark papers as a background for mounting light pictures, or vice versa.

⊗ HARMONY A good layout should also be harmonious. This means that all the elements (lettering, color, materials) must work together to support the basic ideas presented; no one element should distract or capture attention to the exclusion of other important elements.

magic in color

Color is an important element in any successful display. Careful professional study has established what all of us have recognized in a general way: that color plays an important role in evoking moods—in causing us to feel gay or depressed, calm, excited, or restless. In classrooms, color is worth special consideration; wisely used, it is an effective aid to teaching and learning.

A good general rule to follow with respect to color is *keep it simple.* Use a minimum of different colors in any one display, thus avoiding confusion or loss of harmony. Remember that color should be a device to attract attention, to emphasize or contrast, or to create moods. White on black is one of the most effective combinations for carrying power. On any dark background, white, yellow, orange, green, red, blue, and violet, in that order, have great carrying power. On a white or light background, black, red, orange, green, blue, violet, and yellow, in that order, have the greatest carrying power. Neither the use of color nor failure to use it should be habitual or unplanned. Make color *work* for you in your visual presentations.

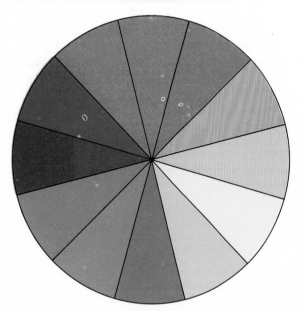

EFFECTS WITH COLOR

On the opposite page are a number of color sketches for bulletin board or poster designs. Note the uses of warm and cool colors. See how the colors create moods. Note how strong colors may either clash with each other or reinforce each other to produce a desirable effect. Recognize that color impact is often created solely by a solid color background with a single color used for accents of design or for lettering. Encourage students to experiment with color. Use restraint in forcing your taste and judgments on them. Sometimes simplicity in the use of color is desirable, but at other times a daring use of color combinations may produce precisely the effect desired.

THE COLOR WHEEL

The color wheel is a practical device to aid in color selection for displays. Consider the following types of color harmony as you design your layout and color plan . . .

Monochromatic. A monochromatic color harmony (a popular one for simple displays) is made up of the tints and shades of any one color. As an example, consider a layout containing only various shades and tints of blue—or pink, green, or yellow.

Analogous. An analogous color harmony consists of colors which are neighbors on the color wheel. As an example, consider a layout plan using yellow, yellow-green, and green.

Complementary. A complementary color harmony uses colors which are opposites on the color wheel. As an example, consider a layout using yellow-orange and blue-violet. Occasionally, an arbitrary choice of complementary colors will produce harsh effects as in the case of using green and red together. However, these colors may still be used if one of them is modified slightly by the addition of white, black, or some other color.

sketching and drawing

Since visual communication is basic to teaching, it may be well to make simple line drawings and occasional cartoons to clarify instruction and create student interest in lesson presentation. But mention *sketching,* and many people begin to defend their inability to draw. "I'm not an artist," they will say. This attitude suggests that the greatest obstacle that discourages teachers from trying to draw is the adult's desire to produce finished, professional-quality art work—or nothing. However, many students have natural talents for drawing and sketching—and so do many teachers! Skillful teachers help their students to refine such talents by encouraging them to communicate through graphics, sketches, cartoons, and other visual presentations. Their sketching and drawing work need not be art in any sense of the word. All that matters is that it communicate effectively. Set standards of performance, then, that establish the levels of quality necessary to achieve that end. With practice, and as skill develops, many teachers are surprised at how well they and their students can draw—and how much the ability to draw can improve teaching and learning.

Here are a few helpful guidelines recommended for developing visual presentations:

⊗ Make drawings simple, bold, and spontaneous.

⊗ Try rough sketching on paper to plan the layout and to test style, legibility, and the message to be communicated.

⊗ Make each planning sketch with attention to final size in which it will be used: as a transparency projected on a screen, as a chalkboard drawing, as a poster, or as a paper chart.

⊗ Don't strive for artistic perfection—maintain informality and spontaneity.

⊗ Don't apologize; just do it, and keep trying.

⊗ For your own presentations, prepare complex sketches before class on large sheets of paper, using light-colored or white paper with contrasting colored felt pens. These sheets may be hung on the map rail above the chalkboard or on an easel. Since they require preparation time, these visuals may be preserved for future use. For additional data or captions, the areas of a chalkboard adjacent to the sheets may be used without spoiling the original visuals.

Faces and figures, which are not difficult to draw, can contribute greatly to chalkboard presentations and to bulletin boards, transparencies, slides, motion pictures, and television visuals. Start with faces, since they can convey many ideas and attitudes. Use a basic simple oval and experiment with step-by-step additions of minimum lines to indicate features and expressions.

Start figure drawings with simple stick figures; they can show action and attitudes with surprising effect despite their simplicity. After you and your students have practiced with basic stick figures, you may then wish to add detail to characters—but without needless trimmings.

You will discover, too, that simple, stylized sketches of real things, such as buildings and automobiles, can establish environments or activities. Examine the sketches below and on the opposite page; then start sketching!

START FACES WITH AN OVAL; ADD MINIMUM LINES.

FOR EXPRESSIONS, ALTER EYES AND MOUTH.

ADD COSTUMES AND OTHER IDENTIFYING DETAILS.

CONSULT PHOTOGRAPHS AND OTHER CARTOONS.

USE STICK FIGURES ALONE OR AS A BASIS FOR MORE DETAILED FIGURES.

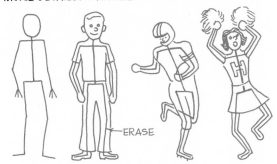

OR START WITH SIMPLE FORMS AND ADD DETAILS.

BUILD OBJECTS AND SCENES FROM SIMPLE SHAPES.

HANDS ARE EASY IF YOU USE YOUR OWN AS MODELS.

EXAGGERATE!

lettering for displays

Other important requirements for creating effective displays are legible and attractive captions, labels, and titles. An otherwise well-planned display may be ruined by unnecessarily poor lettering. There are many ways to achieve satisfactory lettering. The simplest is to develop some skills in freehand lettering with chalk or with felt-tip pens. The use of certain films and other resources can guide your practice and accelerate your learning. For some purposes the typewriter is adequate and convenient. Some typewriters have special type, including large letters; some, with interchangeable type, provide letters in a variety of sizes and styles.

Many different kinds of available lettering aids can simplify preparation of captions, titles, and transparencies. Some of these are shown and explained on the following pages. As you prepare various instructional materials, try your hand at using some of them. Also have your students try the different lettering techniques in their projects and evaluate the effectiveness of the results produced by each of the several methods and evaluate the appropriateness of each style for the media being presented. Communication is the first touchstone for evaluation.

LETTERING METHODS

With practice, most people can do freehand lettering, and there are a number of references and films that can help one learn this useful skill. Freehand lettering can be done rapidly and simply, but when high-quality results are justified, both teachers and students can use precut letters and other lettering aids. These reduce the chore of making a display and speed activity toward a neatly finished product. A number of lettering aids shown here indicate the resources available in art and sign supply stores and from audiovisual dealers.

Lettering guides. From inexpensive kits with plastic lettering guides and brush-type pens, accurate and easily read letters may be produced.

Felt-tip pens. The versatile felt-tip pen can produce professional lettering. Many kinds are available, from the small squeeze bottle with a felt tip in the top to the metal pen with an ink chamber inside. A variety of both tip shapes and quick-drying ink colors makes this a useful tool.

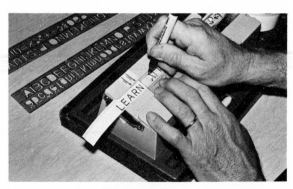

This economical guide permits felt-tip pen lettering to be applied directly to pregummed white tape. The tape can then be adhered to a display to suit the layout. Guide strips of several letter sizes are available, with appropriate felt pens.

Mechanical lettering sets. Template, scriber, and pen guarantee unvarying lines. The template is set against a straightedge, above. The tracer pin of the scriber is used to trace letters in the guide groove of the template. The pen in the socket arm of the scriber (ink-filled) reproduces the letter above.

Dry-transfer letters. Transfer letters, numbers, and symbols are available in many styles and sizes. Use a burnishing tool or any smooth, hard device to transfer the letters to the display material.

Lettering pens. Metal points for lettering pens, fitted in standard penholders and used with india ink, produce, with practice, professional-looking results.

Precut letters. Some precut letters have an adhesive backing which saves time in pasting them to other materials. These letters are also suitable for tracing letter shapes.

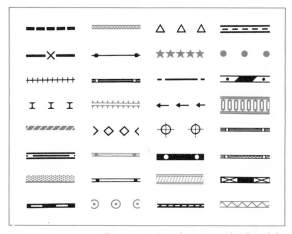

Transparent and opaque plastic stick-on tapes can be used in developing designs for functional and colorful projection transparencies and charts.

Raised letters. Raised letters may make a display especially effective. They are available in many styles and shapes, and practically all can be re-used.

While numerous techniques, skills, and admonitions have been presented in this chapter, remember that perfection in design and display are not expected from either teacher or students. A loose, informal presentation is often completely appropriate and acceptable. However, to be effective, visual communication should be based upon well-thought ideas, well-planned presentations (whatever the medium), and should apply the principles of design, lettering, coloring, and form in a manner that captures attention and leaves the message firmly in the minds of the viewers.

summary

Among the simple, available visual communication resources are the chalkboard, magnetic boards, bulletin boards, and displays—including three-dimensional objects and materials.

Teachers themselves should develop visual presentation skills and ability to judge the visual effectiveness of materials sold commercially or produced by their students. Students should be encouraged to use these same visual communication techniques to develop their own displays and other presentations.

It is not necessary that students and teachers who use display techniques be accomplished artists, but all should develop some skills in using the simple principles of balance, shape, emphasis, contrast, harmony, and color. By working from instructional objectives and proceeding through an orderly plan to a finished product, teachers and students will be able to develop displays that make the classroom environment dynamic, colorful, and stimulating.

The principles of effective design for visual communication discussed in this chapter apply in developing many different media—transparencies, slides, motion pictures, television, and printed materials, as well as a variety of bulletin boards and teaching displays. Some of the basic techniques discussed here involve layout and lettering, cartooning and sketching, and meaningful use of color.

The next chapter, Graphic Materials, continues and expands the discussion of the use of visuals for communication.

6

GRAPHIC MATERIALS

chapter purposes

⊗ To enable you to distinguish the characteristics, advantages, sources, and uses of five different types of graphic materials: (1) graphs, (2) charts and diagrams, (3) cartoons, (4) posters and signs, and (5) maps and globes.

⊗ To provide criteria by which you may evaluate graphic materials.

⊗ To suggest a variety of ways you and your students may use graphic materials effectively to attain learning objectives.

⊗ To assist you in selecting activities for students so that they may develop ability to analyze, understand, and enjoy the different types of graphic materials.

⊗ To encourage you to develop skill in designing and producing graphic materials for your use in teaching and to assist your students in learning how to produce them.

Walter Herdez, published by Hastings House

Graphic materials serve as a universal shorthand to help readers understand the torrent of information with which they are deluged. Much of the content of daily papers, news magazines, technical and artistic publications, and public display boards is conveyed through graphic media. Messages communicated are intended to promote sales, improve health, increase attendance at musical or sports events, or achieve literally thousands of other purposes. In almost every field of knowledge, graphics contribute significantly to communication. Therefore students should know how to prepare and interpret them.

A wide variety of materials has been developed to provide graphic images for diverse purposes. In this chapter we present and describe five of them: (1) graphs, (2) charts and diagrams, (3) cartoons, (4) posters and signs, and (5) maps and globes. Because these five types of graphics have in common a number of instructional advantages, it is reasonable to expect that every teacher will know how and for what special purposes each may be selected, developed, and used.

A most creative use of graphics in an active learn-ing situation is undoubtedly student production of the graphic material. Three different levels of activity, any of which may be valuable for particular individuals, are possible: (1) copying a graphic chart, diagram, or map is permissible if the copy is intended for use in the classroom by the student, but it is a relatively low-level learning activity; (2) more original—and more indicative of the quality of individual performance—is construction of a new graph, chart, or map based on a combination of data from several different sources; (3) complete originality in the development of the graphic is ideal —comparable to original invention in thought processes involved.

Board of Education, Borough of North York (Ontario)

LEVELS OF CREATIVITY WITH GRAPHICS

Levels of creativity were discussed in Chapter 4. In these pictures, three levels of creativity are illustrated. Which represents the imitative level, the adaptive level, and the creative level?

San Diego County (California) Schools

The New School, University of North Dakota

graphs

Graphs show numerical or proportional relationships which can enable readers to grasp quickly and accurately the specific meanings of masses of complex data. Nearly any list of figures can be made into a graph that is understandable and interesting to even unsophisticated readers.

The many possible variations in graph types can be reduced to three basic forms: circle, bar, and line. Many special subtypes have been devised, but each is based on one or another of these forms. For example, the modern pictorial graph uses symbols with one of them—most often the bar graph. Scales used in graphs are of two kinds: those with only one scale of measurement, principally pie or bar, and those with two. Some specialized graphs have more than two scales, and these are of importance in more advanced teaching. Careful instruction in reading graphic symbols is an important aspect of teaching with graphic materials.

line graphs

Line graphs belong to the large family of two-scale graphs. The identifying feature of this family is two scales, called the axes, placed at right angles: one measure is vertical, one horizontal. Each point drawn on the line graph has a value on the vertical as well as the horizontal scale.

ANNUAL PRODUCTION

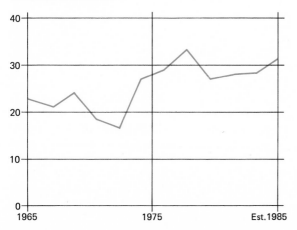

Straight lines connect points representing each measured quantity. Sometimes these points are joined with smooth curves rather than straight lines. If junction points are clearly indicated, the curved-line graph is quite easy to read.

The example of a simple line graph shows how the quantity of something varied from one particular time to another. It gives a clear picture of how production for the most part *dropped in the first few years* and then *increased rapidly, fell off again,* and then *increased* somewhat *more slowly.* In this type

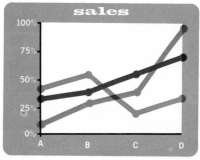

Scott Education Division

of graph the horizontal axis is used to measure *time,* the vertical axis to measure *number* or *quantity.*

One study at the University of Wisconsin led to the conclusion that bar and circle graphs worked equally well for presenting percentage data, but bar graphs, both horizontal and vertical, proved better than line graphs for evaluating and comparing specific quantities.

bar graphs

In comparing the magnitude of similar items at different times, or in showing relative sizes of parts of a whole, bar graphs are particularly useful. The simple *linear bar graph* is usually easy to read correctly. It is particularly useful for students who are just beginning to learn to read graphs.

The *subdivided bar graph* is probably the next easiest bar graph to read and understand. With this type, each bar shows 100 percent of the item being graphed, with different segments of that single bar showing different amounts of the total for similar data.

The bars of bar graphs are generally darkened. Subdivisions or segments of a bar graph should be distinctively colored or crosshatched. Bars may sometimes run both ways from a base and thus indicate changes in two dimensions.

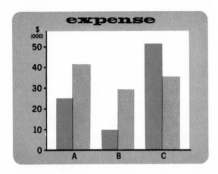

San Diego County (California) Schools

DISTRIBUTION OF UNITED STATES POPULATION BY REGIONS

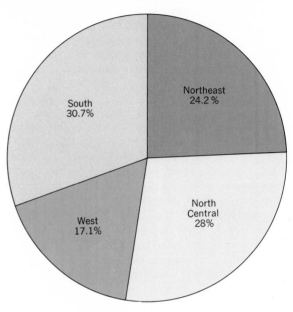

circle or pie graphs

Circle or pie graphs are particularly recommended for showing parts of a whole, such as the tax dollar—its sources or how it is expended—or the consumer's spending dollar. Circle graphs are considered easy to read when used for such purposes. Investigators have found that young children can read them as easily and accurately as they can read line or bar graphs.

pictorial graphs

Pictorial graphs are used in many types of publications and displays to which students will refer during both instruction and study: in government and business reports; in wall charts for classes in economics, political science, and geography; and in exhibits and displays in museums, libraries, and fairs, which include interesting methods of animation and lighting. Since pictorial graphs are widely used for presenting information, students should be given ample opportunity to develop facility in reading and understanding them. One of the best ways to do this is to let them obtain some experience in producing their own.

U. S. cotton production and exports

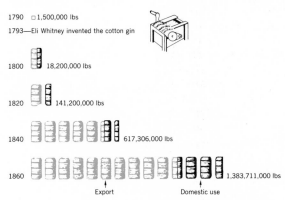

1790 □ 1,500,000 lbs

1793—Eli Whitney invented the cotton gin

1800 18,200,000 lbs

1820 141,200,000 lbs

1840 617,306,000 lbs

1860 1,383,711,000 lbs

Export   Domestic use

From Harold F. Kaye and Bertram M. Wainger, The American Adventure

The following guidelines will help you and your students produce readable pictorial graphs:

⊗ Make symbols of pictorial graphs *self-explanatory*—one for farmers, for example, should be readily recognized as a stereotype for their occupation.

⊗ Keep them as simple as possible; make comparisons using a number of different picture units of the same types and sizes. Avoid using single symbols that vary in height and volume, since the eye cannot easily or accurately estimate numerical changes shown in gross sizes of nongeometric figures.

⊗ Group pictorial symbols in easily interpreted sets—of five or ten, for example—and show clearly on the graph itself the value of each unit.

⊗ If feasible, include figures giving the numerical value of each pictorial graph entry directly after the last symbol in the line.

⊗ If necessary, split the last symbol to represent the partial number it represents in the total.

⊗ Recognize that most pictorial graphs will depict only approximate amounts and that they cannot be read exactly. They should not be used when exact readings are required.

reading graphs accurately

To construct graphs that give true representations of data sometimes presents a problem. The crucial need for the selection of an appropriate scale is shown in the accompanying pair of charts. These graphs show the same data with different treatments. A change on either the horizontal or the vertical scale of a graph alters the appearance of the graph plotted on the scale. Such changes can give a radically different impression of the same set of facts.

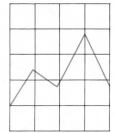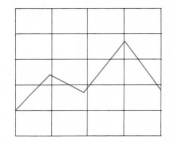

Perspective graphs such as the one below are extremely hard to read, and the deeper the perspective, the greater the problem. Examine this graph carefully before deciding which is the larger section, A or B. Were you fooled? They are supposed to be equal in size. Although you can figure this out from your knowledge of perspective drawing, the important initial impression is deceptive.

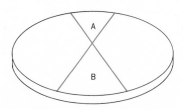

These are but a few of many problems that sometimes make graphs difficult to interpret. In teaching students to read graphs, such special problems should be explained, and attempts should be made to analyze published graphs for the validity of their representation.

SYMBOLS IN GRAPHIC COMMUNICATION

Symbols are becoming a universal language. Those on this page, from the book by Henry Dreyfuss, *Symbol Sourcebook* (McGraw-Hill, 1972), illustrate how the language of symbols evolves. Many valuable experiences with graphics can be developed for students by having them work with symbols. For example, have them study some basic symbols, then ask them to develop graphic displays that communicate solely by symbols. Suggest that they undertake to show progression in numbers, or intensity, by symbols alone. Ask them to modify a single established symbol to change the meaning.

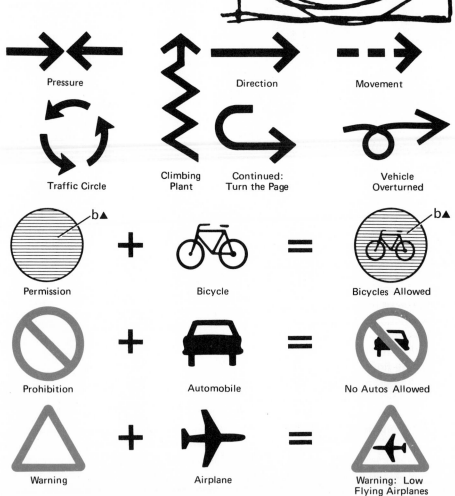

Pressure   Direction   Movement

Traffic Circle   Climbing Plant   Continued: Turn the Page   Vehicle Overturned

Permission + Bicycle = Bicycles Allowed

Prohibition + Automobile = No Autos Allowed

Warning + Airplane = Warning: Low Flying Airplanes

charts and diagrams

An important purpose of many charts is to present visually ideas or concepts which are likely to be difficult to understand if presented in oral or written form. Charts can also highlight important points of presentations.

Effective charts tend to be composed of a mixture of several different types of graphics: pictures, drawings, cartoons, graphs, diagrams, and verbal materials. However, remembering some principles of display covered earlier in Chapter 5, weigh carefully the impact and clarity of the message likely to result from the variety and complexity of materials you include. By all means, be selective. Consider the purpose of the chart. Is it intended for use as an introductory overview? If so, it may suggest, through pictures tied to only a few key words, the principal topics to be covered. Is it to raise problems? If so, it may need to contain a challenging but quite unstructured visual of a problem-raising nature. Or is it for a summary? If so, it may need to be quite specific and to present information in a sequence related to instruction. Are the types and amounts of materials to be displayed too much for one chart? If so, you may need to develop a series of charts to be presented one after the other or perhaps to be used as a wall display.

Remember, charts must be designed for the conditions under which they will be used: in print, either in books or on flip charts; in projection media; for transparency or for slide projection; or for use on TV.

time and sequence charts

The time and sequence charts you make or use may be of several different types. *Time charts* themselves can be chronological listings similar to tabular charts, just discussed. On the other hand, *time lines* may be used to plot relationships of events and time—normally on a long, narrow bulletin board. A single time line may occupy the space available— say 10 feet—with each foot representing 100 years. If this is done, points on the chart must be num-

bered consecutively to show centuries. Pictures or small charts are sometimes inserted on such time lines to highlight important events of the period. Separate time lines, all of them made on the same scale, can also show birth and death dates of a number of famous musical composers. When placed on the same chart, again proceeding from the earliest date on the left to the most recent date on the right, the time lines reveal (1) the life-span of each composer, (2) which composers were contemporaries, and (3) which composers preceded or followed others in history. The same technique can produce interesting results when used for other chronological events—life-spans or periods of time— that overlap.

SOME OF THE PRINCIPAL SIGNERS OF THE DECLARATION OF INDEPENDENCE

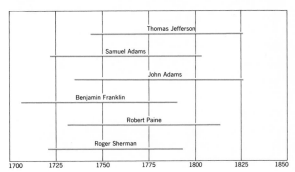

Murals painted on long sheets of paper and placed at one side of a classroom are sometimes developed as illustrated time lines. In a study of the westward movement of the pioneers, for example, your students might make murals of the trek from Missouri to Oregon or California. If they were to draw each stage of this event in a semblance of chronological order, the result would be essentially an illustrated time line. Scale is usually not strictly followed in such murals; human figures are frequently portrayed symbolically rather than realistically. The painting of murals gives your students opportunities for artistic expression and you an opportunity to make some observations as to the accuracy of learning reflected in the painted representations.

strip charts and flip charts

Strip charts and flip charts are two devices used to present data in sequence. The first is constructed as a single chart, with sequential parts covered by strips of paper which can be removed at appropriate times to disclose points of a presentation. The flip chart helps to present a sequence of information which would be difficult to show on a single sheet. To make flip charts, take several sheets of newsprint or other paper available in sheets of the same size, and place appropriate visual or written materials on each. Fasten the sheets together at the top with thin metal or wooden strips—one on the front and one on the back. Mount the supporting strip at the top of an easel. Some schools provide classrooms with commercial easels, which may serve multiple purposes.

When teaching with charts, turn over each sheet as you need it; use a pointer to explain details, and be sure not to block the view of your audience. For review or reference, flip-chart sets may also be separated and displayed on the wall as a series.

classification and organization charts

Several types of classification and organization charts are particularly useful with mature students. The *flowchart* depicts the flow of a process, or it may trace responsibility or work relationships among various administrative sections of a large organization.

The *tree chart* is typically used for genealogy, to show the character, composition, or interrelationships of generations of families or classes.

A *stream chart* is the opposite of tne tree chart. On it, several small sources or tributaries finally unite to form a single stream. A tree chart may be used to present effects of an important invention, such as nuclear fission. A stream chart may show all the raw materials needed in manufacturing a finished product, such as an automobile, and the processes through which they pass. Normally flowcharts or tree charts do not contain numerical comparisons of time, amount, or rate, although occasionally they may.

FLIP CHART

Often used to present topics for discussion or to list procedures to be carried out in sequential steps, as in a laboratory, flip charts are both economical and easily prepared.

FLOWCHART

As shown in the example below, the flow chart is particularly useful to illustrate functional relationships, such as the sequence of events in the judicial process, "Protection of the Accused."

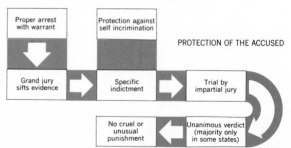

commercially prepared
chart materials

Commercially prepared chart materials, now available for many subjects, are highly useful in teaching. Frequently they represent good art work, reflect dependable research, and make functional use of design and color. An outstanding example of such work is an anatomical chart. Not only for anatomy but for biology, for all of the physical sciences, and for health, literature, history, safety education, and economics, there are charts made by dependable, established companies. Buyers of these charts can be confident that their content is accurate.

The examples shown below are excellent teaching materials. "The Microscopic World of the Molds," left below, from Charles Pfizer & Co., Inc., can give a clear and quick conception of the strange beauty of microscopic fungi—a view ordinarily disclosed only by the use of a microscope. The botanic demonstration chart, one of a series prepared for school use by Denoyer-Geppert Company, shows the characteristic features of the sweet cherry in brilliant colors, with all details clearly visible even from a distance.

Similarly useful charts are sometimes available at little or no cost from suppliers of agricultural equipment and biological products and from manufacturing processors and land and real estate developers. Such materials are discussed later in Chapter 15.

Charles Pfizer & Company, Inc.

Denoyer-Geppert Company, Chicago

MODELS

To supplement wall charts, commercially developed models such as this plastic head and torso are available. They provide tactile learning experiences in such fields as botany, physiology, home economics, physical sciences, family life, and sex education. How would you most effectively use the fact that this torso-head model has ten removable parts, comes in different skin colors, and is available in a talking version by means of a cassette tape? See also Chapter 13, Real Things and their Models.
Denoyer-Geppert Co.

Denoyer-Geppert Company, Chicago

⊗ *Scientific data,* such as ocean currents or geological formations

⊗ *Social or cultural data,* such as population or language patterns

⊗ *Political data,* such as boundaries between states or countries, types of government, or election results

⊗ *Historical changes* in political subdivisions or boundaries often shown in a series of maps or globes

⊗ *Economic data,* such as industrial production, agricultural products, or international trade

As explorations in space provide increasing amounts of information, maps and globes are designed to provide convenient access to the data collected. The lunar globe above was produced from NASA photographs taken by Lunar Orbiters and Apollo astronauts. For advanced students, maps and globes should be available for independent study as well as in forms for lecture-demonstrations by instructors. And map and globe study, correlated with experiences in planetariums, provide stimulating, practical activities.

Still another map form, the pictorial map, combines realistic detail, and often caption materials, that would be difficult to present in any other way.

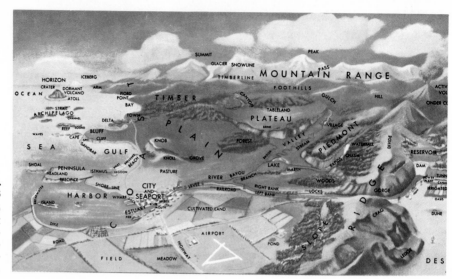

PICTORIAL MAPS

This typical pictorial map is designed to assist young children in recognizing and naming geographical features. *Denoyer Geppert Co.*

problems in map and globe instruction

The task of learning to read maps and globes presents a number of difficulties. In the first place, maps and globes employ symbols to depict geographical features, and these symbols usually bear little or no relationship to things in real life.

A second problem with maps and globes is a bit more complex. Unfortunately, flat maps are made up of a variety of symbols—grids representing lines of latitude and longitude, elevations, surface features, and other data—which have little or no relationship to reality and do not always mean the same things to the different people who read them. Thus, the use of maps is unduly complicated. More will be said about this problem and how to assist students in learning to use maps with assurance.

A third problem that arises in connection with learning to read maps is the fact that map projections are always somewhat distorted. This is chiefly because *they* are flat and the world's surface is round. Because the globe is the only relatively accurate map of the roundish world, instruction in the use of flat maps must be closely correlated with instruction in reading the globe. The traditional Mercator projection distorts distances, especially in high latitudes. An illustration of this disadvantage is the familiar comparison of the sizes of South America and Greenland as shown on a globe and as shown on a Mercator projection map.

But the fact that flat maps must be distorted does not mean that we should stop using them. Because flat maps are needed in many kinds of everyday activities, schools must provide opportunities for students to learn to read and use them.

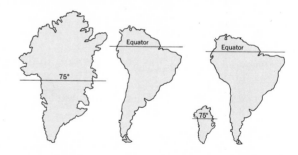

Drawn on the Mercator projection, Greenland looks larger than the continent of South America. On a globe, the true relative size of Greenland, however—one-ninth the size of South America—is as accurate as in an equal-area projection.

CONTOUR MAPS

The ability to visualize three-dimensional land formations from data supplied in flat contour maps is a skill which advanced students must master. Assembling three-dimensional maps supplied in kit form provides experience to develop this skill.

FILMSTRIPS FOR MAP INSTRUCTION

Many map and globe concepts can be studied through filmstrips such as those illustrated here. Filmstrips of white-line outline maps permit projection on dark chalkboards for tracing. *Scott Education Division*

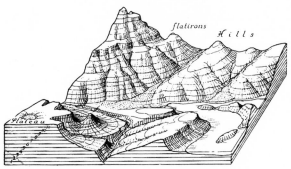

From Irwin Raisz, *General Cartography*, McGraw-Hill

The World Known in 1492

Here different colors are used to show how high the land is above sea level. What color stands for the lowest land?

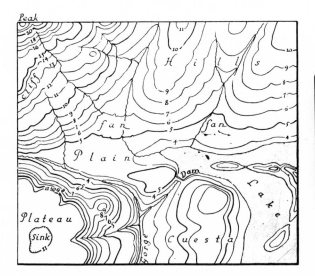

Do you see why the sun's rays feel hotter to us in June than in December?

selecting and using maps

Flat maps are available for a variety of uses in instruction. When selecting maps, a number of design characteristics should be considered, such as size, colors, legibility of details, and suitability for student use. Several types of flat maps in common use are described below, each of which serves a particular instructional function while students develop map-reading skills.

⊗ WALL MAPS Wall maps are designed to provide opportunities for a number of students to study geographic data simultaneously. Most wall maps range in width between 40 and 64 inches. Too much detail makes them hard to read. Small inserts to be used for individual reference and study are sometimes placed in the corners of large wall maps. However, too many inserts create clutter and make the map difficult to read. Be sure such inserts do not infringe upon the identity of the main

map—the major area to be shown. The great value of wall maps lies in the tremendous variety available. Physical-political, relief, or simulated relief maps can be obtained for almost any portion of the world. Colored literary or language maps pictorialize the writings and languages of different countries. For advanced map study, specialized maps may be obtained for each classroom. Know the maps available in order to select those most appropriate for your classes.

⊗ CHALKBOARD OUTLINE MAPS Chalkboard outline maps have a surface that may be written upon with chalk and then erased. They have clear outlines and few colors, usually black or light green,

with blue oceans. Temporary chalkboard outline maps can be made by using templates. (Refer to chalkboard template techniques described in Chapter 5.)

⊗ ATLASES Flat maps for individual student reference and study come in books of maps called atlases. Atlases cover a wide range of materials—the world, regions, historical periods—or one very specialized subject, such as rare metal deposits, often in great detail.

⊗ OUTLINE MAPS The outline map for individual student use contributes to the development of map-reading skills by providing a medium on which to record and fill in geographical data.

FLOOR MAPS

Large, activity-oriented maps can encourage student involvement while teaching many abstract concepts— geographic, economic, and cultural. Students can apply paint, yarn, crayons, papier-mâché, or other materials directly to the map surface. *Denoyer-Geppert Co.*

selecting and using globes

There are several important criteria to be applied in selecting globes.

⊗ SIZE AND LEGIBILITY Suitable globes range from 8 to 24 inches in diameter. Most globes used to study political geography should be at least 12 to 16 inches in diameter. Even larger sizes are desirable if budgets permit. However, legibility is determined by the amount of detail, the size and style of lettering, and the combinations of colors used. Each characteristic must be evaluated.

Globes upon which markings can be made should be as large as 24 inches in diameter. Much of the information contained on globes is suited only to individual or small-group work, because items may be visible but not legible to most of the persons in the class. Only larger features and the properties of a globe itself can be used to teach large groups of students. Whenever inexpensive globes can be provided, it is ideal for each student to have one.

⊗ DETAIL AND SIMPLICITY Achievement levels of students should determine types of globes to be purchased. For beginners and children in primary classes, a simplified, markable globe with a minimum of detail provides opportunity for the children to fill in information as they learn political or geographical names. Words learned one day may be erased to be replaced by those to be learned later. For older groups, globes containing additional detail are appropriate, but there should never be so much detail as to obscure the essential information portrayed.

⊗ COLOR SYMBOLS While color may make globes attractive, its chief purpose should be to distinguish political divisions or to show elevations and land-water differences. Although there are no absolute rules about color, there is widespread, international use of blue for water; white for the continental shelf; green for lowlands; and yellow, red, orange, and brown for progressively higher elevations. Globes and maps made for beginners have only two or three colors. Intermediate globes and maps have four or five; regular globes have ten or twelve merging shades of these colors. If colors are unduly brilliant, the globe is harder to read than if softer, more subdued hues are used.

MAKING GLOBES

The process of assembling a paper globe clarifies the differences between orthographic and Mercator projections. *San Diego (California) County Schools.*

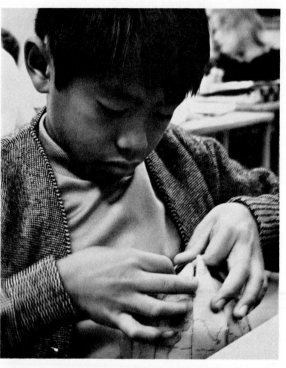

WRITE-ON GLOBES

The world can become more than a symbolic idea with a globe that can be written on. Modeling clay can be made into mountain ranges, and chalk or tempera can be used to label places and land forms. For what stage of globe study would this write-on globe be most useful? *Denoyer-Geppert Company, Chicago.*

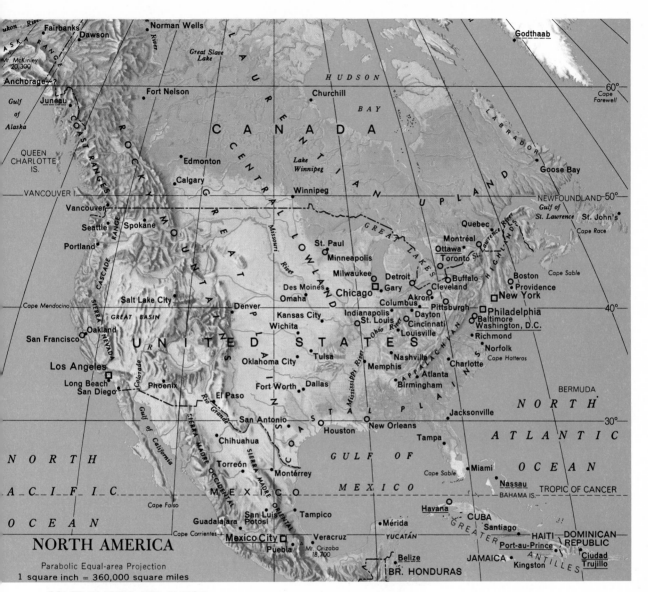

NORTH AMERICA

Parabolic Equal-area Projection
1 square inch = 360,000 square miles

COLOR IN MAPS AND GLOBES

Color in maps and globes is symbolic, normally without much relationship to the real colors in natural features. One exception to this general procedure is the treatment by cartographer Hal Shelton in this map of North America. He uses colors to portray variations in surfaces as they would appear to an imaginary observer in space. For example, this natural-color treatment differentiates between the appearance from the air of the savanna of southwest Florida and the tropical forest of British Honduras. Legends need to be checked carefully for any key to color and other special features. *Jeppesen and Company.*

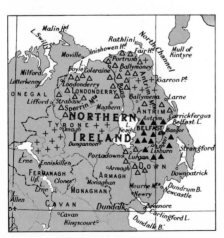

using globes and maps to evaluate learning

To test map-reading ability, develop assignments which can be solved only by working with a map or globe of appropriate type and size. Outline maps can be used to record answers in graphic terms. Also, consider the following possibilities:

⊗ Use commercially prepared transparencies of outline maps. Show them on the overhead projector to the entire class. Use cardboard cutout masks to cover selected areas and test students by asking questions about the obscured parts. Uncover the areas during the posttest discussion or paper-correcting session.

⊗ Mount maps on cardboard sheets. Attach questions that can be answered only by properly interpreting map legends—the scale of miles, the direction orientation, or other data provided. Pass the maps from person to person; ask each to record his or her answers on a special sheet provided for the purpose.

⊗ Use wall maps, flat or pull-down, or globes for testing. Identify questions by numbers affixed to the map or globe, or write out questions on cards and place them near or attach them to the map or globe. Some teachers place question cards on a table in front of a map and run strings from them to relevant map areas.

⊗ Use an opaque projector to project maps in books or on loose-leaf sheets. Use such projections as the basis for making copied enlargements. Omit map elements during the copying process; add questions, numbers, or identifying directional markers, as desired.

DETAILS IN MAPS

Here is a typical teacher's choice: To the left are three maps of the same area, but note the additional data in the second version and the increase of informative detail in the third.

teaching
with road maps

Probably one of the most immediately practical ways for students to learn map-reading is to use road maps. With road maps, many interesting projects for teams or groups of students can be designed. To enable young children to become acquainted with their neighborhoods, the projects can be simple and appropriate; and as students advance, they can expand their knowledge of their community, the county, state, and ultimately the areas of the world—all by road maps.

8- by 10-inch sections, or on two wooden rollers with loops for hanging on hooks. The ease with which the map can be used when needed is an important concern. The spring-roller type mount—ready for immediate use—is considered exceptionally convenient.

graphics and
visual literacy

The need for literacy in the use of graphics is apparent because people should be able to read maps. However, when graphics of all kinds are considered, only a little reflection is needed to realize that graphic literacy involves skill in reading many types of symbols in addition to those used on maps and globes. Fortunately, because students are used to seeing and reading graphic representations, they usually enjoy actively working in school with posters, graphs, charts, diagrams, cartoons, maps and globes, and informational directional symbols. The endless supply of graphic materials in the daily press, weekly news magazines, books and reference volumes, plus the numerous displays for advertising and safety and directional signs—and even graffiti—provide limitless examples for reference and study. Incidentally, taking advantage of all types of available graphics to motivate student interest in important concerns of society should be an easy teaching task.

storing and
displaying maps

A map series should ordinarily be installed in a multiple mounting—on a tripod, a pedestal base, or some type of map rack. Tripods and pedestals are movable, whereas most map cases or racks are attached to wall mounts or display rails in the front of the room. Individual maps on spring rollers should be removable for versatile use. Single wall-size maps can be purchased in various mounts—with spring rollers to fit the map case, folded in flat

summary

Five types of graphics have been discussed in this chapter: (1) graphs, (2) charts and diagrams, (3) cartoons, (4) posters, and (5) maps and globes. Each shares with the others a reliance upon certain forms of symbolism to convey meanings, and each requires skill in reading or interpreting.

Because both cartoons and posters are almost universally utilized as communication media, to use them in school is not only reasonable but necessary for visual literacy. The propaganda potential of both these graphic media is very high, and students need to learn to read them with an analytical point of view.

Although the graphs in common use today provide unique opportunities for condensing otherwise complex data and presenting them in simple visual forms, they also raise a number of problems of interpretation. The most common types of graphs are (1) circle or pie, (2) bar, (3) line, and (4) pictorial. Distortions in graphs—intentional or unintentional—can lead to misinterpretation.

Globes and maps of many different kinds are essential aids to instruction. Map-making activities are effective means of helping students to organize data they have studied as well as teaching them map-reading skills. Such learning experiences are no longer relegated to the development of geographic concepts alone: they belong in all the social studies and in English, foreign languages, and other areas of the curriculum as well. Therefore, the professional education of every teacher should include instruction in the interpretation of various types of map projection; the meaning of the color symbols; and the importance of map size, legibility, and details according to the purpose for which the map is to be used. Teachers should be aware of maps in forms suitable for projection by slides or transparencies.

Finally, through the production of various types of graphics, students learn to think in visual terms and gain skill in organizing and editing their ideas into capsulated, clear, forceful statements. They may apply graphic skills to almost every aspect of communication and to the use of many visual media. They may apply these skills in written reports, presentations and demonstrations, and in productions of slides, films, and TV programs. Probably most important, they are involved in learning by doing.

7

TRANSPARENCIES FOR OVERHEAD PROJECTION

chapter purposes

- ⊗ To encourage you to use the overhead transparency projector and to capitalize upon its well-demonstrated effectiveness for a variety of instructional purposes.
- ⊗ To acquaint you with numerous techniques for using the overhead projector, and to encourage you and your students to experiment with them.
- ⊗ To assist you to locate and to evaluate commercially produced transparencies that will contribute to your instructional program.
- ⊗ To help you acquire specific skills needed to design and produce various instructional materials for use on the overhead projector.
- ⊗ To suggest class activities in which your students can learn to use overhead projection techniques.

3M Company

Most teachers and students recognize the advantages of overhead projection techniques. When using the overhead projector, you completely control your materials at all times. The presentation is yours, with your timing and your choice of when, how, and why you show *what* you show.

Overhead transparencies are also popular because they permit a high degree of enjoyable creativity in the preparation and presentation of ideas, both verbal and visual. With transparencies, you may use a number of methods to disclose ideas in colorful sequence; you may take advantage of many opportunities to use them to create suspense and surprise and to stimulate student attention. Using overhead transparencies in your teaching will reward you with a feeling of success in communicating ideas and in encouraging the participation of every student in your class.

Now, turn the situation around—what about your students? What are the advantages to them of your using overhead projection? In the first place, it will be easy for them to see what you have put on the screen, even in a lighted room. Under proper conditions screen images can be made large enough to be seen by even nearsighted students in some back row. In addition, as in all projection, there is the bright focal point of the lighted screen upon which your images are projected. Action supplied by you as you write, point, underline with colors, or manipulate masks or overlays captures and holds student attention in ways that encourage and facilitate learning.

When students use the overhead projector for their own presentations and reports, they engage in a highly motivating activity and also master a very important communication skill. As they become proficient in creating and using transparencies, they can add vividness and interest to their reports in any class—perhaps even give assistance to an instructor who wants to make a visual presentation.

In the process of designing visuals, students learn to organize ideas for effective communication and make displays that present their ideas with force and clarity. Since many students will be employed in careers which will require communication skills, competence in designing and making transparencies may be a practical vocational advantage. Thus, student practice in using overhead projection techniques can have long-range values as well as adding pleasure to the preparation of communication projects.

A further advantage of overhead projection is that transparencies can be prepared ahead of time, presented exactly when appropriate, and quickly removed when they have served their purpose. If valuable for future use, you may file them away in a drawer or binder and use them again.

Besides the interesting effects that can be achieved with transparencies, the overhead projector offers the advantage of being widely available and simple to operate. There are almost no

San Diego County (California) Schools

schools in the United States or Canada where a teacher or lecturer cannot have immediate access to facilities for overhead projection. Many schools even provide the projectors as permanent equipment in practically every classroom and lecture hall. And collections of prepared transparencies are among the usual resources in many media centers.

creating transparencies

It is easy for both you and your students to create transparencies for the overhead projector. A number of methods have been developed, and the necessary supplies and tools are sold by audiovisual dealers.

Images and designs on a projection screen can be controlled by transparencies which are made by the following techniques: (1) the use of opaque materials to *block out* the light and thus produce combinations of shadow and light; (2) the application of colored, translucent inks or plastic sheets which alter *color* of the light and consequently of the resultant image; (3) the placement of glass trays of mixed colored liquids on the stage of the projector to produce psychedelic effects; (4) the use of plastic materials to *polarize* the light and to suggest motion; and (5) the construction of elements that can be manipulated, such as dials and machine parts. To produce a variety of effective visuals, combinations of these techniques may be used.

Some teachers use the overhead projector chiefly as a substitute for the chalkboard. With a felt pen or other appropriate marker, they write on cellophane or other clear plastic material very much as they write on a chalkboard.

Charles Beseler Company

But many other processes have been developed for creating transparencies; to stop with using a felt pen and a piece of plastic is to miss some of the most exciting experiences possible with the overhead projector. Additional processes range from simple techniques needing no special skills at all to those requiring special equipment and considerable proficiency in using it.

But, before discussing these specific techniques, there is one general principle having to do with transparency format that we wish to bring to your attention.

Transparencies for overhead projection should normally be oriented horizontally. There are several reasons for this. One is that classrooms often have low ceilings; thus, if transparencies are oriented vertically, lower portions of the images are likely to fall below the normal viewer's eye level and be lost in a forest of heads in front of the screen. Further, the horizontal format is less subject to keystoning—a screen image wider at the top than at the bottom. The keystone effect can be corrected by tilting the top of the screen forward to ensure that the projected image falls on the screen at a 90-degree angle.

If the original materials you use, such as those from printed pages, do come in a vertical format, there is one good way to convert them to use as overhead transparencies: divide the material into *two* transparencies by cutting apart and remounting or relettering the material in a horizontal format. Keep this principle in mind as you continue now to explore various techniques for preparing transparencies.

transparencies by direct processes

Various simple markers may be used for the production of transparencies:

⊗ FELT PENS AND PENCILS The easiest way to make a transparency is to write or draw directly on a sheet of clear acetate using one or more colored felt pens. Fine-tipped felt pens generally produce the most satisfactory transparencies because the width of lines and the thickness of letters can be controlled. The fluid in many of these felt pens is *water-based* and can be removed with moistened cotton or cloth. If you use felt pens containing *permanent* ink, you must also use a plastic cleaner or a solvent such as lighter fluid to remove markings. Patience and a certain amount of luck may also be important!

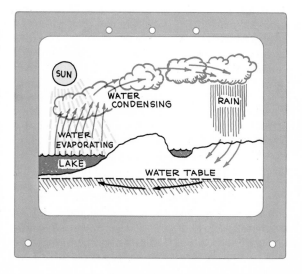

Since many felt pen inks are translucent—i.e., they permit light to pass through them—each stroke or overlapping mark produces color. The result, when projected, may appear uneven and smeared. It is therefore better, when coloring areas, to apply small dots or crosshatched or slanted lines, producing a stippled or textured effect, than to attempt a solid color.

Wax-based pencils, also called *grease* pencils, can be used for writing and drawing on acetate. Some of these pencils can be sharpened. Some are opaque, producing black images on the screen; others make colored markings that project in color. All wax pencil marks can be removed by rubbing with a dry cloth.

To supply presentation details and to add color to transparencies, both felt pens and audiovisual pencils can be used. Although many companies now produce both water-based and permanent translucent felt pens and pencils in a variety of colors, especially for use on overhead transparencies, it is highly desirable to test these articles before purchasing them in quantity to be sure that they perform to your satisfaction with the type of plastic you use. Dealers in audiovisual materials and equipment usually supply pencils and felt pens of satisfactory quality.

When you use transparencies repeatedly, adding and removing inks or wax pencils, your markings may tint the plastic permanently or produce abrasions that spoil the original material. To avoid this problem, place on top of a valuable transparency a clear plastic sheet and make all your marks on that. Later, you can clean this sheet or dispose of it, thus keeping the original transparency in perfect condition.

Your overhead projector may be equipped with an attachment that carries a roll of clear plastic upon which you may write or draw with inks or grease pencils. This roll may be advanced or rewound by means of hand-operated cranks. Prepared transparencies may be protected by placing them under the movable plastic; with this arrangement, your writing may be superimposed on the image of the basal plastic without damaging it. Some teachers find it a time-saving step to make sequences of illustrations, graphs, or outlines on these plastic rolls and to preserve them for repeated use.

This was typed on an elite typewriter.

This was typed on a pica typewriter

This was typed on a primary typewriter

IBM EXECUTIVE

MID-CENTURY TYPE FACE

MID-CENTURY EXPANDED

IBM SELECTRIC

ORATOR TYPE FACE ON 10 POINT

ORATOR ON 12 POINT

⊗ USING THE TYPEWRITER A large transparency containing a simple outline, an identification of a series of steps in some process, a list, or other verbal information can be produced easily with a typewriter. One way is to first type the information on white paper and then reproduce it by one of the processes described later.

However, a special clear acetate called *Type-On* film may be used directly in the typewriter. When an impression is made on it, the result is a sharp, etched image that projects well. Besides typing, impressions of ball-point pens and other writing and drawing instruments used on this film imprint and project well.

When typing or lettering transparencies, making the lettering sufficiently large is very important. Experience has shown that for good reading, lettering on the transparency itself should be at least $\frac{1}{4}$ of an inch high. Whatever kind of lettering you use, be sure to test it for legibility under conditions of actual use and also by reading the original material at a distance of from 5 to 6 feet.

⊗ MASKING AND SILHOUETTES Any opaque material placed on the lighted stage of an overhead projector will show as a shadow on the screen. The shadow itself may be useful either as a mask during a presentation or as a silhouette to represent the subject of a presentation. Such shadows, used as masks, cover portions of a transparency to be temporarily concealed from view. When you wish to make progressive disclosures of tables, drawings, or printed or typed materials, lay a sheet of paper across the transparency and gradually move it downward to reveal the covered material. Onionskin paper is good for this purpose since, though it casts a shadow on the screen, the material on the transparency can be seen through it.

Used simply as a silhouette, a projected shadow may also become the subject of a presentation. Leaves, fabrics, machine parts, or insect body parts, for example, suggest a few of the real things whose shadows may be studied in this way. Other real objects, such as insect wings, may reveal both opaque and translucent structures which, when projected, can be examined in detail.

⊗ TRANSPARENT SHAPES AND OBJECTS Colored plastic cutout pieces or geometric shapes can be moved around on the stage to form various arrangements. This student manipulates translucent plastic shapes to illustrate color relationships.

San Diego (California) City Schools

Translucent arrows, stars, brackets, and special numbers, as well as colored tapes, either solid or patterned, may be purchased for such use.

Transparent rules, gauges and scales, and meter dials can be enlarged on an overhead projector to permit everyone in a class to observe readings at the same time. Also, transparent dishes can be placed on the projector stage, and chemical reactions or color-mixing experiments can be performed for all to see. These things suggest that the projection surface of the overhead projector can literally become a stage on which all types of materials and objects may be manipulated.

indirect, intermediate processes

Several techniques for producing large transparencies require the use of intermediate materials. These indirect processes permit you to reproduce more than one copy of a transparency and to achieve numerous presentation effects. For example, you can develop sequential displays—*overlays*—in which additional transparencies are superimposed over a base. Note also, in the following descriptions, that directly and indirectly prepared materials may be combined in various ways.

⊗ HEAT-PROCESS TRANSPARENCIES Original materials—typed, written, drawn, or taken from a printed page—may be transferred to film by using a Thermo-Fax copy machine, a Masterfax unit, or any similar thermographic (heat-process) machine having an infrared light source to expose the film to original material. In only a few seconds the original, when placed in contact with a sheet of heat-sensitive film and run through the machine, will produce a projection-ready transparency. Depending on the film used, the image may be (1) black on a clear background, (2) black on any one of a number of colors, (3) in negative form (clear lines on a black background), or (4) colored lines on a black background.

However, one characteristic of the heat process must be kept in mind. The marks you make on the original must be added with heat-absorbing material —india ink, a soft lead pencil, such as a test-marker pencil, a good typewriter ribbon, or black printing ink. These marks will absorb heat from the infrared light of the machine, and the resulting temperature increase will affect the film to form the image upon it. *Ordinary ball-point pens, colored printing inks, or spirit-duplicator copies (purple impression on paper) will not reproduce by this process.*

An important advantage of this heat-process method is its speed. Once the original material is ready, only a few seconds are required to make the transparency. Also, the original is not affected, and you may prepare as many copies of transparency as you need. The process is as simple as putting the master and the plastic sheet in the machine.

San Diego (California) City Schools

San Diego (California) City Schools

A number of techniques provide numerous effects with this process. Material may be excised simply by using a sharp knife to cut out unwanted portions. New or different material may be attached to the master by double-faced plastic gummed tape, but be sure the tape does not extend to the edges of the inserts. Various illustrations printed in black ink can be used to add visuals to heat transparency masters; any type of line drawing material printed with carbon ink will serve. Clip art may be purchased for this purpose. Basic transparencies can be supplemented by sketches, symbols, swatches of color, or underlines in color. The process is illustrated in the picture story to the right.

HEAT-PROCESS TRANSPARENCIES

Copying machines originally designed for office use may provide simple, convenient means of producing projection materials. To make transparencies, special transparent film is obtainable for heat-process duplicating machines.

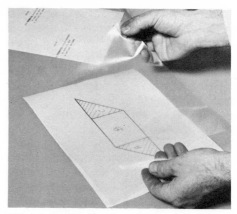

Place the projection film with the notch in the top right corner on top of the original drawing.

Original materials for the heat process may be printed pages or, as shown here, illustrations drawn with carbon pencils.

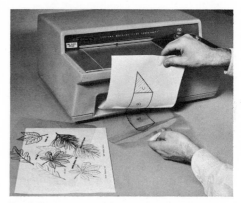

Feed the two materials together into the top of the machine. Separate them as they come out the bottom. The transparency is now ready for mounting (*below*) and projection.

Turn on the copy machine. Set the timer control on "low" or "slow" speed. (In some models an indicator may be set opposite a white mark for making transparencies.)

⊗ SPIRIT-DUPLICATOR MASTERS At times when you prepare duplicated paper materials to distribute to your class, you may also wish to make a transparency of them. The procedure involved is a simple one. The transparency is made from the same spirit master as the paper copies. (See Reference Section 2, "Duplicating Processes.") After first running several sheets of paper through the spirit duplicator to moisten the master, feed in a sheet of finely etched (frosted or matte) acetate plastic, with the etched side up. The acetate will pick up the carbon marks from the master, just as paper does, producing a translucent transparency on the plastic in the colors of the original duplicating master. To improve screen brilliance of the image and to fix it after it is dry, spray the etched film surface with a thin coating of clear plastic, such as Krylon. Remember that, to project properly, the material must fit within the frame-mount opening. Also with such items, give special attention to meeting standards of type size and general legibility.

⊗ DIAZO TRANSPARENCIES The diazo or ammonia process, as it is commonly called, is a versatile intermediate-step process for making transparencies. With this method, no special technical knowledge is required to prepare effective color transparencies or transparency overlays. The basic procedure follows:

First, prepare the master copy of the material on high quality *tracing paper* or on a clear or matte acetate which is *translucent.* It is essential that light pass through the copy. The markings you put

on the paper must be opaque: use india ink, *some* black felt pens, heavy pencil marks (again, test-marker pencils are especially good for this), or other markings that prevent light from passing through areas where screen images are expected to occur.

Second, select a sheet of diazo film of appropriate color. The film, coated with a dye (not visible until after development) is available in many colors.

One company even makes a single diazo film that is sensitive to a number of different colors. Place the diazo film against the drawing on the tracing-paper master, and lay the two in an ultraviolet-light exposing unit in this order: (1) light source, (2) tracing paper or acetate master, (3) film. Set the timer and expose to the ultraviolet light. Exposure time is

critical. Length may be determined by instructions accompanying the machine, instructions accompanying the diazo film, or by making a test strip (or several test strips if necessary) before printing an entire acetate. Make test strips about 1½ by 5 inches cut from a full sheet. Even when instructions are clear, the use of test strips saves money and materials and helps you achieve satisfactory results; this procedure is equally useful when using the heat process described earlier.

Although sunlight may seem to be a good source of ultraviolet light for exposing the master and diazo film, it is not actually very reliable for critical exposure. Various types of exposing units are included in the normal complement of production equipment in school and district educational media centers.

Third, develop the exposed sheet of diazo film in ammonia vapor (chemical ammonium hydroxide) in a large jar or in a diazo machine. Opaque areas and marks on the master drawing block the ultraviolet light from reaching the film; now, in the ammonia vapor, the colored image will appear in those same areas. For the development step of the diazo process, an ordinary wide-mouthed gallon pickle jar, with a small plastic cup in the bottom to hold two or three tablespoons of ammonia liquid, can be substituted for more expensive equipment. Be sure to keep the jar covered except when you insert or remove the film.

San Jose (California) State University

After development is complete, the transparency is ready to be mounted and used.

reflex copying processes

Early equipment for reflecting images from paper to foils for transparency production used a wet process, but more recently developed dry-process machines are much more convenient and are generally available.

Of the several ways to reflect images from your original art or lettering, or from printed materials, the electrostatic process is the principal one. The Xerox Corporation and A. B. Dick Company are among the manufacturers of electrostatic equipment. Exact instructions for equipment operation are issued by each manufacturer, and the instructions must be followed with care. Further, there are numerous manufacturers of transparency materials, not only for the reflex (reflecting) process, but for heat-transfer, diazo, and other processes. The

evolving technology related to electrostatic and other duplicating processes indicates the need for you to check continually on new developments and new sources of both materials and equipment.

With the electrostatic reflex process, a specially coated, electrically charged, light-sensitive film is used. The paper or page to be reproduced is exposed to light, which is reflected from the blank parts of the surface (but not from the printed or drawn lines and words) to the electrostatic film. Where the light strikes the paper, the coating on the film is discharged. The remaining charged areas—where the image will be—then collect a charged toner, a fine black powder, which, upon deposit, results in a visible opaque image. This process produces a good-quality transparency. Most office-type electrostatic copy machines can be used to prepare projection transparencies.

picture-transfer lift-process

Transparencies can also be made by a transfer process from pictures printed on clay-coated paper. (Because this process destroys the printed page, be sure to use only expendable pictures.) To determine whether the paper is clay-coated, first rub a dampened finger on the clear page margin. If a white, chalky residue remains on your finger, you will be able to lift the picture from the paper. There are several ways to do this, and one that requires no mechanical equipment. In the latter case, place the picture face down on the adhesive side of clear, adhesive-backed acetate shelf paper. Con-Tac brand, found in most houseware stores, works well. To ensure a tight seal between picture and acetate, press the two together with a hand roller or rub them with a comb covered by a handkerchief or with some similar implement. Then soak the picture and acetate in cool water. After a few minutes, peel away the (picture) paper from the acetate, upon which the ink of the picture will now remain. Remove any remaining clay residue from the picture by swabbing it with wet cotton. Dry the transparency thoroughly, protect the inked surface, and spray the image with clear plastic to make it more translucent. Mount the transferred picture just as you would any other transparency.

Either a *laminating machine* or a *dry-mount press* may be used to make picture lifts. The laminating machine consists of hard-rubber rollers that exert strong pressure on materials passed between them; it provides an excellent means of putting pressure on adhesive-backed acetate placed on the face of a picture. Another device, a dry-mount press, which exerts both heat and pressure on flat materials, may be used with special film (Transparafilm) to make lifts. Once the acetate or film is sealed to the picture, the procedure for making a transparency follows that described above.

THE DRY-MOUNT PRESS

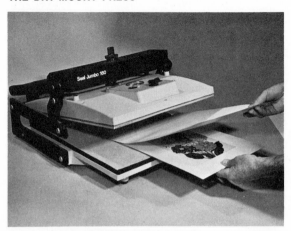

THE LAMINATING PRESS

LIFT PROCESS: STEP BY STEP

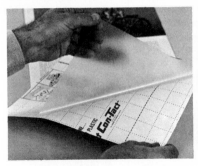

1. In this example, clear, adhesive-backed shelf paper is used to make a transparency.

2. Place the acetate on the face of the picture and rub the surface to ensure good adhesion of the acetate to the picture.

3. Then, soak the picture acetate in water; adding a tablespoon of enzyme detergent to the water improves the action.

4. After a thorough soaking, peel the paper from the picture.

5. With moist cotton, rub off the clay coating that adheres to the picture.

6. After the picture lift dries, spray it with clear plastic spray to make it more translucent and to protect it. It is now ready for use.

other special techniques

Several special techniques are suggested as means of helping you to improve your proficiency in producing and handling large transparencies, including (1) changing their size, (2) adding color, (3) combining images in overlays, (4) selecting a preferred format, (5) preparing handouts to accompany transparencies, and (6) filing and storing them.

⊗ CHANGING SIZES OF TRANSPARENCIES In the preceding discussions of transparency production techniques, it has been assumed that original materials have all been of the proper size for projec-

tion. But this is not always the case. Useful materials are sometimes too large or too small to be transferred directly to acetate and projected. To overcome this problem, another process step is required: enlarging or reducing original materials through a photographic process. But because most teachers may not have ready access to photographic equipment and materials, the work may have to be done through the school or school district educational media facility, or by a commercial photographer. In either case, give clear directions about what you want done. Remember that, when ready for projection, the working area of a large transparency should measure not more than 8 by 10 inches.

As an example, below, the small diagram on the upper right of the page would be useful to show a class. To test the size of the small diagram in relation to the entire transparency area, lay a mounting frame over the page.

A high-contrast photographic negative, below, is made of the diagram. Note the negative on the right which shows the result of opaquing to eliminate unwanted printing. This negative is used to make the enlargement for the transparency.

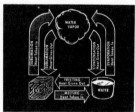

Of course, appropriate colors symbolic of heat transfer (heating or cooling) may be added by using either felt-tip pen techniques or overlays of pieces of transparent colored film.

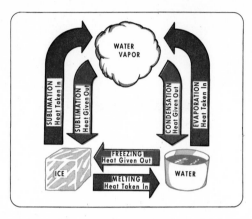

In changing the size of original materials, the photographer will be able to select from a variety of film types according to your need; high-contrast film for black-and-white line drawings or lettering; continuous-tone films for photographs or other materials with a wide range of textures or colors. Note, however, if you ask for a large transparency in *full color,* get a preliminary cost estimate for they are very expensive.

Sometimes only one item to be shown in a transparency needs to be changed in size to match other material that is satisfactory; if so, it is necessary to photograph only the item that must be changed in size. Then either an enlarged or a reduced print of the photograph or art may be pasted into the layout that will be used as the reproduction master.

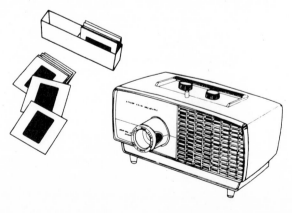

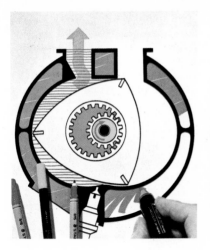
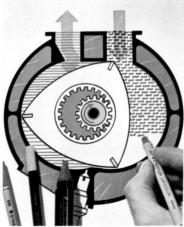
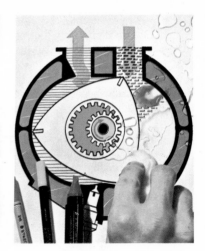

J. S. Staedtler Company, Inc., Montville, N.J.

⊗ USING COLOR Color in transparencies makes them attractive and interesting and generally adds to their value by emphasizing or differentiating areas, content, or categories represented in them. Special audiovisual felt pens and pencils in various colors have been recommended for this purpose. Shown above are colors applied with felt pens—both permanent and washable—which may be used on a single transparency to provide markings for effective communication. The blue and red are permanent colors; the green is washable and can be removed with water and a piece of cotton.

Remember, it is not always necessary to produce works of art to communicate effectively; adding underlines, numerals, asterisks, dots, or other eye directors may be all that you need to improve your transparencies. Lines in colors can be added to enclose areas for discussion or to trace significant elements in a complex diagram. Colored numbers or letters, conspicuously placed, will help to guide viewers during discussions or to identify areas referred to in test questions.

Almost as simple to use as felt-tip pens and pencils are the several types of adhesive-backed films and translucent letters, lines, and symbols that are available in black or in various colors. The sheets can be cut to desired sizes and shapes and adhered to large or small areas as needed. Although these materials require some manipulation in order to be attached to acetate, they add to the professional quality, legibility, and attractiveness of the finished product. Audiovisual dealers and art supply stores usually have them in stock.

And, of course, the acetate for printing base transparencies and overlays may be selected from a wide choice of colors. When several colors are required for a single transparency, different sections of the visual may be made on different colors of acetate and the finished acetates assembled in proper register as a sandwich for projection. Also there are special processes that you may wish to investigate which permit making several colors appear on the *same* acetate. Projected images from negative transparencies—those with black backgrounds and clear lettering or lines—can be tinted in colors by using liquid dyes that are easily applied.

⊗ OVERLAYS: COMBINING IMAGES Among the many advantages of transparency projection is the possibility of using the overlay technique. In such cases, the base acetate carries initially required information and is shown first. This base is usually mounted on one side of the mask, which rests on the projector stage. Subsequently, additional acetate sheets, each containing more information, are placed over this base. Thus, the screen projection shows new information or directive symbols as the sheets are flipped over, in sequence.

To achieve accurate registration with such overlays, position each sheet so that it lines up properly with those beneath. In preparing your original drawings, use a separate drawing for each overlay. Line up the sheet for each overlay by matching corners, or by putting X marks outside the projection area on each master before processing it. Be sure the film used to make the actual transparency matches the corners of the original drawing, or that the X marks are in register.

Overlays are normally attached to a transparency frame (mask) with short strips of masking or cellophane tape; audiovisual dealers also provide a mylar foil tape of great durability for transparency-overlay hinges.

The effects of transparent overlays can be further enhanced by using simple masks, appropriately cut and hinged, to cover or to reveal transparency areas, thus putting control of audience attention in the hands of the operator.

Still another related technique—*progressive disclosure*—may be aided by using either hinged strips or a sheet of paper which may be pulled down to control information exposure step-by-step. For some material, a sheet of paper with a single hole of proper shape and size may be pulled down or across a transparency to expose only one word, phrase, or symbol at a time, as shown on page (131).

MAKING OVERLAY TRANSPARENCIES: STEP BY STEP

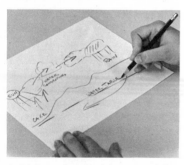

1. Make a sketch of the content for the transparency; decide which parts will be the base and which will be used for each overlay.

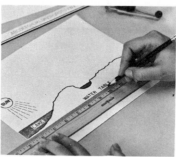

2. Prepare a master drawing for each separate part.

3. In two corners on each master, make register marks that match marks previously put on the sketch.

5. Mount each sheet in register, with the base under the frame and the overlays on top.

4. Prepare the transparency from each master.

6. The finished transparency, with base and two foils in position for projection. Red and blue are hinged on left and right; black foil is mounted on back of frame. *From* Transparency Masters (*to accompany this text*), *McGraw-Hill Book Company, 1977.*

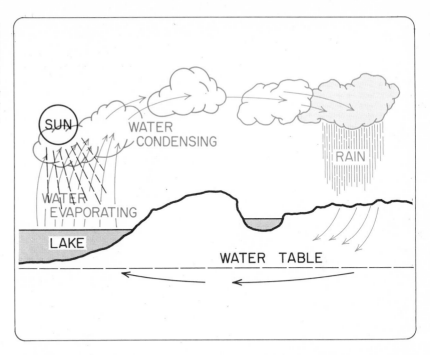

⊗ FILING AND STORING TRANSPARENCIES Transparencies are normally mounted in frames with 10-by-12-inch outside dimensions. This size fits drawers in standard filing cabinets. If such cabinets are not available to you, you may use boxes or transfer cases of approximately the same size. Transparency masters can be filed in manila folders, available for use when new transparencies are required or when copies are to be duplicated.

Organize your transparencies under appropriate subject headings, unit titles, or lesson topics; put dividers, marked appropriately, at the beginning of each category. Do whatever you can to facilitate locating transparencies.

As your collection grows, make a 3-by-5-inch index card for each transparency. Give a brief description of its content and intended use, and make note of activities or comments that may accompany showing the transparency.

⊗ HANDOUTS TO ACCOMPANY OVERHEAD PROJECTION With the duplicating processes which are readily available and easy to use, you may find it convenient to provide your students with duplicates, perhaps modified, of materials shown on the overhead projector. This arrangement will permit them to take accurate notes—perhaps on the same sheet where the reproduced transparency appears.

Charles Beseler Company

TRANSPARENCIES THAT MOVE

The illusion of motion can be created by applying light-polarizing materials directly to the surface of the transparency. The polarizing material selected, and its orientation, determine the pattern and apparent direction of motion that will be seen on the projection screen.

A rotating disk of polarizing material is placed in the light beam to cause the effects which can suggest turning wheels, flowing fluids, moving clouds, or, as shown below, radiating sound waves from a tuning fork to a human ear.

3M Company

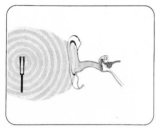

Representations of types of patterns for polarized motion

Simulated samples of polarized material and production kit by American Polarizers, Inc.

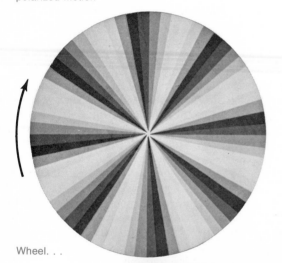

Wheel. . .

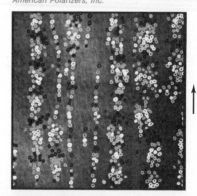

Rising bubbles. . .

Linear motion. . .

**CREATING
TRANSPARENCIES**

Teachers who make and test their own transparencies demonstrate the ease with which the work can be done and the motivation that is developed in the process. *3M Company.*

TRANSPARENCIES THAT CAN BE MANIPULATED

Many transparencies that permit manipulation can be constructed with simple techniques, as illustrated directly below. The abacus and the two-cycle engine on the right are examples of commercially manufactured transparencies. Other typical commercial examples (not shown) are compass rose, slide rules, and calipers.

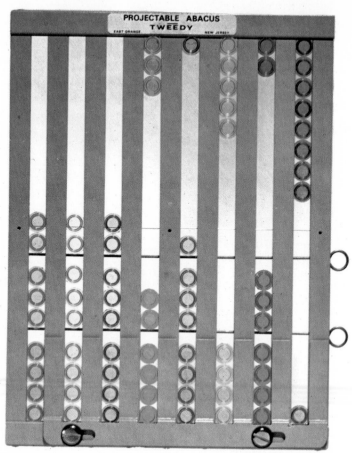

Tweedy Transparencies

3M Company

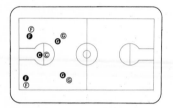

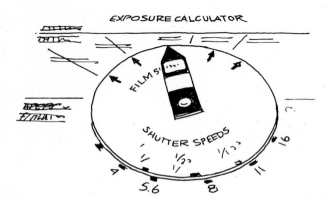

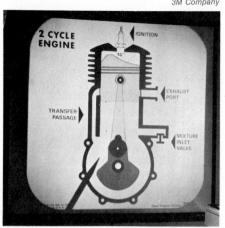

commercially produced transparencies

Many publishers have developed transparency sets for almost every curriculum area and for all grade levels, including college. Some publishers of textbooks produce transparencies correlated with their books and with other audiovisual materials to be used for group or independent study. In the fields of science, mathematics, geography, and vocational education, extensive sets of transparencies are available. In fact, such a wide selection of transparencies is produced by commercial companies that one entire volume of the NICEM catalog is devoted to transparencies useful in teaching.

LaMesa-Spring Valley (California) School District

forms of prepared transparencies

Transparencies are available in finished, mounted form, either in black and white or in color, complete with overlays or masks, as required. Some producers have designed special hinged frames and other devices to facilitate using their transparencies, especially those with overlays.

Transparency masters can be purchased in printed form in packages or in loose-leaf binders. With these master sheets, you can prepare transparencies in your own school by taking advantage of one of the several different reproduction processes that are discussed in this chapter. Teachers sometimes modify these prepared transparencies to adapt them to their own special needs: they excise information to simplify them or add data or captions on the masters or on locally prepared overlays. Commercially prepared masters have the advantage of having been designed by experienced teachers; they have been tested in use; and large numbers of complete transparencies can be prepared from a single master and used throughout the school or school system.

No matter what the original source for transparencies, it is highly desirable that the media center of a school maintain a transparency file that is suitably cataloged and easily accessible to both students and teachers for class use or for independent study.

criteria for choosing prepared transparencies

It is your responsibility as a teacher to apply such criteria as the following in recommending the purchase of prepared transparencies, in addition to the more general criteria mentioned in Chapter 4.

⊗ Does the treatment of the subject lend itself especially to the large transparency form, as opposed to a poster, chart, mounted picture, slide, filmstrip, or other medium?

⊗ Does the display encourage utilization of transparency techniques described in this chapter?

⊗ Are the transparencies technically satisfactory: with bold graphics, lettering of adequate size and style, appropriate coloring, and secure mounting?

⊗ Will the transparency design encourage student participation during the lesson?

⊗ Is too much information crowded into the image for effective communication? Could the content be divided into several more simple transparencies?

⊗ Would overlays for sequential presentation of the content improve the effectiveness of the transparency?

⊗ Would the transparency encourage students to make transparencies for their own presentations in your class or in their other classes?

COMMERCIAL COLOR TRANSPARENCIES

Publishing firms and other companies that specialize in producing transparencies provide materials for the overhead projector for practically every subject taught in schools. Such transparencies range from simple printed materials on a base film to elaborate color presentations with several overlays. *Scott Education Division.*

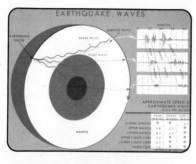

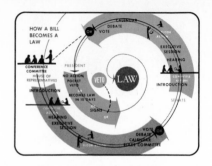

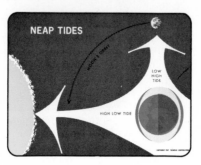

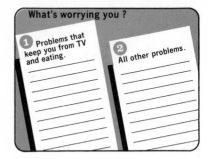

teaching with transparencies

Several general pointers about teaching with large transparencies may improve your performance with them:

⊗ When showing pictures or diagrams, consider using a pointer (a sharpened pencil will do) to direct attention to details. Your pencil will appear as a moving shadow on the screen. It is a good idea to place the pointer, or pencil lead, directly *on* the transparency; otherwise even slight hand movements will appear on the screen in exaggerated scale.

⊗ Add details to transparencies with felt pen or wax-based pencil—before or during projection. But, as mentioned earlier, unless you intend that these additions be permanent, be sure to apply only marks that can be removed with soft cloth or special transparency cleaner. To be absolutely sure no damage is done to a valuable transparency, you may wish to cover it with a clear plastic sheet before adding such marks.

⊗ Turn off the projector when you change transparencies or when you are not projecting at all; avoid a glare that will distract the viewers.

⊗ Use your own overlays to add new material to commercial transparencies. You may make a transparency with properly located arrows, underlines, words, or any other symbols you wish, and then place this overlay over the original.

⊗ When you purchase printed masters designed to be converted into transparencies locally, you may exercise any of several options to make additions to or to eliminate materials from them. For instance, if you find that the vocabulary used on a technical transparency is too advanced for your students, simply change the words to ones you prefer. Or, if you wish to add underlines, arrows, circles, or other guides directly on the base transparency, use carbon ink or a carbon (test marker) pencil and add them to the master. You may also wish to cut up the master and to rearrange the material, to make two transparencies from one master, or to eliminate parts not relevant to your purposes. In such cases, if you do not wish to deface the original, first use a copying machine to make a paper duplicate and then do your editing work on that.

⊗ When selecting printed masters that you propose to process locally for your classes, consider whether the overlays recommended by the producer are actually essential. Sometimes, instead of using separate overlays, you may prefer to devise ways to incorporate items of information on the master, hiding them with masks which you may later remove as appropriate during your classroom presentation. This procedure may save material and reduce processing time and cost without in any way reducing the teaching value of the original transparencies.

A few *specific* applications of overhead projection at various school levels and in various subject areas provide further evidence of the versatility and value of the system:

⊗ In *primary grades,* simple opaque objects such as leaves or cutout paper shapes can be placed directly on the projector stage and enlarged on the screen to stimulate imaginations and encourage discussion. Children themselves can make silhouettes and project them as they tell stories or share experiences.

⊗ In *English* composition lessons, student themes or writing exercises can be reproduced on film by means of the heat or photocopy process and projected. The teacher and students can then analyze the writing in relation to the objectives of the assignment.

⊗ In *arithmetic,* students can present homework for analysis by writing their problems on acetate, using grease pencils; or if they write on paper with soft lead pencils, the heat process may be used.

⊗ In *geometry* and *trignometry,* two- and three-dimensional diagrams can be presented step-by-step, using transparencies composed of color overlays. Geometric theorems and even more complicated problems can be separated into components and presented systematically in segments. In mathematical and technical subjects, translucent plastic rulers, compasses, and slide rules are used to clarify the points of visualized explanations.

⊗ In *physical education,* a group may analyze plays and game procedures through the use of colored plastic or opaque symbols that are moved about on a transparency containing an outline of the playing field.

⊗ In activities simulating *political* or *organizational work,* a secretary can use a cellophane roll (accompanying most projectors) or blank acetate sheets to write nominations, lists, motions for consideration, or important discussion points for all to see and discuss.

⊗ In *art* classes, by using high-quality, full-color commercial transparencies of works of art, clear plastic overlays, and a grease pencil, students may show lines of movement or areas of especially interesting composition.

⊗ In a *music* class, by the use of overlays, a teacher or a student may project three-part music—each part individually or parts in combination.

Finally, with a collection of prepared transparencies filed and cataloged in the media center, students and teachers may construct presentations that will be both instructive and interesting. The collection should include materials from all subject areas in the curriculum. Among the transparencies, commercially prepared masters of appropriate content can be included, along with the equipment and materials for processing them.

San Diego (California) County Schools

summary

The overhead projector has wide acceptance at all levels of teaching. Combining as it does the advantages of a large and brilliant screen image (even in a lighted classroom), relatively easy and inexpensive local production of transparencies, and flexibility of use, it seems assured of expanding applications.

Many publishers of text and reference books offer correlated packets of transparencies to facilitate classroom presentations. These commercially prepared transparencies range from simple verbal materials to a carefully prepared sequence of overlays of graphic or photographic visuals. In some instances, producers provide packages of reproducible masters that permit local production of transparencies and the addition of symbols or other information desired.

A particularly important advantage of overhead projection is the convenience with which materials can be produced. The many techniques range from direct processes, such as projecting real objects on the screen or making simple illustrations with felt pens or grease pencils, to indirect processes which involve the preparation of master foils and overlays by heat, diazo, electrostatic, or lift methods. Files of transparencies, as well as of original masters to produce additional duplicate transparencies, are found frequently in school resource centers as well as in the classrooms or offices of individual teachers. Basically, the system of overhead projection is a flexible medium that invites clear communication by the use of light and thereby provides creative satisfaction for both students and teachers.

Finally, please remember that other sections of the text have material relevant to this chapter: Chapter 5, Displaying, Chapter 6, Graphic Materials, and Reference Section 1, Equipment Operation.

8 PHOTOGRAPHY

chapter purposes

⊗ To interest you and your students in participating in photographic activities for both learning and enjoyment.

⊗ To encourage imaginative applications of photography in the school curriculum.

⊗ To highlight the importance of the language of pictures and show ways photography may improve the visual literacy of your students.

⊗ To present fundamental procedures that you and your students may follow to plan and produce simple still-photo and motion-picture subjects.

⊗ To suggest ways in which specialized techniques, such as copying and close-up photography, contribute to teaching and learning.

National Geographic Society Educational Services

Teachers who have experimented with photography in their classes often talk enthusiastically about their experiences. Students like photography; it leads to numerous creative activities. Still- and motion-picture cameras which are both easy to operate and inexpensive have put photography within reach of nearly everyone—at almost any school level. Many motion-picture activities in schools and colleges are done with video tape. Numerous photographic techniques are applicable in either film or video production, and therefore much information in this chapter will not be repeated in Chapter 12, Television.

Those who have used photography in ways de-

San Diego (California) City Schools

getting acquainted with photography

Before actually taking pictures, you and your students can do many things to get acquainted with photography. Some teachers have organized unit studies about visual communication to help students learn how to *talk with a camera* and to build proficiencies of visual literacy generally. Such activities as the following have been recommended:[1]

⊗ Devote some time in class to discussing the subject of visual communication to see how much your students already know about it. Identify aspects in which they display greatest interest. Use this information in planning a series of visual communication activities, such as those that follow.

⊗ Obtain and use some of the visual literacy materials in filmstrip and film form distributed through the Association for Educational Communications and Technology.

⊗ Provide opportunities for students to add meaning to or increase the instructional usefulness of pictorial materials by writing picture captions or developing interpretive or creative statements about them. In carrying out this exercise, project 2- by 2-inch slides, or hand around a number of black-and-white photoprints, that relate to a single subject or theme, and perhaps limit the time to be spent in writing comments for each. After the exercise is completed, compare the statements and analyze what they seem to reveal about comprehension, interpretation, and the interest generated by the pictures.

⊗ Ask several student teams to select topics for a project and to design and produce pictorial charts, photographic collages, or similar products that represent meaningful visual stories. Pictures can come from magazines and newspapers, old books, old postcard collections, calendars, or any convenient source.

scribed in this chapter do so for several reasons. They find that:

⊗ Taking good pictures is an activity that is stimulating to teachers and students alike. But covering an event with either a still- or a motion-picture camera for the purpose of creating a picture story is even more exciting and inevitably improves visual literacy.

⊗ Photography provides a good basis for achieving other important long-range instructional objectives. The *thinking* that must be done to solve problems related to photographic projects, for example, is especially valuable.

⊗ Providing students with opportunities to express themselves through photographic activities permits them to develop many new skills and to indulge creative interests in constructive ways.

⊗ The subjects students choose to photograph and the comments they make about them often reveal their attitudes and interests.

⊗ Finally, student and teacher photographic activities sometimes result in the production of instructional materials that have value for school use, such as single-concept films on local subjects, slide sets for orientation of new students, and photographic notebooks to guide independent study activities.

---

[1] A number of the suggestions included in this section have been freely adapted from materials originally developed by William R. Hanson, Medicine Hat, Alberta, Canada.

Edmonton (Alberta) Schools

⊗ Obtain and use copies of basic publications produced by various commercial organizations, such as Eastman Kodak Company. See especially such titles as *Producing Slides and Filmstrips, Planning and Producing Visual Aids, Movies with a Purpose, Improve Your Environment—Fight Pollution with Pictures,* and *How to Make Good Home Movies.*

⊗ To teach fundamental visual literacy skill, obtain and use some of the highly effective "Photo-Story Discovery Sets," each of which contains thirty or so small pictures to be arranged in various story sequences.

⊗ Invite local photography club members to your class. Ask them to demonstrate simple photographic equipment of types you and your students may expect to use and also to demonstrate the fundamentals of taking good pictures.

⊗ Ask students to bring to class some of their own photoprints; arrange them as a bulletin-board display, attaching to each a number for purposes of identification. As a learning exercise, ask students to vote for the ten best pictures in the display, applying criteria previously discussed.

⊗ Develop and use cardboard viewfinders as a preliminary to asking students to use real cameras. Cut a rectangle about 1½ by 2 inches, or a square about 1½ inches in each dimension. Invite students to practice holding viewfinders at different distances from one eye, and through this process to discover some of the basic elements of composition—framing, balance, center of interest, inclusion, exclusion, relative size, foreground, and background.

⊗ Accomplish similar goals by inviting students to cut rectangular holes (4 by 5 inches or 8 by 10 inches are good dimensions) in cardboard sheets and use them to frame magazine pictures for best

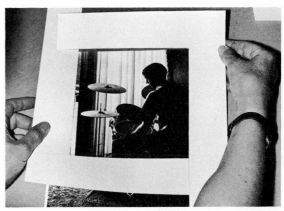

San Diego County (California) Schools

visual effects, taking into consideration balance, interest, and similar qualities.

⊗ Give students the precamera experience of drawing stick figures or other shapes on etched plastic film (similar to 16mm film leader). Principles of animation may be learned this way. Show students how to repeat basic figure shapes in several 16mm frames, varying leg or arm positions in each to suggest walking or dancing. When projected, the effect of motion is created. Other visual experiences of a nonmotion nature can be obtained by drawing and projecting filmstrips or 2- by 2-inch slides. (See further information on this process in Chapter 9, Still Pictures.)

Still other activities with real cameras are recommended as useful preliminaries to serious photography. For example:

⊗ Compare results of taking pictures—especially movies—when holding the camera in the hand rather than placing it on a tripod or monopod; experiment with results obtained at varied shutter speeds.

⊗ Try overexposing and underexposing pictures, or opening up the lens iris while simultaneously increasing the speeds of exposures.

⊗ When taking pictures of moving subjects, compare results of exposures made from oncoming and diagonal views with those made at right angles.

Even before you have developed skill, however, you may find ways to use photography to advantage. With access to a simple still camera—perhaps a Polaroid—you may use pictures to identify:

⊗ The "Citizen of the Week" for a bulletin-board display.

⊗ Students responsible for various weekly room chores.

⊗ Facial appearances and names of each member of the class.

informal photography

The second level of photographic activity in which you and your students are likely to engage (after the very elementary activities just mentioned) may be characterized as *informal,* or extemporaneous. With informal photography, you shoot without specific preparation, seeking to capture useful, interesting, or beautiful pictures with minimum advance preparation. This kind of photography permits you to *select* what you shoot and to think as you go with only general purposes in mind. It permits you to photograph events or situations as you encounter them without the restrictions of a formal script or shooting plan. In this way you have opportunities to shape picture taking to the emerging character of your production: a geographically organized travelogue; highlights of a day at the beach; a chronological field trip record; unanticipated effects of a sudden windstorm; the beauties, landmarks, and people encountered on a walking trip; or a selective documentation of events during an afternoon at the ball game.

Students provided with still- or motion-picture cameras and film may have photographic experiences like the following if they are given the assignment to shoot only what interests them:

⊗ A teacher used a field trip as the basis of an interesting informal photography experience. One Polaroid camera was distributed to each of the five groups in the class. Members of each group were to share one camera. The assignment was to shoot an experience record of people, animals, things, or activities of interest. Later the five sets of pictures were arranged as segments of a large bulletin-board display. One person from each group described the content of their pictures and explained the reasons why they were taken. The class then discussed the quality and style of the pictures displayed.

⊗ A somewhat more sophisticated application of informal photography was made by several members of a high school biology class who were called into service to aid waterfowl trapped in oil escaping from a foundering tanker. There was no time for planning what to shoot, and no script was available. Therefore, the photo crew simply capitalized upon opportunities by selectively shooting action as

it occurred. They took many close-ups of oil-soaked birds, including some that had already suffocated. They showed facial close-ups of students animatedly discussing the tragedies they were witnessing. As they proceeded, there was conversation about how the film might be edited to portray a logical sequence of events. Back at school, they shot several card titles and portions of a map to plot the place where the action had occurred. Then they showed the completed film in their own and other classes and also used it as part of a program at a local luncheon club. Eventually, they placed the film in the school media center for possible use in other biology classes. It was obvious that although the production was developed informally—that is, with only a very little prior planning—the finished film was useful, and the experience of producing it a valuable one for the students.

Paul Fusco from Multi-Media

Photograph by Peter Vadnai

structured photography

In contrast with the informal type of photography just discussed, structured photography follows a plan to achieve some predetermined purpose or goal. Planning, shooting, editing, and presenting 8mm and 16mm films, 2- by 2-inch slide sets, or paper photo-print series or photo essays provide unusual opportunities to develop functional, interesting experiences with structured photography. To produce something that visualizes a preplanned, structured story or demonstrates how to perform some procedure, you must follow through on a number of things:

⊗ Select a subject.
⊗ Describe the individuals who will use the production or to whom it will be shown.

⊗ State as clearly as possible what is expected to occur as a result of using the production.
⊗ Choose the production medium.
⊗ Plan the production content and develop a treatment storyboard.
⊗ Shoot required pictures, and record audio content if sound is used.
⊗ Edit the production.
⊗ Use and evaluate the production.

select a subject

At all levels and in all areas of the curriculum, good ideas for structured photographic subjects regularly occur to creative teachers and students. They stem naturally from classroom assignments, news events, community issues and problems, and field trips. The following represent a few such subject treatments that have proved successful:

⊗ METERS IN OUR FUTURE A sixth-grade class chose the title "Meters in Our Future" for a slide production to summarize a unit on the metric system. The question was: What changes will occur in our town when we start using the metric system? Will things be better or worse? Will we have problems? Photographs showed how new measurements would be made and old ones converted, what new signs would say, and some of the things people would do to accommodate to metric requirements.

⊗ THE HISTORY OF OUR TOWN Several members of a community college social studies class developed a slide presentation from copies of early newspapers, pictures of old buildings, cars, other relics of the past, and "then and now" photographs of early residents, all skillfully woven together to provide historical continuity. Most of the older pictures (a few in slide form) were contributed by residents of the community who appeared in them.

⊗ OUR SCHOOL—AND WELCOME TO IT A high school photography club accepted the assignment to produce a series of 2- by 2-inch slides about personalities, physical features, and the curriculum of their school. These were projected at an orientation assembly for freshmen and explained by students who were in each case involved with the activities and events portrayed.

⊗ WHAT I WOULD SAY IF I COULD SPEAK ENGLISH Three sixth-grade students produced a 2- by 2-inch slide presentation showing several characteristic views of each of a dozen or so animals, birds, and snakes of the local zoo. The sound track, recorded on tape, contained the voices of the student producers who role-played each animal shown. They identified each species, its original home, age, and favorite foods, and said a few things about its favorite zoo visitor—what it liked best about people.

⊗ POLLUTION CITY A high school group produced an 8mm film showing how the environment has been despoiled through extensive pollution.

**DOCUMENTING THE
ENVIRONMENT**

Society for Visual Education

Society for Visual Education

Included in it were sequences on air pollution (smokestacks, smog, commercial open burning, residential open burning, exhaust of automobiles and trucks, and exhaust from jet planes); water pollution (dead fish in a litterstrewn stream, trash along a river's edge, muck emerging from a storm sewer outlet, unchecked erosion, with mud being carried off after a sudden rainstorm); solid waste (unattractive auto graveyards, unscreened wrecker establishments, overturned garbage cans, examples of unauthorized dumping, highway littering); visual pollution (confusions of unattractive highway billboards, electric poles and wires galore, commercial signs in profusion); and noise pollution (visual scenes accompanied by documentary sound recordings of jackhammers at work, branch and clippings choppers in use, gigantic trucks, railroad whistles, fire engine and police sirens). The film received plaudits in a regional film festival and was an effective means of calling to the attention of the people in the community the need to improve conditions leading to pollution of life in the area.

consider the audience

Before proceeding too far with a media production, careful consideration should be given to the expectations, present understanding, and other characteristics of the members of an audience for whom the production is intended. What do they already know about the subject? How skilled are they in performing any tasks that are to be demonstrated?

What is their attitude toward the subject? Will it be necessary to motivate them—to cause them to see why the subject is important and worth their attention? Or may they be assumed to be highly interested?

What is likely to be the viewer's oral vocabulary level? Reading level? Are terms considered for use in the production likely to be misunderstood by viewers? How much will the audience know about various production techniques, such as those of multiscreen media—with combinations of motion and still pictures, audio tapes, and perhaps live narration, for example?

Are there other important questions that should be asked about the proposed audience?

STRUCTURED PHOTOGRAPHY

Formally structured photographic projects commence with a determination of purposes and proceed through stages of storyboarding, photographing, final editing, and showing.

College of DuPage (Illinois)

South Hills Catholic High School, Pittsburgh (Pennsylvania)

College of DuPage (Illinois)

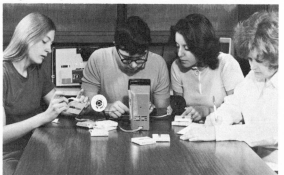

state objectives of creating and using the production

A structured photographic production is usually designed to achieve one or more purposes. To a significant extent, these purposes determine what the visual elements of the production will be and how they will interrelate.

Good reasons for engaging in structured photographic productions come readily to mind. You may simply wish to try something new—to stir interest among your students in an important but unfortunately dull or complex subject. A group project to produce a slide presentation or flat picture display may be just what you need to create a research focus for the study of local government or of recreational services provided in your community.

But while your own reasons for doing the production are significant and must of course be reflected in the arrangements and activities, there is the additional need for students to work with you in determining what they are expected to learn from the experience. For example, in the process of producing, presenting, and evaluating a slide set or other pictorial sequence, will they learn specific facts about the subject? How to think visually? How to *design* communication? What specific skills will they be expected to develop or improve? To what levels of competence?

South Hills Catholic High School, Pittsburgh (Pennsylvania)

PRODUCTION FORMATS

Any one of several different photographic production formats may be used to communicate visual messages.

San Diego (California) City Schools

San Diego County (California) Schools

Green Bay (Wisconsin) Schools

choose the production medium

The choice of medium to be used in your production will be based upon the audience, the subject matter to be included, and the conditions under which the production will be presented. You may choose any one of several forms: (1) *ordinary paper photo prints*—in various sizes, from contact (negative size) prints to enlargements up to 16 by 20 inches; (2) *2- by 2-inch slides,* usually in color; (3) *8mm* or *16mm motion pictures;* or (4) video tape. You may also add sound or provide accompanying printed comments on each. Techniques for producing pictures in all four of these forms are basically similar. Each provides similarly valuable experiences for your students—and you.

To make a firm decision about the format to be used for production, you will need answers to a number of questions:

⊗ Who will use the production—individuals, small groups, large groups, young students, adults?

⊗ Where will the production be used—in a classroom, a hallway, an auditorium, on television?

⊗ Do you have the special equipment, expert assistance, or financial resources you need to produce the subject in the format you choose? Or must you accept some feasible, even if less suitable, alternative?

⊗ If you decide to make your production on film, will you choose the 16mm or 8mm size? With the 16mm size you will spend more money on raw film, but your plan may require it.

Albert Bailey

⊗ Must the production be in color? Productions in color are more expensive than in black and white; color also requires additional skill on the part of camera operators and those responsible for lighting. But color heightens interest, adds realism, and often improves communication effectiveness.

Try to visualize what the effect would be upon the reader if the picture above had been printed in full color rather than in black and white. What specific additional learning might be expected from a color picture? When, if ever, might black and white be better than color for the same pictorial subject?

⊗ Will sound be needed? If so, you have a number of choices. You and your students may plan to present narration and sound effects live, in the form of a tape recording, or as a magnetic track accompanying the film. The techniques of any one of these arrangements are fairly simple; each teaches what is involved in supplementing and extending the communication power of photographic productions by adding appropriate voices, music, or sound

effects. (Technical details of sound recording are given in Reference Section 1.)

⊗ Must pictures be enlarged considerably to provide a suitable image size for large audiences?

⊗ Will the production be used on television?

⊗ Will one print (the original) be all that is required? If more will be needed, how many?

plan production content and treatment

Individuals or groups making films, slide sets, or photo-print series will be helped by doing advance planning *on paper*. Planning should save time and energy and help to ensure that all details are considered, that nothing significant is overlooked.

Production planning usually begins with developing an overview of the subject to identify its *parts*— its elements, the main segments with which you must pace the flow of the story. Such planning must also determine the sequence or arrangement of the several production segments as well as the order of information within each.

A good way to achieve these ends is to develop a storyboard of your production. Write down on separate index cards (4 by 6 inches is a good size)

details of what you have in mind for each still or motion-picture shot. Draw a rough sketch (or stage and make a Polaroid shot) of each key view or

---

Shot
No. ____ Sequence_____

               Shot content:

Special directions:

---

action to be photographed. Add notes to indicate (1) scene number; (2) type of shot—LS for long, MS for medium, CU for close-up, ECU for extreme close-up; (3) special directions for the photographer or director; and (4) the possible nature of any narration, dialogue, or sound effects you expect to add later or, in some cases, to record on the spot. Include separate cards for any titles or graphics that will appear in the production. Rearrange, delete,

SCRIPT SHORTHAND

In making a short script, as described in the text, you can save time and space by using some of the symbols below to indicate various shooting techniques.

L.S. — long shot

M.S. — medium shot

C.U. — close-up

$>$ — fade-out

$<$ — fade-in

— dissolve

H — high angle
(camera looking down)

L — low angle
(camera looking up)

R — reverse angle (camera looking in opposite direction from previous shot.)

revise, or add cards as appropriate until your full story flows.

You may decide to convert your storyboard cards into a shooting script. A shooting script is usually

Edmonton (Canada) Separate School Board

made in at least two scene-numbered columns—one indicating the type of shot (long, medium, close-up, extreme close-up), the other describing the content and arrangement of the scene. If the picture sequence is to be used with a live or tape-recorded sound track, add two other columns—one containing the dialogue or narration, the other directions for types of sound (music and sound effects).

San Diego County (California) Schools

course, if you do this, a fast shutter setting is required.

⊗ Remember that raw film is often the least expensive item in a production budget. To provide an adequate choice of pictures or motion-picture footage at the final editing stage, bracketing of exposures is recommended. Take two or three shots of the same scene, each with some variation in lens openings or shutter speed.

⊗ When shooting motion pictures, use the camera to best advantage. Avoid excessive panning; follow action; pan slowly, with your camera on a tripod. Take scenes of proper length. If your storyboard item for the scene is a good one, the footage indicated for it should be about right.

⊗ Remember, too, that to increase audience interest, you should frequently shoot long, medium, and close views of the same action or scenes. Close-ups are particularly essential; but amateur photographers often overlook them.

Further guidance in planning for the shooting of both still and motion pictures is given in Reference Section 3 on photography in the back of this book. There, you will also find information on how to choose the right film, how to load and focus your camera, how to use a tripod, and a number of other technical suggestions. (Refer also to the bulletins on photography listed in the bibliography.)

shoot the pictures

For production convenience, pictures can usually be shot in random order rather than rigidly following the sequence of the storyboard.

Select picture or scene content, footage, or numbers of different shots, camera angles, and action to obtain the negatives, film footage, or transparencies required by your plan. Check off each storyboard card as you complete the work it calls for.

By observing a few pointers, you and your students will take better pictures:

⊗ If you include people in your pictures, encourage them to act naturally. Catch them in unguarded moments. Have them practice movements or actions without freezing—that is, without stopping a movement just before the shutter snaps. Of

edit the production

Because plans do not always work out exactly as intended, there may be differences between the pictures you planned and those you actually shoot. Thus the need for editing.

The first thing to do in editing is to arrange the still pictures or motion-picture footage according to the sequence of the storyboard cards or the shooting script. Cut apart the footage, or group your slides or paper prints by sequence. After first eliminating all that are improperly exposed, poorly composed, or otherwise unsuitable, arrange scenes or pictures by number, using the storyboard cards or the shooting script as a guide. Splice film footage together in proper sequence and project it for study. Group and view slides in a similar manner. Make editing notes to indicate scenes that must be shortened, rearranged, rephotographed, or eliminated. Then edit to conform to these notes, and reproject to study effects of such changes.

Finally, smooth the edited product into finished form by adding titles, maps, or art work that may be needed. As a minimum, give each production a main title; subtitles are sometimes used to signify major changes of subject or to provide emphasis at the beginning of important sequences.

add sound

To provide synchronized sound on a motion-picture film itself, it will first be necessary to time each scene in the finished silent version. Then you must tailor the sound script (narration, dialogue, documentary sounds, or other sound effects) to conform to the time limits of each scene.

In many respects, the simplest arrangement for providing synchronized sound to accompany a photographic production is to provide it live for each presentation.

Sound recorded and played on an audio machine can be effective, but it is much more difficult to synchronize with motion pictures than sound that is recorded and played back on optical or magnetic-stripe tracks. Still, audio tape recordings have the advantage of being simple; they are convenient and easy to produce locally, and they can be recorded, erased, and rerecorded as many times as necessary to achieve quality results. If either optical or magnetic-stripe sound is to be added at the laboratory, a final, timed tape recording must be supplied for that purpose.

In providing sound to accompany motion pictures or slide sets of the types described earlier:

⊗ Develop a close relationship between the narration and scenes. Talk about only what is *shown*.

⊗ Select appropriate music and/or sound effects if

3M Company

you decide to use them. Percussion instruments, banjos, and guitars record especially well.

⊗ Reduce the commentary to a minimum. Allow the sound effects, the music, and the slide or picture content or the film action to carry as much of the communication load as possible.

⊗ Vary voices. Use different student or adult voices, as appropriate, for narration, dialogue, or crowd effects.

⊗ Talk to your audience on its own level.

use and evaluate the production

A systematic approach to developing a visual production requires that, upon completion, it be evaluated to determine how well it functions in achieving its objectives.

In the case of the short film demonstrating how to operate the spirit duplicator, described at right, such evaluation might be relatively simple. Selected individuals from groups for whom the film was produced (in this case, local teachers and students who did not already know how to operate the duplicator) could be invited to learn to do so by using that film in conjunction with practice on the equipment. Student observers could administer a checklist to rate how well nonoperators performed before and after seeing the film. They could also note the continuing problems of those who were unable to operate the equipment effectively despite seeing the film. Results might suggest defects in film content or organization and changes needed to make the film teach as planned.

Evaluation of imaginative slide, motion-picture, or photo-print productions might be expected to be carried out with less attention to learning of facts or to improving skills than to the affective results of their use. Questions such as the following might be asked: Did the production generate lively discussions? Did students *say* they liked it? That as a result of seeing it they were stimulated to think about the subject in new ways? Did they change in any way of which they were conscious their opinions about the subject portrayed?

case example: a film on operating a spirit duplicator

To test your grasp of the processes of planning a visual presentation, as discussed in the preceding section, assume that you and your students are to produce an 8mm film to help individuals in your school learn to operate and care for a spirit duplicator. Conditions are that the film will be shown on an 8mm silent cartridge-type projector, and the running time cannot exceed four minutes.

Assume that the successful spirit duplicator operator is required to:

⊗ Prepare the duplicator (clean it, attach the fluid dispenser, and adjust the paper pressure level).

⊗ Turn the duplicator drum so that the master clamp is up, and open the clamp.

⊗ Remove protective sheets from the master; attach the master to the drum; close master clamp.

⊗ Load a proper amount of paper into the paper tray.

⊗ Start the motor and test run; examine the quality of a few copies.

⊗ Make any necessary adjustments (bottom, top, or side margins; impression strength; fluid flow).

⊗ Set the run for the number of copies desired; start the machine.

⊗ Run the copies, and monitor their quality.

⊗ Stop the machine; open the master clamps; remove the master and cover it with a protective sheet.

⊗ Remove finished copies; remove the unused paper.

⊗ Close the clamp; set pressure lever to zero; remove the spirit bottle; clean the drum; set the handle to down position; replace the dust cover.

As you develop your storyboard, keep a record of decisions you are required to make about: (1) types of scenes (long, medium, close-ups, extreme close-ups), (2) lengths of scenes (taking into account the four-minute limit), (3) camera angles selected, and (4) need for titling and/or use of symbols or numbers to aid communication.

presentation techniques

Two special presentation techniques may be used to add functional variety to productions you and your students develop. They are (1) the twin-screen slide presentation, (2) the multiscreen presentation (slides only, or slides plus films). Both will appeal to students.

the twin-screen slide presentation

Twin-screen slide presentations involve the use of projectors (preferably automatic, remote-control types) to throw two pictures at a time on a single large screen or two screens arranged side by side. A single operator may handle this procedure without difficulty, keeping both slides in focus and changing, holding, or reshowing slides on either screen as desired.

The twin-screen arrangement is especially effective for presenting certain types of visual messages, such as:

⊗ Panoramic views (broad scenes taken in slide pairs that match reasonably well at the center line —which is concealed, in part, by divisions between screens).

⊗ Illustrations on one screen; related verbal points on the other.

⊗ Close-ups (or extreme close-ups) on one screen detailing an area of the slide shown on the other.

⊗ Several slides shown in sequence on one screen to illustrate graphic concepts held on the other.

⊗ Contrasting shots (a well-laid table on one screen; poor fare on the other).

⊗ A view of a laboratory setup which illustrates principles of water purification paired with a picture of a purification plant in action.

⊗ A map of a geographic area on the left; several views of the area on the right.

EXAMPLES OF TWIN-SCREEN TECHNIQUES

Medium shot   Close-up   Contrasts   *Daniel J. Ransohoff*

Verbal supplement   Panorama

Pan American Airways

multiscreen presentations

While in some respects presentations performed with three or more screens resemble those done by the twin-screen procedure just discussed, their capabilities for variation are considerably greater.

Development of the script for a three-screen (or more) presentation will be greatly simplified by using a form such as the one shown below. In the

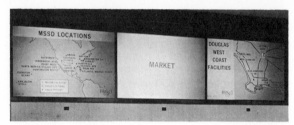

Douglas Aircraft

MULTI–SCREEN PRODUCTION PLAN

Scene No._____
Time _____

SCENE DESCRIPTION:

Medium _____ Medium _____ Medium _____

VISUAL AND TIMING CONTINUITY:

NARRATION, MUSIC, AND EFFECTS:

Arizona State University

screen areas, sketch the visual content to be portrayed; describe each scene in words under "Scene Description"; describe in the "Visual and Timing Continuity" section the order and length of exposure (on the screen) of each scene; and identify elements of narration, music, and sound effects in the bottom rectangle.

By referring to these storyboard cards when shooting, editing, or selecting slides, film clips, audio segments, or other elements of the production, you can ensure continuity and will reduce the complexities of presenting the finished production.

There are numerous opportunities to capitalize on the special advantages of multiscreen projection. For example, you and your students may:

⊗ Project combinations of slides and motion pictures.

⊗ Create a score for and record a coordinated audio track using music, sound effects, and narration.

SIMULATING MOTION WITH MULTISCREEN STILLS

Photograph by Peter Vadnai

Richard Szumski

⊗ Vary sequences of projection by (1) using first, second, third, and fourth screens in order, as in the case of the boy running (left), and hold them all on at once; (2) using slides one at a time, turning each off as the next appears, or dissolving one into the next, across the screen, to give an illusion of motion; or (3) bringing several slides on simultaneously, as for a dramatic panoramic view.

⊗ Organize a group to present a multiple screen presentation, complete with recorded music and sound effects, but without automation. Station a trained student team at each pair of projectors, one team for each screen to handle slide and tray changes, hand dissolves, focus, and other details. Have all groups follow a single script, much like a conductor's score. Students will soon develop the skills necessary to produce a smooth and effective performance.

photographic copies and close-ups

Teachers and students find it advantageous to obtain and use photographic copies, frequently in the form of 2- by 2-inch slides, of various uncopyrighted items of print or art work as well as close-up views of real items or tabletop displays. Here are a few examples of such resources that may be made more readily accessible or instructionally useful through copying or close-up photographic processes:

⊗ SOCIAL STUDIES Maps (rare, color, special types); photographs from books, magazines, newspapers, travel folders; foreign travel scenes from posters or calendars; close-up shots of individual student projects; views of museum materials; prehistoric implements, primitive cloth, tintypes, and old furniture

⊗ INDUSTRIAL ARTS Extreme close-ups of cutting tools, screw types, or bolt heads; blueprint section copied as a slide for group projection and study; close-up color slides to show details of various kinds of wood graining

⊗ SCIENCE Close-ups of flowers to identify parts; extreme close-ups of bugs and insects, cross sections of fruits or nuts; views taken through the microscope of bacterial cultures; views of organs of dissected animals; copies of student scientific drawings; charts from books or magazines

⊗ ART Close-ups of various styles and sizes of brushes, samples of lettering styles and sizes, a color wheel, photographic records of children's

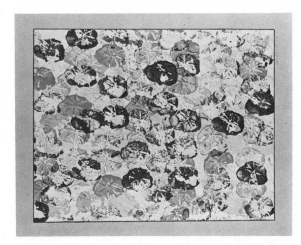

paintings, copies of professional paintings for projected study, close-up views of children's soap carvings.

Technical details of the processes and equipment recommended for use in copying and close-up photography appear in Reference Section 3 ("Photographic Equipment and Techniques"). Special precautions with respect to observing copyright laws are noted in Reference Section 2, "Duplicating Processes."

Eastman Kodak Company

COPYING AND
CLOSE-UP PHOTOGRAPHY

*Encyclopedia Britannica
Educational Corporation*

Encyclopedia Britannica Educational Corporation

organizing your slide collection

You can do several things to organize, protect from damage, facilitate locating, and improve the usefulness of slides you and your students photograph or collect:

⊗ Sort slides into functionally related modular units. Each module should treat one subject or

set. Multiple notes or recordings can be made for different purposes for the same set of slides—one for use with elementary students who have very little background in the subject and a second for ad-

contain a group of slides that would be studied together.

⊗ Place a thumb mark in the upper right-hand corner of each slide as it is viewed in correct position for projection, which is upside down. A very easy way to do this is to make dots, using the flat eraser head of a new lead pencil that has been

vanced students who are prepared to react to suggestions that stimulate reasoning and respond with sensible answers to audio and visual cues.

⊗ Use a uniform modular container—a special inexpensive box that can hold a number of slides, a Carousel tray, a plastic sleeve, or some other unit that fits your projector. Label each unit carefully.

⊗ Use a light table or special slide sorting stand when arranging and rearranging slides.

Again, remember that your school media center staff will usually be able to help you and your students with these tasks and will also make your photographic products available to others for use in your school, school system, or college.

inked on a stamp pad. Use pen and ink to number slides.

⊗ Prepare one or more sets of written notes or audio-tape recordings to accompany each slide

summary

Because photography is a popular hobby, many teachers and students will be aware of ways of using it in school and college. With the development of simplified, semiautomatic and automatic cameras, quality results with photography are increasingly easy to achieve.

Experiences with photography enable students to increase their visual literacy and their ability to communicate with photographic visuals. Today, many students learn about photography by using video systems; the skills they develop are basic to both video and film formats. With either medium, students must select and use lenses and must know about composition, lighting, and direction.

Informal and structured photography are considered in this chapter. With the former, frequent opportunities occur for students to experiment with the camera; the emphasis is on capturing a mood, a situation, an event, or an object in its most favorable or useful aspects. Usually, incidents photographed in an informal mode may never be repeated, and quick recognition of the opportunity and the significant content is essential. With structured photography, however, the intent is to achieve some predetermined purpose, which requires the photographer to follow a carefully developed plan. Often the plan of the project will include a storyboard and a prepared shooting script.

Further important considerations about photography include editing photographs or motion-picture film, or video tapes, and developing sound—speech, music, or effects—to coordinate with visuals. Also, in photographic productions, skills and techniques of displaying are used extensively for visuals, backgrounds, and titles.

Special presentation techniques described are multi-image, twin-screen, and multiscreen presentations. The purposes for such presentations generally put them into the category of structured photography, regardless of the combinations or the number of media used—slides, motion pictures, sound. And, finally, the values of copy work and close-up photography are explored for extensive values in education.

Much of the information in this chapter will be of background value and will be expanded in the next four chapters, Still Pictures, Audio Materials, Motion Pictures: Film and Video, and Television.

9 STILL PICTURES

chapter purposes

⊗ To identify forms and characteristics of still pictures: flat opaque pictures, 35mm filmstrips (silent and sound), 2- by 2-inch slides, $3\frac{1}{4}$- by 4-inch slides, and stereo slides in reels.

⊗ To suggest ways of using still pictures so that students can learn to understand visuals and to improve their visual literacy.

⊗ To indicate criteria and processes to be used in selecting still pictures that will contribute to achievement of objectives.

⊗ To provide the information and guidance needed to prepare flat opaque pictures and slides and filmstrips for use and for accessible storage.

⊗ To suggest how still pictures of all types can be used to advantage in multi-media packages with other learning materials.

National Geographic Society Educational Services

For most students, contact with the real world beyond their doorsteps is made through pictures. One example is the fact that millions of individuals in the United States have never seen the Pacific Ocean—or any ocean—except in photographs. This chapter is about *still pictures,* which, of course, are available in many forms: in print or in slides for projection, in black and white and in shades of gray, or in full, rich color. Whatever the type or source, pictures constitute a large portion of the various experiences by which people know the world. By reading pictures, by working with pictures, and by making picture stories, students become literate in visual terms. Teachers have a great responsibility in helping students achieve this visual literacy.

169

As students use still pictures, they can learn many abstract ideas. Their past experiences provide clues for them: If they know that flowers bloom in summer, they have a start toward understanding a picture of a flowering lily on a pond.

still pictures and visual literacy

Flat pictures may be used to reinforce impressions, to offer new facts, or to form a base on which to develop meanings of abstractions. Students and teachers benefit from exercises directed toward reading the meanings of pictures. Study the picture of the satellite view on page 172. Then turn back to this paragraph and continue reading.

What specific details of the picture do you now recall? Did you recognize India? If so, how? By knowledge you obtained through previous map study? By verbal descriptions you have read? Through other means? Did you know that Sri Lanka, a separate country, is so close to India? Does your knowledge of the fact that the picture was taken from a satellite help you to better understand that the earth is round? If so, what gave you this clue?

How did you proceed in viewing the picture? Did you simply skim over it, without pausing to study in detail certain parts of it? Did the picture caption contribute valuable information or provide assistance in your examination of the picture? Or did it divert your interest from the study? How much information can you *infer* from the picture alone? Did the caption direct your perception of the meaning of the picture? How did your reaction to the satellite picture compare with your reaction to the pictures of the seamen battling waves or of the water lily? Of the boy laughing? Of the Sphinx, or of the converging rail line? How could *you* use these or similar pictures as an exercise directed toward improving the visual literacy of your students?

Consider the possibility of developing a collection of still pictures with which you can experiment in teaching the skills of visual literacy discussed on page 173. You will find the effort to be well repaid in student interest.

MAKING ABSTRACTIONS REAL

Flat pictures may be used to reinforce impressions, add new facts, or provide meanings to abstractions. They deal with experiential matters and enlist the viewer's own capacities for recollection and recall.

"One sound . . . to waterfalls."

". . . straight, and narrow . . ."

"O heavenly colour . . ." through
tones of light and shadow.

Chill, Hays from Monkmeyer Press Photo Service, poet, Keats; color, Standard Oil Company of New Jersey, poet, Meynell; sound, Creative Educational Society, Inc., poet, Emerson; joy, American Red Cross from United Nations, poet, Byron; speed and motion, Eastman Kodak Co., poet, Thomson; straight and narrow, Western Maryland Railway; time immemorial, Trans-World Airlines.

". . . th' Atlantic surge"—its speed and
motion felt . . .

"Ah, bitter chill it was!"

"Let *joy* be unconfined."

". . . from time immemorial . . ."

You're looking at India, as viewed from a Gemini spacecraft. You're also looking at one of the world's largest classrooms, as "viewed" by a broadcast relay satellite. A NASA satellite can relay educational broadcasts directly to television sets in thousands of remote villages throughout India and Ceylon. This spectacular experiment in mass communication provides education in such subjects as family planning and improved agricultural techniques. It's a concrete example of how our space technologies can help bring mass communications and practical education to the underdeveloped countries of the world. Think about the possibilities. *The Boeing Company.*

The exercise makes clear that learning to read still pictures is somewhat like learning to read print: Skills are involved that can be improved with practice. It is important, therefore, that practice in reading pictures be included in the experiences of every student.

Recognition of the importance of visual literacy has given impetus to many activities involving pictures. In this chapter, attention is directed principally toward pictorial materials that are available ready-made from many sources. In the chapters on photography and television, student production activities are discussed. In these pages are ideas that markedly affect the artistic, educational, and intellectual decisions necessary to make pictures and picture stories, with or without captions, with or without sound, and for a variety of purposes—all relevant to promoting visual literacy.

the skills of visual literacy

Several types and levels of picture-reading skills—each related to visual literacy—have been identified. In looking at a *single* picture, different viewers may:

⊗ Recognize objects and call them by name;

⊗ Sort out details within the picture and describe them; or

⊗ Study details of the picture, *interpret* visual cues, and *infer* probable or actual facts about past, current, or future actions and relationships of people, objects, or events portrayed.

Individual students, of course, will reveal differences in their ability to make such judgments or responses. Family activities, social climate of the home, and other influences affect a student's visual literacy.

Even greater differences in ability to make visual interpretations will be noted when students work with *groups* of pictures in ways which require more advanced picture-reading skills. For example, with a number of photographs depicting steps of a process, or several scenes related to some historical event, different viewers may:

⊗ Infer or create totally new ideas by changing the sequence of or making new relationships among the pictures;

⊗ Arrange the pictures in appropriate and meaningful order to compare or contrast them, to establish events in chronological sequence, or to determine whether there are significant omissions or redundancies among them, and cite reasons for judgments or conclusions;

⊗ Recognize that the set of pictures represents a group of related pictorial images or a sequence of actions such as these:

Eastman Kodak Company

Since everyone needs to be proficient in reading pictures, and since the skill develops only with experience, activities that improve the visual literacy of students should be incorporated at every level of education. For example, very young children are likely to regard all four-footed animals as dogs. But after more firsthand experience, they begin to understand that the four-footed animals in the pictures they study are not *all* dogs; they may be one of the other four-footed animals they see on the streets or at the zoo. And within a few more short years they will also learn to distinguish a sheep from a sheep dog, a donkey from a pony, or an elk from a reindeer—perhaps without ever having seen the real-life counterparts of any of them.

"When green buds they were swellin'."
Anonymous.

STILL PICTURES
STIMULATE CREATIVITY

These still pictures of the seasons stimulated creative student activity in poetry writing at the high school level. What other tie-ins could you find for them in a study unit in literature or language arts? In the study of a foreign language? On the elementary school level, the same pictures might be used in conjunction with music activities. *Winter, autumn, summer, T. P. Lake; spring, A. J. Mueller, Courtesy of Johnson Printing, Inc.*

"A lusty winter, frosty but kindly."
Shakespeare.

Implicit in the achievement of visual literacy is an improvement in verbal literacy as well. As students mature, they can be given a wide variety of experiences designed to increase both. They may select pictures, arrange them in sequences that are logical for them, and compose captions that communicate original ideas. In doing this, they may use their own words, or they may select quotations from Bartlett's or other sources that reinforce, clarify, add emotional tone, or otherwise enhance picture-word impressions.

Captions accompanying the four-color photographs on these two pages illustrate some of the possibilities of relating quotations to pictures.

Try another approach. Arrange the photographs in some new order that is unrelated to seasons. Then write original captions for each so that other viewers—your students, for example—will gain new meanings from them. What does this experience suggest to you about the relationships of words and pictures? Share your opinions with others in your class.

"There is a harmony in autumn, and a lustre in its sky." *Shelley.*

"Very hot and still the air was." *Longfellow.*

types of still pictures

For different needs and circumstances, still pictures are available in various forms: (1) flat, opaque pictures (photographic prints and drawings or paintings); (2) 2- by 2-inch slides; (3) 35mm filmstrips (silent or with sound, in color or black and white); (4) stereo reels; and (5) $3\frac{1}{4}$- by 4-inch slides. The talent, artistry, and scholarship of anthropologists, psychologists, sociologists, and illustrators are combined to produce these varied pictorial materials.

flat, opaque pictures

Flat, opaque pictures may be acquired from many sources. They may be bought in sets from com-

mercial suppliers who specialize in study print production. They may be taken from magazines and used as printed; or they may be cropped to emphasize important subjects or concepts, and then mounted and coated in one of several ways to protect them against wear. They may have labels applied to the back with descriptive information, indication of the source, and filing instructions. Sometimes labels may carry test questions or quotations that can be used when the pictures are shown. The study pictures displayed in the chalkboard tray below may be used for individual study or for group discussion.

Denoyer-Geppert, Chicago

National Geographic Society Educational Services

nomical that they can be available in quantity on many topics. Most filmstrips today are in full color; many of them have narrative accompaniments, and sometimes sound effects (including music) on audio cassettes or disc recordings. Nevertheless, the value of silent filmstrips cannot be denied. Many are still used to advantage.

Ways of using both types of filmstrips are discussed later in this chapter.

35mm filmstrips

One especially significant contribution of technology to education in recent years has been the continuing improvement of the quality and quantity of pictures in natural color which are available in a number of formats.

Among the many formats in which still pictures are available, in black and white and in color, one of the most economical is the 35mm filmstrip. For group study or for individual study, filmstrips are so eco-

Scott Education Division

National Geographic Society Educational Services

still pictures in multi-media kits

Multi-media kits of filmstrips, cassettes, transparencies, and worksheets, often with study pictures, provide realistic learning experiences for study of historical and other remote topics.   These National

Geographic pictures from the 35mm color filmstrips, *The Family Farm, West of the Shining Mountains,* and *Harvesters of the Golden Plains,* portray activities in the day-to-day lives of individuals in various parts of the United States.   The Civil War pictures, taken from Matthew Brady's tinted glass plates made on the battlefield, show what life was like in the trenches, prisons, and hospitals of the time.

National Geographic Society Educational Services

National Geographic Society Educational Services.

National Geographic Society Educational Services

Scott Education Division

research on still pictures

Research findings on the value of still pictures suggest the following implications for teaching:

⊗ Pictures stimulate student interest.

⊗ Properly selected and adapted, pictures help readers to understand and remember the content of accompanying verbal materials.

⊗ Simple line drawings can often be more effective as information transmitters than either shaded drawings or real-life photographs; full-realism pictures that flood the viewer with too much visual informa-

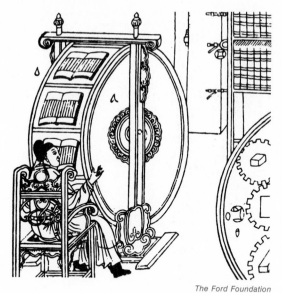

The Ford Foundation

tion are less good as learning stimuli than simplified pictures or drawings.[1]

⊗ Color in still pictures usually poses a problem. Although colored pictures appear to interest students more than black-and-white ones, they may not always be the best choice for teaching or learning. One study[2] suggests that if color is used, it

Merrell Gage

[1] Robert M. W. Travers, "The Transmission of Information to Human Receivers," *Audio-Visual Communication Review,* vol. 12, pp. 373–385, Winter, 1964.

[2] Seth Spaulding, "Research on Pictorial Illustration," *Audio-Visual Communication Review,* vol. 3, pp. 43–44, Winter, 1955.

should be realistic—not just color for its own sake. If only one color is to be added to an otherwise black-and-white picture, the teaching value may be reduced.   But if what is to be taught actually involves color concepts, pictures in realistic color are preferred.

⊗ When attempting to teach concepts involving *motion,* a single still picture (including those in filmstrips) is likely to be considerably less effective than motion-picture footage of the same action. Yet a *sequence* of still pictures, such as might be shot with an automatic 35mm still camera, might reduce flooding brought about by the too-fast flow of live action portrayed in some motion pictures and

U.S. Weather Bureau

thus improve the viewer's grasp of concepts involved.

⊗ Verbal and/or symbolic cueing of still pictures through use of arrows or other marks can clarify —or, possibly, change—the message intended to be communicated by them.

There are numerous opportunities in your own teaching to test these and other tentative conclusions about the values of still pictures for your students.

flat pictures

The concerns of teachers and students about flat pictures (chiefly those printed on paper) range from choosing, teaching, and evaluating with them to finding more of them that are good enough to mount, preserve, and file. Each of these matters is considered here.

choosing flat pictures

Involving students in the process of choosing flat pictures is highly desirable. Student insights provide excellent guides to pictures you may wish to retain for future use.

  Given a topic, reasonably clear purposes, and familiarity with the abilities and interests of students who are going to see a picture, you might expect them to provide positive answers to questions such as the following:

⊗ Is it important to the subject we are about to study?

⊗ Is it interesting? Accurate? Up to date? Well reproduced? Does it hold my attention?

⊗ Does it raise questions in my mind? Will it help us discuss the subject?

⊗ Can we see the picture and comprehend it from where we are sitting?

⊗ Is it large enough or small enough for the purpose for which we intend to use it, such as for a walk-up display?

⊗ If sequential information is desirable or essential, are individual pictures available to present it?

  These questions would be differently phrased, of course, according to the ages and experience of the students.

sources of flat pictures

Flat pictures can be obtained almost anywhere and everywhere! Free pictures for classroom use can come from magazines, newspapers, books, catalogs, calendars, posters, or advertising circulars. Wonderful pictures, picture stories, and picture essays are reproduced with high technical quality in nationally distributed periodicals, which carry colorful pictures on such topics as food, hunger, health, science, geography, urban problems, ecology, oceanography, wildlife, and history. Pictures may also be purchased as sets organized around significant themes in the school curriculum.

  Sources of pictures are actually so endless that you must have your class make an organized attack on the problem of selecting them. As a start, have them see what the school media center has to offer. It probably has a file from which they may borrow pictures immediately. The district or county media center may also have such a file, with fine art prints as well as ordinary flat pictures. And the school or district library is likely to have a collection of booklets listing sources of free or inexpensive flat-picture materials. Some teachers use such source lists for placing personal orders to augment local picture files. For students who can read and write, selecting materials from such source guides and placing orders for them would be an excellent experience.

  Still another source of flat-picture materials may be the students' homes. However, you should be sure that no picture, nor any magazine from which pictures are to be taken, is brought to school without parent permission.

VEGETABLE FARMING

These pictures, selected at random from a commercially produced thirty-one-picture set, are available in 11- by 14-inch dimensions. Together, the full set illustrates these concepts: Vegetables are plants; their roots grow in the soil; we eat parts of plants; sometimes we eat the roots or stems; sometimes we eat the leaves and leaf stalks; sometimes we eat the flower parts and fruits; and sometimes we eat the seeds and seed pods. Are there socioeconomic concepts that can be inferred from these pictures? *Hi-Worth Pictures.*

teaching with flat pictures

Flat pictures can be used in many stages of the instructional process: to introduce and motivate study of new topics, to clarify misconceptions, to communicate basic information, and to evaluate student progress and achievement. Here are some ways to use still pictures, together with suggestions for ensuring maximum student benefit from using them:

⊗ SELECT ONLY MEANINGFUL PICTURES Skill and insight are a teacher's guides in selecting pictures. But student reactions to pictures are final bases for selection and utilization.

⊗ LIMIT THE QUANTITY OF PICTURES USED Ordinarily, individuals among groups of students viewing pictures should not be forced to hurry to keep up. Therefore, if pictures are to be shown one at a time, limit the number; don't ask students to see more than they can read. Simultaneous showing of several pictures proves to be appropriate only when individual students have the opportunity to move quickly from picture to picture and grasp general ideas; retrace for special study, contrast, comparison; move ahead once more; and thus use the pictures as a basis for integrated learning.

⊗ USE VERBAL CUES WITH PICTURES Students learn directly from pictures when they infer meanings or actions from the visual cues portrayed and when appropriate questions focus their attention on

Brigham Young University

the relevant cues. Members of a fifth-grade class studying the westward movement examined several pictures of prairie schooners on the trail. To start the search for visual clues, the teacher asked some questions: What kind of animals are pulling the wagons? What are the wagons made of? How are the animals steered? The students took up the questioning and raised some of their own. Then, interpretive questions were asked by the teacher, again to start a direction of student exploration. Why are so many wagons traveling together? How do you suppose they got the big wagons over steep, rough mountain passes? On the basis of such questions and others raised by the students, library activities were initiated and groups of students began seeking answers, with the ultimate purpose of presenting a pictorial report to the entire class.

⊗ STIMULATE CREATIVE EXPRESSION Pictures not only provide bases for answers to factual questions but they may also stimulate a variety of creative expressions. Showing a single picture depicting a crucial occurrence in the relationship of a small boy and his pet, for example, may trigger the writing of short stories by several students—each with a different plot, action, and ending.

⊗ USE CONTRAST, COMPARISON, AND CONTINUITY Ask students to look for *contrast* within pictures—for differences among the people, objects, materials, or conditions depicted. Contrast the new

San Diego County (California) Schools

Caterpillar Tractor Company

United Air Lines

front of the class, commenting upon or asking questions about each and later passing them for individual study; (2) place them in a file or box where students can examine them individually and in detail; (3) arrange them on bulletin-board or chalkboard-tray displays, perhaps attaching numbers or strings to each to code its relationship to an accompanying set of program notes or questions; or (4) project them in an opaque projector for group viewing and discussion.

Following are several other examples, from specific subject fields, to illustrate different ways in which flat pictures may be used in teaching:

⊗ Members of a class in *home economics* may collect pictures to be used in scrapbooks on the subject of home decorating. Pictures are classified, sorted, weeded, and arranged in suitable order. Each completed scrapbook reflects the personal tastes of the individual who prepared it.

⊗ Several students are assigned the task of locating pictures of flowers for use during a *botany* unit study. They display the pictures—one each day—under the caption "Can You Name the Flower?" Students try to name it independently, and they check their accuracy by lifting an answer flap (a sheet of paper covering an answer) located beside the display.

⊗ A *geography* teacher uses three unidentified, unfamiliar flat pictures in a quick-check test. The pictures are projected, one at a time, through the opaque projector. Students are asked to deduce, from geographical clues in the pictures, the probable location of each scene. They give reasons why they believe their choices are accurate.

⊗ An *industrial arts* instructor mounts on a board a series of still pictures explaining four essential steps in operating the shop planer. The board is hung directly over the machine, and each student is required to consult it before adjusting the planer or starting the motor.

⊗ A *music* class collects still pictures from a variety of sources to illustrate a set of charts prepared to outline events in the lives of four important French composers. Each of the finished charts is displayed in the room at the time the particular composer's life is discussed and his works are played.

with the old, the near with the far, or the known with the new to be learned. Also, ask students to *compare*—the totality of one picture with another, a portion of one picture with another, a particular item with another item portrayed in the same or different pictures. Develop the skill of detecting or inferring *continuity* by using two or more pictures, each showing a different stage of a process, development, or event.

⊗ USE FLAT PICTURES IN SUITABLY VARIED WAYS Depending upon your purposes and the learning outcomes you seek, you should be able to find many different and satisfactory ways to use flat pictures. You may, for example, (1) display them in

evaluating with flat pictures

To give students opportunities to test their knowledge and their ability to interpret pictorial information, flat pictures can be used in numerous ways. Here are a few:

⊗ Display pictures on chalk trays, tables, counters, or any other appropriate location. Number each picture to correspond to a space on an answer sheet or cards. One or more questions may be asked about each picture. When the students have answered questions at each station, ask them to check their answers while reviewing the displays.

⊗ Supply a sufficient number of pictures for each student to have one. Each picture carries a number. Ask each student to write a series of questions for one of the pictures. Students may then exchange pictures and answer the questions, passing the pictures along until each student has answered all questions. Student committees evaluate the questions and check the answer sheets. In this process, it is often helpful to have the students use lines that lead to some part of the illustration, each lettered or numbered, to guide the viewer to the subject of each question.

⊗ Give committees of students randomly packaged pictures of technical processes or other subjects that can best be understood by a series of photos. Have them arrange the pictures in proper sequence to explain the subject accurately.

⊗ Arrange a display of pictures with questions under lift-up cover sheets. The pictures are numbered, and answers are checked or written on an answer sheet. For student confirmation, answers may be listed on a covered sheet at the end of a sequence of pictures, or, for instant feedback, on answer sheets under the pictures.

mounting and preserving flat pictures: principles

Many pictures serve their purpose with only a single use, and then they are discarded. Some may be used more than once, and these may be filed in loose form. Other pictures will deserve to be mounted, protected, and saved for repeated use.

For those that you and your students decide should be treated in this way, the following procedures are suggested:

⊗ Use standard outside dimensions and shapes insofar as possible. For optimum convenience, mounted pictures should fit into a letter-size file drawer or a box of similar proportions.

⊗ Use good quality mounts of stiff mounting board which will not fray or curl. Neutral tones are usually most satisfactory; avoid loud, garish colors.

MOUNTING PICTURES WITH RUBBER CEMENT

Pictures may be easily mounted, with no equipment, by using rubber cement. To make a temporary mount by this method, spread cement on the back of the picture and then set it on a piece of cardboard before the cement dries. To make a permanent mount with rubber cement, follow the steps shown here:

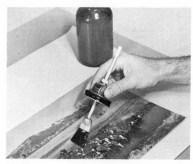

Apply rubber cement to the back of the picture and to the mounting board. Be sure to put the cement on the *back* of the picture. Apply it evenly and with smooth strokes. Set the two sticky surfaces aside until they are dry.

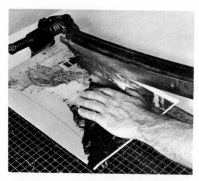

Position the picture on the cutting board and cut it to the proper size to show the content you wish to use.

Then place two sheets of waxed paper on the mounting board to cover the picture area. Align the picture with the guide marks and slide the wax paper sheets out.

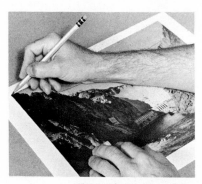

Mark the location of the picture on the board on which it is to be mounted. These marks will guide you in spreading cement and in placing the picture in the correct position.

Rub the picture for good adhesion and to remove air bubbles. Remove excess cement by rubbing around the picture. Erase the guide marks. For protection, the picture may be covered with plastic lamination or spray.

dry mounting flat pictures

Collections of flat pictures that have been selected for use in teaching should be processed in ways that protect them and make them convenient to use.

Essential steps involved in dry mounting flat pictures are:

⊗ First, place a sheet of dry-mount tissue over the back of an untrimmed picture, with the tissue overlapping the edges.   Attach the tissue to the back of the picture with the tip of a tacking iron.

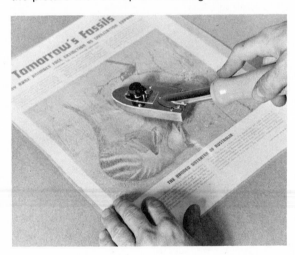

⊗ Next, turn the picture and tissue over, and trim both to the desired size.

⊗ Then tack the tissue to the picture mounting board at two opposite corners.

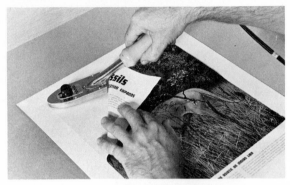

⊗ Place the picture assembly in a mounting press set to the correct temperature, and leave it there for six to eight seconds.

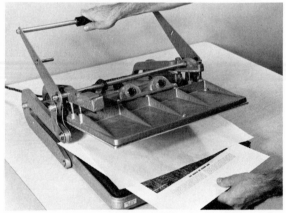

⊗ To protect the surface, apply plastic lamination or spray.

protecting flat pictures

Flat pictures that are mounted for continued instructional use should be protected against unnecessary wear and tear. Several things can be done:

⊗ Cover pictures with plastic spray, as shown in the previous section. Remember that this substance, which comes in pressurized cans, should be handled with great care.

⊗ Consider laminating especially valuable pictures. Many schools now provide laminating machines and supplies for this purpose. School media center personnel—media production technicians, for example—frequently offer this service.

Leon County (Florida) Schools

⊗ Use a special machine to round the corners of mounted pictures to reduce any chances of damage.

Alameda County (California) Schools

filing flat pictures

Once flat pictures are processed for use, a well-organized file, equipped with dividers, is essential for storing them and having them readily available. If no regular steel file is provided, cardboard boxes

George Cochern

or box-type transfer files may be used. Rather than filing all pictures by titles or subjects, consider organizing them on the basis of study units or learning modules. Maintaining such files reduces the loss or destruction of good pictures, makes finding a suitable picture relatively easy, and provides for an orderly expansion of the picture collection.

At the beginning of the year, it is a good idea for students to become acquainted with the content, organization, and purpose of the picture file and to be invited to help keep it growing. With teacher and students discussing ways in which they all may make use of the pictures, they may also decide on ways that more pictures are to be collected, sorted, selected, mounted, and filed. A drop box for pictures may be suggested. Specific times may be set aside for students to look through contributed magazines to select valuable pictures. Collecting appropriate pictures, mounting them for filing, filing them under topics, and maintaining the file becomes a class project, with all students participating and learning as they work.

2- by 2-inch slides

This section emphasizes the kinds of 2- by 2-inch slides (chiefly in color) that may be bought from commercial suppliers and organized and distributed by media resource centers. In Chapter 8, the procedures for planning and producing slides for projection are treated in detail. Much that is mentioned here, however, about the applications of slides in instruction and learning is relevant, whatever the sources of slides.

teaching with 2- by 2-inch slides

There are several advantages in using 2- by 2-inch slides: They may be arranged and rearranged as the instructional activity dictates; some may be added and others dropped; the flexibility in using them makes them especially appropriate for either teacher or student presentations.

San Diego County (California) Schools

When showing slides to a group of students, teachers use techniques that are quite similar to those used with filmstrips, which are discussed later. But since slide sets usually do not have a caption on each slide, what the teacher presents as commentary must be carefully planned. Particularly for young children, continuity in the sequences must be made clear; further, guidance in seeing what is important to the lesson must be provided. Thus a plan is desirable for the presentation. Viewing the slides before the presentation makes possible for the teacher to prepare notes about what must be said. Even more important, techniques used should stimulate questions as students view the pictures; or, as appropriate, as each slide is shown, students can describe what they see as important to the lesson. To have students test their ability to find what is shown and implied in pictures, sets of slides may be used to provide visual cues for evaluation procedures.

Other group activities may include the assembly of slide presentations by students to report on their research findings or their reactions to books they have read or to show historical relationships found in pictures of art, architecture, and other human enterprises. When slides and slide-projection equipment are easily available, many creative ideas for using them will occur to both teachers and students.

handling 2- by 2-inch slides

Today, the trays, carousels, or cartridges used with automatic 2- by 2-inch slide projectors provide a way to keep slides both in correct order and in position for projection; they also serve as storage boxes. For ready-reference storage, they may be filed in labeled boxes or in special slide-storage display cabinets, usually located in media centers. Well-managed slide collections are classified, labeled, and supplied with guide materials. The student below is using a slide sorter, an advantageous aid in assembling slide sets.

other projection media

Two other still-projection media should be mentioned: stereo (Viewmaster) reels and 3¼- by 4-inch lantern slides.

stereo reels (Viewmaster)

Another classroom resource is the stereoscopic three-dimensional slide reel which may be observed in a hand viewer or projected, as a regular slide without the third dimension, in a special projector. Stereo reels contain pairs of pictures, each mounted precisely to provide normal eye-width viewing by means of a special hand viewer which fuses

Hans Noecker

them into a single image to yield a surprising illusion of reality.

Stereoscopic reels are now available in a number of fields—world geography and travel (to most of the states of the United States and to principal countries throughout the world), stories of high adventure, Christmas stories, fairy tales, and various scientific subjects (studies of animals, flowers, insects, fish, and trees).

Advantages of stereo reels lie in their initial low cost and the ease with which they can be studied by individual students. Some school systems stock reels and viewers in central collections, either at individual schools or at district instructional materials centers, to be loaned to classrooms for the duration of units of work. Viewers and reels are commonly kept at reading corner tables.

3¼- by 4-inch slides

Improvements in 35mm (2- by 2-inch) slide production and projection equipment, as well as film, have sharply reduced the demand for 3¼- by 4-inch lantern slides. Occasionally, however, they may be used in the fields of medicine, science, and art. Special projectors are required for them, but projection techniques are essentially the same as those for regular 2- by 2-inch slides.

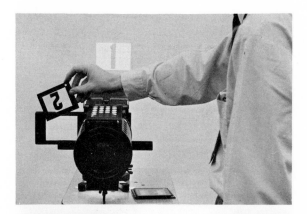

35mm filmstrips

The special values and qualities of 35mm filmstrips make them suitable for independent study as well as for small-group, medium-group, and even large-group instruction. The present discussion relates to criteria that ought to be used in selecting them, where to find the ones that are applicable to your requirements, how to teach with them, and how to use special techniques to increase their usefulness. Sound filmstrips—filmstrips accompanied by sound provided by reel-to-reel or cassette tapes or recording discs—will also be considered briefly.

San Diego County (California) Schools

special values of filmstrips

Generally speaking, the values of filmstrips cannot be considered unique because they are in many cases similar to those claimed for other media types. However, a number of advantages of using filmstrips are worth mentioning.

⊗ Filmstrips are convenient and flexible to use. When projected, they may be viewed effectively by groups of nearly any size or by a single student.

San Diego County (California) Schools

Edmonton (Alberta) Schools

⊗ Both projectors and table-top viewers are available and are relatively inexpensive. These units encourage the use of filmstrips in classroom learning centers and in media centers.

⊗ Since filmstrips present pictures in a fixed sequence, they provide a structure for the subject.

⊗ Many filmstrips are captioned and numbered.

This special treatment often proves advantageous to the learner, especially when one or two students use them in independent work. And numbering makes it possible to locate a frame to be reshown; the number of the frame can be noted in the lesson plan.

⊗ Filmstrips provide an economical means of presenting information. An example is the filmstrip that contains approximately fifty frames, in color, that sells for less than $10. Were the colored pictures contained in even the average filmstrip to be published in book form and sold in comparable numbers, the cost would usually be prohibitive. Filmstrips are cheaper, too, than separate 2- by 2-inch slides.

Usually, circulating filmstrips from remote central collections, such as those maintained at county and district school offices, is not considered economically feasible. Because they are so inexpensive, at least compared with motion pictures, single-school filmstrip collections are common. It seems, therefore, that most of the filmstrips you and your students use will come from your own school media resources center. That being the case, the problem of choosing them will be simplified, and you should find it easy to involve your students in the process.

group teaching with filmstrips

As with other instructional materials discussed in this book, there is no one best way to teach with filmstrips. Much depends upon the purpose you have in mind and the conditions under which you use these materials. But for most situations involving group teaching, a generalized pattern is recommended: (1) *prepare yourself* to use the filmstrip, (2) help the class *develop readiness* to see it, (3) encourage *student participation* during the showing, and (4) *follow up* with appropriate activities. Details of these steps include:

⊗ PREPARE YOURSELF Determine your purposes. Preview any filmstrip before you show it. Very few if any of them are likely to contain *exactly* what you want to show, but this preliminary survey will help you determine how to *adapt* those you do decide to use to your purposes and to the characteristics of the group with which you will use them.

During the preview, note the organization of the filmstrip content and read the accompanying guide,

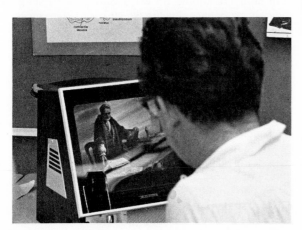

if one is available. Make a list of important terms that are used in captions or on the sound track. Decide how to call attention to them. You may also want to list several questions, perhaps taken from the guide, about information the filmstrip provides. Consider ways you might introduce these questions —such as asking them during a preshowing discussion, asking them at intervals during the showing itself, or placing them on the chalkboard as assigned viewing problems to be discussed after the showing is completed. Note also certain key frames of the filmstrip which merit further class study or later explanation by you. Think ahead, as you study the filmstrip, to plan a few worthwhile student activities that might be expected to grow out of the showing.

⊗ DEVELOP A MEANINGFUL CONTEXT FOR STUDY OF THE FILMSTRIP Just showing a filmstrip and doing nothing more with it is likely to

contribute very little toward learning. However, there are a number of things to do to ensure its making a contribution. The preliminary questions suggested above will help. If you were about to show a filmstrip on Mexico as part of an introduction to a study of Latin America, you might ask such questions as: Have any of you ever visited Mexico? If you *were* to visit this country, what large cities would you prefer to visit? Why? What is the distance from the northern to the southern borders of Mexico? Based on your present information, what is there about the geographic features of Mexico that influences the activities of the people? The showing, in this case, might then be a means of verifying original predictions or correcting misconceptions. Were this same filmstrip to be used as a review or summary of unit study, however, quite different questions could be asked: In what ways does the filmstrip *not* agree with what we have already learned about Mexico? Is the filmstrip coverage fair to the country? If not, in what respects is it unfair?

Another technique of preparing students to see and study filmstrips involves (1) telling them that they are about to see a filmstrip and giving the title and a brief overview of the content; (2) making clear how the filmstrip relates to what they have been studying; (3) giving them a list of major points or questions to which you wish them to pay particular attention during the showing; and (4) indicating that there will be a follow-up discussion, a check test, or both at the conclusion.

These are only examples of different ways you may develop class interest and readiness for seeing and studying filmstrips. By all means, experiment with other techniques and avoid a stereotyped approach. Here are a few additional suggestions:

Ask your students to make up questions as they see a filmstrip, and after the showing have them ask each other these questions.

Tell your students, "When we finish seeing the filmstrip, you will be asked to make a list of the main points it emphasizes."

Prepare study-guide sheets and hand them out

TEACHER'S MANUAL
FOR THE SOUND/COLOR FILMSTRIP SET

Spain and Portugal:
European Neighbors

MANUAL PRICE 50¢

ABOUT THE FILMSTRIPS

There are 6 sound/color filmstrips in this set, **Spain and Portugal: European Neighbors**. The titles are:

Geography of Spain and Portugal (46-1)
Agriculture of Spain and Portugal (46-2)
Spanish Economy and Industry (46-3)
Spanish Cultural Heritage (46-4)
Spanish Historical Background (46-5)
Portugal (46-6)

With ever-increasing demands being made upon the curriculum, the importance of supplying resource materials and teaching the skills to use them become a necessary function of the teaching process.

When there is only a short time allotment in which to develop concepts concerning a culture, such as Spain and Portugal have enjoyed through centuries of history-making background, it is important that independent study be encouraged.

These filmstrips are inquiry-oriented...concept-structured ...designed for class use and for individualized instruction. Correlating closely with the latest editions of leading social studies textbooks, these visual, narrative, and text materials form a transition from traditional teaching to the new.

INDIVIDUALIZED INSTRUCTION

One of the major goals of the social studies program is to provide children with the skills they need for independent gathering of information, classifying and categorizing it, evaluating it, and then using it in ways that have special motivation for them.

If the emphasis is placed upon this method of teaching, the interests and needs of each student will vary and a topical approach might be useful. These could be on an assigned basis or on a personal choice basis. Such topics as cork, olives, fishing, gypsies, matadors, and the like might be used.

A number of ideas suggested in the Student Study Manual will help the teacher to more accurately evaluate both the quantity and quality of the concepts being developed concerning Spain and Portugal.

LEARNING CENTERS

Filmstrips, like books, can open up a world of information. But, like books, the greatest value lies in the ability of the user to make the most of this learning tool.

With the appropriate filmstrips as the focal point, the study can be directed to specific topics such as the "Occupations of the People" (Strips 46-2, 46-3) or "What We Have Gained From Spain and Portugal" (Strips 46-4, 46-5).

By providing supplementary reading materials, exhibit materials, and the Student Study Manual a strong discussion group can be stimulated.

The Student Study Manual has been prepared to help the individual to pursue his personal interests, but it also contains many suggestions for things which need to be discussed. Because it is geared to both individual and group aids it is suggested that a Student Study Manual be a prominent part of each Learning Center related to Spain and Portugal.

WIDE RANGE OF USES

Student abilities. Skill development is needed at all levels of learning. The students may be "slow learners," "gifted" or any of the stages in between. There is always need for moving forward toward the greatest potential of each. This

Encore Visual Education, Inc.

before the showing. These sheets may contain word lists, key concepts to be stressed, specific questions to be answered, a map of the locale of the filmstrip content, and possibly other items.

⊗ ENCOURAGE STUDENT PARTICIPATION DURING THE SHOWING While it is neither possible nor desirable to lay down hard and fast rules about what your students should do while viewing a filmstrip, they will probably enjoy the filmstrip and learn more from it if you do some or all of the following:

Assign different students to carry out a number of specific tasks during the showing. In presenting a filmstrip on England, for example, ask some to note styles of architecture, some the general appearance of the countryside, and still others the types of farming and manufacturing shown.

On some occasions, encourage students to jot down notes about what they see, read, and hear during the filmstrip showing. Because filmstrips need not be projected in total darkness, there will usually be enough light for note-taking.

Invite students to ask questions as the showing proceeds. If questions arise relative to frames which have already been passed, you can turn back and review them.

Have one or more students operate the projector. With the story type of filmstrip for young children, call on other students to read captions.

Stimulate thinking and guide viewing by asking questions or calling for comment about various frames.

⊗ FOLLOW UP THE SHOWING Activities following filmstrip showings should be aimed at clarifying unclear points, evaluating the extent to which original purposes were achieved, and helping students to clinch what they have learned. Follow up filmstrips by activities such as these:

Ask students what they discovered that answered preshowing questions. Discuss each question in turn.

Have students ask each other questions about the filmstrip they have just seen.

If the filmstrip demonstrated techniques or skills which can be applied in the classroom, take time immediately to practice them.

Ask students to write down several main points treated in the filmstrip.

Assign additional research to be done on the filmstrip topic. Students may go to the learning resource center for more information and report back to the class.

Reshow all or parts of a filmstrip as many times as necessary to clarify points. Some teachers, on reshowings, ask different students to explain the significance of individual pictures.

Use a narrative filmstrip to stimulate writing by stopping the showing at a crucial point and suggesting that students complete the story in their own words and according to their own ideas. At a later time, show the rest of the filmstrip and compare endings.

special filmstrip-showing techniques

Two additional filmstrip teaching techniques will be mentioned here. The first is to use a simple cardboard mask, manipulated by hand, to produce flash exposures. This technique is especially useful for

speed-recognition exercises and for identification quizzes. With this procedure it is unnecessary to resort to the use of special tachistoscopic attachments on your projector.

The second technique is to arrange the sequence of filmstrip frames. It is possible, with very little trouble, to skip ahead or to review earlier frames without showing any of the intervening ones. When skipping frames, place your hand in front of the lens and count frame clicks. When you reach the desired frame, quickly withdraw your hand to project the image. Practice a few times to improve your skill in synchronizing the movements involved.

San Diego County (California) Schools

uses of filmstrips
by individuals

Largely because of the availability of simple, economical filmstrip viewers and projectors, extensive use of filmstrips by individuals and by pairs of students has become common practice in classroom learning centers and in media centers. Here are some suggestions for obtaining maximum value from the use of filmstrips in independent study:

⊗ Arrange with the media center to set up a filmstrip learning center with a collection of filmstrips about special topics within study units. In some

LaMesa-Spring Valley (California) Schools

situations the filmstrip center can be in your classroom. Direct students to the collection to find information and ideas needed for reports.

⊗ Prepare study guides for the students to use, or, if available, adapt the guides that are supplied by the producer.

⊗ Rather than permitting students to limit their analysis to identification and description of the content, develop questions that require students to think and to reason as they get visual cues from the filmstrip frames.

⊗ Assign one student to study a sound filmstrip in detail and write a new narration for it which may be read aloud to the class or presented as a recorded tape. Suggest including points that would be important to the class.

⊗ Have a student study a silent filmstrip, or a sound filmstrip without the sound, and write a narration for

an audio tape which will also include music and sound effects, as appropriate. The student production can be presented to the class when the subject treated is timely.

⊗ Record your own narration for a filmstrip to adapt the content of the original sound to the special needs of your group and to meet their specific objectives. You may ask questions or include activities for them to work on during the viewing.

3M Company

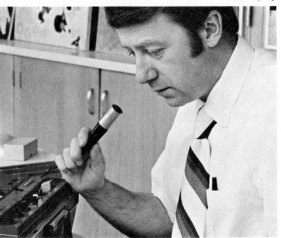

producing handmade filmstrips

Have you considered inviting your students to make their own filmstrips without a camera? You can now. The product which makes it possible is a special 35mm nonphotographic film on which one can write or draw. (See Reference Section 5, Directory of Sources.) The film has a dull side for copy or art work; and because the film has conventional sprocket holes, the finished strip can be projected.

It is easy to draw or write on the film. The dull (etched) surface accepts black pencil, translucent colored pencils, or felt-tip pens. To make the finished work permanent, the film may be coated with clear plastic spray.

Some nonphotographic film will accept typewriting made with a carbon ribbon. For best results, place a sheet of paper under the film before putting it in the typewriter.

Also, 2- by 2-inch slides can be made with this film. After doing lettering or art work in slide format, cut and mount the film in cardboard, plastic, or glass slide mounts.

This write-on film material is also available in 16mm size. Show your students Norman McLaren's handmade animated films, *Hen Hop, Fiddle Dee Dee,* or *Begone Dull Care* (National Film Board of Canada). Many delightful student productions have been made by using this technique.

DO-IT-YOURSELF FILMSTRIPS

Using carbon or translucent audio-visual pencils or pens, students may create their own filmstrips. The framing device shown here ensures that drawings will fit screen dimensions.

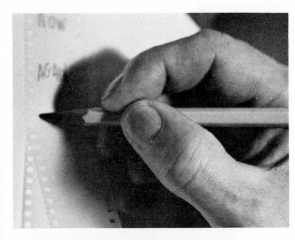

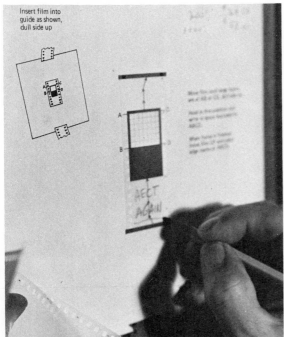

summary

The several forms of still pictures discussed in this chapter include: flat opaque pictures printed on paper, 35mm sound and silent filmstrips, 2- by 2-inch slides, stereo slides in reels, and 3¼- by 4-inch slides. With the exception of the latter, continuing improvements in the quality, availability, and adaptability of these media have stimulated their increased use in education.

Of the several media in the still-picture category, flat opaque pictures and 35mm filmstrips with sound are the most widely used in schools.

Still-picture resources are particularly advantageous because of their low cost and availability, the versatility of their use with either groups or individuals, and the low cost of projectors and viewers.

Useful and valuable flat pictures should be protected from wear and tear; the processes for mounting and coating them are simple, the materials and equipment easy to use, and the results excellent.

Recommended procedures for teaching with still pictures, especially flat opaque pictures and filmstrips, are similar to those recommended for other media forms. But special advantages may be gained from using still pictures by enlisting student participation in processing them and using them in a variety of learning experiences and projects. Examples of such applications are in preparing bulletin boards and table and wall displays, developing picture collections for use in an opaque projector, writing creative stories as assignments growing out of picture study, and participating in visual literacy activities. Still-picture materials are also valuable as the basis for nonverbal, pictorial tests to measure aspects of achievement that are otherwise difficult to evaluate.

10

AUDIO MATERIALS

chapter purposes

⊗ To inform you about the great range of available audio resources and to stimulate your interest in using them.

⊗ To assist you in learning how you can help your students improve basic listening skills.

⊗ To show you how to teach with audio materials in large-group, medium-group, and independent study situations.

⊗ To assist you in developing confidence and skill in making recordings and in learning how to encourage your students to learn through making them.

⊗ To emphasize the continuing importance of radio broadcasting and simulated radio broadcasting as learning activities.

⊗ To suggest the potentialities of telelecturing, telewriting, and telephone teaching in education.

⊗ To stimulate interest in tape exchanges and international correspondence activities.

3M Company

Among the media resources that give students incentive for learning, the audio materials discussed in this chapter have proved to be both practical and effective. Of the many different materials described in this book, those in the audio category are among the least expensive, the most readily available, and the most versatile in application. In addition, the technical equipment required to record, play back, or edit audio communications is easy to operate and transport and sufficiently sturdy to withstand the heavy use to which it is usually subjected. Besides the usual audio resources, innovations in the use of

Mesa-Spring Valley (California) Schools

instruments that transmit and receive radio and telephone signals, including those that write at long distance, are being applied creatively in education.

ready-made audio materials

The many audio experiences which are available in interesting and varied forms include disc recordings, cassette tape recordings, reel-to-reel tape recordings, and recorded cards. There is such a wealth of prerecorded materials, and on such a range of subjects and levels, that the chief problems are how to become familiar with sources of information about them, how to select those of most worth, and how to find sufficient funds to pay for them.

range of prerecorded audio materials

Consider, for a moment, the scope and quality of ready-made audio materials and experiences:

⊗ The thousands of excellent musical, literary, and documentary materials recorded on discs and tapes

⊗ The great variety of foreign language audio and audiovisual instructional materials.

⊗ The many educational and informational radio programs that continue to be broadcast to classrooms

⊗ Study kits of many different kinds that combine audio and visual materials to permit students to pursue studies independently and at their own pace.

As discussed earlier in Chapter 4, your search for suitable audio materials should begin in your own school media center. There you may find many valuable items, including, perhaps, some that will bring to your students voices of past or current political figures: Adolf Hitler, Eleanor Roosevelt, Harry Truman, Martin Luther King, Winston Churchill, Andrei Gromyko, and many others. Other recordings may communicate the tenseness and significance of re-created historical events, helping listeners to be audio witnesses to the murder of Julius Caesar, the rout of the Spanish armada, the death of Joan of Arc, or countless other memorable events.

In literary fields, you may choose freely from recordings of poets such as Dylan Thomas or Robert Frost reading and interpreting their own works. Dramatizations of young people's classics, such as *Treasure Island* or *The Three Musketeers,* may also be available. Or you may introduce the Old Vic Theater Group in its interpretation of *A Midsummer Night's Dream* or *Macbeth.* Entire plays, such as *Death of a Salesman* or *The Cocktail Party,* may be on hand.

Specially produced teaching tapes are now available commercially on hundreds of subjects and for all grade levels: spelling and sight vocabulary drills; grammar and punctuation (*Paul the Period, Intro-*

United Nations   *United Press International*   *United Nations*

duction to Syllables); library skills (*How to Find a Book in a Library, Dewey Decimal System*); listening and reading skills (*Audio Reading, How to Read Line Graphs*); social studies (*Visit to Germany, Zones of the Earth, Our Shrinking World, Revolutionary War*); and science (*Animal Classifications, Human Heart, Our Bird Friends*).

You can do a number of things to become acquainted with the content and usefulness of pre-recorded audio resources and to recommend the best of them for purchase by your school media center. Of course, final selections ought to reflect the usual cooperative media-selection practices and policies of your school, as discussed in Chapter 4.

Jeffrey Norton Publishers, Inc.

But it will also be advantageous for you to audition as many items as you can and to write for audio catalogs (usually free) from major producers. (See the list in Reference Section 5.)

compressed speech: a listening innovation

Many teachers have become interested in rapid, or speed, listening—sometimes called compressed speech. Technological developments have made it possible to record speech at a normal rate—i.e., 100 to 125 words per minute—and to play it back at faster speeds without changing the pitch of the sound. This shortening of replay time is accomplished by automatically editing out the minute time intervals of silence within or between words.

By recording in this way, it is possible to increase presentation speeds up to 400 words per minute without distortion. However, with the delivery rate thus increased, listening with understanding is more difficult than at a normal rate. Yet, successful listening at accelerated rates of speed can be developed by nearly everyone.

Rapid listening at the upper end of the speed scale seems to be best suited to the review of familiar content. Moderate speed increases appear to improve concentration upon materials heard, whether or not the listener is familiar with them. Some speech compression recorders also have the capability of lengthening time intervals between words and word portions, an especially valuable feature.

educational tape-recording project (NCAT)

A particularly important source of tape recordings you and your students will want to use is the National Center for Audio Tapes (NCAT), at the University of Colorado (Boulder), associated cooperatively with the Association for Educational Communications and Technology (AECT) and the National Association of Educational Broadcasters (NAEB). Under the plan of operation for this project, various agencies throughout the country—college, university state departments of education, government, commercial, and public school—contribute master tapes to the National Center collection. These, in turn, are classified, cataloged, and reproduced (duplicated) upon direct order from educational institutions.

National Center for Audio Tapes

ncat
ncat
ncat
ncat

NATIONAL CENTER FOR AUDIO TAPES

CATALOG $4.50

UNIVERSITY OF COLORADO
348 Stadium Building
Boulder, Colorado 80302
TELEPHONE 303 443-2211,
EXT. 7341

Programs available through NCAT range in content from *American Tradition in Art, Adventures in Music,* and *Critical Issues in Education* to *Driver Training, Old Tales and New, Children of Other Lands,* and *South Africa's Racial Problems.*

Program content appeals to various grade levels, and, in reel or cassette form, the programs of the organization, supplied by the Center, are always recorded on high-quality tape. NCAT policy permits those who use its services to reduplicate for noncommercial educational purposes as many copies as desired of its more than 14,000 programs.

Service costs, which vary according to program length, are well within reach of the materials budgets of most schools. A catalog which classifies and annotates the entire NCAT master tape collection is available from the Center. Check with your school media center to see whether a copy is on hand there.

student listening skills

You and your students can ask yourselves a few questions about your own experiences with listening: What skills do we consider to be most essential to listening? Do we believe that our listening skills can be improved? If so, how? How much attention has been given in our school careers to help us improve our listening skills?

research about listening[1]

Research and experience, together, tell us much about listening. We know, for example, that:

⊗ As much as 70 percent or more of the average adult's working day is spent in verbal communication—with 45 percent of that time used for listening.

⊗ Nearly 60 percent of the time in elementary classrooms, and 90 percent of the time in high schools and colleges, is likely to be spent in listening.

---

[1] Adapted from Sanford E. Taylor, *Listening,* National Education Association, Washington, D.C., 1964.

Aurora (Illinois) Public Schools

Oregon State Department of Education

⊗ Students retain surprisingly little (perhaps only a fifth to a third) of what they hear. Even mature persons retain only 50 percent, on the average, no matter how hard they try. Two months later, they will be unable to recall even half that.

⊗ People write, on the average, about 25 words per minute, speak at about 100 to 150, and read or think silently at about 300.

⊗ The rate of speaking appears to influence the listener's learning, but only when it falls short of or exceeds broad limits of the range between 125 and 200 words per minute. While faster listening rates may be maintained for short periods, listeners prefer a rate of between 150 and 175 words per minute.

⊗ Attention and concentration are important to listening results. Attention will wander if what is listened to is either too difficult or too easy to follow and understand, or if the room environment is not right (too hot, too cold, acoustically poor, badly ventilated, or filled with visual or aural distractions).

⊗ Improving listening skills must be planned for and developed by thoughtful application and practice.

⊗ Improvement of aural literacy skills (of knowing how to listen efficiently) permits students to extend the range of their contacts with the world, and thereby, to increase their capabilities to experience, to learn, to apply, and to enjoy.

improving listening skills

Teachers and students are interested in finding out how effectively they listen and what they should do to improve their listening skills. To determine the present status of your students with respect to these skills, you may:

⊗ Administer a standard listening test for which norms have already been developed.

⊗ Develop and refine through further use a listening test of your own. Try to include in your test means of measuring the listening skills you consider to be most important.

Typically, listening tests will emphasize measurements of the abilities of students to (1) direct and maintain attention toward what is being listened to, (2) follow directions, (3) use auditory analysis, (4) obtain meaning from context, (5) distinguish between relevant and irrelevant information, (6) find and recall main ideas and details, (7) listen critically, and (8) listen for appreciation. The guidelines which follow suggest class activities that permit measuring and improving listening abilities.

⊗ DIRECT AND MAINTAIN ATTENTION Find opportunities for students to measure or practice their ability to direct and maintain their attention while listening to you or others read aloud interesting descriptions or dialogues. Ask them to close their eyes for a few seconds, to open them, and to prepare a list of words identifying every sound heard in the interval. Or read short paragraphs aloud and ask them to count the number of times they hear particular words.

⊗ FOLLOW DIRECTIONS Provide practice that will help improve student abilities to follow directions by listening to you or a tape recording and entering appropriate signs or comments on a prepared answer sheet. Oral directions may state: "Put an X on . . .; circle the . . .; fill in the rest of the picture under number 7; estimate the distance in miles, between points A and B of the map." Or ask students to listen to and repeat aloud a set of directions typical to those that might be given to a traveler. Make a point of not repeating directions. Stress the importance of paying attention and getting the message the first time.

⊗ USE AUDITORY ANALYSIS Read a series of nonsense syllables (or use foreign-language phrases) and ask students to repeat them. Play sound-effects records and ask students to identify sound sources.

⊗ USE CONTEXT IN LISTENING Read aloud sentences containing unfamiliar words; determine the accuracy of student interpretations of their meaning; discuss clues to meanings provided by the context. Read aloud open-ended, incomplete sentences and ask students to finish them in ways that make sense, using only the contextual clues.

⊗ DISTINGUISH BETWEEN RELEVANT AND IRRELEVANT INFORMATION Read a paragraph containing at least one out-of-place sentence; ask students to identify it. Read sentences containing poorly chosen or inappropriate words and ask students to identify them. Read a short speech containing several appropriate and one or more inappropriate arguments; ask students to identify those in both categories and to give reasons for their selections.

⊗ FIND AND RECALL MAIN IDEAS AND DETAILS Read aloud a story containing descriptions of a number of events. Ask students to restate them—in their own words and in the order of their occurrence. Read aloud a very short story, scrambling the order of some events; then ask students to retell the story, putting events in their proper order.

San Diego County (California) Schools

teaching with audio materials

In teaching with disc or tape recordings, as with most instructional materials discussed so far, there appears to be no single best formula or procedure to follow. You will recognize the utilization suggestions that follow as extensions of those of a more general nature supplied earlier. But here, they are adapted specifically to audio media.

prerecorded audio materials in large- and medium-group instruction

The familiar five-step procedure for using educational media with large- and medium-sized groups is but one of several promising approaches to using audio materials. It requires that you:

⊗ PREPARE YOURSELF To use audio materials with medium- or large-sized groups, you must, of course, do advance planning. Consult available printed materials or program notes and listen to recordings before you attempt to teach with them.

Take notes on important points; decide how to build interest, what to discuss, and how to test comprehension or appreciation. Give special attention, too, to ideas which may prove difficult for students to grasp and which should be explained before listening begins. With radio programs, decide whether to listen to the immediate broadcast or to record and use it later. Decide, too, whether one or more student committees will be assigned any special responsibilities in connection with the listening experience.

⊗ DEVELOP STUDENT READINESS Bring your students to the point of readiness to listen. Stimulate and focus interest through preliminary comments and questions.

⊗ LISTEN TO THE PROGRAM Lead into the listening experience with proper timing and minimum delay. Perhaps you will want to dim room lights or

3M Company

inaccurate? Check the ability of your students to use the key words and phrases that were identified earlier. Correct any misunderstandings identified through discussion. Decide, too, whether it will be necessary to listen again to all or part of the material to clear up such problems. In many cases, the postlistening discussion may complete the listening experience. In the majority of cases, however, it is expected that students will be motivated to want to learn more about some specific aspects of the subject through library reading, studying the text, seeing films or filmstrips, hearing other recorded programs, or engaging in other appropriate learning activities.

⊗ LISTEN CRITICALLY Ask students to listen to a recorded speech with several questions in mind: (1) What are the speaker's apparent motives or purposes? (2) What emotionally toned words or expressions does he use? (3) Do his views seem to be based on fact or on opinion? (4) Does he use propaganda techniques or logic?

⊗ LISTEN FOR APPRECIATION Use any of the multitude of excellently prepared recorded materials related to effective use of phraseology, such as poetry readings, dramatizations, or monologues. Ask students to note cadence, inflection, emphasis, and other skills of delivery evidenced in the presentation.

partially draw curtains. Encourage good listening by asking students to (1) listen quietly and courteously; (2) concentrate on the audio material, upon what is said, how it is said, and what it means; (3) listen with a willingness to hear a point of view although it may differ from their own; (4) consciously relate what is heard to problems and questions set up during the prelistening period.

⊗ FOLLOW UP THE PROGRAM It is nearly always appropriate to hold some type of postlistening discussion. Move back to the list of preliminary questions. Which were wholly or partially answered? Do students agree with viewpoints expressed? In what respects do they differ with them? Were characterizations valid or were they overdrawn or

Clark County (Nevada) Community College

audio-card instruction

A particular advantage of audio recording is demonstrated by magnetic audio-card readers of the types shown above and below. These devices permit one or more students to listen to words and repeat them, and at the same time to see the words in print. The practice responses of students can be recorded simultaneously and played back immediately for comparison with the card model or for teacher evaluation.

Other creative uses of this device may occur in teaching foreign languages and mathematics,

Audiotronics Corp.

especially in making effective combinations of pictorial and verbal materials in self-check items.

tape teaching in small groups

Tape teaching in small groups provides opportunities to divide your class to meet several types of learning needs at the same time. For example, you may assign two or three students to complete yesterday's recorded lesson involving worksheet calculations on simple fractions while you carry out reteaching activities with several others whose previous work showed lack of understanding.

San Diego County (California) Schools

In another situation, you may give tapes with different assignments in a subject to groups of three or four students, each with different achievement levels. Although they work side by side, the students do not disturb each other; each concentrates on completing his or her own assignment.

If desired, you may also include on your tapes the information the students need to check their own answers, and even point out reasons why certain answers would not be correct.

If you have just one tape recorder, and use it only as a group teaching device, you are not taking full advantage of its potentiality. Add to its usefulness by feeding its playback output into a connector box containing jacks for multiple earphone sets. As many as a dozen or more sets of earphones can be

Clark County (Nevada) Community College

Dallas Community College District

used to listen to the output of a typical reel-to-reel recorder; even portable cassette machines will usually handle several sets of earphones without additional amplification.

self-instruction tapes for independent study: listening only

One reason for the increased utilization of self-instructional tapes is the improvement in quality and the reduction in cost of audio cassette equipment. Recorder-players are now sold at prices many students can afford. Furthermore, the equipment for duplicating cassettes has been developed for personal use so that students may now go to the media center circulation desk, choose the particular tapes they want, and duplicate them in only a few minutes, usually at no charge. They may listen to such tapes in the center or take them home or elsewhere to

study at their own convenience. When the tapes have served their purpose, they may be erased and reused.

Growing use is also being made of wireless playback systems, the short-range signals of which are picked up by receivers built into headphones, thus freeing listeners to hear while moving about within a limited area.

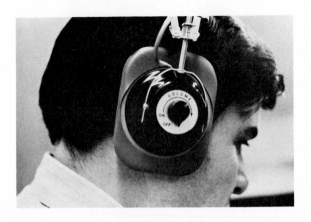

self-instruction tapes: audiotutorial learning

Creative teachers and students alike find it interesting and rewarding to explore uses of audiotutorial instruction in classrooms, laboratories, and school media centers. As mentioned earlier, in Chapter 2, the process involves the effective mix of print media with real things and audiovisual resources, and guidance from accompanying audio tapes. Many audiotutorial units are commercially produced.

Hagerman (Idaho) High School

They usually come with a teacher's guide, a set of disc or tape recordings, and a student workbook. Sometimes, realia, inexpensive devices, filmstrips, or other audiovisual media are also provided.

3M Company

Dallas Community College

The principles of using either commercially produced or locally made audiotutorial kits are essentially the same. Yet if you plan, produce, use, and refine some of your own, you probably will learn much about teaching and learning that you can use in other ways. A few examples of audiotutorial applications will suggest their usefulness:

⊗ In a high school chemistry class, a teacher recorded introductions to each laboratory experience; pictures, printed with locally available, low-cost offset processes, were provided in student binders to be used as the tape was played. Students were instructed in the terminology and proper arrangements of equipment and in the procedures and objectives for required activities. As lessons proceeded, they were directed (by tape) to perform manipulations; to go to special stations to study slides or short film clips or loops; or to visit an exhibit, in the corner of the classroom, which visually presented elements of the process not revealed in real time or with real things. During instruction, the teacher was available for conferences, and also observed student study techniques and sought ways to improve instructional processes.

⊗ For a home economics course, each project was introduced by means of a tape-slide lecture-demonstration prepared by the teacher. Each such presentation contained built-in audio quizzes and

reviews and gave students opportunities to use real things in demonstrating what they have learned.

⊗ For a high school economics course, the instructor prepared several audio study guides—one for each reading assignment. Using either their own cassette recorders or units borrowed from or used in the learning resources center, students listened to the instructor as they completed out-of-class reading assignments. On the tape, the instructor introduced the reading assignment and provided a set of directions for items of interest and importance to which special attention was to be given. The student was frequently asked to stop the tape, read only part of a page or a paragraph, then turn on the tape again. The instructor then commented on the material read or asked questions about it.

You will find it rewarding to try other things on your own. For example:

⊗ Prepare a series of exercises and place them in ordinary three-ring notebooks. These may be developed through use of pictures, maps, charts, diagrams, or text material cut or reproduced from magazines or books or typed from your own notes. Record your directions, questions, and comments on tape. Guide the study of accompanying visual materials by placing identifying numbers or letters at appropriate locations on the pages. Provide, in another part of the notebook, self-check answers to problems.

⊗ Draw (or cut out) pictures of common sound-producing machines, animals, or other things children are likely to encounter in their environment. Tape record, in random order, the sounds actually made by these things, preceding each with a spoken number or letter. Ask students to use these numbers or letters in matching sounds with pictures.

⊗ Record spelling lists by reading words and providing examples of their uses in sentences. Leave a sufficiently long time interval between words to permit students to spell them without having to stop the playback unit. Include an oral end-of-list verification.

⊗ Read and tape record a story, or information section, from a book. Ask students to listen to the story and, at the same time, to follow it in the book.

⊗ Develop table-top number games with directions for each narrated in your own voice.

3M Company

evaluating with recordings

Recordings provide a variety of materials with which to measure student skills and abilities, understanding, and appreciation. Following are several methods for making such evaluation—in large- or medium-group or individual study modes.

⊗ HAVE STUDENTS LISTEN TO A DRAMATIZED RECORDING OR A RECORDED LECTURE Ask definite questions of fact or interpretation based entirely on what was presented. Note and evaluate the significance of variations in responses.

⊗ PLAY A PORTION OF AN UNIDENTIFIED DRAMATIZATION OR SPEECH Ask students to identify the speakers, the occasion, the time, events preceding or following, and the thoughts expressed in the selection.

⊗ PLAY ALL OR PART OF A SPEECH OR DRAMATIZATION Ask students to evaluate critically what they hear by giving attention to opinions or ideas expressed, quality of the dramatization, the speaker's enunciation, emphasis and expression, or length of the speech.

⊗ LISTEN TO PART OF A PROBLEM-STORY PRESENTATION, BUT STOP SHORT OF THE ENDING Ask students to produce their own endings based on applications of principles they are supposed to have learned.

⊗ PLAY ONLY A DRAMATIC ENDING TO AN UN-FAMILIAR STORY Ask the class to develop, crea-

Folkways Scholastic Records

tively, the basic elements for events which could have taken place prior to the ending they have heard.

Another use of the tape recorder in evaluating student learning involves an oral test. With this technique, a full list of discussion or essay-type questions may be developed and handed out to the class in advance. Instructions indicate that, on the day of the examination, each student will be asked to speak extemporaneously on just *one* of them —but that his or her particular question will be selected by means of a random drawing. At the time of the examination, and with the entire class present, the first student draws a question, takes a minute to review what he or she will say about it, steps to the microphone, and begins. While this is going on, the second student draws a question and plans a statement, and so forth. Each student thus receives a question he or she has already seen (along with perhaps thirty others), has a minute or so to organize an answer to it, and delivers that answer orally into the recorder microphone while others listen. The instructor later listens to the resulting tape, reviews each performance, and analyzes and evaluates it at his or her own convenience and with as much repetition as is needed.

recording as learning experience

In the relatively short time they have been available, audio tape recorders, particularly of the cassette type, have become highly valued and widely used to promote applications of locally made recordings and recording activities. The following examples suggest a few of the many ways of developing such opportunities:

⊗ Because each student may listen to his or her own cassette, place drill materials on tape for individual instruction. Then, while your students respond to problems, or carry out whatever activities are assigned, circulate among them to observe their work and give assistance.

⊗ Assure consistency in repetitive activities, particularly those related to drill exercises, tests, or the giving of directions.

⊗ Rearrange audio materials, by shortening, changing the order, or adding materials from one or more sources. You may juxtapose materials from different program sources for comparisons or contrasts.

⊗ Record and preserve sounds for future use and analysis.

⊗ Plan and edit in order to time audio materials exactly. Because tape recorders run at standardized speeds, the amount of time required to play a certain footage of tape can be computed accurately in advance of its use.

Oregon State Department of Education

⊗ Automatically synchronize the sound and pictures of a sound-slide presentation. Special tape recorders, or accessory devices for standard recorders, permit recording of pulse-cue signals to operate various types of projection and playback equipment.

⊗ Use opportunities to evaluate your performance as a teacher and the performance of your students. Making and studying *before* and *after* tape recordings will help your students to hear for themselves how they have improved or failed to improve audio performances such as speaking, discussing, or singing.

⊗ Make useful alterations of sounds for analysis or explanation. Slowing down a tape recorder while playing bird calls originally recorded at higher speed, for example, produces sounds which, although unlike those of nature, provide clear audio clues to their structure.

⊗ Transport sounds from one place to another. Congress has recognized the instructional value of tape recordings by permitting them to be mailed like books, at low postal rates.

⊗ Make duplications of sounds easily and economically. An original tape recording can be reproduced in any number of copies, each with practically the same quality of sound as the original.

⊗ Perfect the sound materials—through trial, editing, retrial, alteration, deletion, and rerecording —until they meet your standards and objectives.

Applications of tape-recording techniques extend across the full range of the curriculum, as suggested by the following selected examples:

FOREIGN LANGUAGE Use special language laboratory equipment or creative combinations of ordinary tape-recording devices to provide students with audio-active foreign language practice. One tape player may be used to present the language model, for example, while a second is used to record responses. In many schools, the media-center staff provides audio lesson-duplication services, supplies duplicated lesson tapes, and, on request, transfers masters to the students' own tapes, to be used at their personal convenience.

⊗ SPEECH Give students opportunities to evaluate their own speech—enunciation, tone, pronunciation, speed, expression—and to recognize proficiencies as well as errors or inadequacies to be corrected. Make tape recordings at the beginning and end of the semester to provide a useful basis for evaluating change and progress.

⊗ MUSIC Use tape recording to provide objective bases for self-evaluation and improvement in music classes.

⊗ BUSINESS EDUCATION Use prerecorded tapes to carry necessary introductions, directions, and post-use evaluation suggestions for specific individual or group assignments such as those for alphabetic drills, keyboard-review drills, carriage-throw drills, or common-word drills. Use tapes with shorthand classes to demonstrate desirable timing, pronunciation, or emphasis.

⊗ ENGLISH Teach discussion and listening skills through tape recordings. Use the tape-recorded sound track of a problem-type film, for example, to practice (1) identifying what was said, (2) determining the meaning of what was said, and (3) assessing its significance. Oral book reports may also involve functional uses of tape recording. Also, in English, or for almost any other type of instruction, you may save your own time and that of your students, by recording your critiques of written assignments. As you review student papers, record your comments

3M Company

on short tapes (five minutes on a side, for example), one tape for each paper on which you write brief marginal reference notes to guide students as they listen to your tape.

⊗ PHYSICAL EDUCATION Record personal notes on the field—orally—for later reference in the locker room. The handy off-on switch on the hand-held microphone facilitates such recording, conserves tape, and helps avoid recording irrelevant sounds.

⊗ ART Use a portable cassette tape recorder while walking about an exhibit of creative student work, making comments to be referred to later in critique sessions with the artists.

learning by editing tapes

Editing tapes is a process that requires skill and integrity—skill in handling alterations to be made and integrity in respecting the rights of those whose voices are recorded and avoiding any distortion of their meanings.

However, with such skill and integrity, acceptable and desirable alterations can be made to tapes through editing. For example, it is possible to do the following:

⊗ By judicious excising and bridging, reduce the overall length of a program.

⊗ Rearrange the order of events by placing one sound sequence ahead of or behind its original position on the tape, thus providing bases for comparisons, contrasts, or emphases.

⊗ Insert new materials to round out taped presentations.

⊗ Intermix speech and music, using music for transitions and for setting mood.

⊗ Add pulse-cue signals at desired locations in the tape to control synchronized sound-picture presentations.

Most of these program modifications can be achieved electronically—i.e., by playing one tape recording into another recorder and onto a new tape without cutting the playing tape. Alternatively,

FINAL DRAFT

    T A P E   S C R I P T :   G E O G R A P H Y
This tape will help you select courses for your
~~geography~~ major.

~~As we proceed,~~ *B̶e* sure you have a current copy
of the San Jose State University GENERAL CATALOG
FOR UNDERGRADUATE STUDY. If you do not ~~have one~~
~~with you now,~~ STOP THE TAPE, find one, and START
AGAIN.   ~ PAUSE ~

At San Jose State University, the study of Geography
involves many things--from the traditional approach
of looking at life styles to more technical matters,
such as working on environmental or urban problems.
In Geography you may learn to make maps--which
now uses computer technology and draws on air photos
and satellite photography. You may also become
involved in land use mapping.

but less desirably, with reel-to-reel tapes, editing can also be accomplished mechanically by using scissors or tape splicer and splicing tape. Reference Section 1 provides detailed technical information about editing tapes.

To emphasize how easy it is to edit tape recordings, here is a brief report of an actual incident:

Several years ago, a nationwide ninety-minute broadcast of *Hamlet* was announced. A high school English department asked the school media director to arrange to tape it so that students who wanted to could hear it later.

It was obvious, however, that it would be impossible to use a full ninety-minute recording within the usual fifty-minute period. And, if the recording were divided into two forty-five-minute parts, the effect would be impaired.

As a way of overcoming this problem, two students volunteered to edit the tape and to reduce it to just half its original length—to forty-five minutes. This is the way they did it: First, they duplicated the original tape by playing it on one machine while rerecording it on another. Second, they played and replayed the ninety-minute version while following a regular printed copy of the play. They located and marked each speech on the tape. By the time they had run through the play twice and had read the printed version, they were quite familiar with *Hamlet!*

After some discussion, the two students decided which speeches to leave intact and which to excise entirely or to shorten. Of course, their primary concern was to preserve the thread of the story, the most important passages, the flavor, and a sense of continuity—all as best they could. After marking the printed text to identify the portions to remain in the tape, they decided what comments or musical bridges were needed to provide smooth, understandable transitions between excerpts. These comments generally described the thread of the action and led nicely into the next excerpt.

Then, with the marked text in hand, the two students began to rerecord the edited version. They played the original ninety-minute tape on one recorder and recorded it on a second, stopping and starting the second to drop out unwanted portions. They developed considerable skill in fading the sound, bringing in microphone comment and music, and bringing up the *Hamlet* dialogue to provide smooth transitions. Whenever appropriate, they shortened muscial sequences. By these means, the students produced a shortened forty-five minute version of the play that retained the

original story substance and quality. Finally, they made a duplicate of the short version to protect the original of the edited version.

This report emphasizes several points: (1) that it is relatively easy to work with tape recordings, (2) that tape recordings can be shortened and edited, (3) that recordings can also be added to, and (4) that programs can be recorded from the radio or, for that matter, from a television receiver, another tape recorder, or a disc record player. It also suggests that one does not actually need to be an expert to make or to edit good tape recordings. And it emphasizes that student ingenuity is a priceless commodity available, in some measure, in every class and in every subject.

The learning outcomes of this experience with tape recording were numerous. The two students involved were allowed to *work* with a play. While doing this, they developed insight into the plot and characterizations as well as an appreciation of Shakespeare's artistry. They made judgments about important ideas and incidents, and they contributed original commentary to bridge gaps in omitted materials. They learned a good deal, too, about the versatility of tapes, about how to do effective editing, and about how easy it is to distort meanings unless one is particularly careful in the process. The outcome of this particular project was also a resource that contributed to the learning of other students who borrowed and listened to the recording in subsequent years.

recording discussions

A simple but usually rewarding activity is to tape-record discussions verbatim—and to analyze them without editing. Results of doing this may be an enlightening surprise. Try this procedure: Next time you lead a class discussion, arrange to start your tape recorder; let it run without interruption. Later, listen to what happened. Did the preponderance of the discussion really come from the students? Or from you? Did you handle student contributions fairly and encouragingly? Were most student comments simply "yes" or "no" statements? Did the discussion wander aimlessly?

Were there monopolizers of the discussion who prevented others from contributing? Does the recording suggest the need for you to find ways to stimulate improved class participation in discussions?

A tape recording of a discussion session permits analysis of the participation pattern of an entire class. Your students will gain considerable insight about themselves and others taking part by tabulating the number and kinds of discussion participation and the extent to which their own ideas were agreed with and accepted, or disagreed with and rejected, or reinforced or altered by others. Each may also discover patterns of behavior, such as who are the ones who change subjects frequently, who are sticklers for detail, arguers, or negative critics. Such discovery may be a help to later improved participation.

Teachers interested in demonstrating the extent to which their classes improve in ability to discuss problems sometimes make tape recordings—one near the beginning and one toward the end of the school year or term. By comparing these two tapes, students can begin to gauge their progress—or their lack of it.

a good discussion . . .

With your class, develop criteria that may be used in evaluating a discussion. Then ask the class to record one of their discussions so that they may evaluate it. After the recording, play all or parts of it for evaluation. Together, decide whether:

⊗ The discussion is based on a suitable topic—that is, one that cannot be answered by yes or no, and that is clearly defined and delimited.

⊗ It involves the participation of everyone in the group and is not dominated by only a few individuals.

⊗ It moves from point to point, often with an acknowledgement by participants that points of disagreement may not have been resolved but have been sufficiently examined.

⊗ Participants use available facts, whenever possible, rather than only biases and opinions.

⊗ Participants welcome, examine, and respect different viewpoints.

⊗ The discussion proceeds to some logical conclusion.

recording interviews

Another good source of audio instructional materials is the recorded interview. The following suggestions will assure a smooth procedure when students carry out this activity:

⊗ In advance, inform the person to be interviewed about the questions that will be asked. Let the person know the identity of the audience and why he or she has been selected for a recorded interview. A telephone call, a note, or a meeting prior to the appearance will help to clarify what is to be discussed and establish mutual understanding.

⊗ Conduct the interview in a suitable environment. Avoid areas having disturbing noises such as clanking machinery, noisy air conditioning units, or loud talking by others. However, on-the-spot documentary sounds are often good for backgrounds and can add realism to the interview without distorting the recording.

⊗ Make the interview friendly and relaxed—and well organized. Use the background information you have about the person interviewed and ask questions about his or her known interests; play down the mechanics of the recording operation itself.

⊗ Don't dominate the interview. Avoid making too many elaborations and observations except, perhaps, to refocus the interview or to provide opportunities for the interviewee to rephrase, to reemphasize, or to elaborate upon a point of particular interest.

⊗ Keep your listening audience in mind. Inject into the interview references to the listening audience or identify questions as those that were asked by particular members of that audience.

Interview tapes are valuable as learning resources in many different curriculum areas. In social studies, for example, tape-recorded interviews will provide for group or individual listening to the comments and observations of community members who may differ in their views regarding local issues; interviews with the mayor, the town clerk, the sheriff, and other community officials can provide insights into the nature and importance of their work. Interviews with local scientists or business or professional persons add interest to studies of vocational opportunities and requirements.

Portland (Oregon) Community College

courtesy and ethics in recording

Courtesy and ethics are important in making and editing tape recordings. Here are a few suggestions:

⊗ Request permission before recording any speaker or discussion.

⊗ Be as inconspicuous as possible and make arrangements beforehand for setting up equipment, placing the microphone, and finding a place at which to record.

⊗ Monitor the recording as the event proceeds.

⊗ Be ready to change the reels or cassettes when needed.

⊗ Avoid allowing recorders equipped with end-of-reel signal systems to disturb proceedings.

⊗ When shortening, juxtaposing, or bridging tape content, be ethical in preserving the integrity of the original.

educational radio: the hidden medium

A somewhat surprising development in the field of radio has been the continued upswing in the number of licensed radio stations in the United States and Canada, especially in the FM (frequency modulation) range. Large numbers of them have been licensed for educational purposes. The school audience as well as the general radio audience continues to grow.

School-owned and nonprofit stations are also increasing their activities, largely because of the growing interest in FM broadcasting. Several types of stations now operate in this category: college- and university-owned stations, those owned by school systems, and privately owned nonprofit stations. All of these types of broadcast units make frequent use of program materials distributed through cooperative efforts of the National Association of Educational Broadcasters (NAEB) and National Public Radio (NPR).

Pistor Jr. High, Tucson (Arizona)

Educational radio broadcasting is a useful means of providing learning experiences for large numbers of students. It has been an especially helpful resource for remote, isolated schools; yet school systems of many metropolitan areas of the country also use radio to enrich in-school instruction.

In the past, radio broadcasts for classroom use were handicapped by a number of difficulties. Schools found it troublesome or undesirable to use radio programs that came at the wrong time of the day or term. Departmentalized schools could not use broadcasts for all sections of a course without changing class schedules. But tape-recording equipment now helps with both these problems, enabling teachers and students to record and repeat programs at will, to save them for delayed use, or to edit or shorten them.

Still another significant contribution of radio to learning occurs when students are involved in planning, writing, rehearsing, producing, and sometimes recording and editing radio programs that are publicly broadcast. Many such programs are produced each year in cooperation with commercially owned or district-owned stations.

KXCV, Maryville (Missouri)

simulated radio broadcasts

The tape recorder and permanently installed public address systems are useful devices for simulating live radio broadcasts. Simulated radio activities appeal to students of all ages. To discover their inherent values, give your students the experience of planning a dramatization, writing and reading a script, or writing and performing music or producing sound effects for taped dramatizations. Here are a few suggestions:

⊗ CHOOSE A WORKABLE SUBJECT A good subject is one that is interesting and at the same time sufficiently limited in scope and complexity to fit the capabilities of the group proposing to do the production.

⊗ DECIDE UPON A STORY OUTLINE Ask each student to write a complete story outline—a start-to-finish description of the story in synopsis form. Then assign a class committee to choose the three or four outlines which seem to offer the most promise, and from them choose the final outline.

⊗ CAST THE PLAY AND ASSIGN RESPONSIBILITIES In addition to actors taking part in the finished dramatization, the simulated project will require other personnel: a director, possibly an assistant, and one or more sound-effects technicians. Someone must also be responsible for providing musical excerpts; and student recording technicians must be assigned to operate the recorder, to place microphones, and to edit tapes.

⊗ PRACTICE AND REVISE THE SCRIPT Try out the script one or more times under regular rehearsal conditions. Make whatever revisions are necessary to give it the quality the group desires.

⊗ MAKE A FINAL TAPE RECORDING When everything is ready, make the final tape recording.

⊗ EXPERIMENT WITH SOUND EFFECTS Records with nearly every kind of sound effect needed for radio dramatizations can be purchased. But homemade effects will also be surprisingly satisfactory.

STUDIO LANGUAGE

Watch me for your cue to start.

You're on; start your lines.

Move closer to the microphone.

Move back from the microphone.

Speak more loudly.

Talk faster; speed up action.

Fade out (music or sound effects).

OK, or fine.

Stop, or cut!

Speak less loudly.

telelecturing, telewriting, and telephone teaching

Telelecturing, telewriting, and telephone teaching represent three related and relatively recent additions to the fund of media resources useful to education. All three have in common the fact that they employ telephone lines to transmit voice, music, or other sounds; telewriting provides a further stimulus —long-distance writing—by means of an electronic, remotely controlled pantographic stylus.

American Telephone and Telegraph Co.

telelecturing

In the case of telelecturing, special pickup and amplifying equipment (furnished at a rental fee by the local telephone company) is used to transmit a lecturer's remarks from one location to several classrooms, or even to individuals, in widely separated locations. Telelecturing has a two-way communication capability which permits listeners to query the lecturer and to hear his immediate response. Students in one location hear, and can also query, those in other locations who are on the line. A relatively inexpensive adaptation of this same procedure involves use of a regular telephone which is attached (again at a fee) to a smaller, less complicated loudspeaker-microphone unit. With this arrangement, students may call individuals in their own communities or, for that matter, anywhere in the country or in the world where telephone service exists and interview them during class time. Con-

versations can be private (between caller and respondent only) or may be amplified to include a number of students.

telewriting

Telewriting involves an oral lecture, explanatory comments, and questions, accompanied by long-distance writing or sketching. Usually, it is an adjunct to organized telelecturing services. The

American Telephone and Telegraph Co.

American Telephone and Telegraph Co.

lecturer using the telewriting device (sometimes called a remote blackboard) may or may not be presenting materials simultaneously to a live class in his own institution. The telewriter receiving unit incorporates overhead projection to permit a group of students to see the visual images.

Extensive work with instructional uses of telewriting has been undertaken by the University of Missouri in St. Louis, New Mexico Highlands University, Stephens College, and New Mexico State University.

telephone teaching

Telephone teaching, under arrangements that permit free participation in classroom discussions, is also provided for shut-ins who cannot come to class. The New York City schools, for example, operated an experimental program of this type for students who had never been in a regular classroom. In this case, the telephone teaching program enriches the regular work the student is doing under the supervision of a home-study instructor who meets with him or her in person, approximately four hours a week. Each housebound student is equipped with

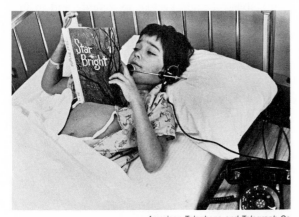

American Telephone and Telegraph Co.

a private unlisted telephone, complete with headset and desk speaker. Reading material is supplied for study, followed by assigned listening, and, finally, participation by telephone with other students on the same line.

Pupils speak to pupils around the world

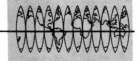

INTERNATIONAL TAPE EXCHANGE

tape exchanges

Because sounds, voices, and music can be captured on tape and readily boxed and shipped anywhere in the world, audio tape recording provides excellent opportunities to further the international exchange of personal communications. This capability is of particular importance to elementary and secondary school students who find it interesting and instructive to record and exchange tapes with audio correspondents elsewhere in the world.

Whether you and your students are producing tapes for exchange within the United States or Canada or seek contacts in foreign countries, several cautions and suggestions apply:

⊗ Before preparing a tape, write a brief letter to express the interest of your group in a tape exchange arrangement and to invite your potential correspondent (an individual or a group) to express willingness to participate. In it, give necessary technical details of the tapes you propose to exchange. Tell whether the tapes are cassette type or reel-to-reel type; give playing speeds, inches-per-second or appropriate metrical system measurements as applicable, and reel sizes. Specify also the track pattern to be used.

⊗ When the tape exchange is agreed upon, produce an introductory tape. Students undertaking the program often make this a program tape, opening with some music, preferably live-group singing, for example. Then a student announcer may identify the school name, grade levels, subject,

World Tapes for Education

formation about services provided by the exchange.

organizing your tape collection

In organizing your tape collection, you are chiefly concerned with (1) standardizing, if possible, on either reel-to-reel or cassette recordings, (2) having a marking scheme that is easy to use and leads you quickly to the tapes you want, and enables you to find the exact places at which to start or stop listening. Here are some suggestions:

⊗ Consider standardizing on cassette tapes as an optimum choice.

⊗ Purchase cassettes of different running times: C (for cassette) 15, C30, C45, C90; the number represents the total playing time of both sides.

⊗ Time your programs and select a cassette that will contain as much of both sides of the tape as possible.

⊗ To simplify playing instructional programs, continue recording the material on the other side of the tape to eliminate rewinding.

⊗ Establish a code system to identify and to file tapes. Keep the system simple, but functional. Perhaps a combination of a letter and numbers will suffice. Include a code to indicate whether the programs are on cassettes or on reel-to-reel tape.

town, state, and teacher and describe the general location of the group. He or she might then give an overview of the tape—its purposes and the main points to be covered or demonstrated in sound. The announcer may provide bridging comments, perhaps with a background of music in the sections following. Toward the end of the tape, a brief summary would be appropriate. Once more, the announcer might identify the group making the recording, invite the correspondent to respond as soon as possible, and mention several specific things fellow students would be interested in hearing about on the return tape.

⊗ Be original in your use of sounds and be sure to identify them. Consider using school musical groups—chorus or folk singers—or the sounds of school buses leaving at the end of the day, the ring of class bells, sounds in the hallway between classes, in the cafeteria during the noon rush, at a faculty meeting in progress or during an interview with the principal, sounds of a typical class in session, of an original poem as read by its author, of a choral-speaking group in action, of an interview with the local chamber of commerce director, or other sounds typical of your environment.

The International Tape Exchange, launched several years ago (by Mrs. Ruth Y. Terry, 834 Ruddiman, North Muskegon, Michigan 49455), has sponsored direct tape exchanges between class groups in this country and those abroad. Write for further in-

⊗ Label each cassette case and each cassette with the identifying code number. Also provide essential information such as date, event, program title, participant(s), location, and running time. (With tapes on reels, include playing speed.)

⊗ Notes about tape contents may be typed on paper or gummed labels and placed either inside or on case covers.

⊗ To systemize your collection, use special storage racks for tapes and cassettes or book-type holders for cassettes—available at music stores.

⊗ Prepare a catalog or index of your collection; provide simple procedures to add new titles.

⊗ To prevent the accidental erasure of program material, remember to remove the safety tabs on cassettes that are circulated for replaying.

Audiotronics Corporation

developing an audio performance test

The following script for an audio recording performance test includes tasks for listening, recording speech with a microphone, using a record player with the tape recorder to record music and speech, and editing with splicing tape.

You should have a reasonable time to complete the test. Note that the test does not require use of long selections that would consume too much time and make checking your work difficult.

To do this assignment you will need a reel-to-reel recorder, a reel of unrecorded tape, a record player, and a disc recording, and all required accessories. Follow these directions:[2]

⊗ Load your recorder with the unrecorded tape and run 5 feet onto the takeup reel. Set the counter at 0.

⊗ Run the tape forward to 35 on the counter. Record the following script: "In making this recording, I am using a take-up reel that is the same size as the full feed reel. I tried using a very small reel, but it didn't work."

⊗ Play back this recording to check your work.

⊗ Run the tape forward to 60 on the counter and record: "I tried using a very small reel for take-up in the beginning, but it didn't work."

⊗ Use the tape splicer and splicing tape to eliminate the words ". . . in the beginning" by excising them and rejoining the tape. Play back and check your edited sentence.

⊗ Advance the tape to 90 on the counter. Connect a record player to your tape recorder and record a very short musical excerpt preceded by your announcement: "I will now play _____ performed by _____." (The music should last at least 10 counts on the counter.) Then fade out the music and say: "That concludes this music lesson. Your narrator has been _____."

⊗ Rewind and listen to your recording.

---

[2] Adapted from John Hofstrand, *Preparing Test Items in Educational Media,* Educational Media Institute Evaluation Project, San Jose, Calif.

summary

The world of sound provides an exceptionally useful variety of resources for learning. Various technical improvements in sound recording and reproduction are being made. The range and quality of audio materials to be used with them are also being improved. While there is a wide assortment of listening and recording devices available (including dial-access units and electronically equipped carrels, and satellite transmitters), the portable cassette player-recorder attracts more and more users among students and teachers alike.

As with other educational media discussed in this book, the value of audio media is determined by the uses to which they are put and the creativity of the users. Here we have stressed the importance of identifying purposes to be served through listening or recording. We have discussed listening skills in general, and how they may be improved. The special skills required to record discussions, record interviews, and edit tapes have also been identified and suggestions have been given about how to conduct these activities.

The continuing influence of educational radio broadcasting and its utilization have been mentioned, as have the opportunities to provide interesting learning experiences by simulating radio broadcasts. Further uses of audio facilities for telelecturing, telewriting, and telephone teaching were suggested. The exchange of tapes (as well as letters) with correspondents in other parts of the country, and of the world, was presented as an especially useful way of extending the range of student learning experiences by capitalizing upon the unique qualities of audio equipment and resources.

11 MOTION PICTURES: FILMS AND VIDEO

chapter purposes

⊗ To remind you of the many important contributions motion pictures can make to help your students learn.

⊗ To examine several common forms in which motion pictures are produced and distributed.

⊗ To offer guidelines to help you locate, choose, obtain, and teach with motion pictures.

⊗ To provide background information about several motion-picture techniques that will improve understanding and enjoyment of motion pictures studied.

⊗ To provide information about ways to organize study of motion pictures as creative art.

⊗ To review some of the contributions to education made by motion pictures produced primarily for entertainment.

Teaching Film Custodians, Inc.

Great Plains TV Library

Motion pictures are rich resources on almost any imaginable topic. A survey in 1974 indicated that there were well over 80,000 16mm instructional films, nearly 20,000 8mm loops and cartridges, and nearly 12,000 video tapes for instruction; these commercially available titles totaled over 112,000. By now, the figure is much larger.

Of course, in order to fill special needs not met by commercially produced subjects, many hundreds more motion pictures on film and video tape have been produced for instruction by school systems, school districts, and colleges and universities. In interesting contrast, in the early 1950s the total number of instructional motion pictures was estimated to be from 9,000 to 12,000 titles. At present, video tapes show the most rapid growth among all instructional media, substantially more than films. All of this suggests that motion pictures are a very basic, easily available medium for instruction.

You may wish to debate the differences between instructional material on films and video tapes, and perhaps argue the logic of calling them by one term, "motion pictures." However, for the purposes of this discussion, please accept the fact that motion pictures used in classrooms and auditoriums or in study carrels are a one-kind medium: films and video tapes may be with or without sound, in black and white or in color, and the images from them seem to move. Both present to viewers images of living things or of art work—animated or static, graphic or photographic.

The equipment and the image carriers of motion pictures differ in film and video methods of access; perhaps another principal medium may soon be used—for example, clear plastic or paper discs. But, motion pictures, regardless of technology associated with them, are a widely used and continually important instructional resource at all levels of education and may be useful in all modes of instruction. Thus, to eliminate repetition and perhaps confusion, the term "motion pictures" will be used.

University of Nevada, Las Vegas

special values of motion pictures

The chief and most often cited advantages of using motion pictures for instruction are:

⊗ They help to bypass some of the intellectual barriers to learning. They communicate effectively and directly without requiring much word-reading skill. The student who experiences difficulty in comprehending such terms as "electricity" or "nuclear fission" through the verbal print medium alone, for example, will usually be helped to achieve such understanding through viewing sound motion pictures on these subjects.

⊗ They aid in overcoming certain physical barriers to human experiencing. Special motion-picture techniques—microphotography, photomicrography, telephotography, and animation—permit the viewing of actions in motion which the unaided human eye would be incapable of perceiving.

⊗ They present action in continuity as it occurs, or as it can be purposely changed to provide some special visual experiences essential to understanding. For example, motion pictures can show action as it takes place normally, and they can speed up, slow down, or freeze the action and show a still picture.

⊗ They enable us to re-create real or imagined events, actions, or processes that have occurred, that may possibly occur, or that may not even be capable of occurring in real life.

⊗ They can often compensate for the differences in background among members of the same group. By presenting an experience that is common to all, the motion picture can be the basis for a discussion that will lead to exchange of ideas and improved understanding of points of view.

⊗ They may often be useful in evaluating students' knowledge or their ability to analyze. Because a motion picture may be stopped immediately, at any appropriate time, a showing may be interrupted and the students given time to answer questions or to ask them.

As you become acquainted with motion pictures of different types and on different subjects, you will find many opportunities and reasons to use them.

MOTION PICTURES:
A SOCIAL RECORD

Among the contributions of motion pictures is the record preserved for society of its history, its problems, and its progress. Through motion pictures, the past has been reconstructed, interpreted, and preserved. Supplementing, and sometimes transcending, the record in writing, in art, and in still pictures, films and video programs today are extending the knowledge, understanding, and appreciations of people of all ages, colors, and creeds. The examples on this page suggest but in a small way the amount and scope of motion-picture material on the past and the present of society.

North American Indian (*Contemporary Films / McGraw-Hill*)

A Time Out of War / *Pyramid Films*)

FILM GUIDE

Film 1. A HISTORY OF THE NEGRO IN AMERICA—
1619-1860: OUT OF SLAVERY
(80 minutes—black and white)

OTHER TITLES IN THE SERIES
Film 2. A History of the Negro in America—1861-1877: Civil War and Reconstruction
Film 3. A History of the Negro in America—1877-Today: Freedom Movement

FOR USE IN
Junior and Senior High School courses in United States History, and in Intergroup Relations.

Church, Industry, Labor, Interracial groups and other adult and youth groups, particularly those of the "inner city."

ADVISER
Dr. Benjamin Quarles
Professor of History, Morgan State
Baltimore, Maryland

PURPOSE OF T
To revi

development of the African slave trade and the growth of slavery in the American colonies. It then presents the Negro's part in the American Revolution, slave labor as the foundation of Southern wealth, and the everyday life of the Negro-slave and freeman, North and South. The Negro's constant resistance to slavery and his important role in abolitionist movement are traced. Film 1 eve of the Civil War.

BEFORE SHOWING
Make it clea
of hi

Contemporary Films / McGraw-Hill

Self-Incorporated (*Agency for Instructional Television*)

MOTION-PICTURE CONTENT

Motion pictures show things, events, and ideas of the world—as they were, as they are, and as they might be.

Indigenous art forms preserved . . .
National Film Board of Canada

Political history re-created . . .
National Film Board of Canada

Welfare of people through health described . . .
National Dental Association—
Wexler II Productions

Protection of endangered species encouraged . . .
National Film Board of Canada

films and video forms

Images can be produced in numerous film sizes, including 70mm, 35mm, 16mm, and 8mm; often productions are undertaken in 35mm, then reduced to 16 or 8mm and distributed for educational use. Video tape production may be on tapes 2 inches wide, 1 inch wide, $\frac{3}{4}$ of an inch wide, $\frac{1}{2}$ of an inch wide, or even smaller. Film programs may be transferred to tape, and tape programs may be transferred to film. For optimum quality, production is usually undertaken in the larger sizes of film or video tape, and reduced in size for distribution and use.

16mm films

The most widely used film in the field of education during the past forty years has been 16mm, with an optical sound track—i.e., a sound track on one edge of the film exposed as variable dark and light areas or densities. Through this track passes a beam of light focused on a photoelectric cell; the pulsing light is converted to sound by electromechanical processes.

During the past ten to fifteen years, motion-picture projectors have been improved, are more efficient and convenient to use, and are available in a variety of forms. Cartridge-type sound and silent projectors and viewers have been developed for 16mm film, but are more frequently used for 8mm film. For large groups, 16mm film is the most satisfactory for viewing; but Super 8mm film is increasingly used for groups of moderate size and for individual and small group study.

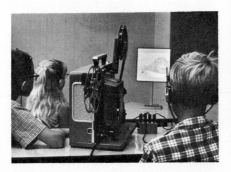

8mm films

Although 8mm film has been used for many years for amateur film production and for home entertainment, it was not until comparatively recently with the development of Super 8mm film that film smaller than 16mm was extensively used in education. Super 8mm cartridge systems usually require no complex threading or rewinding; the cartridge in which the film is loaded is simply inserted into a slot in the projector, and the machine is turned on. Some types of Super 8mm equipment—either front- or rear-screen projection—rewind and turn off automatically at the end of a film and leave the cartridge ready for immediate reshowing. Some Super 8 projectors incorporate automatic programming devices to stop the film on a cue to permit student activity and, at the push of a button, proceed with the showing until the next stop.

video tape formats

Video tape systems, including video cassettes, for presenting motion pictures were developed during the 1970s and have gradually standardized on a few formats and tape widths. Though $\frac{1}{2}$-inch reel-to-reel equipment is popular for inexpensive and convenient local production work, the $\frac{3}{4}$-inch video tape cassette has become nearly a standard for distribution and utilization of video tapes for presentation of motion pictures. Other widths of video cassette tape, or reel-to-reel tape, include $\frac{1}{4}$, $\frac{1}{2}$, 1, and 2 inch; the latter two are used principally for production work.

When considering a rapidly changing technology, keeping up to date with trends in the field as well as current status is a necessity. For example, a 1975 report on Video Disc, a new development, suggested that this medium for the distribution of motion pictures might ultimately replace video tapes and films. Video Disc is discussed in Chapter 12.

Some states and provinces have made efforts to standardize all motion-picture materials on one medium, either film or video cassette. As another approach, many producers are willing to supply their motion-picture products in any format desired by the user. However, consideration of the format of material used is secondary in importance to the procedures for locating, choosing, and using the motion-picture content that is suitable for instruction.

Sony Corporation of America

locating, choosing, and obtaining motion pictures

Locating, choosing, and obtaining the motion picture you use with your classes may present some problems. But if you seek the help of the professional media personnel of your school, the process will be easy.

sources of motion pictures

The sources of motion pictures you use will vary according to the types you need and the specialized services of the many distributors.

Most of the instructional motion pictures you use will come from your school district media resources center. A relatively small number of large secondary schools, and many fewer elementary schools, maintain their own single-school motion-picture libraries. However, with the advent of more titles in the 8mm format and the increased use of film of that size, the number of single-school libraries is growing. In addition, great numbers of motion pictures are still rented directly to schools through college, university, and commercial rental libraries. Even though some state institutions are required to limit their services to schools and organizations within state boundaries, others are not. To supplement their own local libraries, schools regularly use the motion-picture loan services of institutions such as the University of Iowa, Indiana University, New York University, the University of Washington, the University of Arizona, and Syracuse University.

There are numerous sources that specialize in providing programs in video formats, though in each case one should also inquire about the film formats. The Great Plains National Instructional Television Library, located in Lincoln, Nebraska, is a principal source. The National Instructional Television Center in Bloomington, Indiana, is another. The Western Instructional Television Center in Los Angeles, California, is a source for both purchase and rental of programs especially produced for education. The Public Broadcasting Service (PBS) operates the Public Television Library to make available programs from the PBS. Check your media center for other sources.

**GREAT PLAINS NATIONAL TV
LIBRARY PROGRAMS**

One of the pioneering video program libraries, the Great Plains National Television Library distributes programs produced by ETV stations and school districts, colleges, and universities. Generally of high quality, programs are used for in-school motion-picture material and for broadcasting to school and college audiences.

Classrooms in Vermont and South Dakota during showings of the Great Plains programs.

The Metric System: ". . . the professor" and junior scientist Harvey discuss changes in units of measure.

Alistair Cook (center) and Charles A. Siepman in the "Communications and Education" series.

Entertainment motion pictures and others intended for school or film society use are customarily rented from commercial film libraries. Rental rates vary according to the appeal, quality, and recency of subjects offered, and sometimes according to the size of the group applying for the film. Many feature-entertainment motion pictures have been edited and adapted especially for school use by Teaching Film Custodians. The Teaching Film Custodians collection is circulated by the Indiana University Film Library. Among other sources of entertainment motion pictures is Films, Incorporated, in Wilmette, Illinois.

Free or sponsored films produced and distributed by industrial or business film libraries add to the fund of film resources available to schools. State and federal agencies also produce and distribute large numbers of educational and informational films, often without charge. In many cases, their products are also available to schools, on a loan basis, through regional depositories such as the college and university film libraries just mentioned.

information about motion pictures

Once you are familiar with the sources of motion pictures, whether on film or video tape, you will need to know the availability of titles that deal with the topics and purposes you have in mind. You will need to know what such motion pictures portray, by whom they were produced, what techniques were used, how long they are, and the level of sophistication of the audience for whom they are intended. If the motion pictures you consider using are not available through your own school media center, you will usually rely upon printed information about them, including that in the school media center files, which may contain card catalog information about titles that are available elsewhere or detailed data about the source of titles that are frequently used in your school. Also, you may have available the district catalog or catalog cards, sometimes distributed directly to individual teachers or to the school media center or the administrative office. And do consult other teachers who have used the motion pictures.

When you are seeking motion pictures in film, video tape, video cassette, or other formats, you can get much of your information from catalogs of motion-picture distributors or producers. Here are some of the catalogs you will find in most media centers:

⊗ Catalogs of regional, state, college and university, or commercial motion-picture rental libraries that distribute on a loan or rental basis; their fees usually support costs of their services,

⊗ Catalogs of various suppliers of free motion pictures such as the American Petroleum Institute, American Iron and Steel Institute, major automobile manufacturers, airline companies, telephone companies, and many other manufacturers and distributors of products and services.

⊗ Catalogs of film producers—frequently available in the media center or from the producers—such as Encyclopaedia Britannica Educational Corporation, Film Associates, McGraw-Hill Films, Great Plains National Television Library, the National Instructional Television Center, Films Incorporated, National Film Board of Canada, Pyramid Films, and many more. (See Reference Section 5.)

⊗ Omnibus-type film and video catalogs, such as those produced by the National Information Center for Educational Media (NICEM) at the University of Southern California; the *Index to 8mm Motion Cartridges,* published by R. R. Bowker Company; the *8mm Film Directory,* published by the Educational Film Library Association; *The Educators Guide to Free Films,* published by Educators Progress Service; *U.S. Government Films,* published by the National Audiovisual Center, Washington, D.C.; and *Videoplay Program Source Guide,* published by Tepfer Publishing Co. in Ridgefield, Connecticut, which lists suppliers of video programs in ¾ U-matic video cassette format.

⊗ Journals of state, national, and regional associations of teachers and media specialists whose columns carry reviews and critiques of new films and video tapes, besides other media used in instruction.

⊗ Study guides that are often published by producers for the titles they sell. Your media center may have a file of guides often used by your school. To the right are excerpts from a guide for a film on biology. Note the headings and the picture from the film.

THE LIVING CELL: AN INTRODUCTION

A 16mm Sound Film, 20 Minutes

BIOLOGY

In collaboration with
HEWSON SWIFT, Ph.D.
University of Chicago

EDUCATIONAL OBJECTIVES

• To identify the cell as an independently functioning unit and to introduce its major features.

• To provide a brief historical background to the discovery of the cell and to relate this discovery to the development of the microscope.

• To suggest theories and stimulate thought concerning the possible origin of cells.

SUMMARY OF CONTENT

Early Renaissance anatomists, using only the limited surgical techniques known at the time, had already succeeded in describing in detail the organs and tissue systems of the human body. But no one had suspected that the integrated structures and functions of complex organisms were results of processes occurring within invisible cell-like subunits of the body. With the invention of the microscope in the seventeenth century, however, all living things were discovered to be composed of cells or cell products and to function according to the same principles of reproduction, inheritance, molecular composition, and metabolism. *The Living Cell* introduces students to the fundamental principles of cell biology, examining the various structures and biochemical processes that occur in all living cells.

Although cells are the elementary units of all living things, they are structurally and functionally complex. Each cell is separated from its outside environment by a thin, semi-permeable membrane that controls the flow of molecules into and out of the cell medium. Within the cell medium itself are several distinct structures that regulate the biochemical activity of the cell, the most prominent of which is the cell nucleus. Within the

FOR DISCUSSION

1. Briefly describe the structure and function of the cells shown in the opening sequence of this film. How is *Didinium nasutum* typical of all living cells? How does it differ from other cells?

2. What invention led to the discovery of the cell? How did the discovery of the cell influence the study of living organisms?

3. Describe several internal structures common to most cells.

4. What function does the cell membrane serve?

5. How do most cells reproduce? What role does the nucleus play in cell reproduction?

6. What is the hereditary material of the cell? How is this material distributed during cell division?

7. Why is the ordinary light microscope limited in its usefulness to microbiologists?

8. Describe the nucleus. How is the nucleus able to communicate with the rest of the cell? How does the

FOR FURTHER STUDY

1. The resolving power of a microscope may be defined as the smallest distance of separation (measured between centers) at which two objects can be just distinguished. Pairs of objects closer together than the resolving power will appear as a single blur. Resolving power thus determines the maximum useful magnification of a microscope. Use an encyclopedia or other reference sources to discover how the resolving power of a lens system is related to the properties of light. Would X-ray microscopes be more efficient (provide more useful magnification) than ordinary light micro-

*Encyclopaedia Britannica
Educational Corporation*

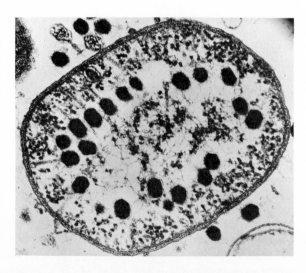

ordering motion pictures

In a school well organized to provide educational media services, you may order films and video tapes somewhat according to the following pattern:

⊗ First, decide, with the help of your media specialist, which motion pictures you want. Many school districts offer one- or two-day services for locally owned films. Unfortunately, in some cases it will be necessary for you to place orders for films well in advance of the dates you wish to use them. In extreme cases, you may need to order them for as much as a year ahead, although this practice is discouraged.

⊗ Second, give your list of desired films or video tapes to the school media director, and, if necessary, indicate sources and rentals. Alternative dates or titles may be required.

⊗ Finally, await a confirmation notice that the motion pictures have been reserved for shipment to you on the dates indicated. File confirmation slips with your lesson plans for the dates the motion pictures are expected to arrive.

motion-picture evaluation files

School media center personnel sometimes maintain files for frequently used motion pictures which include such items as the following for each title: producer's brochures describing the production, a study guide also prepared by the producer, copies of printed reviews that may have appeared since the production, copies of local previewers' comments or evaluations, local teacher and student comments and criticisms, and teacher-prepared tests related to the item.

Among the most important records about a motion picture are the reactions of students. Seek ways to record and report student comments and questions in terms that will be helpful to other teachers who are planning to use the same production. Make notes on ways the picture has been utilized—for class presentations, for independent study, for motivating discussions and written composition, for individual student analysis, and in student presentations. Keep copies of your notes with your curriculum folders and guides.

**FILM AND VIDEO
LIBRARY OPERATIONS**
The operations of film and video libraries, whether small or large, are complex and require efficient procedures and management. *Top center, Agency for Instructional Television; far right, Green Bay (Wisconsin) Schools.*

types of motion-picture action

For optimum enjoyment in viewing motion pictures, and improvement of visual literacy, students can learn how various types of action are created, including the following:

⊗ *Time lapse* sequences, which involve taking large numbers of single still pictures at intervals, and over an extended period of time, and later projecting them at normal speed. This technique permits the viewer to watch a seed germinate and develop into a plant—all in a matter of seconds, despite the fact that the real time action was days or weeks. The effect of continuous motion achieved with this technique involves placing a motion-picture or a television camera before the germinating seed and recording the action one frame (picture) at a time (taken, perhaps, at one-hour intervals), until the plant reaches its full height. When shown at normal speed, pictures of the growing plant show it rising quickly, twisting and turning as it follows the light source.

⊗ *Slow motion,* which involves taking motion pictures at a higher than normal speed and showing them at normal speed. With this technique, such actions as the wing movements of a honeybee in flight, which in real time are too fast to be perceived by the human eye, can be slowed down for detailed study.

⊗ *Stop motion,* which permits the study of movements, or the results of movement, to be frozen, or shown as a still at a given moment. This technique involves making multiple exposures of single frames, which are usually taken as live action. Stop motion may be used to illustrate such events as the change in shape of a golf ball as it is struck by a club, or similar occurrences involving ultra-high-speed real-life motions.

⊗ *Animation,* which is usually performed through the process of filming stills of numbers of drawings, each of which is changed slightly from the one preceding it to produce, when projected, an illusion of motion.

Three other photographic techniques may also be learned by student film viewers:

SPECIAL MOTION-PICTURE TECHNIQUES

Underwater photography, used here in an 8mm film loop for teaching swimming, provides instructional information that would be difficult to communicate verbally. *McGraw-Hill Films.*

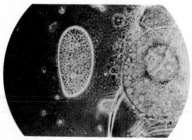

Microphotographic processes are often used for science instruction. *From the Encyclopaedia Britannica film* Protozoa: One-celled Animals.

⊗ *Microphotography* (or photomicrography), which produces ultra-close-up, enlarged views of microscopic life. Motions of single blood cells moving through the capillaries, the process of cell division, or activities of bacteria filmed in this way are easily studied by individuals or by large groups.

⊗ *X-ray photography,* which uses x-ray film and equipment to record actions (such as actual movements and articulation of skeletal parts of joints) that are ordinarily invisible.

⊗ *Telephotography,* which uses long-range telephoto lenses to bring distant scenes close, sometimes with variable focal lengths, as in the case of zoom shots that give the illusion of the scene being transported toward or away from the viewers.

Through filmographic techniques, still prints, photographs, paintings, or drawings can be used to create illusions of motion. *William Cahn, Metropolitan Museum of Art; McGraw-Hill Films.*

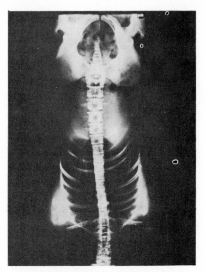

X-ray photography is one of many special film techniques that broaden the scope of the classroom screen. *From the Encyclopaedia Britannica film,* The Spinal Column.

Animation techniques bring motion to static drawings *From the National Film Board of Canada film,* A Christmas Cracker.

Zoom lenses bring the subject close or move it far away. Have you discovered any difference in results between using a zoom lens and moving the camera with one focal length lens toward or away from the scene? *National Film Board of Canada from* The Red Kite.

research about motion-picture utilization

For many years numerous studies have been made on the values of film utilization in instruction, and in recent years similar studies have been made on television utilization. Of the latter studies, many refer to classroom uses of television in modes similar to those of the film studies. The studies have been widely reported and may be found classified under "film research" and "television research." From the research results, here are a few representative generalizations:

⊗ Motion pictures may sometimes be used as the sole means of teaching some kinds of factual materials or performance skills.

⊗ One showing of a motion picture dealing with a complex skill is usually not sufficient. Repetitive showings are recommended, such as may be conveniently provided by loop films or video cassettes. Repetitions are especially valuable if it is possible for students to practice the skill immediately and, at the same time, to refer repeatedly to the motion-picture model. Mental practice—imagining one is actually going through the motions portrayed—also seems to aid learning.

⊗ Note-taking during showings is generally not to be encouraged, because it interferes with attention and hence with learning. Where appropriate, a suitable substitute for note-taking is stopping the program occasionally to permit viewers to ask questions or to make notes.

Great Plains National TV Library

⊗ Increased learning will result from a motion-picture showing if viewers are told in advance what they are expected to learn from it and that they will be checked for what they have learned. Common sense and experience indicate that repetitious use of this technique for all motion-picture showings often produces negative feelings toward motion-picture viewing generally and thus reduces the effectiveness of the instruction.

⊗ It is important for students to know in advance any special terminology or nomenclature they must learn in order to grasp the full meaning of the motion picture they are about to see.

⊗ Motion pictures that have built-in viewer participation activities and planned redundancy, or repetition of key points, appear to produce greater learning than those which do not. Therefore, teachers can improve learning from motion pictures by providing for participation or repetitive experiences related to content.

⊗ Learning from a motion picture may be increased by providing a verbal or written introduction, stating the purpose and the importance of the showing and explaining how the content pertains to the study already under way or about to be undertaken. Also, learning can be increased by repeated showings as well as by pretesting and posttesting.

⊗ Ability to learn from motion pictures improves with practice in viewing them.

⊗ Motion in pictures appears to improve types of learning that involve speed, action and reaction, directionality, changing viewpoints, serial ordering, and progressive changes.

⊗ Color is, of course, important, even essential, to learning some types of content, but students do not learn and retain more merely on the basis of color versus black and white. In some subjects color may actually detract from learning. For some subjects, if presented in color, what is learned may be remembered longer, and color appears to be especially useful in student achievement of affective objectives involving feelings, moods, and attitudes.

uses of motion pictures in teaching

In classroom situations, moving pictures fulfill a variety of purposes: to communicate information, to change or strengthen attitudes, to develop skills, to whet interest, to raise problems, to invoke moods, to emotionalize learning. Sometimes they can be used to test the abilities of students to apply principles to problem situations. Sometimes they may be shown more than once. On some occasions the sound may be turned off, providing an opportunity for either students or teacher to do the commenting, allowing the pictures alone to carry the message. Teachers sometimes show only selected parts of motion pictures, using excerpts that apply to particular topics of immediate significance. Or the picture may be stopped to freeze the action at a predetermined scene to invite class discussion or to check understanding of the points demonstrated.

In similarly varied ways, students may use motion pictures outside the classroom for independent study or for small-group learning experiences. They may use them more or less as they would use printed reference materials, or as parts of assigned program modules. Under certain circumstances, viewing motion pictures may be regarded quite appropriately as a leisure-time or recreational activity, somewhat in the nature of free reading.

Students seem to be tuned in to the new techniques that are less linear than conventional motion pictures, the content of which is based upon a logical or chronological sequence. They readily see the meaning in productions which tend to give a stroboscopic presentation of a series of events from which the significance must be inferred. Because these productions do not have a direct instructional emphasis, they may be used in a wide variety of ways by both students and teachers, and in a number of subject fields. Films of this type include *Help! My Snowman's Burning Down, Fiddle-de-dee,* and *N.Y., N.Y.*

Eastman Kodak Company

using motion pictures in group instruction

Whenever a motion picture is to be used in group instruction, joint planning by both teacher and students is desirable. Both should know the reasons for using the picture at a particular time and what results to expect from it.

A recommended motion-picture utilization pattern, but certainly not the only one, suggests that you as the teacher should (1) take into account the purposes for using a particular production, (2) prepare yourself by becoming thoroughly acquainted with the content, (3) prepare your students to be ready to profit from seeing it, (4) plan appropriate ways to involve students actively in learning from it, and (5) with the students, plan activities to follow up the showing. Further details about these steps are provided in this section.

The only really effective way to *prepare yourself* to use a motion picture for a group presentation is to preview it with pencil in hand, taking notes as you proceed. Read the program guide; check the sample discussion questions; substitute your own questions or ideas if you consider them to be more appropriate than those in the guide. Ask one or more of your students to preview with you. They will almost certainly see things you will overlook —and from their special points of view. They are also likely to notice new or unusual words or situations that ought to be explained before the motion picture is shown. If the production is not of the straight expository type, students may be particularly helpful in judging its value and suggesting the most desirable ways of adapting it to group use. Or they may be good judges in deciding that it should not be used at all! If the decision is made to use it, plan activities and, if needed, prepare a quiz covering the content. Obtain other instructional materials likely to be needed after the showing—reference books, atlases, globes, maps, specimens, flat pictures, filmstrips, recordings, or magazines. Make arrangements for technical equipment if your

classroom does not have the necessary permanent facilities. Arrange to have a student technician handle the showing, or be ready to handle it yourself.

You can help to *develop readiness* by making clear to your students why they are seeing the production and *what* they are expected to learn from it. But no set of routine steps need be followed invariably. Depending on the age and the experience of the class, creative teachers develop class readiness for seeing motion pictures in many different ways, as suggested by the following:

⊗ Discuss what is already known about the subject of the motion picture and lead into what might be expected from viewing it.

⊗ Introduce key words by listing them on the board. Develop class familiarity with them by explaining their meanings in the content of the motion picture to be shown.

⊗ Develop a list of questions to be at least partially answered by the information contained in the motion picture. List these questions on the board as a guide to viewing.

The usual procedure is to show a motion picture all the way through, without interruption. If an interruption in viewing will cause a serious break in

**PREPARING TO USE
MOTION PICTURES**

A teacher previews . . .

and develops a lesson plan.
Sony Corporation of America.

Students give their judgments and suggestions following the film preview with the teacher. *San Jose State University (California)*.

Students take responsibility for preparing for the showing of films.
Clark County (Nevada) Schools.

thought for viewers, the full, uninterrupted showing is quite desirable. But there may be times when only one *part* of a motion picture will be needed. In such cases, show only the part or parts you need —even if they last only a minute or two. Instructional motion pictures sometimes are intentionally produced to be shown in segments so that a production may be stopped at convenient intervals to permit a discussion of problems raised or to practice the skills demonstrated.

There are still other reasons for varying the procedures for showing motion pictures. For example, you may have obtained a film with a narration which is above the level of understanding of your class. If so, turn off the sound and give your own narration, or, as an alternative, ask one of your students to do it.

Finally, develop *follow-up activities*. As students think about and discuss motion-picture content, you should be able to detect any misunderstandings they may have. A second showing of the picture —in whole or in part—may be necessary to clear up any confusion. Other appropriate follow-up activities (only a few of which will be used at any one time, of course) may be employed as follows:

⊗ Divide the class into small groups after the presentation of an open-end motion picture (one which presents a problem but does not provide a solution). Ask each group to propose a solution to the problem. Regroup to compare results and, perhaps, to produce a consensus.

⊗ Take a field trip for which the motion picture serves as a preparatory introduction.

⊗ Practice the skills demonstrated in the motion pictures.

⊗ Administer a written or oral check test covering the major points treated.

⊗ Ask the class to obtain more detailed information about the main ideas.

⊗ Assign one or more student groups to develop a bulletin-board display to clarify and elaborate upon principal themes developed.

For teachers or students who are trying to decide whether or not to use a certain film, the written comments and evaluations of students who have seen the film are helpful guides.

special uses of sound tracks

Copyright permitting, you may sometimes wish to record the narration, dialogue, music, or documentary sounds of motion-picture sound tracks and to reuse them later as additional learning experiences for your students. To make such a recording, place your microphone in front of and about a foot from the speaker. Preferably, locate the speaker well out of range of any machine noise. To edit out any portions of the track you don't need, start and stop the tape recorder, at the same time leaving the projector or video player on.

Use this audio tape to provide your students with a review of a motion picture after it has been returned to the media center. Ask them to give special attention to ways in which the actors delivered their lines; listen again to the music and natural sound effects and discuss how they added meaning and reality to the production; study other audio signs and judge how they contributed to an understanding of the subject portrayed.

Richard Szumski

using motion pictures in independent study

Using motion pictures in independent study may be limited only by your imagination and your determination to obtain maximum benefit from the medium in your teaching situation. This list illustrates the kinds of uses now undertaken in many schools:

⊗ Place films or video cassettes in carrels with appropriate equipment. Provide study guides or project sheets. Ask students to write out answers to programmed questions or to follow directions for completing proposed experiments or projects.

⊗ Arrange a schedule of repeated motion-picture

among themselves, and to prepare a report on it to present to the class.

⊗ Remember that just as students may read a book for pleasure, they may choose to see a motion picture for the same purpose. Such an experience might inspire some further activity—in art, construction, writing, or the social sciences, recording interviews, or taking pictures.

⊗ Encourage students to view and re-view a skill motion picture as many times as they choose and to practice the demonstrated skill along with the pictured demonstration until they are satisfied that their acquired performance meets the acceptable standards, just as the students below are doing.

Hans P. Noecker

Hewlett Packard

showings at the school media center to which some students may be invited to attend while the remainder of their group is involved with other activities.

⊗ Use a cartridge film or a video cassette showing how to operate a piece of equipment (such as a spirit duplicator). Without further help from you, have students learn to operate it simply by following and practicing the steps and procedures of the demonstration.

⊗ Offer students opportunities to go to the media center on their own for specific purposes and to decide which motion pictures they will view individually and what they will do, if anything, as a result of studying them. In some schools, students are allowed to order motion pictures they wish to see.

⊗ Assign a student committee to go to the media center to view a film or video cassette, to discuss it

Clark County (Nevada) Community College

evaluating with motion pictures

Motion pictures are sometimes used to evaluate the ability of students to apply or use what they have learned through instruction. A few examples will illustrate the possibilities:

⊗ Show a complete motion picture of an experiment, leaving the sound on until a crucial point in the action. Then turn off the sound and leave the picture on until the experiment is completed. Ask students to explain the significance or meaning of the action shown and to give reasons why it occurred.

⊗ Show only part of an experiment (the ending, which pictures the result, for example). Ask students to describe the action which probably preceded or followed the action they saw.

⊗ Use a motion picture your students have never seen before. Provide your own narration as you project it silently. Ask students to evaluate the accuracy of your narration, explanations, and interpretations of the action they view.

⊗ Show two motion pictures, both biased but at opposite extremes and dealing with the same social problem. Ask students to evaluate critically the positions taken by applying criteria learned earlier in the course.

⊗ Show a motion picture dealing with a social problem, but stop it short of the solution or resolution. Ask students to state their stands on the problem and to support their views by citing facts or principles developed through previous study.

⊗ Show a motion picture portraying a dramatic story with which the class is unfamiliar. Stop the film before the story ends. Ask students to create their own endings in written form. Then proceed with the rest of the picture to compare the conclusions reached by the class with those of the actual story.

⊗ Within well-designed motion pictures, some visuals can often serve as an excellent review and as a basis for evaluating student knowledge of the facts that led up to the visual. Such an example is this test/stop frame, above, from the Chem Study Series; animation has been used in the film to facilitate the understanding of molecular spectroscopy; by plan, this frame—with the projector on "Hold"—is used with test questions.

⊗ Show a motion picture in normal fashion, but without introduction or interpretive comment. Follow immediately with questions which can be

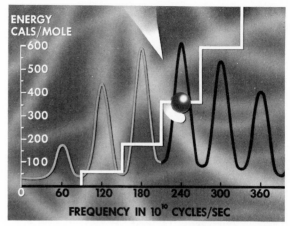

The Chem Study Project

answered only by recalling and interpreting the content of the film. *Determine in this way differences in the abilities of students to see, understand, or remember what they see.* Use the results to decide what to do to help them improve these abilities.

what about free motion pictures?

Many organizations and industries produce sponsored motion pictures which are usually loaned to schools without charge, except for return shipping. Here are samples of catalogs of free films and video tapes.

Though the majority of free motion pictures are designed to advertise products or services, or to promote the point of view of a special interest group, many are attractive, useful, and entirely appropriate for instructional use. When you are deciding whether to use a free motion picture in your teaching, you should consider the following several cautions:

⊗ The use of free, sponsored motion pictures may or may not be approved for your school. Check your school policy.

⊗ Sponsored motion pictures are often produced for adult audiences. Consequently, they may be longer and more sophisticated in their treatment of subject matter than would be suitable for your specific class.

⊗ When time is taken into account, no motion picture should be considered free. If you use one film or video cassette that contributes little or nothing to your students, you will be wasting valuable time.

A motion picture is neither good nor bad just because it is sponsored or loaned free of charge. Rather, like other instructional resources, it must be judged on its own merits. If it contains useful material which is presented in good taste, with truthfulness and style, and if it is suited to the ability and experience levels of your students, use it—if school system policy permits. But be sure not to use a motion picture just because it is free.

short subjects

A special type of motion picture under a general classification of "short subjects" may often be useful in the instructional program. These have been produced for many years for filler material in motion-picture theaters and for standard time units for television. Often of high quality production, typical short subject materials include travelogues, short science features, pictorial surveys of progress on major construction projects, reviews of some social problem of general interest, industrial or agricultural developments, and reports of activities of persons with unusual talents or hobbies.

The selection of illustrations in the accompanying picture story indicates the variety and scope of short subject motion pictures. These were provided by Films, Incorporated, one of the distributors of short subjects and theatrical motion pictures for school and general public use.

**SPECIALS AND
SHORT SUBJECTS**

. . . the "Primal Man Series."
Wolper Productions.

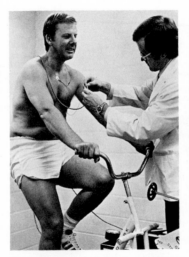

How We Adapt. From the CBS-TV series, "The Human Journey."

Miss Goodall and the wild chimpanzees. *National Geographic Society.*

motion pictures as creative art

The visually literate person understands and appreciates the techniques of film production and the language of motion pictures generally. He or she is able to analyze, compare, and contrast motion-picture styles and techniques and to talk and think about motion pictures in their literary, social, or historical contexts.

Motion-picture criticism extends through a full range of plot and story-line treatment, directing, animating, acting, cutting, editing, schools of production, photographing, and providing musical accompaniment and real-life or manufactured sound effects. It is based on some knowledge of film techniques and outstanding productions of the past, development of perspective on the film as art, and appreciation of similarities and differences among film productions of many different countries throughout the world.

Some of the details of film language as they relate to understanding or producing special effects have been provided in Chapter 8.

The remainder of this section describes a number of developments relative to the study of the film as a creative art and the extension of its influence in society.

film societies and film study

The many film societies of the United States and Canada, made up of individuals who seek to study, to understand, and to foster the preservation and development of superior examples of film art (including avant garde types), often have headquarters in public libraries. But it is not unusual for such groups to develop in secondary schools and colleges, often as outgrowths of activities in English or creative writing classes. If several such organizations exist within the same community, there may be occasions to develop cooperative relationships among them to help with the financing, promotion, and administration of series showings. Motion pictures on video cassettes are appearing in library collections and are opening new opportunities for the selection of titles by the film societies in their activities of study and research. Small dues or series ticket sales may be used (if necessary and

permissible by policy) to finance film rentals. Programs may be organized to present several different examples of film art, such as the following:

⊗ *Film classics* among American or foreign films judged outstanding productions of the past (sometimes extending back to earliest days of motion picture production) are perhaps the most commonly used type. Typical examples are *Potemkin* (Russian); *Bicycle Thief* (Italian); *Mayerling* (French); *Snow White and the Seven Dwarfs* and *Fantasia* (American); *Mr. Hulot's Holiday* (French); *Grapes of Wrath* (American); *Birth of a Nation* (American); and Charlie Chaplin's *City Lights* (American).

⊗ Also used in programs sponsored by school film societies are a growing number of *experimental* film productions which provide interesting examples of the film maker's art. These films may be on serious social topics or explorations of what film or television cameras can do with people, things, light, and space.

So-called "underground films," often the products of school and college experimental film production units, represent another useful program resource for film study groups.

information and materials about film study

A number of sources of information about film societies and film study generally are helpful. The Educational Film Library Association (EFLA) has for some years (since 1958) conducted the annual "American Film Festival," which attracts more than a thousand short film entries. From among these are selected first- and second-place winners in each of thirty or more categories, including nature and wildlife, music, dance, mental health, ecology, film as art, social documentaries, and stories for children. The *Festival Film Guide,* issued annually by EFLA in conjunction with each festival, provides production and content data pertaining to the productions judged best for the year. It is an especially useful publication for individuals seeking information about film study programs.

A number of other information sources and publications related to film study are presented in Reference Section 5 ("Classified Directory of Sources").

American Film Institute

Established for the principal purpose of "preserving the heritage and advancing the art of film and television in America," the American Film Institute was developed as a nonprofit, nongovernmental agency under a grant from the National Endowment for the Arts. Additional funds provided by the Ford Foundation, the Motion Picture Association, and various individuals and private corporations have assisted the Institute in its work of

⊗ Preserving important films for future use

⊗ Providing guidance to teachers generally and to those specializing in the film as a form of art and creative expression

⊗ Conducting advanced study and research in film production and use

⊗ Operating a national repertory theater

⊗ Publishing film periodicals and reference works of many different kinds

⊗ Encouraging the work of young film makers

The Center for Advanced Film Studies of the American Film Institute provides a unique course of instruction in film making, film research, and seminars which brings film professionals into close contact with new film makers. The film collection of the Center now numbers nearly 10,000 American films that are preserved for current study and future use as examples of America's film heritage.

More than a hundred young film makers have been aided through the Institute's grant program to advance their careers in the field. Special help to teachers who teach film appreciation as well as aspects of film production and utilization is provided by the Institute through a membership arrangement: dues for educators and students are reduced to encourage membership.

National seminars are conducted for teachers of motion-picture studies. Surveys of college film and television courses are conducted. The biweekly journal of the American Film Institute, *Filmfacts,* provides useful information about current film releases.

Other countries have organizations to sponsor research and production in the film field; one notable example is the National Film Board of Canada.

National Film Board of Canada

A model for other countries in its promotion and use of films in the public service, including bilingual education, the National Film Board (NFB) of Canada was established in the early 1940s. John Grierson, pioneer Scottish documentary film producer, was the first director of the NFB. A mandate of the NFB is to use film and other audiovisual media "to interpret Canada to Canadians and to other nations." Prolific in its production activities, the NFB is widely respected for the high degree of creativity and the superior production quality of its work. Films from the NFB are accepted and circulated widely throughout the world.

Norman McLaren, one NFB producer-director, brought early attention to the NFB. Drawing by hand on film, McLaren created unusual color films, with sound, that continue to delight audiences; some examples of his films are *Hen Hop, Fiddle-de-dee,* and *Neighbors.* Other types of McLaren productions include *Pas de deux, Canon,* and *Lines.*

But the range of productions of the NFB has included a wide diversity of subjects and film types such as documentary films, educational films, and film reports for government agencies of Canada.

documentary films

The "documentary film" is a special approach to communication with motion pictures. In more than fifty years, many excellent documentary films have been made that are considered classics. Documentaries depict essentially true stories about real-life situations and real people. They also reflect the viewpoint of the film maker, and poetic narration, authentic music, sound effects, and dialogue are often directed toward building moods to strengthen the message. Documentaries do not avoid shocking or profane substance, if such qualities are indigenous to the subject, and, if, in the viewpoint of the film maker, the truth is told. Documentaries continue to be produced for television specials and for general showings. There are many references on documentary films, their producers, and their techniques, but the best way to study documentaries is to see them. On the right, a few examples are listed and several typical scenes are shown.

DOCUMENTARY FILMS

Many documentaries stress strength of personal character. Above, *Leo Biderman* (*Centron Films*): Below, *Nanook of the North* (Robert Flaherty, director; *courtesy of McGraw-Hill Films*).

Above and right, *Solo* (Mike Hoover); bottom, *The City* (Ralph Steiner and Willard Van Dyke). *Courtesy of Pyramid Films.*

Other documentaries recommended:
The Drifters; Night Mail (John Grierson)
Building a House (Julien Bryan)
You Don't Have to Buy War (American Documentary Films)
One Spring Day (H/K Films)
The Plough That Broke the Plains (Pare Lorenz)
Juggernaut (National Film Board of Canada)

entertainment films in schools and colleges

Among the numerous films that have been produced for commercial distribution are many that have great visual, musical, and literary value. In educational institutions at every level, these productions have had ready and continuous use; they are now augmented by numerous productions made for television both in the United States and abroad. From various organizations that specialize in theatrical films, films may be rented or leased, and many schools and colleges present a series of entertainment motion-picture programs of merit and interest as a regular service to students. In many cases, the selection of the programs is based on a careful coordination of titles with the instructional program in literature, music, history, science, and the arts. And, of course, the study of the motion picture as literature is included in many curricula. Listed here are some of the important reasons for using entertainment films as part of the instructional program:

⊗ To provide students with opportunities to develop standards for literary criticism and to apply them to motion-picture productions either on theater screens or television receivers

⊗ To give students pleasure from seeing superior-motion-picture productions

⊗ To influence students to support and attend film programs and to watch television programs of superior quality

⊗ To motivate students to read books and plays from which productions are made and to seek background information on the subject matter presented

⊗ To enable students with limited reading ability or cultural background to experience important literary works in modes compatible with their present abilities and interests

Complex works of theater art, such as *Hamlet* and *The Red Shoes* (see top right of this page), are easily comprehended and enjoyed by many individuals of all ages when the plays are well portrayed with music, decor, color, and great acting. Such

United World Films

United World Films

productions seem to have continuing attraction as classics, and well serve educational goals.

Numerous entertainment films have been edited to shortened versions especially for school use. Originally produced by Teaching Film Custodians, representing a consortium of major motion-picture producers, these films were selected and edited with the guidance of national organizations of teaching groups such as the National Council of Teachers of English and the Social Science Teachers Association. Sequences with special educational values were excerpted from selected films. Many of these films are available from the Indiana University Film Library at Bloomington, Indiana; a few others may be obtained from Films, Incorporated. The scenes, right, are from films that are useful in social studies and English classes: (top to

bottom) *Up the Down Staircase, Becket, The Friendly Persuasion,* and *Billy Budd.*

In promoting literacy, study programs about documentaries, special short subjects, and motion-picture classics deserve the serious consideration of those who plan curricula, organize instruction, and teach. The range of resources is great: Many motion pictures, including those mentioned above, have been based upon important books; one classic example is *The Good Earth,* written by Pearl Buck and made into a memorable motion picture by MGM. And, below, is the 19-inch-high giant, King Kong, a popular star in a film enjoyed and respected for over forty years. Major technical achievements were developed during the production of *King Kong* by RKO which immediately proved to be entertaining. This film continues to be a popular one in the long list of motion-picture classics that are available for study.

Orville Goldner and George E. Turner,
The Making of King Kong

summary

As the output of 16mm and 8mm films and video programs steadily increases, so does the adaptation of these media to school use. Both developments emphasize the value of motion pictures as a medium for instruction and learning. By reducing or overcoming certain intellectual and physical barriers to learning, motion pictures permit students of varying, and even limited, abilities to have meaningful experiences which might otherwise be unavailable to them.

Teachers have responsibilities in locating, ordering, and using motion pictures, and in doing so they must use care and insight. Motion-picture evaluation files located in learning centers may be expected to provide help with these processes. While the familiar formula of "prepare, present, and follow up" can be applied to motion pictures as well as to most, if not all, of the media discussed throughout this book, in the utilization of motion pictures endless functional variations may be developed as creative substitutes for such formalized activities.

Free motion pictures must be used with great caution. While there is no reason to avoid using a film simply because it is supplied free of charge by an interested sponsor, neither is there reason to use it only because it is free. Such decisions should be based on the same criteria that are used for selecting any motion picture. Does it do the job that needs to be done—in good taste, without improperly overemphasizing a product that it advertises?

The motion picture as a creative art and an aesthetic experience also attracts an increasing amount of attention among students of all ages. The interest of film societies in presenting films of many different kinds—including film classics, documentaries, and underground types—grows each year, suggesting that the serious study of film art should be included in the school curriculum. And, today, the term "motion picture" suggests that television programs are beginning to be included in the same field of study.

12 TELEVISION

chapter purposes

⊗ To present some of the many uses of television in education.

⊗ To suggest ways that student viewing of television at home can be related to classroom instruction.

⊗ To indicate some of the methods by which television can be incorporated as part of classroom instruction.

⊗ To report some of the trends in television development that have implications for education in the future, and to indicate ways that students can keep informed about these trends.

⊗ To help you and your students make successful and varied uses of television camera and recording systems in the school program.

University of Colorado

The word *television* represents many different things to different people; a single definition —beyond a purely technical description—is impossible. In education, there are many interpretations of what television is. To some it is a lecture series produced on video tape and used for instruction on closed-circuit channels; to others it is portable equipment, taken to the field to record and observe and study athletes or student teachers; for some it is a receiver provided to watch programs in a media center or a teachers' lounge. Many think of television as a medium for students to use for creative communication; to some television is a source of home entertainment and news; and others consider television as an instrument for presenting motion pictures in the classroom.

Thus, in the evolution of television in education there has been a continual modification of viewpoints and a constant expansion and differentiation of the functions that television actually performs. Not the least cause of these changing concepts has been the explosive developments in the technology of television itself.

One trend serves as an example: With the advent of video tapes and video cassettes, they were rapidly accepted as means of presenting instructional programs for in-class viewing. With these media, teachers and students have control of the time when the program is used, how it is used, whether it is stopped and restarted or repeated. This control is similar to that made possible when using motion-picture films in the classroom. *The subject matter and how it is used is much more important than what medium is used for the presentation of moving images in black and white or in color, and whether with or without sound.* Therefore, for entirely practical purposes, in this book we have ignored any differentiation among video tape, video cassette, video disc, and films for showing motion-picture images in the classroom, under the control of teachers and students, either by groups or by individuals. Video-based instructional programs and film utilization are included in Chapter 11, Motion Pictures: Film and Video, along with a discussion of the sources of instructional programs on films, video cassettes, and video tapes.

This chapter concentrates on the characteristics of television which cause it to have many potential values beyond that of serving as a medium for in-class presentation of motion pictures. Television technology permits:

⊗ Transmission of live programs or motion pictures by cable, microwave, Instructional Television Fixed Stations (ITFS) systems, satellite, or conventional broadcasting.

⊗ Schools or colleges or other institutions to view instructional programs when they are transmitted, or to record them for subsequent use.

⊗ Local production of instructional units that are not available from any other source; such units may be used to extend student learning options, to provide several levels of subject difficulty, or to present content of purely local interest.

Richard Szumski

⊗ Students to produce motion-picture programs to increase their visual literacy or to communicate their ideas and observations to others for pleasure and as practical learning experiences for themselves.

⊗ Students to benefit from a variety of materials of educational value derived from programs that they see at home—programs that are broadcast from commercial, public, school, school system, or college and university broadcasting stations.

out-of-school television viewing

Some teachers consider out-of-school television viewing to be unwelcome competition with homework and reading. Probably more teachers, however, see a constructive challenge in television: How may it be used effectively and in correlation with formal instruction? What are the viewing habits of students, and what programs are popular with them? What they view at home may markedly influence their value judgments and their behavior patterns in school; these influences cannot be ignored. Occasionally, classroom topics may be introduced by discussing television programs that students have watched outside school. And an alert teacher will encourage students to view well-produced programs presented after school hours.

The inherent instructional values of ordinary television programs (including those broadcast over educational stations) must be discovered and capitalized upon by design. There are a number of things you can do:

⊗ BECOME A DISCRIMINATING TELEVISION VIEWER YOURSELF Discover the numerous fine programs that are available. Take the trouble to look for them. Remember that your own interests may be quite different from those of your students, and quite naturally so, but, as a teacher, you also know student likes and dislikes and respect their choices. Thus, helping them to discover values in different programs is one of your challenges. So watch television; find bridges between the television viewing choices of your students and their work in your classes.

⊗ BE WELL INFORMED ABOUT THE PROGRAMMING OF YOUR LOCAL COMMERCIAL AND PUBLIC BROADCASTING (EDUCATIONAL) STATIONS Arrange with local stations to receive advance listings of their offerings. Stations and networks may provide lists of programs judged to be important for educational purposes. Get in touch with the chairmen of film and television groups in local organizations interested in education: the PTA, the Legion of Decency, the League of Women Voters, or others; ask them for program information and ratings. Subscribe to publications of your local public broadcasting stations.

⊗ VERIFY WITH YOUR SCHOOL OR DISTRICT OFFICE WHAT PROGRAM INFORMATION MAY BE AVAILABLE FROM LOCAL EDUCATIONAL AGENCIES Determine whether your administration office communicates with managers and educational directors of local television stations. If they do not, get acquainted with them yourself. They may be very much interested in your request for program information and receptive to ideas about the types of programs that ought to be provided for schools.

⊗ INVITE TELEVISION STATION MANAGERS TO VISIT YOUR SCHOOL OR YOUR CLASSES, OR HAVE STUDENTS CONDUCT TAPE-RECORDED INTERVIEWS WITH THEM TO DISCUSS PROGRAMS AND CAREERS IN THE TELEVISION FIELD Help students to evaluate the significance of station policies and presentations; encourage them to suggest programs they believe ought to be put on the air and to comment on programs that they enjoy and find useful in their schoolwork. Perhaps a class may wish to maintain a bulletin board calling attention to recommended television programs. Encourage students, according to their ability levels, to discuss, evaluate, and occasionally write criticisms of programs they see. Help them to discover programs that appeal to various student interests: science, social science, literature, or hobbies. Ask them to encourage wide student participation in recommending programs for the bulletin board.

⊗ ESPECIALLY BECOME WELL ACQUAINTED WITH YOUR LOCAL PUBLIC BROADCASTING STATION To ensure receiving current program information, become a subscriber. You will discover many programs of general educational value, as well as some that can be directly correlated with your curricula. Arrange for station personnel to visit your classes, and make field trips to the station. Involve students in reporting on and discussing the programs they follow. Help them to develop interests in a variety of program types: science, ecology, sociology, politics, homemaking, literature and the arts, current events, and community problems.

⊗ ASSIGN HOME VIEWING WITH CARE When you suggest home television viewing, family problems may result—especially when there is only one receiver in the house. Schools that have television receivers in the media center may overcome this problem. And, you may also arrange to have certain programs taped for students to see later.

There is every reason for you to consider the valuable contributions television can make to your students and to remember that television is informing them and influencing them whether or not you make any effort to use their viewing experiences constructively. Thus, you are encouraged to look upon television as a potential ally in your instructional program, and not as an enemy! What *you* do may make the difference.

PBS: PROGRAMS FOR HOME AND SCHOOL VIEWING

The Children's Television Workshop, an activity of the Public Broadcasting Service, continues to extend the success of ''Sesame Street,'' ''The Electric Company,'' and other programs of merit for home and school viewing. While ''Sesame Street'' was designed primarily for home viewing by preschool children, ''The Electric Company'' was designed to teach basic reading skills to children between the ages of seven and ten. The history of these programs is well known, and their effectiveness has been well tested.

SPANISH AND ENGLISH
WORDS TO LEARN
PALABRAS EN INGLES Y ESPAÑOL

HEAD

¿Cómo te llamas? What's your name?

COAT

HAND

LEG

SHOES

BOY
NIÑO

To the right are the recently developed multi-media kits for in-school reading development based upon the learning and motivation children have received from watching ''Sesame Street'' and ''The Electric Company.'' Produced by Addison-Wesley Publishing Company in conjunction with Children's Television Workshop, the kits include sound filmstrips, audio cassettes that are used with activity books, comic books, minibooks, and games; a Teacher's Resource Book of alternative ways to use the kits is also provided. All elements in the kits were planned and evolved with formative evaluations and in correlation with basal reading curricula. Such programs are of special interest as examples of comprehensive, coordinated, multi-media instructional planning. Courtesy: *Patricia L. Harrison, Director, Media Services, Addison-Wesley Publishing Company.*

Children's Television Workshop, Ted Kurihara, and Addison-Wesley Publishing Company

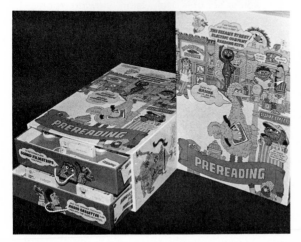

in-school teaching with television

If your moving-picture programs are on video cassettes or in other video formats, the procedures you will use are directly comparable with those you use for motion-picture film viewings. However, if the direct broadcasting of television programs to classrooms is on a fixed schedule in your school system, you will have special problems. For any program:

⊗ Are you thoroughly informed about the objectives and content of the program? Will it fit into the sequence of learning activities of your class?

⊗ Can you arrange your class schedule to be ready when the program is broadcast? Are there alternative possibilities for class or individual viewing? Are other showings scheduled? After school repeats? Can you have the program video taped to show at a more convenient time? Is there a rebroadcast at night?

⊗ Can you preview the program, preferably with several of your students, to prepare for the viewing by the entire group or by only those in the group who can best benefit from the subject? Do you have study guides or data from other teachers who may have seen the program?

⊗ What activities should you plan to introduce or to lead into the televised lesson you intend to use? Will there be opportunities for class participation during the program? After seeing the program, will you clarify answers to questions, appraise outcomes of the experience, guide students toward related classroom activities?

⊗ What will you do to evaluate what students learned from the program? How can you be sure that your students recognize what they learned? Help with such questions may come, in part, from printed materials dealing with the program provided by your school or school district or by the agency that broadcasts the program.

As with any instructional medium, the success of a television lesson transmitted into the classroom depends on the skill and advance preparation of the teacher. The fact that a program comes from an outside source makes using it more difficult.

ITV Center, Broward County (Florida) Schools

viewing the program

There are other considerations for television viewing:

⊗ PHYSICAL ARRANGEMENTS Arrange seating and place equipment properly for optimum viewing. Plan to have some daylight and/or artificial light in the room, but avoid distracting reflections on the television screen image. Stagger chairs, if necessary, and close in the arrangement as much as possible to provide a satisfactory viewing angle and distance. If you have more students than can see a single receiver well, provide additional receivers.

In some schools, television viewing rooms or areas are provided, including earphones if required. Programs with which students are required to participate actively—such as art or mathematics lessons—may require tables. Be sure to provide for the essential working materials.

⊗ CHECK EQUIPMENT Before the program begins, test the receiver and make necessary adjustments. If the program is to be played on a video tape recorder, test and adjust the equipment system (recorder-player and receiver) before you start the program.

⊗ PREPARE STUDENTS Give a suitable lead-in before the program starts. Base your comments or directions on your study of printed materials describing the program or on your preview notes, as mentioned earlier. Help your students approach the viewing session with anticipation of what they are to see and what the presentation may contribute to their understanding of the subject under study.

postprogram activities

Following a television program, in order to assemble information to use in a report to those who planned and presented the program, you will want to: (1) determine what the program contributed to understanding the subject presented and what unanswered questions the students have and (2) assess the strengths and weaknesses of the program, from both your viewpoint and that of your students.

The first purpose will usually be achieved through a brief question and discussion session that is related specifically to the program content. Often, the second purpose may be achieved by responding to a program evaluation sheet, supplied from the school district office, containing such questions as these: (1) Did the program deal with a significant topic related to what your class was studying? (2) Was the program well done, interesting, and understandable to your students? (3) Should changes be made in future productions of similar programs? (4) How can the teacher's guide be improved?

ITV Center, Broward County (Florida) Schools

successful scheduling of television instruction

When early efforts were made to broadcast television for in-school instruction, no problem discouraged educators more than scheduling programs to meet the needs of large numbers of classrooms. Here, however, is a case study about how the scheduling problem was solved. The solution resulted from the evolving technology of television, a fresh approach, much common sense, and a determination to cooperate. This is a brief description of procedures used in the Clark County, Nevada, School District, where programs are now distributed to the schools for convenient utilization.

Motion pictures are delivered to the schools of Clark County by telecasts over four 4-channel ITFS (2500MHz) stations (in effect, private channels) and by school district delivery vehicles. The schedule for program delivery is based upon a priority system which is under continual analysis by computerized records of program requests and use. Programs are repeated each week day, and at different hours each day, giving optimum freedom for teachers to select the best time for their students to see them. Or, for optimum freedom in scheduling, each secondary school and some elementary schools have a video tape recorder so that broadcasts can be taped for delayed or repeated use through the school master antenna system. When film programs of low demand are requested, the film itself may be delivered to the school. Thus, the system is quite nonrestrictive in the use of motion-picture materials.

All programs used in Clark County are selected by committees of teachers working in cooperation with the district curriculum office; most of the selections are directly correlated with the official, teacher-planned curricula of the district. Motion-picture films used are either leased or purchased from educational film producers, with special fees arranged for use of the films on the television system; films are transmitted over the ITFS private channels and cannot be received by the public. Video taped programs, many in series, are obtained from four principal sources: The Great Plains Instructional Television Library, National Instructional Television Center, Western Instructional Television, and productions undertaken by the local district.

Clark County (Nevada) School District

The District ITV Center often records pertinent programs that are received during the evening hours from the Public Broadcast Service; these programs are then rebroadcast during the day for use in classrooms.

Local production is undertaken to meet only very special aspects of the curricula for which outside productions are not available; and some productions are undertaken because of the advanced design of the instruction which they support, such as courses in which learning objectives and criterion-referenced tests are established and materials and procedures for instruction are prescribed. Again, teachers are closely associated with all production activities, and maximum involvement of both students and teachers in planning and developing instructional programs is standard practice.

Further, to maintain close cooperation between the school staff and the ITV Center, two ITV Resource Advisors (certified teachers) spend three-fourths of their time in the schools assisting teachers in planning for and using the programs; feedback about program effectiveness is continually verified, and any modification in programming is guided by the flow of information. Teachers are urged to recognize that use of the television programming is optional; however, every possible incentive is given to have students use television lessons either individually or in small groups. For every program, teachers are provided with study guides and in-service workshops to familiarize them with ways to use the television programs and all related types of instructional resources.

With a firm basis of curriculum planning and by wise use of technology, flexible, relevant, carefully selected programming with films and video tapes is made effective.

alternative routes for television transmission

The number of alternative routes by which television programming can be transmitted is a significant element in the various modes by which schools have made successful use of the medium for instruction. First, the broadcast channels VHF (very high frequency) and UHF (ultra high frequency) now cover most of the population of the United States and Canada. The channels assigned for general educational purposes, under the United States Public Broadcasting System, numbered 236 in 1974. Many of these stations devote all or most daytime programming to instruction, and many include some formal instruction during the evening hours. Major stations in the system are particularly concerned with public education in the general sense.

Instructional Television Fixed Stations (ITFS) systems

To provide needed multichannel television opportunities for education, the Federal Communications Commission reserved channels in the area of the 2500MHz band for school-system use. This low-power, limited-range system, less costly than VHF and UHF installations for transmitting programs, makes possible television services to schools in areas approximating large school districts. Its signals, in effect, are private because home television sets cannot receive them. Schools, however, may have special receiving systems that convert the ITFS signals for distribution on their closed-circuit channels.

Today, a number of school systems have installed and are operating these ITFS systems, often with several channels. Most of these operations use at least four or more channels; as many as twenty channels are used in some to serve the various school subjects, school levels, and district administrative requirements. In general, large cities and large parochial administrative units are the principal users of ITFS systems.

Though principally designed to facilitate the use of television for educational purposes, ITFS, for many reasons, has not been widely adopted. Why?

cable television (CATV)

Recent, rapid development of cable television (Community Antenna Television, or CATV) is of special interest to teachers. Each CATV system either *initiates* television programs on tape, film, or live or *receives* them from regular broadcasting stations and redistributes them to cable service subscribers in its region. High quality and reliability of reception are two characteristics of these services; but, of even greater interest to education, the cable service can carry as many as thirty or more separate programs simultaneously, thus opening the possibility of using the CATV technology to solve persistent problems of program distribution. In 1970, the National Education Association called for reservation of at least 20 percent of all CATV channels for educational uses (an action reminiscent of the successful effort to obtain broadcasting channels for education twenty years earlier).

A few of the special advantages of cable television for schools are that it could:

⊗ Permit redistribution of open-circuit instructional television broadcasts over additional channels at a variety of times and on schedules useful for classroom utilization.

⊗ Facilitate electronic distribution of films or other motion/sound media via CATV channels from a central library of materials for teacher previewing, class showings, or rerecording for later use.

⊗ Provide special audiences with the programs they desire, such as in a second language, preschool education, and school-home counseling.

⊗ Permit the transmission of teacher education programs for in-service instruction for use after school or at other convenient times, making it unnecessary for participants to travel to colleges, universities, or other special training centers.

⊗ Connect clusters of schools through existing cables to permit exchange of programs that students produce.

⊗ Permit teachers to observe demonstration teaching sessions and participate in other in-service teaching activities.

⊗ Provide data transmission capabilities for administrative or counseling information and other facsimile transmission.

⊗ Offer a means of transmitting adult education programs directly to homes.

In spite of the fact that cable service has a vast capacity to provide multiple channels to many people for many purposes, there have been many political, economic, financial, legal, and regulatory obstacles to its rapid and orderly utilization. Later in this book we will examine some of the advantages of cable television that we may see developed in the future.

Investigate to determine the status of cable services and their use in the schools in your area and throughout the United States and Canada. Analyze the problems that have been solved, and those that require further study and resolution in order that cable may become a valuable, widely utilized medium of program transmission, and possibly one for interaction among members of the community.

by hand and by parcel post

Ironically, in spite of the development of telecommunications both in theory and in practice, a large portion of the instructional programming from television production is delivered to the user by shipment through the mail or by hand-carrying processes. And the medium shipped may be in any motion-picture format needed by the user: video tapes, video cassettes, or films. Earlier in this book we discussed using films and video tapes in the classroom. Later, in the last chapter, we will discuss some of the prospects for using telecommunications when all the relevant parts can be brought together to improve education through systematic planning.

a growing edge: television in nontraditional education

Trends in the 1970s indicate extensive and effective development of nontraditional patterns for post-secondary education, often with television as an important element. Early efforts in the United States in continuing education include "Continental Classroom" (in-service education programs for science teachers and many others interested in post-Sputnik physics and chemistry); "Sunrise Semester" (CBS-TV and New York University); "University of the Air" (State University of New York); and the Chicago TV College.

In Great Britain, the Open University started in the early 1970s; it was later adopted in the United States, sometimes with substantial adaptation. Some consider that the courses presented by this pioneering effort are traditional, except that with the use of television, radio, and multi-media kits of materials for home study they reach a large population desiring advanced education. It seems probable that early developments in television instruction in the United States influenced Open University, and it is quite clear that Open University has markedly influenced more recent developments in nontraditional education in the United States. For your further investigation, here are some of the significant developments after the mid-1970s (see also Reference Section 6):

⊗ S.U.N. (State University of Nebraska) program for a nonresidential curriculum. S.U.N. courses are characterized by systematic development processes by a team of specialists including evaluators, content specialists, instructional designers, and production talent. Each course plan includes multimode instruction in modules, interrelated by plan and using multi-media packages: television units, audio cassettes, 35mm slides and a viewer, record albums, and printed materials including a study guide and a text, the latter of which is the principal print resource for the course. In daily newspapers covering the region reached by the course, a feature article is published on the lesson of the week.

S.U.N.'s courses differ in pattern, with resources and methods selected by careful design to meet instructional goals. With this example to show an evolutionary trend, here are other programs worthy of study:

⊗ Southern California Regional Consortium for Community College Television.

⊗ Project Outreach (University of California, San Diego, and the Coast Community College District).

⊗ Maryland College of the Air (offering some Open University courses and others developed locally).

⊗ Miami-Dade Community College, Florida, and the University of California, San Diego, engaging in a joint effort to utilize "The Ascent of Man" series. Miami-Dade Community College planned resource materials so that students without science backgrounds could benefit from the programs, while the University of California, San Diego, developed materials for upper division, four-year college students and adults.

⊗ The Bay Area Consortium of Community Colleges (in the vicinity of San Francisco, California).

One characteristic of all these efforts is the concept of cooperation, through consortiums or other working arrangements, to permit effective planning, sharing of talent, producing, and distributing high-quality instruction for postsecondary students and an even greater number of persons who do not have convenient access to traditional formal instruction. The trend is encouraging to those who have faith in and hope for an open university for all who need it.

THE OPEN UNIVERSITY

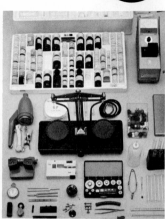

using portable video tape equipment

Video cassette and video tape recorders and players are becoming standard school resources. Earlier, in discussing motion pictures, the use of video systems for self-instruction and group viewing was presented. Now, however, because activities with television equipment can be conducted in

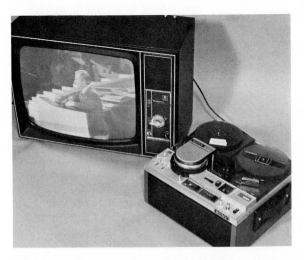

San Diego County (California) Schools

many areas of study, attention will be directed to possibilities for using video systems for creative production work by both teachers and students.

video tape and video cassette recorders

The number of economical, portable television camera/recorder systems now available makes possible many valuable classroom television production activities. Some school systems in the United States, Canada, and other countries provide one or more video tape recorder-players for every school; other school systems provide equipment only on a demand basis or circulate television recorders, cameras, monitors, and accessory equipment on schedule. Teachers are used to having equipment readily available, and the ease of operating it provides both teachers and students with opportunity to make practical application of the *instant production* capability of television. The trend toward standardization to make possible the interchange of program tapes from one brand of video tape recorder to another is further stimulating student and teacher use of the equipment. See Reference Section 1, Operating Audiovisual Equipment, for technical operating instructions; this chapter is devoted principally to the instructional uses of equipment.

what can you do with simple television equipment?

Some call the use of low-cost, portable television recording systems "instant television production." It is true that the development of simple equipment has brought to each classroom the potential of making satisfactory quality productions of a variety of television recordings. Only a few of the activities that can be undertaken are described here, but exploration of the literature reporting successful use of this equipment will be of further help in your own teaching. The following are some sample applications of instant television:

⊗ TEACHING DEMONSTRATIONS RECORDED FOR REPEATED USE The television camera—with several lenses or with a zoom lens—can record steps in a demonstration, often from the point of view of the manipulator, for correct orientation to a subject and for reuse when required. The recorded demonstrations can be remade until perfected; upon presentation, there will be no unnecessary delays. Likewise, the recording will have eliminated the time usually invested in repeated setups by teachers or technicians. The recording of the demonstration can be used and reused at any time in lecture halls, laboratories, or carrels. When necessary, recorded demonstrations can be updated quite easily.

Below, a student television technician records a demonstration to be used with several gym classes.

Edmonton (Canada) Catholic Schools

⊗ STUDENT PERFORMANCES RECORDED FOR REPLAY AND EVALUATIVE ANALYSIS Whether in sports or in public speaking, in typing or in role playing, in music, in drama, or in auto mechanics, playing back the video-recorded performance for evaluation by the student alone or by his or her peers or instructor can lead to improvement. But recording and playback of performance are not sufficient in themselves. For instructional success, each performer must accept a commitment to attempt to attain clearly understood standards of achievement.

⊗ YOUR OWN TEACHING PERFORMANCE RECORDED FOR PRIVATE REPLAY AND ANALYSIS Just as students can benefit from seeing themselves perform, you too may benefit by seeing yourself as your students see you. Ask a student to operate the video tape recorder-camera system as you teach a lesson or present a demonstration. Have him follow you as you use the chalkboard, the overhead projector, real objects or models; suggest how to record your performance as you lead a discussion, picking up occasional views as the students react (or don't react). With pictures and sound together, you then will have an opportunity to examine your speech and actions, and you can draw inferences about your influence as a motivator of students as well as a presenter of information or instructions.

⊗ MICROTEACHING In teacher education programs and in some in-service programs, you may participate in *microteaching*. In its simplest form, this is a procedure by which you will teach a small class of students a short lesson with specific objectives and teacher-performance criteria, after which the replay is evaluated by you, by the students, or by an expert teacher or professor. The usual procedure then calls for replanning and re-presenting the lesson for another recording and evaluation.

Recently, a media director in a high school set a goal to list twenty-five uses of portable television production equipment for each curriculum area in the school, and in the library, and for the administration. How many uses can you list for your special field? Here are some examples of possible special uses, in addition to those mentioned above:

⊗ Use a portable video tape recorder on a field trip to record points of interest and important information obtained during the trip for replay and restudy and for a pictorial report for other teachers and students interested in the same field trip.

⊗ Produce information programs for career counseling about local businesses and industries. A representative for each organization could give a guided tour of the company operations, pointing out typical jobs and explaining pay, opportunities, and educational requirements.

⊗ Make motion-picture reports about situations of critical interest in the community that can be used in class presentations or as reference material in the media center for independent study. The topics for such reports might include urban renewal, waste disposal, minority groups and their cultures, and water supplies.

⊗ Make video recordings of guest speakers—with their permission—that can be used on subsequent occasions.

⊗ Prepare a series of video-taped instructions suitable for use in training student technicians to become competent audiovisual and television equipment operators. A program on the appropriate conduct of a technician on assignment in a classroom may help to maintain standards of projectionist performance.

⊗ Make orientation programs for new students on aspects of the school program: use of the library, school policies, services to students (health office, bookstore, cafeteria), student government and its activities, and local agencies useful to students, such as the public library, city recreation department, and public parks programs.

⊗ Initiate a program of communication activities, including the production of a weekly television newscast for closed-circuit or in-class showing: Student and community events could be reported, guests interviewed, and sports events briefed.

With this as a start, make your own list of what can be done with simple television production equipment to create interesting learning experiences.

student television productions

All over the country, students of all ages are using television equipment to produce interesting, creative programs.

Primary grade students record their presentations—the show and tells, the story reports—and look at them with great interest, continuing the long, long process of evaluating their own self-images and personal skills.

Clark County (Nevada) Schools

Intermediate students write, script, build models and nearly full-scale sets, create costumes, learn lines, perform, record, and present full-length motion pictures made with low-cost, portable television equipment and with reusable video tape.

College students of television journalism write, edit, and present news from the press wires and from news they have gathered—photographed and recorded, edited, and narrated—a real-world level of experiences with practical communication.

The projects undertaken by students with television equipment are as rich in variety as the imagination and confidence of youth—and, as the wisdom, patience, and skills of their teachers.

If you want to incorporate television activities into the lives of your students, here are several questions you might consider:

⊗ To what extent should students of any age be required to learn the technical fundamentals of television production before they are permitted to use equipment or to produce any programs on their own?

⊗ Is it possible that technical information and skills should be taught after students have had the opportunity to experiment with television equipment? Would this procedure impede or enhance their interest? Their ultimate skills?

⊗ How can you provide opportunities for students to try various aspects of television production in order that they will acquire a broad background for making judgments on production quality and will learn to respect the jobs of others who are working on the production?

⊗ Are there ways in which studies in visual literacy, experiences with still-camera photography, and film and television program analysis can be used to prepare students to undertake production of motion pictures by using television equipment? Are there dangers involved in having them cling to stereotyped techniques? How can you arrange for them to learn the similarities and the important differences between television and film production?

These are but a few of the questions teachers may well consider if they want students to benefit fully from experiences with television.

Since this text does not present an in-depth discussion on any one production medium, the questions above may help you make some deductions from the selected examples shown here. In each example consider what the teachers did to prepare themselves and their students for the experiences shown. What point of view toward the television medium would the teacher hold? What criteria would the teacher set up to measure the value of student effort? What criteria would the students be encouraged to establish to judge their own work?

Think of the skills students can learn as they plan and produce television programs: reading, writing, organizing, cooperating with others, speaking, directing, performing, assuming responsibility, drawing, lettering, photographing, editing, recording (pictures and sound), advertising. Remember the list in Chapter 1, "Activities Leading to Learning"? If you, as a teacher, have confidence and a little knowledge about how to put together a simple video system, and have tried it for yourself, you should be ready to let your students go to work in an environment that is ready-made for learning.

SILVER CITY—1880
MOVIE OF THE YEAR
With minimal television equipment, a maximum of enthusiasm, and teacher support, a sixth-grade class at Jo Mackey school in Las Vegas, Nevada, produced a ninety-minute (!) historical western. With models, life-size sets, a story, and a plan: What do *you* think the project meant to the students?

Clark County (Nevada) Schools

SCHOOL STUDIO PRODUCTION

For students interested in the television industry, studio experience in the techniques of television production and the associated skills of photography, graphics, engineering, and management can best be learned in a facility such as this at South Hills Catholic High School, Pittsburgh, Pennsylvania.

For example, allow students with special talents to share their interests with others, and, at the same time, give the student television production crews valuable experience. Students in a Clark County (Nevada) school teach chess on a taped program for library viewing.

Production experience seems to be appropriate for individuals of any age, and production activities serve many instructional purposes. Here, at Virginia State College, students undertake extensive work in the television studio.

Clark County (Nevada) Schools

At North Street School, Greenwich, Connecticut, students use the Philips Mini-Studio equipment, which includes small but complete studio facilities: two cameras, controls for cameras and effects, audio control, multiplexer, and video recorder.

School production experience may be used later in public library television production facilities, such as with this Sony Video-Rover portable system available at the Huntington, New York, Public Library.

video tape recorders and visual literacy

Many teachers are using visual literacy activities to encourage students to open their eyes to the world about them. Some students will use still and motion-picture cameras to record their ideas and reveal the world as they see it, but others find the portable television system an attractive choice. The instant replay feature of video equipment and the possibility of erasing and reusing video tape until results are satisfactory add to the value of television as a convenient recorder to show interpretations of things seen. Because of the erase-rerecord feature, the economy of using video tape instead of film also adds to its attractiveness: with the video tape, students feel free to experiment and create, with no obligation to follow traditional rules for film making. And, since immediate results can be evaluated, some students who are working with motion pictures may wish to make video recordings for analysis before making final productions on color motion-picture film.

image magnification with television

Image magnification is a very popular and practical use of television equipment. With a single inexpensive television camera connected with one or more receivers, demonstration activities may be magnified for a classroom, a laboratory, or a large-group instruction area. Usually the camera is mounted on an adjustable stand over a table; receivers connected to it are distributed at convenient viewing locations. Students can then observe demonstration materials in proper perspective and see them in enlarged detail. Experiments with this technique have proved valuable in teaching biology, typing, electronics, English, psychology, jewelry making, and many other subjects in which close-up viewing is essential. Observed benefits include improved learning by students, reduced time for demonstrations, and economy of effort in preparing lecture-demonstrations. The art instructor below arranges materials before his demonstration.

San Jose State University

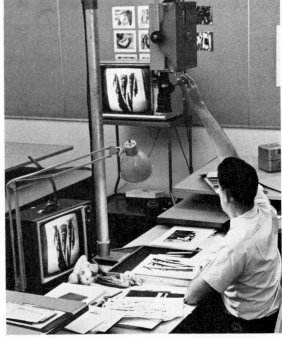

simple television production techniques

With almost startling speed, it has become possible for both students and teachers to engage in television production. Notable, too, is the trend away from elaborate studios and formal production techniques toward simple, portable, relatively inexpensive equipment used for informal program preparation. This trend has extended the opportunities for students to learn many ways to prepare effective programs and to simulate professional creative work and for teachers to produce instructional units especially needed, but not otherwise available.

Earlier in Chapter 8, Photography, information was given on basic procedures for planning both formal and informal motion-picture photography. That information will not be repeated here. However, additional points should be stressed to emphasize the important advantages of using the television equipment system for school productions. The system permits:

⊗ Immediate observation of results, and the opportunity to repeat scenes until they are right.

⊗ Flexibility in program development by recording pictures and sound simultaneously, by recording sound first and then adding pictures according to plan or need, or by intermixing the procedures as suggested by the requirements and circumstances.

⊗ Facility in editing, which is essential to permit maximum utilization of the television medium.

⊗ Use of equipment suitable for productions such as those just described and for simple performance evaluation by recording and playback of sports activities, role playing, dramatizations, public speaking, and microteaching.

These suggestions are especially helpful in producing programs for later and repeated use—programs such as laboratory demonstrations, office machine instruction, presentations by guest speakers, or a summary report of a field trip. These suggestions will also motivate students to do successful creative work as they work with television equipment.

As a general criterion for television production, start with this point: Any television recording should permit viewers to see what must be seen and to hear what must be heard. With small format television production—as with ½-inch video tape or ¾-inch video cassette equipment—it is essential that one learn the advantages and the limitations of the equipment.

. . . to see what must be seen

For television projects, in addition to your usual procedures for planning for any production—study of audience needs, the objectives of the project, and the content—you should also study the conditions under which the program is to be made: by whom, where, and the skills and equipment required. Here are a few suggestions:

LIGHTING There must be adequate light, of the right kind and in the right places. In reasonable weather, outdoor lighting is seldom a problem. Indoors, lighting is often more complicated. But in any case, keep lighting arrangements as simple as possible. Have some kind of portable lighting kit. Use it to light faces, especially, and also whatever else may be an important subject in the production—a carburetor, a sewing machine, or an assembly of glassware, each of which will give you a different problem. Set up ahead of time, and make test recordings; then analyze the results. For pictures of people, keep in mind the triangle system for lighting, illustrated below and right. And don't shoot toward windows or into bright lights.

IMPROVE YOUR SKILLS: PRACTICE, AND STUDY RESULTS

A number of photographs from the how-to programs by Reynolds and Pensinger, San Jose State University, illustrate points that can improve productions.

USING THE CAMERA For a very important starter, whenever possible use a good tripod and a good quality pan and tilt head on which to mount the camera. This equipment provides the best way to obtain smooth camera movement. But if your work necessitates that you hand-hold your camera, then think of yourself as a tripod. Practice being smooth and steady as you use any camera. Don't move the camera at all, unless you have to. Work with a wide-angle lens (with small-format television 10 to 12mm); if you use a zoom lens, move it to *wide* and leave it there unless the program demands that you zoom. Otherwise, instead of zooming, consider doing a walking dolly shot, but be smooth.

Focus problems are reduced with a wide angle lens, but always check your depth of field, especially when you are doing close-up work; and always check your focus and any distortions that may detract from your purposes.

Give continual attention to the camera position in relation to the subjects being photographed: From what perspective do you want the viewer to observe? As if participating in the action? Or as a spectator, from an audience view? Do you need a wide orientation shot to let the viewer see what surrounds the area of the action, and then a close-up or very close view of what is taking place? Can the viewer see what must be seen?

Develop some minimum standards for lighting, camera work, and communication to be either achieved or exceeded for every program. And you can read some good books (some are suggested in Reference Sections 1, 3, and 6) and see some video tapes on how to make good video tapes.

. . . to hear what must be heard

Undistorted, and very clear, understandable sound is essential to any successful program for any audience. Think about how the sound can best be picked up for the specific program you are producing.

MICROPHONES You should have good quality microphones that can equal or exceed the full capability of the equipment to record and play sound. You should also have a lavalier microphone on a long cord, a table, and a floor stand, or a simple, small boom, to position the microphone properly. For field work the microphone in a hand-held television camera will sometimes be sufficient, but you will often need better control of sound. Try to position the microphone 8 to 12 inches from the performer or other sound source. A good stand microphone has a shock-resistant mount and a non-reflective finish which will help to eliminate light flares during shooting.

Remember that in productions of the types we are discussing, there is no need to worry about whether the microphone can be seen or not. Only in a theatrical performance, especially one with historical scenes, would a microphone be out of place.

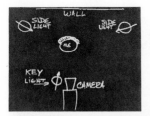

The primary rule is that an audience should hear what must be heard.

helping performers

For informal productions, don't write a script, don't plan every detail ahead of time, don't allow anybody to read anything for the recording, and don't rehearse the spontaneity out of the performers. And don't be highly directive as a director. You need to know what the performers are going to do, how they propose to go about the doing, and, in general, the areas to cover and the angles from which you are to work. Simply block out what is to be done; let the performers do exactly as they would under normal circumstances. Don't let them think they are acting. Let them concentrate on completing the project. Don't permit the environment for your work to resemble that of a studio or formal television production. The more natural the situation you create, the better your production and performers will be.

Start by thinking about what is to be seen. You are making motion pictures of someone doing something that needs to be seen in motion. (If you aren't, then you should be making slides.) Think about what actions will need verbal clarification (because *what* is seen may not be clearly under-

stood by the uninitiated viewer); consider what action may need verbal emphasis (because it is very important for the viewer to know *how* it is done); and explain the implications of the action (*why* it is done in that particular way).

editing

Editing video tape is one of the most important techniques to increase the usefulness and effectiveness of productions.

The primary reason for editing video tapes is to eliminate or to add information. Presenters and experts have a great tendency to want to tell all they know—usually far more than the audience needs to know. Be prepared to edit *out* anything other than that which must be known.

As you learn to use television for practical, interesting projects, make very short but complete productions. Practice editing as you go along: record, check, and re-record if necessary. Later, after all the photography is done, continue with insert editing: cutting, adding, rearranging, changing or adding sound, and making other arrangements of the content for maximum visual communication in the shortest possible time, and with the fewest possible words.

sources of information about television

A textbook cannot hope to report the constantly changing circumstances and developments in educational television. Only continual reference to the current literature can keep you up with the field. Information about television is usually listed under numerous reference headings, including *educational television, instructional television, closed-circuit television, public broadcasting, cable television, Instructional Television Fixed Stations (ITFS)*, and *television research.*

Among the best reference sources are the publications of the National Association of Educational Broadcasters, including its bimonthly *Educational Broadcasting Review* and occasional publications, including *Memos on Instruction.* Other professional education journals, such as *Audiovisual Instruction,* published by the Association for Educational Communications and Technology, print relevant articles.

For information about the programs distributed through the public broadcasting stations, a primary source is the Corporation for Public Broadcasting.

In addition to the journals mentioned above, others of interest to educators include *Educational Broadcasting* (bimonthly), *Educational and Industrial Television* (monthly), and *Televisions,* formerly called *Community Video Report* (published several times a year), which gives wide coverage to community, educational, and commercial broadcasting.

The Joint Council on Educational Telecommunications (JCET) is a useful source of information. It was originally organized to bring to focus the purposes of groups interested in the reservation of television channels for educational purposes. The organization continues to serve the educational community for the support of new telecommunication technology, such as cable and satellites.

Other sources of information are the regional organizations that have formed networks to facilitate program development and exchange:

⊗ Eastern Educational Television Network

⊗ Southern Educational Communications Association

⊗ Central Educational Communications Association

⊗ Central Educational Network

⊗ Midwestern Educational Television

⊗ Rocky Mountain Corporation for Public Broadcasting

⊗ Western Educational Network

⊗ Washington Educational TV Authority

Many large school systems support television programs that may include broadcast stations, ITFS systems, microwave links, tape/film distribution services, and substantial production facilities; their programming may include numerous locally produced series, as well as programs from outside the district. In some areas, school systems are working together with other cities or districts in a consortium to reduce production costs by sharing—and the production quality and scope of programming has markedly improved.

An excellent example is the Regional Educational Television Advisory Council (RETAC) in southern California. Two separate television operations have been cooperating for nearly twenty years: seven southern California counties and sixty of the school districts in those counties, and the Los Angeles City Unified School District. Los Angeles schools operate KLCS-TV (Channel 58), broadcasting more than thirty hours of ITV weekly, including staff development programs and programs for continuing education and the general public. RETAC not only produces programs, but utilizes programs from the program suppliers mentioned earlier and the regional networks listed above.

Another very important consortium is the Agency for Instructional Television (AIT) at Bloomington, Indiana. This nonprofit American-Canadian organization of forty-one participating states and provinces cooperatively plans and produces needed instructional programming.

Wherever you live, check with state departments of education, with large school districts, and with media specialists in your schools to investigate developments in instructional television. Each of the agencies listed above can provide you with guidance to informative sources; and do not overlook your local television stations, both commercial and public (educational), as additional sources of information. For current addresses, consult the latest volume of *Educational Media Yearbook* (revised annually), R. R. Bowker, Publishers.

summary

No matter what you think about television, it affects you and your students, and it may increasingly affect much more of the world than it does now. A teacher must be prepared to use either broadcast or closed-circuit programs that are scheduled for use in class instruction; just as important is to learn to operate simple portable television cameras and recorder systems. If required to do so, you should be able to teach before a television camera as well as make effective and imaginative classroom use of television in combination with other teaching resources and techniques.

Another responsibility you have is to help students learn to evaluate television programs seen at home and to benefit from those that relate to class activities. Teachers of language arts, sciences, and social studies, particularly, have numerous opportunities to direct student attention to out-of-school programs that relate directly to or effectively supplement regular class work. However, until the structure of education is quite different from that of today, what *you* do about television, and *what you plan for your students to do* in relation to the programs they see, will determine in great measure the benefits and the learning they derive.

Further, development of public television stations, of ITFS systems, and of CATV channels allocated for education suggests that amounts and varieties of television programming may be expected to increase. These programs will be designed for direct use in the curriculum at all levels of education, from preschool through graduate studies, and in nontraditional education.

Because of the big investments involved in television production for education, there is a strong trend toward consortiums and cooperative regional associations of institutions and stations to facilitate production and distribution of television programming for education.

Probably the most interesting aspect of television is the opportunity it provides for students to learn to create productions that can improve their ability to communicate, as well as increase their visual literacy. Simple, economical television systems—cameras, monitors, and recorders—can provide many valuable activities. And, as equipment has become less costly and simpler to operate, many programs for instructional use can be produced locally to meet special needs. In these enterprises, both teachers and students may have active, interesting, and creative experiences.

13

REAL THINGS, MODELS, AND DEMONSTRATIONS

chapter purposes

⊗ To encourage the use of real things, in appropriate forms, for learning activities.

⊗ To focus attention on the various forms in which real things may be used: *unmodified* (as they are), or *modified* (simplified, changed in scale, or rearranged in convenient, safe, observable form, or in a form in which they may be manipulated).

⊗ To suggest how students may demonstrate or display real things as a creative activity.

⊗ To indicate ways of using real things or their models that give students incentive to work with subject matter instead of just reading about it.

⊗ To show that student interests in collecting real things can be related to instructional goals.

⊗ To show how both teachers and students may present demonstrations by using real things in combination with other resources, including pictures and graphics.

Hi-Worth Pictures

There are many advantages in using real things in instruction, not the least of which is that students become familiar with objects studied and become aware that these objects are part of their environment and relate to their problems and activities. As with other resources, however, these real things have instructional value only as students themselves become involved in using them to learn. There are many ways in which such involvement may be encouraged.

Do you want an alligator? Perhaps, after your class sees a film on the Florida Everglades, a real, live baby alligator will be just the right resource visitor. Or do you want frogs, lizards, mice, turtles,

269

an ant colony? The W. M. Welch Scientific Company can supply them all—alive and guaranteed; so can Ward's Scientific Establishment and other companies. Perhaps you need a set of real telephones to dramatize the need for your students to learn better communication skills? Check with your telephone company.

All of which suggests that there is no shortage of realia to bring activity to your classroom. They are often free for the asking, or available at very little cost and effort. The challenge is to locate and acquire them and to find profitable ways to put them to work.

San Diego County (California) Schools

real things in the classroom

Definitions of the several types of real things may be categorized under the headings of (1) unmodified real things, (2) modified real things, and (3) specimens.

⊗ UNMODIFIED REAL THINGS Unmodified real things are things as they are, without alteration, except for having been removed from their original real-life surroundings. They have all segments intact; they may operate, work, or be alive; they are of normal size; they can be recognized for what they are. An automobile engine mounted on heavy support stands, a live owl or rabbit, a Confederate flag, or an old-fashioned phonograph are all examples of unmodified real things.

⊗ MODIFIED REAL THINGS A good example of a modified real thing made suitable for school use is a human skull, elements of which have been *separated* and *rearranged* by skilled craftsmen to clarify structure. As one catalog (Welch) describes it: "Exploded skull: . . . each bone mounted rigidly at a distance from its normal position, but relatively placed."

An entire human skeleton can also be obtained for study. Do you want it painted or plain? Painted portions may be used to identify important parts, such as attachment points of muscles.

San Diego County (California) Schools

In many of the different ways in which real things can be used for instructional purposes, each student will study or manipulate the objects, practice with them, and use them to discover their characteristics, operational actions, or behaviors. In others, however, when safety precautions or special skills are required for operation, manipulation, or handling, it may be preferable (or necessary) to demonstrate the object while the students observe. You, as the teacher, may do the demonstration, or you may have a specialist expert do it for you; sometimes, a student may be the best demonstrator. Techniques by which a demonstration is performed will influence the interest generated and the learning gained. Thus, in this chapter we review real things and their models, in various categories and forms, and the ways in which they contribute to instruction. We also describe techniques of demonstrating them to achieve instructional objectives.

Another useful form of modification of real objects is the *cutaway*. Usually, this term applies to mechanical devices, such as engines, through which cuts have been made to permit observation of otherwise hidden parts, either moving or static. Cutaways (sections, slices) are also common in biological or anatomical fields, but these are most often in the form of specimens.

⊗ SPECIMENS A specimen may sometimes be unmodified—simply a piece of the environment.

More often, though, it will be what the dictionary says it is: a part or aspect of some item that is a typical sample of the character of others in its same class or group. Specimens used in instruction are sometimes packaged in bottles, jars, boxes, and other containers to permit direct observation and study. For example, W. M. Welch will provide a Portuguese man-of-war (in a jar), or bottled, preserved earthworms. Specimens may also be embedded in plastic for safe and convenient study.

TYPES OF REAL THINGS

The forms of real things selected for instruction should be determined by the purposes for which they are to be used. The live chicks; representing unmodified real things, appeal to several senses: they are active, warm; they produce sounds; they have characteristic textures. *Creative Playthings.*

Specimens may be left as they are, bottled, or otherwise preserved to facilitate study.

This skeleton has been modified to preserve and protect it, to make it more convenient to distribute, and to aid the learner in studying the relationship of its parts.

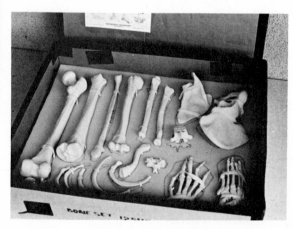

teaching with real things

Having real things available to observe, handle, manipulate, display, discuss, assemble, and disassemble is a means of encouraging students to use such resources of learning. For example, in nature study, the observation of birds in the trees outside the classroom window may signal the start of a genuine interest in bird study. But to study birds closely requires other special sources of information. Illustrated books, pictures from magazines, and excellent motion pictures and filmstrips should also be available. Perhaps local zoos, or some of

your students, may have live birds that could be brought to class. If you are really fortunate, you may be able to borrow specimens of birds, skillfully mounted in lifelike poses and carefully selected to represent general classes of birds. With these, similarities among birds may be observed and their adaptations to habitat and life patterns may be discovered. Comparing a bird's skeleton with the taxidermist's bird, complete with feathers, is a meaningful experience that further promotes an understanding of the reality of birds.

The coordinated use of pictorial and verbal information about birds—including motion pictures—with real birds, both alive and modified, will also give students valuable sensory experiences from which they may derive both specific information and general conclusions about the subject.

real things in school or classroom museums

Every school, no matter what its size, can have its own museum, however unpretentious it must be. The school museum may be only the corner of a classroom or of the school media center where several cases of materials are displayed. Or museum materials owned by the school may simply be placed in cardboard boxes, catalogued, and stored to be loaned to individuals or to student groups.

Facilities for the museum are much less important than the ways a school uses collected materials to help students learn. Students usually profit more from conceiving, planning, producing, and evaluating the effectiveness of their own displays than from observing those in museums.

living things in the classroom

There are subjects in the curriculum for which living things are of major importance as a basis for learning. Although bringing live things into the classroom for such purposes may take a bit of doing, the educational value of the effort makes the task worthwhile.

As a starter, and if needed, you and your students might build your own *classroom aquarium.* Even if it must be inexpensive, it can show much of what is demonstrated in the finest aquariums. Building it will help your students to understand such principles of life science as evaporation, relationships between plants and animals, and the full meaning of the term "balanced aquarium." Even in the early grades the concept of ecological balance can be shown through the controlled reality of an aquarium.

Probably the best type of classroom aquarium consists of a rectangular glass tank formed in a reinforced, caulked metal frame. But other types of containers may be used—preferably glass, however, with a maximum of exposed surface. Note these pointers about the construction:

⊗ Clean the container thoroughly, and put in clean sand (which has been washed several times) to a depth of an inch or two, spread out evenly along the bottom. Fill the tank with water. You will need plants to supply oxygen.

⊗ Carefully anchor selected water plants into the sand. Plants suitable for aquariums include straight-leaved tape grass, arrowhead, fanwort, watercress, and milfoil. Water hyacinths or algae are good surface plants. To improve their chances of survival, be sure to place them at only a shallow depth in the sand.

⊗ Then add a few smooth rocks to improve the effect of the aquarium and to aid in holding down the plants.

⊗ Live specimens suitable for the aquarium include native fish which students may collect and contribute, snails, guppies, freshwater clams, mosquito fish, goldfish, small snails, and tadpoles (be sure to put in a rack which rises above the water surface to allow snails or frogs that will develop from tadpoles to leave the water when necessary). Snails will eat some of the refuse which would otherwise cloud the aquarium water. Insects suitable for aquariums include water boatmen, water striders, whirligig beetles, and larval forms.

⊗ Water for the aquarium may come from a number of sources. Creek or river water may be satisfactory as it is. So is water which has *not* been chlorinated. But if you must use chlorinated water, boil it for ten minutes or so and let it stand for seventy-two hours to reabsorb oxygen. You may add to its oxygen content by pouring it back and forth between two containers or by beating it with an eggbeater.

⊗ Observe a rule with regard to the number of fish in an aquarium: "One inch of fish to one gallon of water," excluding tadpoles, snails, or water insects.

A *classroom terrarium* is also useful for showing interrelationships of plants, animals, and their environment. Plants and animals found in the vicinity of the school will usually survive in a schoolroom terrarium provided moisture and temperature conditions are controlled. On the following page information is presented to guide you and your students in making, stocking, and caring for the living creatures in the terrarium. To prepare a classroom terrarium, follow these suggestions:

⊗ Obtain six pieces of double-strength glass: three of them 8 by 16 inches, for the sides and bottom; two 8 by 8 inches, for the ends; and one 9 by 17 inches, for the top.

⊗ Assemble the glass sections by applying 2-inch adhesive tape. (Colored plastic tape may also be used.)

⊗ To avoid the danger of cuts, tape all exposed edges of the glass.

⊗ Hinge the glass top so that the back edge is flush with the edge of the terrarium.

⊗ Place the terrarium on a solid surface and cover

the bottom with about an inch of fine, thoroughly washed sand.

⊗ In the sand, place suitable potted plants. Any

succulents, such as cactus, will be excellent for this purpose. Add small pieces of driftwood or other items to produce an attractive arrangement.

⊗ Stock with animals—a nonpoisonous snake, a chipmunk, a land turtle, or a garden spider. Clean the terrarium at least once a week, and feed and water the living things according to need.

EVALUATING WITH REAL THINGS

Tests based on real things are often superior to tests that are purely verbal. While testing with real things may seem most practical in technical and scientific subjects, there are many opportunities to use realia to evaluate student knowledge and skills in other fields.

IDENTIFY THE BIRD

Great Plains National TV

evaluating with real things

Models, specimens, and real equipment can be used in evaluation activities. Ask students to identify them by name; to describe their purposes, functions, or composition; or to compare their relative social value.

It would be entirely appropriate and useful to base at least a portion of nearly any test upon real things. Use a table arrangement upon which to place items, each with a large identification number. Include questions in the test that can be answered only by studying items displayed. These items could be real things, models, reproductions, art prints, cutaways, or specimens. The important factor is that students must examine, react to, and make decisions about something real rather than respond to a verbal description which may contain hidden cues or which may otherwise prejudice responses.

Here are a few additional pointers on how to use real things in testing situations:

⊗ To each object attach a numbered tag which carries one or more questions about the object. Ask students to respond to questions, by number, on a separate sheet. If the number of objects does not equal the number of students, exchange objects, on a timed interval basis, until all have finished. In some circumstances and with some materials, pregummed labels of the removable type can be applied to the surface of the object. Type or print the questions and number them clearly.

⊗ As an alternative to the numbered tag arrangement, develop a set of pages made of tagboard or manila file folders on which you place pictures, or drawings of real things. For each item, attach one or more questions to the back or front of the paper base. Number the questions to correspond to numbers on the answer sheet which each student receives.

⊗ In classes where it is appropriate—in home economics, for example—give students several numbered containers, each filled with a different aromatic herb or liquid, and ask them to make a sniff test to identify them. Develop another test by passing around separately wrapped food specimens to be tasted and identified or appraised.

modified real things

In practice, real things in their natural states are not always available when and where they are needed. Even if they were, they might be too big, too complex, too heavy, too costly, or too dangerous to use. The processes of distillation and cracking oils undertaken in a large refinery, for example, are so complex that students would find difficulty in understanding them even if they took a field trip through the plant.

But these same processes can be made quite understandable by demonstrating and explaining them through use of a miniature reproduction or model of the plant. Thus, through skillful modification of reality, large and complex ideas and systems can be made clear.

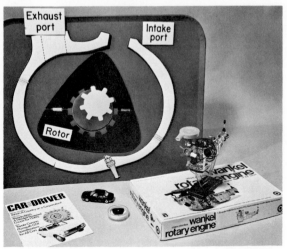

Entex Industries

models

Models are modified real things—reproductions of costly or delicate items that can be provided at reasonable cost and are safe to use. Some models are smaller than real size for the sake of space or economy; others may be extensively enlarged to permit convenient study. Generally, original characteristics of a real object are carefully reproduced in the model.

Beacon News, Aurora (California)

The combination of a working scale model and a simplified cutaway provides an excellent basis for understanding the operation of the Wankel engine shown here. Using such a combination for illustration, students may practice explaining the factual knowledge involved.

Interested teachers, especially those who deal with complex scientific phenomena, have encouraged the commercial production of various kinds of *kits* for use in schools. Model oil refineries that work, flight demonstrators, and model weather stations are available in kit form.

The contents of commercial kits sometimes reflect a systematic approach to teaching. One such kit, for example, provides a comprehensive handbook for teachers that describes the purposes, content, structure, and material covered in a full-year course in general science. This same handbook also offers detailed instructions to adapt the kit to meet unusual course and student requirements that may be encountered. For each instructional unit in the kit, detailed guide sheets for the teacher are provided. These guides include background information, tested procedures and activities, content suggested for student investigations, evaluation materials and test items, and practical hints from expert teachers who have actually taught the course. It is not unusual for kit producers to supply inservice teacher education suggestions as well.

CREATIVITY
AND MODELS

The three interrelated levels of creativity discussed earlier (Chapter 4) may also be discerned in the various model-making activities in which students engage.

Imitative or cookbook—the model kit containing all materials and full directions for assembling it. *San Diego (California) County Schools.*

Adaptive—as with this reflecting telescope, built adaptively, using functional variations of a plan, from a combination of manufactured and handmade parts.

Inventive—as with this water-powered spit, the creative development of a group of Santa Clara (California) County Boy Scouts.

TYPES OF MODELS

Different learning advantages are offered by different types of models: the cutaway (torso) and the life-size mannequin.

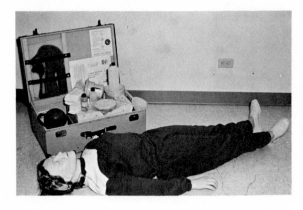

San Diego County (California) Schools

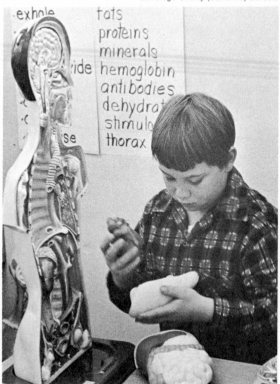

mock-ups

Mock-ups, widely used in industrial training, are also found frequently in schools. A mock-up is a simplified version of reality—a representation of a real thing so constructed as to highlight essential parts or functions and to eliminate unneeded details. As in the case of the balance scales shown here, mock-ups are sometimes produced as working models to demonstrate the essential operations

San Diego County (California) Schools

of a real device. Mock-ups may also be developed with the parts arranged on a flat board as a display, with the various parts maintained in their proper relationships.

creating reproductions of reality

Although real resources that you would like to use may not always be available to you and to your students, there are ways in which you can create some of your own. These self-made resources include miniature dioramas, sand-table constructions and models, and items produced in various plastic materials.

miniature dioramas

Dioramas in three-dimensional displays found in museums of science or natural history are usually produced life-sized. They represent the best of the

George Cochern

Sawyer's

professional designer's art of visual communication in this medium.

School-made dioramas are most often produced as miniatures, however, in proper scale to fit instructional and space requirements. Students who construct dioramas may start with shoe boxes which they paint or cover with colored paper. Ordinary corrugated cardboard or wood boxes will also serve this purpose. Any of a number of different materials may be used to develop the scenes. Oak tag can

serve as backdrops. Clay or plastic figures will suggest reality. Stuffed animals or birds may be shown in their natural habitats. To provide realistic textures, chicken wire covered with plaster of paris, sand, pine needles, leaves, or rocks can serve as a base. Mirrors embedded on a scene floor create the illusion of water. Cardboard or wood boxes may be covered with plastic, clay, or paper to make castles, houses, or furniture. An underwater scene can be portrayed by covering the front of the diorama with blue cellophane. Adding lights may make a diorama look more realistic. If lights are added, bulbs of low wattage should be used, and great care should be taken to avoid lamp contact with combustible material. In some displays, a flasher unit, such as that used to make Christmas tree lights blink, will be both functional and attractive. Finally, cards or labels should be prepared to explain details, to ask questions, or to make meanings clear.

Pairs of students can design and produce miniature dioramas to summarize or highlight some aspect of a unit of study. Have the teams print or type accompanying comments and questions that will invite the participation of viewers as they move from station to station.

sand-table construction and use

The sand table continues to be regarded as an excellent medium to help students learn to construct scale reproductions of real things. To produce an accurate replica of any element of the environment, students must do appropriate research and make a number of decisions about their objectives, their standards, and their procedures. Reproducing to scale a scene of a section of the Nile River Valley, for example, requires knowing such facts as the dimensions of original pyramids, the width of the Nile at the point selected, the distance from the pyramid to the town depicted, and the relative height of trees along the river.

As a skillful teacher, you will also encourage activities that are relevant to contemporary student interests as well as to broad goals of the curriculum. Your students may produce sand-table replicas of sections of the moon surface, for example. Some who are skilled in model building may construct moonmobiles and LEMs; less creative students may find and assemble plastic models of rockets and other important elements of the scene. In every case, library research is needed. You may also review with your class the catalogs of booklets and pictures available free or at low cost from NASA, or similar agencies, and thus help them locate useful pictorial as well as written information.

American Telephone and Telegraph Co.

Harry Haworth

In developing sand-table or other table-top displays, you may produce models from inexpensive materials, including some described in the next section, Special Plastic Recipes. Mount cardboard cutouts on stand-up cardboard or wood bases; paste toothpicks to bottoms of cutouts to suggest pilings. Create figures from pipe cleaners draped with odds and ends of cloth or colored paper. Make trees and shrubs from bits of paper, real shrub twigs, pieces of sponge (spray painted in appropriate colors), dyed popcorn, or small pine cones. Crepe paper or colored sawdust will produce the illusion of grass; make fences from twigs or small cut pieces of wood.

To produce the appearance of a lake, embed a mirror in the sand or put blue paper under a piece of glass. Moss can be used to look like underbrush. Add colors to displays to improve effects: white for snow, green for grass, yellow for dry fields. For structures of various kinds, remember that balsa wood is easy to work and to finish, and it is generally available. This wood comes in rods, beams, and sheets, often in sizes that will be appropriate for the scale of your constructions.

For the base of a sand-table display, you will find that dampened sand is easier to mold than dry sand. But if you do use dampened sand, you should line your box with tin, zinc, tar, or another appropriate material to waterproof it. Pure white sand does not pack as well as common sand, but it is excellent for marking roads.

special plastic recipes

Many of the procedures just described call for the use of plastic mixtures. Several of them may be produced right in the classroom by using recipes listed below. We recommend, in each case, that you first have students make a small test sample of the recipe selected to determine the proper proportions for it, the time it requires to set or dry, and its workability.

⊗ CEMENT MORTAR Mix 1 part dry cement with 2 parts sand. Add enough water to make the mixture workable. The sand helps to prevent cracking.

⊗ CASEIN GLUE Heat 2 cups skimmed milk (using an enameled pan) to 90 degrees Fahrenheit. Remove from the stove and add $\frac{2}{3}$ cup vinegar. As you stir in the vinegar, curds should appear. Remove them by straining the mixture through cheesecloth. Wash the curds in cold water and allow them to dry. Then add water (twice as much water as there are curds) and blend thoroughly. Dissolve 2 tablespoons sodium bicarbonate in 2 cups warm water and add to the casein solution. Stir thoroughly. The mixture will soon become casein glue.

⊗ FLOUR PASTE Mix flour with water, starting with only a little water and 1 cup of flour. Slowly add more water until the mixture becomes smooth and creamy.

⊗ PLASTER OF PARIS Select an enameled container for easiest cleaning after mixing. Pour in the amount of water needed (about half the volume of the anticipated finished plaster of paris mixture). Without stirring, slowly sift plaster of paris into the water and continue as long as it sinks. When enough has been added so that it stays on top without being absorbed, stir it and press out lumps by hand. Use mixture immediately, for it dries quickly.

⊗ FLOUR AND SALT MIXTURE Use 2 cups flour and enough water to mix into creamy consistency. Add 1 cup salt. Hardens when dry.

⊗ PAPIER-MÂCHÉ This can be made in several ways. You may (1) soak torn paper bits in thin paste and mix well; (2) boil paper bits, mix them until they form a smooth mass, squeeze out the water, and add paste, glue, and plaster of paris; (3) tear toilet paper into shreds, boil and beat until smooth, squeeze out water, and add paste; or (4) dip 1-inch-wide paper strips into paste and lay over a torn or wadded paper center to produce the desired form. As an alternative, you may now also buy "instant papier-mâché" (such as Celluclay), which is very easy to use.

⊗ SPRAY PAINT Mix 1 pint white calcimine in water to a thick, creamy consistency. Add $\frac{1}{3}$ pint alcohol or white shellac. Mix thoroughly by shaking. Add more water if the mixture seems too thick. Ordinary white poster paint or paint contained in pressurized cans may be substituted for the mixture.

⊗ PUPPET MIX Ingredients: $\frac{1}{2}$ cup table salt; $\frac{1}{4}$ cup cornstarch; $\frac{1}{4}$ cup water. Mix ingredients thoroughly, then cook over low heat, stirring continually. The material quickly stiffens into a lump. When it is sufficiently cool, knead it briefly. It is then ready for use. To color, add dissolved watercolors, melted crayons, or ink to the original ingredients. This product may be wrapped in waxed paper and stored in a refrigerator until you are ready to use it.

⊗ SAWDUST MODELING MIXTURE Mix 2 cups sawdust, 1 cup plaster of paris, $\frac{1}{2}$ cup dry wallpaper-paste flour (wheat), and 2 cups water. Start modeling before mixture has hardened.

⊗ CONCRETE MIXTURE A good concrete mixture can be made from 1 part cement, 3 parts clean sand, and 5 parts coarse aggregate (crushed rock, stone, or coarse gravel). Mix all ingredients before adding water. A proportion of 6 gallons of water for each standard sack of cement is usually about right for average-strength mixtures. Where less strength is needed, additional water (up to $7\frac{1}{2}$ gallons) may be used. Greater strength can be obtained by making a 1-2-4 or even a 1-2-3 mixture of cement, sand, and aggregate in that order. Concrete should be allowed to set three or four weeks before heavy use; keep surface moist for at least a week. To keep concrete from sticking to forms, oil them liberally. To fill unwanted holes or broken corners, make up a mixture of 1 part cement and 2 parts fine sand moistened with sufficient water to make a heavy, workable paste.

motion in reproductions of real things

Since motion is an element of many real things, you and your students should seek to find or invent ways to introduce appropriate motion in your displays. There are two principal ways to make your objects move: they can be activated by the viewers themselves, or by electric motors or solenoids installed in them. Invent ways to involve viewers or users by attaching levers or knobs to parts of devices that must move up or down or turn around. Levers can be made of balsa wood strips, tongue depressor blades, or short pieces of wire; knobs can be made from heavy cardboard, balsa wood, or furniture drawer grips.

You have seen the moving advertising displays in stores: figures bob up down, swing from side to side, or rotate. Effects produced are a result of the design of the display and the manner in which the small motors are attached. These motors will run for hours, driven by only one or two flashlight batteries. Sometimes a turntable, also operated by flashlight cells, can be used to power the motion of some replica of a real thing. Invite your students to join in the search to locate mechanical (spring-wound) or electrical units for this purpose.

additional ways to reproduce real things

Several other materials and techniques which may be used as means of producing models or replicas for instructional purposes are suggested by the following:

⊗ WORKING WITH METALS Small pieces of sheet aluminum or metal from tin cans can be used to produce small models, such as a radio telescope or a geodome (geodesic dome). A pair of tin snips, a small pair of pliers, and a small vise will be helpful tools for this work. If your students are capable of handling a soldering iron, they may use it to add realistic details.

⊗ MODELING WITH FOILS To simulate metal sculpture, aluminum foils can be molded into different shapes or used to cover cardboard or other materials.

⊗ BUILDING WITH CARDBOARD Various types of

Herbert Breuer

stiff and corrugated cardboard are excellent materials for making models of small-scale replicas of buildings, such as the Elizabethan stage setting (above) which is being readied as background for a student-produced puppet play.

⊗ MODELING CLAY AND OTHER PLASTIC PRODUCTS Molding figures, perhaps to fit into diorama exhibits, can be an especially satisfying experience for students. Use pipe cleaners or pieces of soft wire for the skeletons of figures; use clay or plasticene to form the actual bodies. Students can make an almost limitless number of items from clay or other plastic materials. These range from the giant pyramids of the Nile Valley, a relief map of the British Isles mounted on plywood, or a Navajo hogan of the Southwest, to a replica of the Moon's Sea of Tranquility.

⊗ MAKING CARVINGS Students will enjoy creating small reproductions by carving wood (balsa is particularly good for this purpose) or bars of soap. The basic forms of small replicas of real things, like moon-landing modules or moon buggies, can be carved, and other parts can be made of toothpicks or balsa wood and buttons.

⊗ MAKING PAPER SHAPES Instructional devices can be modeled from various weights, colors, and types of sheet paper.

⊗ MAKING WOOD PRODUCTS Discarded wooden boxes and crates can supplement the schoolroom lumber pile and be used to produce playhouses that simulate frontier homes and stores or a variety of other buildings.

collections
of real things

Students who are collectors may serve as resource people for others. Their hobbies may tie directly to many of the goals of your instructional program. As their teacher, you may help them turn what may be random, unsystematic collecting into a purposeful, selective hobby leading to improved skills in planning, in library research, in classifying and packaging collections, and in mounting, preserving, and arranging displays. They may collect coins, stamps, model cars, bottle caps, or any of hundreds of other items. But the nature of items collected is usually not as important as the ways collections are used to achieve learning.

teacher interest
in collections

In what ways can you tie the collecting enthusiasms of your students to geography, history, economics, art, reading, or languages? Consider for a moment either coin or stamp collecting. Think of the possibilities of having your students share their collecting interests by making presentations that encourage others to initiate collections or to make reports

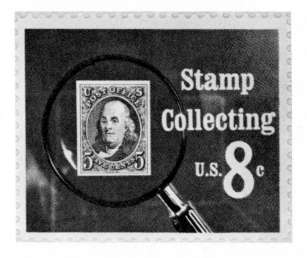

related to a coin of the Civil War period, to a stamp commemorating the evolution of transportation, the anniversary of a great musician, or possibly the statehood of Oklahoma. Usually, interests of students in collecting such things and the goals of the curriculum are but a short way apart; bridging the distance between them is something you can easily do. Along with the significant content and a system for organizing collections, of course, standards of conduct for the collector are also essentials to be taught and learned; everyone must respect the line between collecting and vandalism, or the unauthorized taking of souvenirs.

student interests
in collections

To understand better the students' point of view about collecting, it may be helpful to recall experiences you yourself have had with the activity, even if in the distant past. Bring to class samples of items you have collected: postcards, beads, Kachina dolls, or a few old coins. Indicate your belief in the value of collections—of almost any kind. Perhaps you might also bring in a list of the kinds of real things that have been collected by children you have known. A list of things adults collect may be just as helpful. Remember, "The world is so full of a number of things . . ." that there is something for everyone. A group of college students discussed collec-

PROCESSING AND DISPLAYING COLLECTIONS

For every type of collection there are appropriate processing and displaying techniques. *Ward's Natural Science Establishment.*

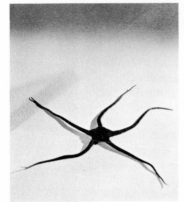

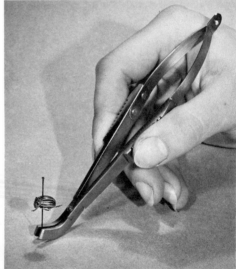

tions of children and adults and made the two lists below. What would you add to or omit from each?

Children

Dolls	Beetles
Butterflies	Jewelry
Stamps	Model cars, planes
Recordings	Bottle tops
Rocks	Maps
Travel postcards	Dishes
Gummed labels	Corks
Pets	Pictures
Coins	Pressed flowers
Miniature objects	Insects
Movie stars' pictures	Autographs

Adults

Insects	Stamps
Books	Costumes
Recordings	Jewelry
Shaving mugs	Ship models
Guns	Old tools
Old furniture	Magazines
Theater programs	Menus
Art prints	Travel posters
Thermometers	Clocks
Coins	Old bottles
First editions	Paintings

processing collected items

Amateur student collectors should learn early the importance of the several steps of processing their collections. Clear and accurate identification and labeling of items come first. In the process of labeling, the whole organization of the collection must be studied, and the purpose of the collection and its value to others who will see it must be made clear. The physical procedures for mounting, protecting, and displaying collected real things offer creative opportunities for students, and an opportunity to use the techniques of expert collectors. If you can guide and encourage young collectors by giving them practical suggestions with which to begin their work of processing, you will perform another important function as a teacher.

⊗ DISPLAYING OBJECTS Objects such as "exploded" component parts of electric motors, radio receivers, tools, or innumerable other items may be mounted for study on veneer board, hardboard, or pegboard. Before preparing the display, the collector should first plan a layout on paper or arrange items in their proper relationship on some flat surface. Then drill small holes at appropriate points and anchor parts with soft copper, iron, or aluminum wire. Twist wire from the rear to tighten into place. Use a painted background to provide contrast in the display; a wood frame will give it rigidity. In some breadboard displays, colored, pregummed tapes may be useful in defining areas or relating components. Labels of appropriate content, size, and style may be required.

⊗ HANDLING CLOTH SWATCHES OR SAMPLES Cloth swatches, pieces of thin metal, specimens of paper (including wallpaper), paint chips, and similar materials can be mounted on uniform-sized cards and used as hand-around items to permit close examination. If it is necessary to see both sides of the specimen, glue or sew along one edge only; some types of materials can be attached to cardboard by invisible cellophane tape used as a hinge. Then the information about the specimen is placed directly on the card, preferably on the front, so that

San Diego (California) City Schools

it can be read as the specimen is examined. Pregummed labels, available in several sizes at stationery stores, facilitate the preparation of labels by hand lettering or typewriting.

⊗ EMBEDDING IN PLASTIC Liquid casting plastic can be used to preserve specimens indefinitely and thus permit handling and examination without damage. Typical of specimens which can be treated in this manner are poison oak or other dried leaves; starfish, beetles, or spiders; small electronic parts; plaster models; small soap carvings; jewelry and coins; and shells.

Several liquid plastic products are available under different brand names. At room temperature they are water-clear and syruplike, about the consistency of thick cream. Once hardened, however, the plastic becomes clear and glasslike. It can be drilled, sawed, carved, or polished to a high gloss—all in the classroom, or possibly elsewhere with the help of the school shop personnel or of parents who have suitable hobby equipment.

demonstrations

It is not always possible for students, as a group, to have direct experiences with real things or their models. For various reasons, including time, safety, and cost of materials, a demonstration may be the most effective way to show how a thing works, of what it is composed, or why it is important. Even though students themselves may not directly manipulate items shown in demonstrations, they may become involved with the subject because they are looking at specific things which hold their attention.

You may present a demonstration, or you may have one or more students plan, prepare, and present one. On some occasions, you may ask experts to come to your class to demonstrate items or processes which neither you nor your students are prepared to deal with.

Giving demonstrations is a particularly valuable activity for students. When they give demonstrations, they gain by participating in a process of communication that involves multiple responsibilities and actions. They may provide oral explanations, for example, and use a variety of media to clarify points. They may put drawings, schematics, or names of parts on transparencies and show them in careful coordination with real things and verbal explanations. Sometimes they may also show slides or pictures taken from magazines as parts of demonstrations. In all such cases, they may learn

Hagerman (Idaho) High School

much about which techniques produce effective communication and which do not.

preparing for demonstrations

Because demonstrations involve two opposite elements—demonstrator and observers—preparation for a demonstration requires careful attention to the needs of both. The observer must be able to see and to hear—and to understand—what the demonstrator wishes to show and to have understood. Lack of proper preparation on the part of the demonstrator may create an end product in which both demonstrator and observer fall between the boat and the pier. Here, then, are conditions that should be respected when you plan a demonstration:

⊗ First, consider whether there are motion pictures, films, slides, filmstrips, or other visual materials available to supplement or to replace all or portions of the demonstration—i.e., materials that can save time or present information better than the live demonstration you are considering.

San Diego County (California) Schools

⊗ Determine the purposes your demonstration is to serve: Is it to demonstrate a skill observers are to learn to perform? Is it only to arouse interest in the content of the presentation? Is it to have observers understand a principle and be able to demonstrate their understanding of it in some specific behavioral way? Define *how,* at the end of the demonstration, you will determine whether you have accomplished your purpose.

⊗ Select the basic real things, models, pictures, films, sketches, or other supporting resources that will contribute to the demonstrations. Will you use flip charts, the chalkboard, slides, or graphic displays to show and clarify elements of the item or process to be demonstrated?

⊗ Arrange the sequence of events and the content of the demonstration; use the real thing or real things, as well as the other media selected, as the basis for your outline. Determine what you will show first, second, and so on. Plan how you will arrange materials in the demonstration area to have them conveniently at hand when you need them.

Santa Clara County (California) Schools

⊗ Arrange the room—at least on paper or in your mind's eye—and consider viewing angles from which observers will watch your demonstration. Can everyone see? Will the real things, models, graphics, and other materials be clearly visible and of a size to be understood completely by even a very nearsighted viewer in the last row? Should you arrange for a television camera and monitors to provide close-up views for every single student? Will the demonstration require that arrangements be made to permit viewers to come up front to get a close view of what is shown? If it is feasible and if it would be helpful for them to do so, will all of them still be able to see what is being demonstrated?

⊗ Are your time limits realistic? Can every step of the demonstration be included in the time available? Have you limited the demonstration to points of importance (in contrast to points that are only of personal interest to you and not really relevant to the observers' objectives)? Have you allowed time for questions as you proceed? For questions at the end of the demonstration? Is there time to repeat any elements of the demonstration that need to be repeated?

⊗ Have you planned the introduction to your demonstration to ensure that the observers know exactly what purposes the event is to serve, why it is important to them, what they will learn from it, what they will be able to do about it, and what will be asked of them after it is over?

⊗ Have you determined whether or when you could be assisted by others during the demonstration?

⊗ Have you planned whether or when to provide the observers with handouts—perhaps drawings, sketches, or photographs of the real thing or model to be demonstrated? Such illustrations in the hands of the observers will facilitate their making notes or their learning key points you want them to retain. Do you want them to take notes? If so, a printed outline might guide them. If they do not need to take notes, it is good procedure to tell them so.

Each of the foregoing suggestions about demonstrations applies to students as well as to teachers. A well-organized presentation by students for their peers often produces superior learning for both, and adds credence to the value of learning by teaching others.

conducting the demonstration

Although pitfalls and hazards are possible in demonstrations, the following positive suggestions, in addition to those just presented, will ensure reasonable assurance of success:

Cajon Valley (California) Schools

San Diego County (California) Schools

demonstrations to evaluate student learning

Demonstrations may also be used as bases for evaluating student learning merely by adjusting the procedures in the presentation. Here are three examples of testing by demonstrations:

⊗ A chemistry instructor performs several operations for the class, mixing several chemicals, identifying each at the right moment, and, at a crucial point, stops and asks: "What would happen if I were to add the contents of this bottle?"

⊗ In a home economics class, students are tested on their ability to recognize correct and incorrect procedures in measuring and combining ingredients to make bread. After the teacher demonstrates a number of clearly defined operations, students determine which, if any, are performed out of order and whether the correct ingredients have been used.

⊗ A static demonstration, such as a breadboard display of a process, is used to present experimental results; students are to explain the process that leads to the conclusions drawn.

⊗ Speak loudly enough so that everyone can hear. When manipulating real things, or when showing correlated materials on flip charts or overhead projectors, remember to speak up!

⊗ Tell or show only what viewers need to meet their goals. Use the "need to know" idea as a guideline for your plans. Concentrate on the essential ideas.

⊗ Keep an eye on your observers. Watch for puzzled expressions or other signs of confusion.

⊗ Develop a sense of showmanship. Build suspense, if this is appropriate; keep the viewers alert.

⊗ Be aware of the pace of your demonstration. Move slowly over difficult parts and repeat them if necessary.

With the technique of time-lapse photography used by film makers, you can demonstrate a sequence of procedures in a relatively short time. If they are nonperishable, materials used to demonstrate the several stages of a process may be treated in advance and brought out when needed in the appropriate sequence.

summary

Real things, in either unmodified or modified form, have unusual value in instruction. In unmodified form, they are available from the community—from the museum, from homes of students, business establishments and industries, and from school and school district resource centers. Modified real things may be purchased from companies specializing in teaching materials, or they may be created by teachers and students themselves. The processes of creating models, mock-ups, or other modified forms of real things have two advantages: students learn while planning and making the modified real things, and they learn from studying them in finished form. Modified real things range from simple to elaborate replicas of their real-life counterparts.

Collections of real things, including specimens, and the activities of organizing, processing, and displaying or explaining them are also valuable teaching tools. Skillful instructors build upon student interests in collecting by relating them to the objectives of the instructional program.

When real things are used in presentations, the results may be twofold: (1) student interest may be stimulated and (2) ideas and concepts may be presented with clarity. Techniques of demonstrations include uses not only of real things but of correlative materials as well—pictures, graphics, or handouts.

A successful demonstration by either teacher or student is based upon a skillful combination of elements: proper selection of the real things or modified real things, careful determination of just exactly what is essential information for the observers, and judicious use of supplementary resources such as slides, sketches, transparencies, chalkboard drawings, and printed handouts. It is especially important that the demonstrator watch the observers to be sure they understand what the demonstration is to convey. To this end, asking the observers frequent questions and inviting questions from them are essential procedures.

Finally, demonstrations are an efficient method of presenting overview explanations of what real things are and what they do.

14

GAMES, SIMULATIONS, AND INFORMAL DRAMATIZATIONS

chapter purposes

⊗ To introduce games, simulations, simulation games, and informal dramatizations as means of instruction.

⊗ To help you select from the activities discussed in this chapter those that will meet the instructional needs of your students.

⊗ To guide you in performing appropriate roles as you work with students while they play games and participate in simulations and informal dramatizations.

⊗ To suggest ways in which you and your students may invent your own games and simulations for useful learning experiences.

⊗ Finally, to encourage you to want to seek games, simulations, and informal dramatizations that can make your classroom an attractive, dynamic place for students to learn.

Pistor Junior High, Tucson (Arizona)

From kindergarten through the university, games, simulations, and dramatizations may serve as instructional resources. People of all ages enjoy them. Both adults and children play charades in many forms. Games are the basis for popular adult television programs. Without guidance, children imitate those about them and invent stories and events which they dramatize, playing as themselves, fantastic creatures, or as people they know—the doctor, the cowboy, or the spaceman. Games that children play inevitably influence their development and therefore may be considered an important aspect of their education.

This chapter introduces a variety of educational experiences which provide entertaining, productive, and interesting involvement for students of any age. For younger students, games, simulations, informal dramatizations, and puppet and shadow plays are discussed. For older students, role playing, games, simulations, and simulation games are discussed as productive activities which may be as sophisticated as simulations conducted with the use of a computer. These activities encourage students to participate in wholesome competition and productive learning, to practice skills, and to make decisions and discover the consequences of them. They provide opportunities for creative play, experiences of interpersonal role playing, and dramatizations that serve student interests and fulfill educational objectives.

In all of these experiences, media may be used advantageously; in all of them students learn as *active participants.*

games and simulations

Let us start with some clarifications: What are the principal or distinguishing characteristics of instructional games and simulations? First, they are *not* purely recreational activities that provide only exercise or just fill time. They are designed to help students to learn, to achieve specific goals or objectives, in an active rather than a passive climate. Many games are advertised as educational, and so they often are. For our purposes, however, we will concentrate on those games, simulations, and related activities that are especially suited to classroom and other in-school use.

⊗ INSTRUCTIONAL GAMES An instructional game is a structured activity with set rules for play in which two or more students interact to reach clearly designated instructional objectives. Competition and chance are generally factors in the interaction, and usually there is a winner. A familiar example of an instructional educational game is the word game "Scrabble." More sophisticated contemporary examples are the mathematical "Wff'n Proof" games (developed by Layman E. Allen), in which

Wff 'n Proof

students learn mathematics concepts and logic by using dice to form equations. Although games are valuable as instructional activities, they do not necessarily attempt to imitate real-life situations.

⊗ SIMULATIONS A simulation is in many respects a model of the real world. Simulation participants are assigned specific roles: they make decisions and solve problems according to specified conditions. As with instructional games, simulations also have instructional objectives. But a simulation is usually more loosely structured than a game; with a simulation there is no winner as such but merely a changed condition or situation to be achieved by participants.

A familiar example of a simulation process used for instruction is the mechanical/electrical/electronic driver-trainer simulator. This system provides realistic simulations of actual driving conditions for each student, although several students may drive simulators at the same time, using the same filmed situational presentations. In such cases, however, each student normally interacts with the situation rather than with other students. Real-life simulation proves a valuable means of aiding students to develop confidence and driving skills while practicing under safe and controlled conditions.

⊗ SIMULATION GAMES There is a third category of gaming/simulation activities, usually associated with social sciences, known as the simulation game. This activity combines decision making and real-life elements of the simulation, but it also includes the clearly specified rules for interaction and competition that are characteristic of games.

The simulation game as a category suggests the difficulty of making completely clear and simple separations among types of games and simulation activities. In a simulation game, for example, the outcome may or may not require a winner or a winning team. One example is "Consumer," produced and distributed by Western Publishing Company. With it, participants interact on a practical economic level—some as consumers deciding if and when they should seek credit; others as credit managers to determine who should get credit, and how much. The objective is for each group to get the best possible terms. The winners, if any, would be those who felt that, through bargaining, they had obtained a "good deal."

Thus, as you see, there is a cloudy borderline between games and simulations, and you may find the terms used interchangeably, or together, as *simulation game.* But don't worry too much about the terms. Although the ability to identify characteristics of games and simulations may help you to select activities for your students, the much more important consideration is, of course, the appropriateness of the activity for the particular instructional needs.

types of games and simulations

Examples of games and simulations for several curriculum areas may help you understand how best to select and use them.

In *mathematics,* for example, games are usually designed to make abstract concepts concrete, understandable, and enjoyable. Many games make practice pleasant and motivate drill on essential basic skills.

⊗ "Take" (Academic Games Associates) is actually a set of board games, developed by James S. Coleman, an early researcher of games and gaming theory. These games illustrate the basic properties of numbers and are readily adapted to young children and adults. The "Wff 'n Proof" games, mentioned above, and "On Sets," developed by Layman E. Allen, are designed to teach mathematical logic and theory. In "On Sets" students build set-theory equations for themselves while preventing others from doing so.

⊗ Other mathematical games, covering elementary arithmetic skills and using cards and boards, are produced by Edward W. Dolch (Garrand Publishing Company, Champaign, Illinois) and Science Research Associates (Chicago). Among a number of Dolch games is "Make One, A Fraction Game," which combines computations of percentages and fractions. Dolch has also developed a simulation called "Pay the Cashier," in which primary school children practice buying in a store.

Games used in the *language arts* are usually directed toward skill development in reading, phonics, spelling, grammar, and vocabulary. You may want to examine some of them:

⊗ A series for primary grades by Dolch includes "Consonant Lotto" and "Vowel Lotto," in which students learn consonants and vowels through matching picture cards with a playing board by reading the sounds; "Match," in which students must find matching pairs of verbs; and "Read and Say Verb Game."

⊗ Other games, such as "Word Power" (Avalon Hill Company, Baltimore, Maryland), in which players are required to differentiate between synonyms and antonyms, and "Scrabble" (Selchow and Righter Company, Bay Shore, New York), available not only in English but also in foreign languages, are aimed at building student vocabulary in middle elementary and higher grades.

Garrard Publishing Company

A *Dolch* PHONICS GAME

Vowel Lotto

IN THE PHONICS GAME SERIES

DESIGNED BY
EDWARD W. DOLCH, Ph.D.

THE GARRARD PRESS, Publishers
CHAMPAIGN, ILLINOIS

Copyright 1956, by E. W. Dolch   Made in U.S.A.

helps children HEAR vowels and vowel combinations

In the *social sciences,* especially in economics, history, business, geography, and psychology, numerous simulation games have been developed. Their popularity is attributed to the interactive nature of the gaming circumstances. With them, person-to-person social, economic, and political situations can be developed; but, to heighten the reality, the impact of simulated business organizations or of nations may also be specified. As players, students of secondary levels and above gain understanding of complexities, conflicting interests, and emotional potentials of problems presented. Here are a few examples:

⊗ "Crisis" (Simile II, La Jolla, California) presents a fictitious situation about nations involved in a mining dispute. Participants use all the mechanisms of international relations (conferences, bargaining, alliances, even threats) to achieve goals. As in many simulation games, situations are continuously fluid and allow for maneuvering; participants are challenged to think hard, to plan strategy, and to anticipate consequences of their actions. The goal of "Crisis" is to obtain a strategic mineral at least possible cost and yet avoid war with other nations. Play is designed to last at least one class period; an equivalent amount of time is needed for teacher-student preparation for it. To gain their objectives, players must communicate with each other, either orally or in writing. In negotiating, students deal with a variety of problems arising out of "balance of power" situations, thereby gaining valuable insights into difficulties involved. As with many other well-designed simulation games, extensive manuals and instruction sheets are provided.

⊗ Another simulation game, "Ghetto" (Western Publishing Company), can be played effectively by individuals who live in a ghetto or those who don't. For city students to whom ghetto life may be normal, this game can mean playing out and analyzing familiar, everyday problems. For students to whom ghetto life is an unfamiliar, remote condition, playing the game provides realistic experience that may enlighten them about situations faced by people in a ghetto.

Many more games and simulation activities are available for the social sciences, such as "Democ-

Western Publishing Company

racy," "The Generation Gap," and "Propaganda."

Simulation games can also be useful in the area of *guidance and career choices.* At the elementary school level, games are available which provide inside views of different occupations as well as broad overviews of entire areas of career choices. One example which helps students understand the many variables to be considered when deciding upon a vocation is "Life Career," developed by Sarane S. Boocock (Western Publishing Company, Wayne, New Jersey).

The number of games and simulations available through commercial sources (many of them developed by teachers and subject matter specialists) increases each year. It is recommended that you search current literature and source directories for announcements of new games. One reference for this purpose is David W. Zuckerman and Robert E. Horn, *The Guide to Simulation and Games for Education and Training* (produced by Information Resources, Inc., Lexington, Mass.). This comprehensive source book, which is updated regularly, includes evaluative comments and a list of games for students and teachers who have had limited experience with the use of games in education.

games in
multi-media kits

Games and simulation activities are appearing with increasing frequency in multi-media kits and learning modules. The kit below, "We Stand Together," published by Listener Educational Enterprises, includes a social science board game, "Survival," and a card game, "Occupations." The kit is designed to help students clarify personal values, to focus on universal human needs, and to understand needs of the family in society. Audio cassettes, study prints, a songbook, lyric sheets, and teacher's manual are also included. Two other multi-media kits correlated with "Sesame Street" and the "Electric Company" include a total of eighteen games.

Listener Educational Enterprises

choosing games
and simulations

Although teachers have used games for centuries, the many new forms of simulation and game materials are not familiar, and new bases for choosing them are still developing. During the past decade research findings about games and simulations have shown conflicting results, but indications of values are available, particularly from the fairly extensive program of the Center for Social Organization of Schools at Johns Hopkins University. A few of the findings are:

⊗ The conditions under which simulation games are conducted are important. Students are much more positive about games when they understand the instructor's purpose in using them.

⊗ Simulation games can teach facts, concepts, and relationships as effectively as conventional instruction which has the same objectives; but, as with many new instructional materials, most research shows no significant difference in results.

⊗ In certain fields such as vocabulary building and math, both simulation and nonsimulation games are better than conventional instruction in producing greater learning.

⊗ Generally, games increase students' motivation to learn. And there are strong indications that in both game and simulation activities, the competition encourages students to help each other with their schoolwork, especially students of different races or different sexes.

⊗ The most conspicuous contribution of games appears to be in the affective area, since games provide motivational support of learning and contribute in some cases to attitudinal changes.

⊗ Because of the wide variety of possible games, simulations, and learning environments, teachers and students must try out new games and simulations carefully in order to select those of value to them.

Experienced teachers have indicated that in evaluating games and simulations, items such as the following are important: Will the time required to use them fit into your schedule and your available time? Is the game you are considering a board or nonboard type, and will it be reusable? Will it fit the size of the class in which you would want to use it, or is it flexible enough to be adapted without losing its value as a learning resource? Is it too elaborate or too expensive for the purposes to be achieved? Does it require a computer in order to be used effectively? Is it a game to be used by individuals or small groups in a classroom or school learning center, or is it better suited for large-group learning situations? Are the directions for use appropriate for your students, or are they too complicated and time consuming to learn?

After choosing a simulation or a game and planning for its use, and before using it with the entire class, having a few students from the group examine it and play it is particularly helpful. And these students can be of great assistance later in playing the game with the whole class.

games and simulations in the classroom

As you plan to use games and simulations for instruction, consider a number of suggestions from reports of experienced teachers:

⊗ An important responsibility of a teacher is to assess the readiness of students to participate in and succeed with any games selected. Start by deciding whether a particular game will suit your

San Diego (California) City Schools

purposes and benefit your students. Anticipate the questions that may arise when you use it in class; determine how much and what information must be given when you start the game; whether the stu-

dents or you should choose players; how to arrange the room; and how to introduce the game. Remember that students should learn from *playing the game,* not by listening to what you *tell* them about it in advance.

⊗ Find ways to *adapt* games and simulations to the capabilities and interests of your students as well as to the physical environment in which you use them. Many of the best and most successful simulations and games from commercial sources are not delivered with every detail planned. When you encounter this situation, it will be your responsibility to decide how to introduce and set the stage for them. Further, in some instances, you may need to produce additional materials.

⊗ Determine, in advance, whether to involve all or only part of a class in a game. Many classroom games involve at least two (and often more) players or call for play by two or more teams. Because of the great amount of activity involved in many games, one part of a class usually cannot be expected to continue studying while others play the game. However, groups of students can play different games at the same time without interference.

⊗ Begin your use of games or simulations with simple but interesting activities. Generally, students are reluctant to listen to long, detailed explanations. Still, they need to know the requirements for participation. If instructions are complex, present only enough information to start the game. Inform students that you will provide additional instructions as required or when requested. Though students of all ages need help in starting play, thereafter only their request for assistance or evidence that the activity is falling apart justifies your interference. To the greatest degree possible, let the combined influences of the rules of the game and the admonitions of fellow players serve to control any dissident student; avoid directive or traditional disciplinary action.

⊗ Remember that, although you *are* a teacher, when you play classroom games or conduct simulations, you must not be one in the traditional sense. Instead, act only as a guide, coordinator, or moderator. Make every effort to avoid being a leader figure. Your students must take the primary roles in such activities. Enjoy the opportunity to observe them as they learn. Restrict your own intrusion into their programs by giving help only where it is essential or unavoidable. And, even then, avoid being helpful too quickly or being too generous with your advice. Since students need time to think, silence may indicate concentration. Only when project development stops will guidance be needed.

San Diego (California) City Schools

⊗ Reserve time for postgame discussions and evaluations of game experiences. Much of the learning acquired in game playing is inductive; often students do not realize that they have learned. For that reason, then, postgame discussions are necessary to help students recognize what they have learned. Under no circumstances, however, should you present a lecture to summarize what you think they should have learned or done. Let them tell you. You may want to have student-to-student or team-to-team discussions, or you may want to set up a few leading, open questions to start the students inquiring for themselves what they have learned; their discussion becomes, at once, both identification of their learning and a summary, review, and reinforcement of what they have learned from the game or simulation experience.

⊗ Be prepared, with games and simulations, to accept a new level of normal noise. Quite unlike a room full of silent readers, the interaction processes of games and simulations involve noise; but this is completely normal. As students gain more experience, they learn acceptable action levels for effectively conducting games. They become increasingly self-directive and self-controlling; they find the bounds of reason that produce results. If your students are *not* active, *not* interacting, look for symptoms of lack of involvement or enthusiasm, a lack of learning; then it is urgent that you seek to identify causes. But, most often, games and simulations create an atmosphere of enthusiastic participation and an end product of learning that both you and your students recognize and appreciate. In considering the climate for successful game and simulation activities, you may find profitable some readings in procedures for the conduct of open classrooms.

some special examples

An excellent example of a simulation game with a somewhat different approach is "Mini-Economy," designed to develop economic literacy among children in elementary school. It is organized to provide active, real experiences in making decisions in a classroom which becomes the scene of an actual economic system. It is based on the idea of scarcity of anything needed in the classroom—books, chalk, paper, or any other item—and provides for discussions of allocation methods at the beginning. From these studies and discussions, payments and rewards, or money systems, gradually emerge and an economic society is evolved.

Dunham Elementary School, Tucson (Arizona)

In actual classrooms this approach has led to the operation of a stock market, banks, insurance companies, real estate offices, consulting firms, and a wide variety of service or marketing businesses.

The third-grade class of Mrs. Helen Condit, in the Dunham Elementary School of Tucson, Arizona, developed a money system, a number of shops, including a Paper Gift Shop, an accounting firm, and a Boutique. In other classes, students have been in many enterprises—music, jewelry, or papier mâché, for example. Governmental functions develop, including the collection of income taxes and the various assignments of civil service employees. Students learn about economic costs, the need for capital, supply and demand, money as a source of power, value of savings, property rights, and similar economic facts. Through the ''Mini-Economy'' game, third graders may learn some of these concepts in a way that many adults in our society have never known and understood.

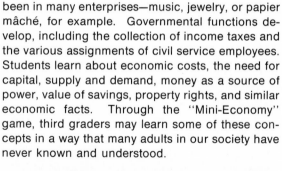

Card games are popular with people of all ages. Margie Golick, a psychologist who has worked with students who have learning problems, says, ''More than any other educational toy I know, a deck of cards has the potential to exploit the child's readiness, through play, to take an active role in the learning process.'' In her book *Deal Me In!* Dr. Golick describes nearly 100 card games that, from her experience, help students with verbal, intellectual, and social skills. Handling cards helps develop manual dexterity, visual efficiency, and contributes to the development of number, space, and time concepts. Highly motivating, card games cause students to play over and over—learning as they play.

There are many opportunities for students and teachers to invent new games that contain content materials useful for the group. The picture story on the opposite page is a convincing indication of the absorption of students as they play with cards.

PLAY CARDS AND LEARN

. . . my deal!

. . . your play!

. . . eeee—I won!

. . . Oh, no! . . . Mmmm! . . . maybe yes?

Dr. Marjorie Golick and
Jeffrey Norton Publishers, Inc.

creation of games and simulations

Should you and your students create your own games or simulations? It is not an easy thing to do, but why not try? You all learn a great deal; but enter the project with care. First, to gain experience, use and test a variety of commercially prepared games with your classes. After this, and perhaps with some further exploration of the literature, have your class try to create a simple game or simulation. Many students enjoy this activity, and some of the best ideas for games and for practical tactics come from them. Students with special interests will learn not only through playing games but also by helping to develop them in this way. The following step-by-step sequence of events should help with the creation process:

⊗ DEFINE LEARNING OBJECTIVES What will students be able to do after having played the game that they were unable to do before?

⊗ SET PARAMETERS What is the time scheme within the game itself? For how long are students to be occupied in playing it?

⊗ IDENTIFY PLAYERS AND THEIR GOALS What role will each participant play? What will each seek to accomplish?

⊗ IDENTIFY AND SPECIFY RESOURCES TO BE USED With what will each player work? Will dice, cards, spinners, or some other means be used to control the action?

⊗ IDENTIFY THE PRINCIPAL RULES OF PLAY AND DETERMINE HOW PLAYERS WILL INTERACT What will determine the sequence of events? What does each player do or not do to reach specified goals? Decide the direction in which players will move. Will players meet obstacles en route? If so, what will determine their actions? How should bonuses or penalties for chance occurrences be distributed? Will specific incentives be provided, either for progress during the game or for the winners?

⊗ ESTABLISH HOW AND WHEN THE GAME IS TO BE WON Will it be won by a single player or by a team? Does winning mean reaching a fixed goal first, or do quality and quantity in meeting goals determine who wins?

⊗ LIST AND DESCRIBE THE MATERIALS AND ARRANGEMENTS REQUIRED FOR PLAY Are required materials readily available, or obtainable at reasonable cost? Must materials be constructed? Are there appropriate facilities available to play the game? Many games devised will require some type of game board. Since interaction on a board game is usually well defined, such games will require detailed development. Even the simplest game boards require time and thought for design. Does the completed plan require a board to organize the movement of the players along one or more tracks? Or may players determine their own movements, as, for example, on a grid? Or should the board be simply a series of zoned areas in which materials are placed according to instructions?

⊗ UNDERTAKE ONE OR MORE TRIAL RUNS Try the game with a typical group of players; work out any problems. This step may save time, inconvenience, and later frustration for you and your students.

⊗ DEVELOP SUGGESTIONS FOR POSTGAME EVALUATIONS Suggestions for postgame evaluations will assist players in assessing what they have learned from the experience and how their future performance in playing the game, and even the game itself, might be improved.

The following case example shows how the process of inventing games or simulations just described was applied in designing a simulation game to help students learn about judicial processes in the United States:

First, learning objectives were defined: After playing the game, students will be able to (1) describe the procedures and the sequence of principal events of a jury trial; (2) tell by examples from the game how courtroom procedures are used to give everyone a fair say about the case; (3) differentiate between factual evidence and supportive testimony, hearsay, or personal opinion; and (4) understand the court procedures that are used to separate these various types of data in determining an individual's guilt or innocence.

The time parameters of the game were given to approximate the duration of a one-day jury trial. (You can't give a whole year to it!)

The players and their individual goals were identified: one judge; twelve jury members, including a foreperson; a defendant; a defense lawyer; a prosecution lawyer; a bailiff; a court recorder; and witnesses for defense and prosecution. The goal of each player was to simulate satisfactorily the duties and responsibilities of a particular participant in a jury trial.

Various resources were prepared which included descriptions of duties and behavior patterns of courtroom participants and the facts of the case to be tried.

To establish interaction and game rules, standard courtroom procedures were specified in writing. The students were encouraged to attend a local court in action or to watch courtroom procedures in television productions.

From the film To Reason Why, *American Bar Association*

creating games: a summary checklist

⊗ Define your learning objectives.

⊗ Set the parameters of your game.

⊗ Identify players and their goals.

⊗ Identify and specify the resources that are to be used.

⊗ Identify principal rules of play and determine how players will interact.

⊗ Establish how and when the game is won.

⊗ List and describe materials and arrangements required for play.

⊗ Undertake a trial run.

⊗ Develop a postgame evaluation plan.

real-life simulation devices

A number of *simulation devices* have found favor in the schools, where they are believed to be capable of making instruction applicable to real-life situations. Examples of such simulation devices include the Link Trainer, driver-trainer units, and telephone company utilization kits.

⊗ LINK TRAINER Available for many years and in a number of forms, the Link Navigation Trainer is designed to teach fliers, under simulated real-life conditions, to solve problems of aircraft navigation in safe surroundings. The Link Trainer is a full-scale, full-size working model of an airplane cockpit with all essential instruments provided. This model moves in response to the pilot's operation of the controls, and the instruments respond normally to the pilot's actions. Given a navigational problem, the pilot simulates flying the airplane, works the navigation problem, and experiences a fairly realistic sensation of flying. Most important, however, is the fact that on a remote table a "crab"—a mobile ink-trailing device—accurately records the performance of the pilot for subsequent evaluation with the instructor.

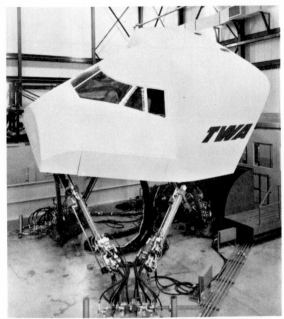

Singer Simulation Productions

A planned sequence of driving experiences is provided—all within the classroom. This Drivo-trainer system provides the teacher and students with information about student driving performance and indicates the need for additional trainer practice—all before practice on the highway.

San Diego County (California) Schools

For instruction and for student practice, simulation devices such as these, though costly, are often superior substitutes for real things. Hazards are eliminated, and savings in time and equipment make them advantageously cost-effective.

⊗ DRIVER TRAINERS Not unlike the Link Trainer, the driver-trainer system now used by schools incorporates working models of automobile controls and uses projected motion pictures which show normal and emergency driving conditions to which student drivers react.

⊗ TELEPHONE KITS Major telephone companies provide kits to teach how to improve the use of telephones and to increase their value in both business and private life. Elementary-level kits contain telephone instruments, motion pictures, filmstrips, paperback texts, and teachers' manuals. The telephones themselves do not work, but they do provide an illusion of reality for student practice in dramatic play form.

For upper grades or secondary schools, more complex kits are available. In some of them, telephones are interconnected through a central control unit. The system produces normal sounds—dial tones, busy buzzes, dial clicks—and transmits conversation. The control unit has additional features—a loudspeaker to permit an entire class to hear a conversation, and an output jack to which a tape recorder may be connected for playback and evaluation of student performance.

hand calculators

Hand calculators, marvels of modern technology, are used in many schools and are justified as appropriate, real-life equipment when computations are required. In higher education, calculators are commonly used in courses in business, mathematics, engineering, sciences, statistics, and in social sciences such as psychology, economics, and geography.

In many elementary and secondary schools, students are encouraged to use calculators as a means of increasing the amount of practice undertaken in computation skills, and at the same time developing competence in the use of a widely adopted resource.

Numerous studies are seeking optimum ways that calculators can be used in education at different levels and are determining how to avoid possible detrimental effects from using them.

As a device for use in games and simulations, the calculator is better than paper and pencil. A number of games have been developed which are built around the calculator itself and its capabilities. Variations in the time allowed for calculations, or in the methods of using the calculator, make the games adaptable for individuals or for teams of players. When time is a variable, the calculator can be passed from one player to another, or games can be played in which each player has a machine and competes directly with the other players. Although a number of games have been developed primarily for fun, the hand calculator can be used with many instructional games and simulations. In fact, calculators may be used with any game or simulation that involves computations.

As with playing cards, numerous stunts can be done with calculators, another means of motivating students to develop computation skills.

More sophisticated, versatile, and programmable calculators are becoming available at lower prices and will make possible the playing of more complicated games. However, complex simulations will require the use of computers.

simulations and the computer

The tremendous capacity of the computer to handle data and to react to alternative trial decisions by showing probable results of each has attracted the attention of educators interested in developing student learning through games and simulations. A potential exists for excellent computer-based simulation games which can help students learn to work with probabilities and the results of alternate decisions. Therefore, with the development of less expensive computers, greater school use may be made of the computer in games and simulations.

Laurence Hall of Science

Clark County (Nevada) School District

Just as business, industry, and government make use of simulations with the computer in training programs and when seeking solutions to real problems, so universities and colleges use computer-based simulations for instruction in various academic fields. As school systems adopt computers, opportunities will increase for using them in various phases of instruction, including simulations and games.

informal dramatizations

While games create a high level of involvement, with simulations and other types of dramatizations individual participants identify with and become involved with the roles they are playing. The value of the outcome is generally determined by the extent to which the participants project themselves into their roles, understand who they are, know why they behave as they do, and feel what they are communicating. In role-playing activities, media are often crucially important.

role-playing activities

Role playing normally involves impromptu dramatizations presented before a class or group. In most situations, learning outcomes are dependent upon discussion or another activity following a performance. The performance sets up a problem, presents criteria of behavior, or otherwise furnishes a basis for subsequent discussion and exchange of ideas. Thus, the role play is an unstructured simulation, and the performance is impromptu. A role may represent an activity in which a student is involved, as in a secretarial class in which students play receptionist roles, or students may simulate a group of scientists discussing ecology from the points of view of industrial organizations, farmers, or lay citizens who are interested in how the issues concern them.

The most effective role-playing situations are those which grow out of problems concerned with people, their actions, and their beliefs. Topics with strong emotional content usually spark performances with the most frank and honest reactions. It is best to start with simple problems and with a minimum number of participants. Two to four roles are sufficient for most problems.

As in preparing for all simulations, the class must have sufficient background information to undertake the role play of the topic selected. Informal discussions and questions from the group may help to set the stage. It is important that all present know who the performers are and what is to be done by the group when role playing concludes.

When preparing for role-playing activities, encourage participants to take their roles seriously and to emphasize positive performance. They will soon learn that to show ten things *not* to do in an interview is easier than to show three that *should* be done. For both the participants and the observers, the desired outcomes may be realized only by positive performance.

Some situations may involve the whole class as participants; for example, the class may represent an adult audience at a meeting where the pros and cons of a major community issue are discussed. In this case, the role play necessitates having several persons assume roles—as leading citizen, banker, business person, or politician. But the rest of the class, as citizens, also become involved by asking questions and making supportive or critical contributions to the discussion that follows. Such role playing can become complex and the ideas diffused; therefore it is wise to try the simpler situations first.

An often used, quite typical role-playing situation is that of employer and prospective employee. In this case, two people take the roles; background information is provided to indicate what position is open, who the people are in the interview, and what behaviors are to be noted as the role play proceeds. After the performance, the group undertakes an evaluation, extracting and discussing evidence of desirable and undesirable behavior from each interpretation.

As students learn to use role play, several developments in skills that should be nurtured will become evident. First, experience should help students avoid stereotyped behavior in the characters they present. In making role-playing assignments, the teacher, too, should avoid making stereotyped assignments—casting girls as "motherly types," for example, or as "beautiful receptionists"; casting boys as "business types," "workingmen," or "athletes" because of personal characteristics of the students selected. Casting should be done in a way which will help students focus on the realities of interpersonal situations and to identify the infinite number of nuances in human behavior that reveal individuality.

Frank dramatizations of contemporary problems of importance to students are usually effective for role-playing assignments, especially at high school and college levels. Often, students who seem minimally motivated need only an opportunity to participate in such performances and discussions relevant to their personal concerns and feelings. Given that incentive, those who may be inclined to be uncommunicative often make significant contributions and reveal interesting and divergent viewpoints.

The value of role playing can sometimes be enhanced by using audio and/or video tape equipment to record the role play and the follow-up activities for later review and evaluation. After hearing or watching the recording, ask those involved in the performance to comment first about the replay, giving their own reactions to the quality and clarity of presentations. This procedure will help guide the other students in their appraisals and you, as teacher, in the comments you make. It will also help to ensure that the feelings and insights of performers are respected and understood before others make evaluative comments. Remember, all simulation and role-playing experiences are learning experiences and must be so considered in every aspect: preparation, presentation, and evaluation.

Tempe (Arizona) Schools

free-play activities

Unstructured free-play activities allow children in early grades to express their individual values—to interpret characters and events and to reconstruct their world as they perceive it. This is completely unstructured role playing. As they study space,

Life and Allan Grant

they relive space flights. As they read about the westward movement, they become pioneers or scouts. To the reality of their imaginative world, they add their own props—a pioneer wagon train (made from their own wagons or paper boxes), a campfire within a wagon circle (sticks over a red light bulb), native huts, or a blockhouse on the plains.

In free play, children do not follow a set story line, nor should they. The teacher gives very little direction. Play develops as the children create incidents, sing, dance, or march. When inaccuracies appear in constructions, props, or characterizations, the teacher may provide fresh information or steer the class to additional reading assignments or to films. But the students are not required to follow definite story lines or literal facts; their primary concern is simply to express values they feel and believe in.

acting out stories

Story acting may be conducted in several ways. In elementary classes, for example, you may assign roles of individuals in familiar stories. Depending upon the familiarity of the story situation and the communication skills of your students, either you or a student may read the story aloud while role-playing students act the story in pantomime. In another approach, the group may discuss the characters to be portrayed, plan scenes to be played, and suggest ideas for lines to be spoken by performers. To combine reading and oral skills in enjoyable practice situations, this technique may be used with either elementary classes or with high school and college students in foreign language, social studies, and literature classes. On the elementary level, as planning proceeds, put key words or phrases on the chalkboard to reinforce vocabulary learning and to serve as a cue sheet when students are later extemporizing lines.

In all such story presentations, it is well to start by keeping scenes brief and simple and to focus on instructional purposes of the experience. At appropriate intervals, discuss what has happened and verify what has been learned. Especially with very young children, scenes should move quickly to satisfactory endings, and they should provide for students' needs for activity by being interspersed with singing, dancing, and marching. Generally, only simple props are needed to help with the action after oral descriptions of the scene are given. Filmstrip projection can provide a setting in front of which the action may take place. Recorded music may be interpolated, and, with that, the term ''instant operetta'' might be used to describe a story that is acted out.

Though the informal dramatization activities discussed here have centered especially on younger people, there are many, many situations in secondary and postsecondary education where important curriculum objectives can be achieved by role playing and free play—in an adult sense. Courses in marketing and labor relations in schools of business, courses in courtroom procedure in schools of law, courses in police and military activities, courses in languages for immediate use are all examples in which the skills of interpersonal relationships can

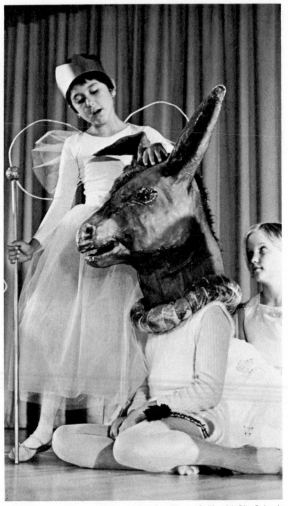

San Diego (California) City Schools

best be learned by various types of informal dramatization and role playing.

Maximum benefit can often be realized if the activities are recorded either on audio or video tape and played back for evaluation. It is suggested that the first critique offered should usually be by the person who has performed; in this way, the performer is encouraged to be objective in making observations and develops skills of self-evaluation. Then, others in the group may make their observations.

puppets and puppet theaters

The most commonly used puppets are (1) hand puppets, (2) glove-and-finger puppets, (3) rod puppets, (4) marionettes, and (5) shadow puppets. Here are a few suggestions about making and using them:

⊗ *Hand puppets* generally consist of a head figure and a loose garment or dress fitted over the operator's hand. The garment covers the operator's wrist and helps to hide it from view. The index finger fits into the puppet's head, and the thumb and middle finger slide into tiny sleeves to form two movable arms. The hand puppet is operated from below if a stage is used.

⊗ *Glove-and-finger puppets* make use of gloves to which small costumed figures are attached. To

San Diego County (California) Schools

make them, cut off the first and second glove finger. The operator uses the index and middle fingers as puppet legs. Puppet bodies can be either flat cutouts or doll-like figures. These puppets are operated from the back of the stage.

⊗ *Rod puppets* usually have jointed bodies made with stiff wire, umbrella ribs, or thin wooden sticks

attached to arms, legs, or heads. Rods can also be used to push animal cutouts, stage furniture, or scenery on or off the stage or to move them while onstage.

⊗ *Marionettes* are flexible, jointed puppets operated by strings or wires attached to a crossbar and maneuvered from directly above the stage. Although they can be almost any size, they are usually between 16 and 24 inches in length. Weights placed in their feet help to keep them upright and in proper working condition. Marionettes are considerably more complicated than puppets to make and to operate.

⊗ *Shadow puppets* are usually formed from pieces of thin cardboard or wood, to which handles are attached to permit manipulation behind (and close to) a rear-lighted white cloth or milk-plastic screen. What is seen from the front is a series of sharply focused, moving shadows.

TYPES OF PUPPETS

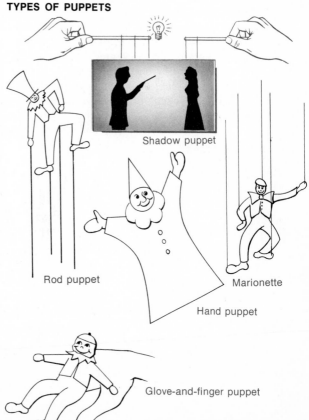

Shadow puppet

Rod puppet

Marionette

Hand puppet

Glove-and-finger puppet

building a puppet theater

A simple puppet theater can be made by arranging a screen on a stand in the front of the room, in front of a window, or in a doorway. This screen should be large enough to hide all the operators. Use a cardboard or wooden box to form the stage itself; make it open to the front and closed on the top and at both sides. It should be at least partially open at the bottom to permit puppets to be moved from underneath.

String a curtain across the front, running pull cords to each side to permit opening and closing from one position. Paint scenic backgrounds either directly on the stage back or on pieces of paper that can be tacked to the stage. Simple furniture or three-dimensional objects (trees, rocks, toy automobiles) can be tied or tacked to the floor out of the way of the operator's hands. An electric light will sometimes improve the general effect and, at the same time, permit partial room darkening during presentations. A filmstrip projector provides an effective spotlight for the theater.

Musical backgrounds and sound effects increase the illusion of reality and add to the fun. It is often desirable to tape-record the entire dialogue, complete with music and sound effects, but only after considerable rehearsal. This procedure frees operators to handle their puppets during presentations.

For shadow plays, a translucent screen should be mounted in front of the stage area. The operators work below this screen, since bright lights are arranged at least 6 feet behind the operators. Reflectors direct the lights, or a slide projector serves as a source of light. Settings are made by painting on the screen with opaque or translucent paints, or by using cutouts of colored plastic sheets. The operators work their puppets from below.

Shadow-puppet silhouettes representing characters or objects (such as airplanes, ships, automobiles, trees) are cut from stiff, opaque paper or tagboard. Silhouettes can be constructed to allow students to move body parts by pulling on strings or fine wires from below.

making puppet heads

There are many ways to make puppet heads; the illustrations on the next page show one good way: (1) Fill a paper bag with crumpled paper. Put a cardboard tube into the bag and tie it by the neck. (2) Mix paste with water to a thick, creamy consistency. Soak strips of newspaper in the solution and wind them around the bag. Add several layers of these strips to cover the bag, smooth out any wrinkles, and allow each layer to dry before adding the next. (3) Create facial features with poster paints. Add hair or headdress. The head may be used on a hand puppet or a glove or finger puppet.

Some students have a natural bent for puppetry,

**MAKING
PUPPET THEATERS**

The first step is to nail stick legs to each corner of a wooden crate that has two sides removed.

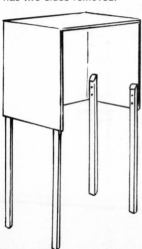

Drape cloth from the bottom of the box and tack it around the sides and front. Operators crouch behind the theater.

Plywood, heavy cardboard, or Masonite may be used to produce a self-standing puppet theater. Refrigerator cartons have been used too.

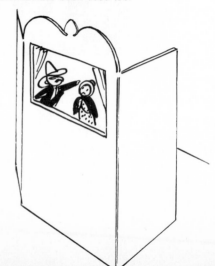

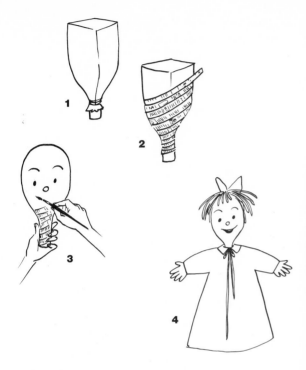

Rabbit in the World" and "The Singing Snake." She spoke the parts, sang songs, and used all kinds of theatrical effects and stagings. Always in demand for performances, Heidi found creative puppetry to be a rewarding hobby. Not all students will go as far as Heidi, but many can experience the joy of communicating through creative puppetry.

as did Heidi Rosen of Palo Alto, California, when she was in Wilbur Junior High school. At an early age her interest developed until she had created a number of skits and stories peopled with her own performers. Two of her skits were "The Fastest

summary

In all games, simulations, and dramatizations discussed in this chapter, the principal common characteristic is active participation in the events by individual students. All these instructional activities are and should be highly motivating and at the same time cause students to seek to achieve the objectives set for each activity.

In using games, simulations, and dramatizations the responsibilities of teachers are not in telling students what and how but rather in setting up conditions under which they can proceed, with minimum guidance, toward enjoyable and spontaneous learning experiences. But great care must be exercised in the evaluation and selection of games, simulations, and simulation games as well as topics for dramatic activities. The concerns to be respected in selecting any medium or resource for instruction must be considered, as well as the appropriateness of the activity to student backgrounds, experience, and readiness to participate in the game or simulation selected.

The variety of commercially produced games and simulations increases each year, as does the frequency of their use. You are encouraged to capitalize on the natural pleasure your students will realize through participating in games, simulations, simulation games, and informal dramatizations and to use them in your teaching.

Through games, students may practice language arts or mathematics; through simulations they may gain insights and knowledge about problems, social processes, and personal responsibilities in contemporary society. Through dramatizations, role plays, and puppet theater activities, students may become more creative and free to communicate, either as themselves or as other personalities whose identities they temporarily assume. All these activities are and must be pleasant and rewarding. In addition, they should be conducted in an atmosphere of freedom and self-imposed responsibility.

15 FREE AND INEXPENSIVE MATERIALS

chapter purposes

⊗ To highlight the many useful purposes to which free and inexpensive supplementary informational materials may be applied in teaching and learning.

⊗ To suggest ways in which students can be encouraged to maintain a constant search for valuable free and inexpensive materials that are useful in their studies.

⊗ To encourage you to seek and obtain appropriate free and inexpensive instructional materials to enrich learning activities.

⊗ To assist you and your students in developing and applying criteria to be used in screening and evaluating free or inexpensive supplementary materials.

⊗ To extend your acquaintance with the sources of information about free and inexpensive materials and how to obtain them.

Free and inexpensive materials that are useful in instruction come in varied forms—pamphlets, brochures, books, statistical reports, charts, films, filmstrips, tape recordings, construction kits, product samples, comic booklets, and even scraps. They are developed and distributed by similarly varied groups—community, state, and county agencies; myriad professional organizations; trade associations and private industry; foreign governments; travel-related organizations; the United States government—and are available locally, sometimes as unsalable items or discards, materials that students may obtain for the asking. To select and use these materials properly as supplements to instruction and learning requires a focused purpose and special attention.

using free and inexpensive materials

Free and inexpensive supplementary materials make several significant contributions to teaching and learning:

⊗ As supplementary resources they provide up-to-date material that often has not yet appeared in regular textbooks, reference books, and other data sources of the school media center.

⊗ They furnish materials that can be edited, or used as is, for bulletin boards or other teaching displays and learning center projects.

⊗ They provide items for students to use in illustrating their own written or oral reports.

⊗ They offer opportunities for students to select, catalog, classify, and edit instructional resources to achieve learning objectives.

⊗ They encourage students to exercise ingenuity in creating a variety of useful learning products from readily available scrap materials.

⊗ They provide motivation for students to explore community agencies that may have materials relevant to projects undertaken in schools.

Alum Rock (California) Schools

get acquainted in your community

Chapter 3 described the community as a valuable source of educational experiences. The institutions, agencies, and private homes of the community are prime potential sources of materials to supplement school media resource collections.

You may be fortunate enough to teach in a school in which the media staff has canvassed the community for places where supplementary materials may be obtained. Perhaps they have prepared a directory of field trip sites, as recommended in Chapter 3, which also lists booklets, films, and samples that may be requested. Or perhaps these items are already on file. In any event, inquire about them. If the resources which you need are not immediately available, ask the media staff to direct you to promising local sources that might supply them. Involving students in the process of finding supplementary materials provides opportunities for them to learn new and useful things about the community in which they live.

other local sources

Additional local materials are often available in publications of nongovernmental agencies such as chambers of commerce, county historical societies, and patriotic organizations. Also consult civic groups (e.g., League of Women Voters), regional trade or promotion groups (e.g., New England Council), gas and electric companies, railroads, telephone companies, water companies, transit lines, and newspapers. Some counties and many of the larger cities publish descriptive materials, annual reports of particular departments or agencies, and other bulletins.

To obtain more information about the media resource potentialities of your community, consider these suggestions:

⊗ Check the yellow pages of your local telephone book. Look under such headings as "Government" or "Associations" or for class names of products or types of occupations.

⊗ Learn the names of your local, state, and national legislative representatives; they will often supply publications without cost.

⊗ Become acquainted with personnel in local travel agencies, consular offices, and transportation companies.

⊗ Above all, don't overlook the obvious! Local, regional, and state instructional media centers, public libraries, the state library, and libraries of nearby colleges and universities often have excellent collections of free and inexpensive materials.

There is still another local source of items that can be used in your school. Do not forget your own students, whose families may willingly, and without embarrassment or compromise, volunteer such needed items as:

⊗ Leftover paint to decorate a playhouse

⊗ A set of fifteen slides about farming in northern New Mexico

⊗ A specially edited recording of the sounds of animals in the local zoo

Beverly Hills (California) High School

⊗ A tape recording of a television opera performance broadcast on a previous occasion

⊗ A book about some topic which is not treated in items in the school media center

⊗ A junked electric motor that can be disassembled and studied

⊗ Loan of a stamp or coin collection to round out the study of a foreign country

⊗ Half a can of plaster of paris left over from a recent bathroom repair job

⊗ A display shelf that could be used for a theme exhibit for a PTA meeting

⊗ An extra slide projector for a super production on parents' visiting day

⊗ Ten glass cider jugs that could be filled with water for a jug band

⊗ Several empty ice cream cartons to make a classroom totem pole

⊗ A dozen empty two-pound coffee cans, with plastic lids, to be used for a musical game

⊗ Several different kinds of wood scraps to be used for a science project

evaluating sponsored materials

School policies pertaining to the selection and use of sponsored instructional materials may range from absolute prohibition at one extreme to no restrictions at all at the other. It is your responsibility to be aware of and to be guided by whatever regulations do exist. The media specialist staff will be a first source of information about those policies and how they apply to you.

If you find that the regulations in force are poorly conceived or overly restrictive, work to improve them—but don't ignore them.

Several important listings of criteria for use as guides in selecting free and inexpensive sponsored materials have been developed. Most such lists start off by giving major attention to these questions: Should materials containing advertising be allowed in the schools? What actually constitutes advertising? At what point does advertising make materials unsuitable for school use?

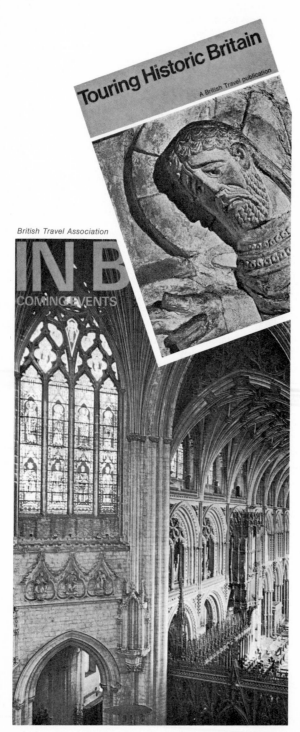

What do *you* consider to be suitable answers to such questions? Is the mere presence of a commercial name or trademark on an otherwise useful piece of instructional material sufficient reason to prohibit using it? At just what point should you draw the line on advertising?

Actually, although some otherwise valuable free or inexpensive sponsored instructional materials do contain excessive advertising or biases toward certain commercial products or partisan points of view, great numbers of them do not. And perhaps many of those which do can be made suitable for student use by editing with scissors and rubber cement.

With federal government publications, on the other hand, as well as those of many national non-profit institutions and organizations, there is usually little question as to the suitability of materials offered. State, federal, and local government publications are frequently prepared by highly qualified experts who write authoritatively and objectively about the topics they discuss. Many publications of state, regional, and national associations portray broad viewpoints of entire industries. National social service organizations (the American Cancer Society, National Tuberculosis and Respiratory Disease Association, and the Joint Council on Economic Education, to name a few) also provide expertly prepared materials that are widely used for instruction.

Finally, even materials which do contain advertising may at the same time be highly acceptable for teaching. A brochure on Madrid, with its maps of the city, photographs of palaces and monuments, and historical notes, for example, may be an undeniably excellent source of authentic information about the city, even though the ultimate intent of those producing it is to encourage tourists to visit that part of Spain.

In examining and evaluating free or inexpensive educational media resources of the types just discussed, several criteria should be kept in mind:[1]

⊗ If the material contains advertising, is it in good taste? Limited in amount?

⊗ Does the material present useful information that

---

[1] Adapted from Ruth H. Aubrey, *Selected Free Materials for Classroom Teachers,* Fearon, Belmont, Calif., pp. 2–3 (revised biannually).

is clearly related to one or more educational goals with which you and your students are involved?

⊗ Is the content of the material honest—objective and unbiased?

⊗ Is the material written or illustrated in such a way as to communicate effectively at the level of the students who will use it?

⊗ Does the material effectively complement—perhaps update—other information presented about the same subject in currently available text-books and other media to which you have access?

⊗ Will the material have other related uses, such as for tearsheets, bulletin-board displays, flat-picture collections, or test material?

professional associations

Hundreds of professional associations—i.e., of business people, doctors, lawyers, educators, engineers, environmentalists, safety engineers—seek to improve conditions in their fields by collecting and disseminating information and conducting research. These organizations usually compile and issue pamphlets, news stories, reports, yearbooks, and magazines about their purposes and activities. Some do much more by producing films and filmstrips, arranging and conducting workshops for educators, and offering scholarships and other awards. Together, professional associations constitute a rich resource for education.

national education association

The National Education Association (1201 Sixteenth Street, N.W., Washington, D.C., 20036), with well over a million members, is of primary importance as a political action voice for the teaching profession. It is the largest professional association in the world.

National affiliates and associated organizations, many of them independent in their operations, relate to such fields as art education; educational communications and technology; industrial arts; teacher education; business education; the teaching of mathematics, science, social studies, music,

MEDIA PRODUCTS FROM PROFESSIONAL ASSOCIATIONS

and home economics; school libraries; journalism; health, physical education, and recreation; school public relations; driver education; school nursing; rural education; and exceptional children.

Several publications available from the NEA Division of Publications offer further information about the many books, monographs, bibliographies, and audiovisual materials available from the organization. Some topics covered in such publications include teacher bargaining, educational accountability, bilingual education, the energy crisis, human relations, education as a career, mainstreaming, and environmental education.

NEA's own journal, *Education Today,* frequently publishes helpful articles on educational media and innovative instructional techniques and practices.

other professional education associations

Every state now has numbers of state, local, and regional professional education associations, many of which are affiliated with the NEA. Similarly large numbers of other professional associations of teachers and educational specialists function independently of the NEA in performing valuable services for their members. Included in this latter group are the National Council of Teachers of English, the Music Educators National Conference, Child Study Association of America, Modern Language Association, American Educational Research Association, and the National Kindergarten Association.

Two organizations are significantly important in a search to improve applications of media and instructional technology in your teaching: (1) the Association for Educational Communications and Technology (AECT), and (2) the American Association of School Librarians (AASL).

One of the AECT catalogs lists publications and other media that suggest the range of interests of that organization: *Individualized Instruction: Its Nature and Its Effects* (filmstrip); *Selecting Media for Learning; Extending Education for Instructional Technology; AV Communication Review* (quarterly journal); *Audiovisual Instruction* (monthly periodical); *Learning Resources* (monthly periodical); *New Media and College Teaching; Hey, Look at Me!* (16mm film on visual literacy); *Technology and the*

PUBLICATIONS OF AECT AND AASL

Association for Educational Communications and Technology

Management of Instruction; and *Educational Facilities with New Media.*

AASL, an affiliate of the American Library Association, continues to emphasize programs to improve school library media services and programs to modernize preparation for specialist personnel. To achieve mutual goals, its members and central staff work closely with such groups as the National Association of Secondary School Principals, National Council of Teachers of Mathematics, Prime-Time Television, and Action for Children's Television.

AASL publications include: *School Media Quarterly; Student Success through Joint Counseling and Media Services; Welcome to the U.S.A.; Evaluation of Alternative Curricula* (for the preparation of library/school media personnel); and *Steps to Service* (school library procedures).

AECT and AASL cooperated in the writing and publishing of *Media Programs: District and School* (1975) which offers guidelines for the development of adequate media resource programs to be used in schools.

professional associations in other fields

Teachers and students seeking up-to-date information about some special professional field will find a likely source of help to be the national headquarters of an association organized by the professionals of that field. Representative of the many such associations that maintain headquarters staffs are the American Dental Association, the American Home Economics Association, the American Library Association, the American Medical Association, and the National Association of Broadcasters.

trade and industry associations

Literally thousands of trade and industry associations with interests throughout the United States publish materials that schools can use, despite the fact that their primary businesses relate to such diverse products as insulation, automobiles, refrigerators, and food products. Although many publicity materials of such organizations reflect special aims and interests, this fact alone does not reduce their possible instructional value. A few specific examples emphasize this point:

⊗ *Telephone companies,* located in all the principal cities throughout the United States, provide schools with numerous free materials. Among the most widely used items are the *Elementary Telezonia Program* (a kit of materials, including films and filmstrips as well as charts and booklets) to promote proper use of the telephone; teletraining materials —teacher aid materials, filmstrips, and activated telephones—for English, speech, and business studies; and a variety of 16mm films on science and general interest subjects. To learn the address of the distribution unit nearest you, call the local telephone business office.

⊗ *The American Iron and Steel Institute* maintains a teaching aids distribution center from which it distributes a large number of materials on the iron and steel industry, most of them for the elementary and secondary levels. These materials include comics and other booklets on steelmaking; magazines on current developments in the industry; and a series of 35mm sound filmstrips in color, with teachers' guides, with such titles as *The Chemistry of Iron, Raw Materials of Steelmaking,* and *The Cradle of an American Industry.*

⊗ *The National Dairy Council,* with headquarters in Chicago, is a nonprofit educational organization of the dairy industry. It operates through a network of

Dairy Council of America

American Telephone and Telegraph Co.

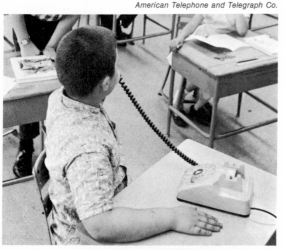

affiliated units throughout the United States. Its purpose is to "promote optimum health through adequate use of milk in accord with current scientific recommendations." To accomplish this, it prepares a variety of health and nutrition education materials, including posters, booklets, dairy farm panorama kits, charts, 16mm motion pictures, 35mm filmstrips, and reference sources for all levels of instruction. Most of these materials are available for purchase by teachers and others from the Na-

Standard Oil Company of California

tional Dairy Council. They may also be obtained from local dairy councils.

⊗ The Standard Oil Company of California, San Francisco, maintains a well-developed information and public relations program directed toward schools and colleges. Many of its educational products deal with factual aspects of the petroleum industry. Also available, however, are a variety of 16mm free loan films on science, travel, sports, education, and foreign countries. Provided free, upon request, to schools in the nine western states are multi-media kits which contain filmstrips, cassettes, duplicating masters, and charts on "Energy and Your Environment."

Here are a few suggestions from sponsors of free instructional materials to help you and your students obtain items you want and need and, at the same time, to maintain good relationships with those who provide them:

⊗ Write on stationery with the official school letterhead which shows clearly the name and school address (including zip code and telephone number and area code).

⊗ If students write the letters, be sure that you countersign them.

⊗ Write only one letter for an entire group, rather than have students write individual letters.

⊗ Request only the number of items actually needed, within limits imposed by suppliers. Sponsors recommend that teachers request one copy of an item for careful review and evaluation before asking for it in class-size quantities.

⊗ Specify the kinds of information and materials desired. Describe clearly the field of interest and the grade level in which the items will be used.

Representatives of private industries and trade associations who maintain public relations contacts with schools through the production of sponsored instructional materials are usually interested in how their products are used and the degree to which they meet educational standards. As for themselves, they regard the recency of their materials and the expert knowledge of the subject matter, which others who prepare instructional materials may not have, as being of special value.

Teachers and students who use sponsored materials will offer a valuable service to those who supply them if they will write giving their reactions, appraisals, and suggestions for other valuable items and treatments in future productions.

**FREE MATERIALS
ABOUT AUSTRALIA**

Australian Consulate General

foreign government agencies

The instructional values of selected information-type media distributed free by offices of foreign governments in this country are well recognized. Most if not all of the nations of the world engage in such distribution to stimulate improved business and political relations and increased travel and foster appreciation of the cultural contributions and aspirations of their people.

Good examples of the range of subject matter and media formats included among foreign government-sponsored items in this category are the offerings of the Australian Consulate General offices in New York City, Washington, D.C., Los Angeles, and San Francisco. Items recently available from such sources include:

⊗ Booklets titled *Australia in Brief; Australia: Handbook; A Look at Australia; Australia: Facts; Birds and Animals of Australia; Flowers and Trees of Australia; Australia: Travellers' Guide; Australian Birds and Animals* (a coloring book for young children).

⊗ Other media are also included: *Illustrated Atlas of Australia; 16mm Films from Australia,* titles which may be borrowed, free of charge, from several sources; *Australia: The Beginning* (chart); reference papers on a variety of topics: Sydney Opera House, Transport and Communications, Forests and Forest Industries; and an exceptional set of full-color picture panels, 36 by 48 inches, printed on sturdy paper, depicting such topics as "Bush Dwellers of Australia" (birds and animals of the back country).

⊗ A letter addressed to the Embassy of _____, Washington, D.C., or to the nearest regional or district consular office of a particular country, will usually be a good start toward obtaining catalogs, materials lists, or materials themselves. Further information about offerings of such agencies will also be found in source books mentioned later.

travel-related organizations

Most foreign and domestic travel-related organizations with offices in the United States also prepare and distribute informational materials of potential interest to schools. Typical examples: Quantas Airlines, Japan Airlines, Cunard Lines, Trans World Airlines, Pan American Airways, French National Railroads, Swiss Air, and German National Railroads. In addition to printed travel brochures and sightseeing bulletins, regional offices maintained by

German Information Center

these agencies in some of the larger United States cities may provide 16mm loan films, books, magazines, and resource persons qualified to speak on subjects related to the countries which they represent.

German Information Center

local ethnic cultural groups

In many communities, the local ethnic cultural groups are often willing to provide instructionally related resources and services of many different kinds. A few examples:

⊗ Invitations to students and teachers to attend their annual (sometimes more frequent) fairs and fiestas that feature handicrafts, music, costumes, dancing, foods, and arts.

⊗ Visits in person to your school or classroom to demonstrate handicrafts techniques, show fiesta costumes, discuss cultural similarities and differences, or lead folk dancing.

⊗ Loans of artifacts and other materials during a period of study of a particular country or area.

United States government sources

An especially rich and comprehensive source of free or low-cost instructional materials is the federal government. Nearly all departments, divisions, bureaus, and agencies of the federal government maintain staffs of specialists to do research and to write and illustrate pamphlets, charts, booklets, and other materials which meet informational needs. Some departments, notably Agriculture and Defense, also maintain film-production units.

Government publications are inexpensive. Numbers of pamphlets are free; many sell for 10, 25, or 50 cents. Large books frequently cost as little as a dollar or two. Because the agencies involved can draw on unexcelled resources, government materials tend to be authoritative, complete, and up-to-date. In fact, they often provide the primary sources of information for commercially produced publications.

Hagerman (Idaho) High School

general information about the federal government

The ramifications of the federal government are so vast and the output of instructional materials so extensive that those who would like to send for available items should know the procedure for ordering them. The key to ordering is your understanding of how the federal government is organized, and how to use the facilities of the Government Printing Office and individual governmental subdivisions. Here are several pointers about beginning the process:

⊗ Consult the *U.S. Government Organization Manual* (revised annually). It describes principal functions and responsibilities of various federal departments; their component divisions, bureaus, and agencies; and the several important independent agencies. The manual is helpful in identifying government offices from which to obtain information. It is available from the Government Printing Office.

⊗ It is not always necessary to write to the main office of a government agency or bureau to request information or publications. Many agencies maintain regional offices where literature can be examined and purchased on the spot. Become acquainted with federal offices in your area; you will find them listed in local telephone directories under "United States."

⊗ Again, you may find that your school media center staff will be able to direct you to appropriate government resources. Many of the vertical file collections of such centers contain selected government documents classified according to local curriculum topics.

Your school media center will probably have a copy of the *U.S. Government Organization Manual,* which is a valuable annotated directory of federal offices from which you may request media products and services. The chart "Federal Departments and Agencies" (opposite page) identifies specific offices of particular interest to educational institutions. To write for directories of educationally related information products and services, from which specific orders may later be compiled, letters may be addressed to the Office of Information of most units.

**THE GOVERNMENT OF THE UNITED STATES
ORGANIZATION CHART**

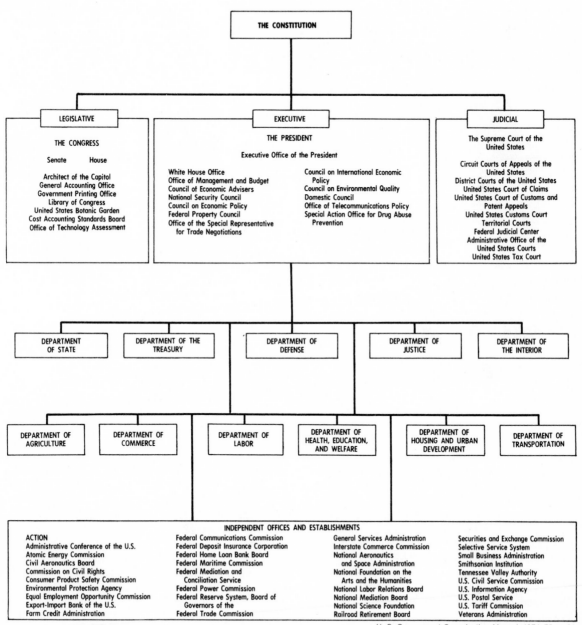

U. S. Government Organization Manual, 1974–75, p. 21

Some of the publications which provide short surveys of United States government materials are the following:

⊗ *Selected United States Government Publications.* Biweekly, free, available from the Superintendent of Documents, U.S. Government Printing Office.

⊗ *Monthly Catalog of United States Government Publications.* U.S. Government Printing Office, $12.50 per year, monthly. Covers books, pamphlets, maps, and serials; arranged alphabetically by departments and bureaus; monthly and cumulative indices by author, subject, and title.

⊗ *Government Reference Books: A Biennial Guide to U.S. Government Publications,* Sally Wynkoop, compiler. Libraries Unlimited, Inc., Littleton, Colorado. An annotated record of more than a thousand United States government reference materials.

⊗ *Federal Government: A Directory of Information Resources in the United States Government.* Sciences and Technology Division, National Referral Center, Library of Congress, 1974. Covers all subjects related to activities of the federal government; cites areas of interest, agency publications; information services provided.

Government Printing Office

The central source for the majority of federal publications is the GPO—Government Printing Office, Washington, D.C., 20402.

Despite the tremendous output of the GPO, the Superintendent of Documents makes it relatively easy for schools to obtain publications promptly and with minimal expense.

You may sometimes wish to examine copies of available materials to evaluate them for use in your classes. The GPO regularly sends copies of all its publications to more than five hundred libraries and other depositories throughout the country. Your nearby university library may be one of these; state libraries and many large urban public libraries are also included in the list. To learn the location of the depository nearest you, ask your school media professional, consult your college or university library, or write to your state librarian or to the GPO.

National Audiovisual Center

The National Audiovisual Center, authorized in 1968, seeks to achieve efficient use of federal audiovisual materials and to furnish information about their sources. *U.S. Government Films: A Catalog of Motion Pictures and Filmstrips for Sale by the National Audiovisual Center* is an impressive collection of materials covering topics of general educational interest. Subjects range from agriculture, education and culture, and human relations to national security and woodworking.

Costs of government-sponsored films and filmstrips are generally lower than those for commercially produced items, chiefly because of the federal subsidy provided to sponsors.

The complete address of the Center is National Audiovisual Center, National Archives and Records Service, General Services Administration, Washington, D.C., 20409.

other government offices

Among the many other United States government departments and independent offices that produce informational materials suitable for schools are the following:

⊗ DEPARTMENT OF HEALTH, EDUCATION, AND WELFARE This department combines the National Institute of Education, Office of Child Development, Office of Consumer Affairs, Office of Education, Public Health Service, Social and Rehabilitation Service, and the Social Security Administration. A principal unit of the National Institute of Education is the Educational Resources Information Center (ERIC), discussed in Chapter 16. The Office of Education maintains the Information and Materials Center, which is concerned with juvenile literature, textbooks, audiovisual materials, and related educational media.

⊗ THE SMITHSONIAN INSTITUTION Widely acclaimed as "a treasure house of our inheritance," the Smithsonian Institution is truly a prime national resource. Actually, it is several institutions in one: Museum of Natural History, Museum of History and

A NATIONAL RESOURCE...

THE SMITHSONIAN INSTITUTION

What do you know about the Smithsonian Institution? Where is it located? What are its principal functions? Who provides funds for its support?

This skeleton, reassembled for the display, represents the remains of an animal that once roamed much of the world. What was its name? Where was it most prevalent? When did it become extinct?

adventures in science at the Smithsonian

Technology, Astrophysical Observatory, National Air and Space Museum, National Zoological Park, and Tropical Research Institute. Many of the Smithsonian publications have direct application to school requirements, including such titles as *Adventures in Science at the Smithsonian* and *The World of the Dinosaurs*.

⊗ DEPARTMENT OF AGRICULTURE (USDA) Traditionally responsible for "acquiring and diffusing useful information on agricultural subjects," the USDA prepares accurate and up-to-date materials in such areas as science, home economics, agriculture, vocational guidance, industrial arts, and social science. Its national publications, as well as bulletins and advice tailored to the needs of local areas, are often available from county agricultural or home economics agents. Extension offices representing both the USDA and the state land-grant universities and colleges will be found in most counties.

⊗ ENERGY RESEARCH AND DEVELOPMENT ADMINISTRATION This agency consolidates federal research and development activities that were formerly managed by the Atomic Energy Commission, Interior Department, National Science Foundation, and Environmental Protection Agency. Its chief purpose is to develop new and improved energy

Atomic Energy Commission

sources consistent with sound environmental and safety practices. The subject matter of booklets, films, and other audiovisual media distributed by this agency range from nuclear power and uses of radioisotopes to career information. Included in its loan collections are many items formerly distributed by the Atomic Energy Commission. New descriptive media catalogs are now available.

⊗ LIBRARY OF CONGRESS The Library of Congress was established to serve the reference needs of Congress, but its present activity is much more comprehensive than that. Its collections of photographs, music scores, maps, phonograph records, and tape recordings are unequaled. Its unique

materials are supplied to other libraries through interlibrary loans for scholars or researchers who cannot go to Washington. It photoduplicates materials and reproduces photographs at cost, prepares special bibliographies, maintains a Central National Union Catalog, issues library catalog cards, and provides braille books and Talking Books for the blind.

⊗ DEPARTMENT OF DEFENSE Each of the three branches of the Department of Defense—Navy, Air Force, Army—maintains public information offices in Washington, in district headquarters, and in other key cities throughout the country. Pamphlets on subjects such as careers in the armed services, specialized training programs, careers in nursing, the service academies, and other topics of student interest are revised regularly. Films and other materials are also available for free loan. Local recruiting offices and regional headquarters of each branch of the services can provide some materials; others must be requested from the public information office of each service in Washington, D.C.

⊗ NATIONAL ENDOWMENT FOR THE ARTS The National Endowment for the Arts makes grants to nonprofit, tax-exempt organizations and to individuals of exceptional talent; it seeks to make the arts available to Americans, to preserve the cultural heritage, to strengthen cultural organizations, and to develop the nation's finest artistic talent. It is interested in museums, film, video, television, radio, theatre, and related activities, and sponsors the American Film Institute.

⊗ NATIONAL GALLERY OF ART Through its Extension Service, the National Gallery of Art produces and distributes audiovisual programs in art education which are loaned free to schools; it also loans films, color/sound slide lectures, and filmstrips.

⊗ U.S. NATIONAL PARK SERVICE The U.S. National Park Service provides information on the National Park System and the National Park Service. It maintains a large file of color and black-and-white photographs which it loans to media users and assists in the preparation of books and articles for magazines and encyclopedias. The National Park Service maintains a center for the production of interpretive exhibits, audiovisual materials, and publications in Harpers Ferry, West Virginia.

how to order United States government publications

Once you have determined the precise United States government publications desired, ordering procedures are relatively simple. A few publications will be noted as being available directly from the issuing agency. Many may be bought over the counter at Government Printing Office field office units and bookstores or other government branch units. Your congressional representatives will sometimes be able to obtain copies of certain publications for you upon special request.

Government Printing Office bookstores which stock the most popular government publications are located in Atlanta, Birmingham, Canton (Ohio), Chicago, Dallas, Denver, Kansas City, Los Angeles, New York City, Pueblo (Colorado), and San Francisco.

In ordering government publications, the rule is:

BUSINESS & INDUSTRY

40C HIGHWAY SAFETY PROGRAM MANUAL NO. 5: DRIVER LICENSING. A guide for States and their political subdivisions to use in developing highway safety program policies and procedures. Topics discussed include purpose, authority, general policy, program development and operation, program evaluation, reports, and local government participation. 1974. 88 p. TD 8.8/3:5   S/N 5003–00195 **$2.30**

41C THE AUDIBLE LANDSCAPE: A MANUAL FOR HIGHWAY NOISE AND LAND USE. Outlines various admin-

Be specific! There are often publications with similar titles, and it is therefore best to use the entry number, the name and number of the series (where applicable), the full title, and, for serial publications, the date.

To pay for GPO materials, you have several choices:

⊗ Check or money order made payable to the Superintendent of Documents. *Do not send stamps.*

⊗ For small orders, check or money order, or either of the arrangements mentioned below.

⊗ Advance deposit accounts of $25 or more can be made with the GPO. Deductions will be made from your account as your orders are filled. When your balance runs low, you can send additional money.

⊗ Special GPO coupons can be obtained from the Superintendent of Documents. These coupons have a face value in varying small amounts under $1. When ordering documents, simply enclose the amount required.

The Government Printing Office address is North Capitol St. between G and H Streets, N.W., Washington, D.C. 20402. Copies of large numbers of government publications may be consulted, before purchase, in one of the many Federal Depository Libraries located throughout the country, locations of which may be obtained upon request from the GPO.

state sources of information

The numerous and varied offerings of the federal government are supplemented by a similar output of publications and other media at state, county, and local levels. Free or inexpensive materials prepared by state and local government offices often contain more specialized or locally oriented information than similar items prepared in Washington. Publications describing the work of these agencies appear in a variety of forms: periodicals, pamphlets, films, maps, charts, and sometimes books. A few state agencies, especially those related to agriculture, have such extensive offerings that they publish lists of available pamphlets. For detailed information about them, consult *Monthly Checklist of State Publications,* United States Library of Congress.

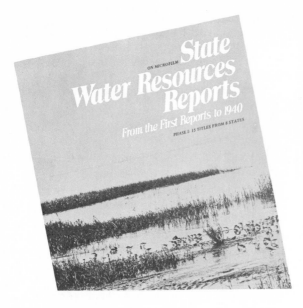

the secretary of state

Teachers who wish special information about some aspect of state government usually write first to the office of the secretary of state in the appropriate capitol city. In most states this office provides special publications about the state—its state flower, song, seal, common birds, and early history and development. For further information, see *The Book of the States* (The Council of State Governments, Chicago, published biennially), which offers up-to-date information about available publications and projects in each of the states.

Also obtainable from the office of the secretary of state is a copy of the organization chart that shows the agencies, departments, bureaus, and offices which handle various phases of state government. The following list includes typical divisions of bureaus at the state level:

Corrections	Parks
Education	Sanitation
Fish and game	Health
Forestry	Taxation
Geology	Water pollution
Housing	Food and drugs

branch offices

Various offices of state government sometimes maintain branch units in principal cities where it may be possible to obtain copies of official publications, to borrow films or other audiovisual resources, and to engage officials to serve as resource visitors.

further sources of information

Your own school media center may have copies of a number of publications in which you will find further information about free and inexpensive materials of the types discussed in this chapter. Look especially for some of the following items which provide information about educational media resources dealing with different facets of the school curriculum:

⊗ Ruth H. Aubrey, *Selected Free Materials for Classroom Teachers,* Fearon, Belmont, California, revised biannually. Sources are listed by subject matter, with items keyed for appropriate grade levels. This especially useful book is indexed and cross-referenced. It contains sections on bases for selection of free materials, what their uses are, and how to order them, as well as a section on local sources.

⊗ *Educators Index to Free Materials,* Educators Progress Service, Randolph, Wisconsin, revised regularly. Free materials are listed by curriculum areas. Emphasis is upon pamphlets, but books, magazines, posters, exhibits, maps, and charts are also included.

⊗ *Free and Inexpensive Learning Materials.* George Peabody College, Division of Surveys and Field Services, Nashville, Tenn. 37203. A compendium of addresses and annotations of educationally related materials issued by various commercial organizations and associations. Revised biannually.

⊗ *Educators Guide to Free Social Studies Materials,* Educators Progress Service, Randolph, Wisconsin, revised regularly. Lists free filmstrips, slides, transparencies, audio tapes, video tapes, scripts, records, and printed materials for use in teaching the social studies.

⊗ *Elementary Teachers Guide to Free Curriculum Materials,* Educators Progress Service, Randolph, Wisconsin, revised regularly. Materials are listed under subject heading, with brief descriptions. Pamphlets predominate, although there are some books, charts, maps, exhibits, and similar items.

⊗ *Vertical File Index,* H. W. Wilson, New York. A monthly publication devoted to listings of inexpensive materials but not intended to be a complete listing of all pamphlet material. Items are arranged under subject headings with a title index, and some entries have short notations. There is an annual cumulation.

Other sources of information about free and inexpensive media resources:

⊗ Various annual listings produced by Educators Progress Service (Randolph, Wisconsin), including such titles as *Educators Guide to Free Films; Educator's Guide to Free Health, Physical Education, and Recreation Materials; Educational Guide to Free Social Studies Materials;* and *Educators Guide to Free Filmstrips.*

⊗ Various lists developed by Serina Press (70 Kennedy St., Alexandria, Virginia 22305) with such titles as *Guide to Foreign Government Loan Film; Guide to Free-Loan Films About Foreign Lands; Guide to Free-Loan Sports Films;* and *Guide to Free-Loan Training Films.*

⊗ Loan film offices of Modern Talking Picture Service, with twenty-five offices in major cities throughout the United States.

⊗ Loan film offices of Association-Sterling Films, with twelve offices in major United States cities and in Quebec and Ontario, Canada.

Teachers use the "materials" columns of their professional magazines to obtain data about new free or inexpensive instructional resources. See especially the periodicals published by the National Education Association and its many affiliates. See also other widely read magazines such as *School Media Quarterly, Scholastic Magazine, Teacher, Instructor, Media and Methods, Educational Technology, Learning Resources,* and *Audiovisual Instruction.*

summary

Quantities of free or inexpensive materials with varying degrees of instructional suitability are available for use in schools and colleges. Some of this material meets highest standards of preparation and provides resources which otherwise would often be unobtainable by schools. On the other hand, much of it is obviously unfit for school use because of its bias or advertising content. Each type—free, inexpensive, with sponsorship, or without sponsorship—presents a special problem of selection.

Some school systems solve this problem simply by prohibiting the use of all sponsored materials in the classroom and by subjecting inexpensive materials to usual selection procedures. Other systems say nothing at all about the use of sponsored materials, leaving the matter to the decision of individual teachers, administrators, or supervisors. The present trend seems to be toward appointing special committees to develop local standards and procedures for appraising and selecting free and inexpensive materials. Once such standards have been set up, the materials are selected in much the same manner as all other types of instructional materials.

Efforts expended in becoming familiar with high-quality free and inexpensive instructional materials are usually well repaid. For that reason teachers and students need to know where to obtain and how to order them, and how to apply practical criteria in choosing them.

16 print, multimedia, and microforms

chapter purposes

⊗ To provide information about distinguishing and useful characteristics of the variety of printed text and reference materials available to schools.

⊗ To improve your performance in choosing, using, and validating textbooks, reference books, supplementary books, microforms, and various types of media packages related to your teaching.

⊗ To suggest how printed text and reference materials may best be used in conjunction with all of the other media resources.

LaMesa Spring Valley Schools

The electronic revolution discussed in this book is often compared in its effects upon humanity with Gutenberg's development of printing with movable type. Sometimes, too, it is predicted that the new communication techniques and devices growing out of this revolution will soon replace books and other printed resources on which education has depended for so long.

We believe that this is unlikely to occur. Rather, printed materials of many different kinds—those now existing and those yet to be developed—will continue to serve educational purposes for years to come.

But probably increasing use will also be made of systematically designed media resource units that combine the best qualities of print and nonprint media. And use of microforms, of the types discussed in this chapter, will be expanded to fill requirements which are met most efficiently and economically through those media.

textbooks

The textbook is a versatile product. It provides convenient and random access to the messages it contains; users control the process by which they gain the information they seek. Paging and indexing permit them to move ahead, to skip, or to dwell at length, to skim quickly, or to read slowly, perhaps with exceptional care. Books are portable; they are economical to buy and maintain. They often contain excellent visualizations of concepts and information with accompanying verbal content to explain them.

Textbooks are used in several ways: (1) as basal texts, and thus as the chief sources of information for students in a particular course; (2) as co-basal texts, with two or more titles adopted and both used, complementarily, in studies; and (3) as supplementing and enriching sources only, and not as a chief source of information or guide to a course.

What, especially, do textbooks contribute to teaching and learning?

advantages of textbooks

Textbooks offer several advantages when used for classroom instruction:

⊗ INDIVIDUALIZATION OF INSTRUCTION Textbooks help to individualize instruction by enabling students to proceed at their own rates, and, to a limited extent, according to what they are interested in studying. With textbooks, students need not all study the same things. They may choose what they require.

⊗ ECONOMY Textbooks are used and reused; their actual per-pupil costs are therefore quite low. If one compares the cost of a textbook containing hundreds of pictures, charts, maps, diagrams, car-

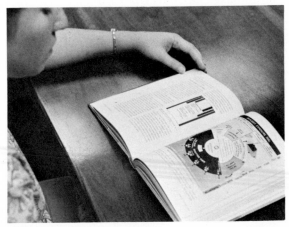

3M Company

toons, and other visuals with the cost of those same items in filmstrip, slide, or still-picture form, books remain the inexpensive choice.

⊗ ORGANIZATION OF INSTRUCTION Textbooks help to organize instruction by providing common reading experiences, suggested activities, recommended readings, and questions. Expertly written textbooks give unity to classroom instruction. Also, they are graded; in introducing new concepts or content, they build upon what has preceded.

⊗ TUTORIAL CONTRIBUTIONS Teachers often maintain that textbooks help students learn how to read better, to study, to weigh evidence, and to solve problems. They disclaim any necessity for memoriter learning when teaching with them. Instead, they advise using the suggested activities, thought questions, and chapter bibliographies to stimulate and guide different kinds of learning.

⊗ IMPROVEMENT OF TEACHING Textbooks are also regarded as helpful in improving teacher skills (especially through teachers' manuals and notes), and in suggesting ways to handle instructional problems.

McMurry and Cronbach used four types of communication—narration-description, prescription, generalization, and theory—portrayed in the chart below, to classify the content typically included in arithmetic and English textbooks. Examine a few textbooks in some field of interest to you. Pay particular attention to the content and to the form in which it is communicated.

Do the McMurry-Cronbach generalizations apply? If not, what new or changed classifications are needed?

In making this examination, also consider the degree to which each of your books meets the criteria enumerated later in this chapter, which should be taken into account when appraising and choosing textbooks.

FOUR TYPES OF VERBAL COMMUNICATION AS USED IN TWO TYPES OF TEXTS

	NARRATION: DESCRIPTION	PRESCRIPTION: DIRECTION	GENERALI-ZATION	THEORY
ENGLISH TEXTS	Works of literature Description of evolution of a particular word Biographies of writers Incidents to arouse interest	Rules of grammar Procedure for using reference books Statements of accepted usage Forms for business letters	"A good paragraph is unified." "Contractions are rarely used in formal writing."	Systematic treatment of grammar and its logic (definitions of parts of speech, for example) Comparison of poetic forms
ARITHMETIC TEXTS	Incidents to arouse interest History of numbers	Model solution for type problem Procedure for attacking problems	Numerical combinations Rule for area of triangle	Concepts distinguishing decimal, binary, and base numbers Definition of multiplication as multiple addition

Text Materials in Modern Education, University of Illinois, Urbana

American Book Company

criticisms of textbooks

There are several common criticisms of textbooks. First, it is sometimes said that by presenting materials in logical, predigested form, textbooks relieve learners of much need to think, to do their own organizing, or to arrive at independent conclusions. This being the case, the critics say, the learning task is too often more reading to remember than finding, choosing, and using data to solve problems or to serve other useful purposes. Coupled with this criticism is one that says textbooks ossify subjects by treating them too sketchily or by providing only minimal information, thus stimulating little interest on the user's part to pursue any of them very far.

The claims that textbooks ignore instructional innovations, and yet that they influence teaching more than they should, are well known and interrelated. Often, these claims are coupled with derogatory comments about "textbook teaching" in

which principal learning activities are characterized as a deadly routine of assigned reading and recitations. It is sometimes argued as well that the practice of making cyclical adoptions of single texts every few years prevents adventurous teachers from encouraging publishers by buying books that deal with subjects creatively.

recent developments

Two recent developments highlight the changes in educational publishing. One is that numbers of publishers now produce instructional materials not as textbooks but as multi-media kits. These kits often contain correlated booklets, filmstrips, and other appropriate materials. Many of them deal frankly with controversial topics such as drugs, race, poverty, marriage, premarital sex, and war.

Another publishing innovation of the last few years stems from the rising interest in the teaching of minicourses, particularly at the secondary level. Typically, a minicourse is a six-week module of instruction. Numerous minicourses are offered as electives. Students may include several of them in a single year's schedule. Printed materials used with minicourses are considerably different from those used in year-long courses. They are shorter, more restricted in coverage, and especially readable and interesting.

Gage Educational Publishing, Toronto

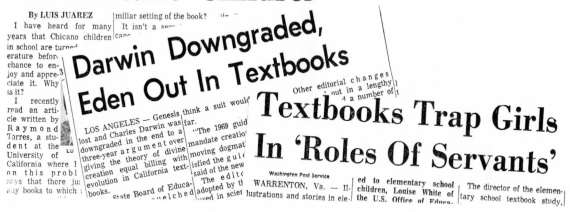

Books Fail To Relate To Chicano Children

By LUIS JUAREZ

I have heard for many years that Chicano children in school are turned ... erature before ... chance to en- ... joy and appre- ... ciate it. Why ... is it?

I recently read an arti- cle written by R a y m o n d Torres, a stu- dent at the University of California where I on this probl ... says that there ju ... any books to which i

miliar setting of the book?

It isn't a ...

Darwin Downgraded, Eden Out In Textbooks

LOS ANGELES — Genesis lost and Charles Darwin was downgraded in the end to a three-year a r g u m e n t over giving the theory of divine creation equal billing with evolution in California text- books. State Board of Educa- ... ... u e l c h e d

think a suit would far.

"The 1969 guid- mandate creatio... moving dogmat... isfied the g u i d said of the new The edit o adopted by tl ... sed in scier

Other editorial changes ... out in a lengthy ... a number of ...

Textbooks Trap Girls In 'Roles Of Servants'

Washington Post Service

WARRENTON, Va. — Il- lustrations and stories in ele-

ed to elementary school children, Louise White of the U.S. Office of Educa-

The director of the elemen- tary school textbook study,

appraising and choosing textbooks

Appraising and choosing textbooks to recommend for purchase are professional tasks you may be asked to perform. Generally speaking, educators appear not to favor state adoptions of basal texts. Instead, they prefer local adoptions and increased participation in all selection processes by teachers and students who will use them.

As a participant on a selection committee, you will be expected to:

⊗ Use open, honest, and carefully developed proce- dures

⊗ Suggest desirable characteristics (criteria) of textbooks to be selected

⊗ Be consistent in your applications of evaluative criteria for the textbooks you assess

⊗ Know individuals, organizations, publications, and lists that provide assistance in finding textbooks to match those criteria

Typically, in the past, textbooks have been recom- mended for adoption largely on the basis of evalua- tors' *subjective personal estimates* of quality and suitability. But there is a growing concern about

this practice among those who believe that, in the selection of any textbook, recommendations must be objective, based upon the single significant ques- tion: *How well do students learn by using it?*

At the present time, the most practical means of approximating a satisfactory answer to this question has been to ask experienced teachers, supervisors, and school administrators, media specialists, stu- dents, and parents to make combined assessments on the basis of whatever data are available upon such important factors as the following:

⊗ LEARNER VERIFICATION How was the book tested and refined during the process of its develop- ment? Is evidence supplied to show that this veri- fication process was valid for the purpose?

⊗ VALIDATION Does the publisher provide evi- dence to show that, since publication, students of certain characteristics who have used the book in stated ways have learned satisfactorily? Are the characteristics of your students and the proposed ways of using the book in your situation sufficiently similar to those of the validation groups to predict that they, too, will learn well by using it?

⊗ TREATMENT OF MINORITIES Are members of minority groups treated in an accurate and dignified

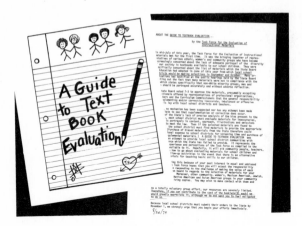

manner? Are there omissions of significant minority-related content? Is the textbook Eurocentric? Do the verbal generalizations and pictorial content about minorities lead to inaccurate stereotypes? Are cultural and other contributions of outstanding members of minority groups recognized and discussed?

✪ SEXISM In this context, "sexism" is meant to denote discrimination based on gender. Are men and women, boys and girls, treated by the textbook primarily as people rather than as members of a particular sex? Are individuals of each sex cast repetitively in sex-typed roles: females as housewives, secretaries, or nurses; males as breadwinners, managers, or doctors? The former as submissive, withdrawn, and dependent; the latter as aggressive, outgoing, and independent?

Other questions to be asked relate to: (1) scope or content, (2) grade or achievement level appropriateness, and (3) physical-mechanical features.

✪ SCOPE OR CONTENT Does the textbook contain information related to topics to be studied in the course? Does the text interpret curricular objectives prescribed by the course of study? (Select a topic in the text and compare the information it gives with that called for by the course of study.)

✪ APPROPRIATENESS OF GRADE OR ACHIEVEMENT LEVEL Are style and vocabulary suitable for the age, grade, and achievement levels of students who will use the text? (Check such items as sentence and paragraph lengths; check vocabulary difficulty, using some approved basic word list.) Can the material be adapted for use by individuals of varying abilities? Will the material appeal to students of both sexes? Are aids provided for slow learners? Are extra challenges provided for superior students?

✪ PHYSICAL OR MECHANICAL FEATURES Is the type clear and readable? Is it sufficiently large? Is there enough leading (spacing) between lines? Are lines of proper length for easy reading? Is the paper of good weight and durability? Is the binding reinforced so that the book is held firmly in its cover? Are pages uncrowded and readable? Are the index and table of contents complete and easy to use? (Note subjects mentioned on one or more pages of the text, and check against index entries.) Are difficult and unusual words defined in a glossary? Do well-organized summaries and reviews appear at ends of chapters and units? Do bibliographies include up-to-date materials, both printed and audiovisual? Are visuals correlated with the text? Do they add meaning to verbal content? (Check reality in color, artistic appearance of pages, size of illustrations, and lack of irrelevant details.)

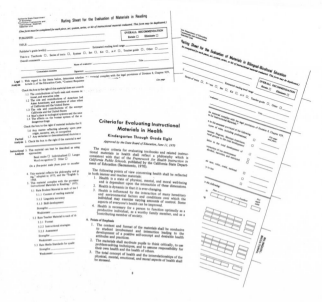

teaching with textbooks

Most criticisms of textbook teaching would not be made if textbooks were used imaginatively. Obviously, there is no need to stop with the read-and-remember routine; other, more functional uses of textbooks should be substituted.

⊗ USE TEXTBOOKS AS TEACHING ASSISTANTS To use textbooks as a course outline is usually a mistake and a temptation to be avoided (with some exceptions, or course). Adapt the textbook to your purposes and to the status and capabilities of members of the groups with whom you work.

⊗ ADD REALITY TO TEXTBOOK ABSTRACTIONS Make textbook content come alive and help students to understand and appreciate real-life applications of what they study in textbooks by showing appropriate films or filmstrips, taking field trips, arranging and explaining displays or exhibits, listening to recordings, or experiencing or preparing other related audiovisual media.

⊗ INDIVIDUALIZE INSTRUCTION THROUGH INDEPENDENT STUDY ASSIGNMENTS Produce your own self-instruction sheets to accompany the textbook. In them, invite students to use particular textbook portions to answer questions or solve prob-

lems; indicate other textbooks, encyclopedia articles, magazines, and pamphlets to be consulted and the purposes to be achieved by using them; list filmstrips or flat picture sets to be studied, tapes or discs to be listened to, or motion pictures to be seen—in the classroom, in the school media center, or elsewhere.

⊗ HELP STUDENTS IMPROVE LANGUAGE AND RESEARCH SKILLS Textbook study can help to improve language and research skills. Ask students to skim for sense, to note paragraph headings, to use indexes or tables of contents to locate specific information. Or give an assignment to express the essence of several pages in a brief paragraph, or to compare the viewpoints of several books about the same topic. Have students look up meanings of new words and use them in writing and speaking; teach them to read footnotes, and to use the library to locate references.

⊗ MAKE FULL USE OF THE VISUAL CONTENT OF TEXTBOOKS Be especially aware of the visual content of textbooks. Charts, maps, graphs, pictures, diagrams, and occasional overlays represent unusual and valuable resources for learning. But the mere presence of these items in textbooks gives no assurance that students will study and understand them or be helped by information they contain. So call attention to textbook visuals; assist students in learning to use them as sources of information.

Lowell School, Mesa (Arizona)

media packages

The continuing interest of educators in individualized instruction, much of it as independent study, has stimulated the production of several types of learning packages, mentioned earlier. Some of them are multi-media kits intended to assist teachers in making effective group presentations; these sometimes contain such varied, but interrelated, items as large transparencies, 2- by 2-inch slides, filmstrips, flat pictures, posters, and graphics. Some may contain only printed materials; others may be truly multimedia, including numerous printed as well as audiovisual resources. Some may be programmed for self-directed independent study; others may be developed as collections of correlated (but not programmed) media to provide for various ability levels.

multi-media packages: some examples

Typical multi-media packages, or kits, contain a variety of self-teaching and self-testing materials. Major portions of many of them take the form of printed booklets, self-check tests, filmstrips, tape recordings, study prints, and 2- by 2-inch slides. From this basic core of materials the multi-media kits are often expanded into other kinds of teaching media. In the area of foreign language, for example, kits sometimes contain motion pictures and textbooks. One cross-media kit for science includes identification charts, real objects (some of them alive), plastic-embedded real objects, overhead transparencies, and motion-picture films.

Two examples of multi-media packages provide yet other materials:

⊗ A learning package on the subject of "Historical Highlights in the Education of Black Americans" was developed by the National Education Association. It includes a twenty-two-minute sound filmstrip, a history booklet, and a three-panel display of still pictures. Both the booklet and the display pictures may be purchased in quantity if desired. The kit was designed to be used in English, social studies, art, or for special events such as Negro History Week or American Education Week.

BFA Educational Media

⊗ A 1972 production by BFA Educational Media, titled *Databox: Fort Bragg,* was developed to provide realistic practice in community study in the classroom. Using the data supplied within the package alone, students are encouraged to experiment with and invent their own solutions to problems it presents. The kit presents a multifaceted picture of Fort Bragg (California). Included in it are (1) eighty student project cards that pose problems to be examined by small groups or individual students; (2) blank cards on which to record other projects thought up by users; (3) twenty copies of a photographic inventory of Fort Bragg—hundreds of pictures of the town including every building on Main Street, signs, vehicles, people at work and at play; (4) twenty copies (in book form) of source materials about Fort Bragg—census data, stories, want ads, and letters to the editor, and city ordinances; (5) a local telephone directory, (6) a topographical map, a nautical chart, an air navigation chart, and a highway map, (7) two filmstrips—one containing historical photographs of the town, the other showing contrasting views of life in a similar town in Norway, in another California town, and among the Indians who first settled the Fort Bragg area; (8) two audio cassettes of recorded interviews with local people related to community problem solving; (9) more than a hundred copies of a relief map of the town, and a reproducible master to permit production of additional copies locally; and (10) a teacher's manual suggesting ways to use the kit to advantage.

programmed media products

Although programmed instruction has an important role to play in the solution of numerous instructional problems, no attempt will be made here to discuss it in detail; there are many useful references on the topic. Here, we wish to mention only a few of the principles of programmed instruction and to encourage you to consider the broader applications of the process of programming in planning any instructional sequences.

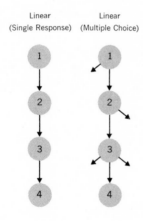

Linear (Single Response)   Linear (Multiple Choice)

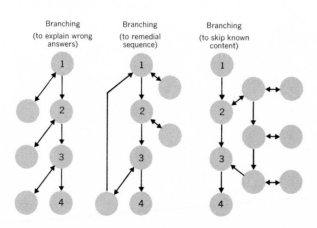

Branching (to explain wrong answers)   Branching (to remedial sequence)   Branching (to skip known content)

Two patterns—shown at left—are used in programming: *linear* and *branching*. Linear programs present the material in small amounts to students who often answer a question after each segment by choosing a yes/no or a multiple-choice answer; if the answer is correct, they proceed to the next question. The branching (or adaptive) patterns, as shown, provide alternative choices, and, according to the response, students are guided to the most desirable subsequent action. Note that known content may be skipped, or remedial sequences may be required; the program seems to be adapted to the needs of different learners.

Programs may be only in print, or they may be accompanied by visual materials or direct learners to laboratory stations or other experiences outside the program. The length of programs varies with the instructional need: short modules permit rapid mastery of a block of content, or they may be longer, more complete modules, or even complete courses. A most important contribution made by the developers of programmed instruction is the process itself, strongly reflected in the procedures for designing instruction as described and charted in Chapter 1, with emphasis upon learner-centered instruction, upon mastery of clearly specified content to a specified level of competence, and at a pace suited to each learner.

supplementary books

Supplementary fiction and nonfiction books are used to enrich class learning and to accommodate individual differences among students. They are frequently brought to classrooms from the school media center to form a decentralized loan collection. Such collections are sometimes augmented by books students bring from their homes.

teaching with supplementary books

Experienced teachers find many ways to use supplementary books. During the exploratory or initiatory phase of teaching, for example, supplementary books constitute a rich resource to build class interest in a subject and to meet needs of individual students as they work alone or divide into small groups or committees to carry out cooperative projects. During the developmental or work-activity phase, supplementary books furnish day-to-day help and stimulation in completing assignments and preparing reports. Showing a motion picture may be the occasion to encourage several students to read and report on a number of different supplementary books.

Portland Community College

San Diego County (California) Schools

Supplementary books may also be used as the basis for assignments that require students to practice using library aids (card files, guides), book study aids (indexes, tables of contents), and reading techniques (skimming, detailed studying). Students interested in making tape recordings to dramatize incidents in history, for example, will have reason to consult library books and, at the same time, a functional reason to practice library skills.

Many teachers dramatize supplementary book collections by obtaining the cooperation of school media-center personnel in arranging book-jacket displays and giving book reviews.

Teachers should create opportunities for their students to exchange ideas, opinions, and comments about supplementary books they have read or used.

paperback books

Paperback books by the thousands play increasingly significant roles in the education of students. Paperbacks read by elementary and secondary students are principally in the area of literature, including not only fiction, but travel, biography, science, and politics. They offer a welcome variety and change from a steady diet of textbooks. Through book clubs, such as the Teen Age Book Clubs promoted by *Scholastic Magazine,* the student use of paperbacks has been encouraged.

Perhaps as much for their unforbidding size and eye-catching covers as for their content, paperbacks are well regarded by students. And their low cost, as compared with the relatively high cost of hard bound books, makes them attractive as components of the modularized, self-instructional, and individualized learning packages discussed earlier.

Paperback book displays are easy to arrange, and they stimulate student reading. For this purpose, the mobile display units, shown below, are especially convenient.

J. Roy Barron, Santa Barbara (California) City Schools

encyclopedias

Independent study activities frequently require two other important aids to learning: encyclopedias and similar reference books. The stimulus to search for information in these sources may be the unanswered question after seeing a film, the need for data to augment a field trip, or a missing fact to be depicted in a classroom bulletin-board display. It is essential that students become familiar with the contents and arrangement of many different kinds of reference books in the process of locating information to solve such problems.

choosing encyclopedias

As with textbooks and supplementary books, teachers and students are frequently asked to assist in recommending encyclopedias for school purchase. In making recommendations, they will often seek answers to the following questions:[1]

⊗ PRELIMINARY CRITERIA What are the characteristics of the students who will use the sets? If the school now owns encyclopedias, what has been the experience in using them? Based on classroom tryouts with competitive sets, from which do students get the most benefit? Which do they prefer?

⊗ ORGANIZATION Are articles arranged alphabetically, or are they classified? Are articles detailed and long, or short and of a summary type? Are there extensive cross-references with articles? Is there an index with each volume, or in a separate volume? Is the index easy to use? Does it supply supplementary information?

⊗ STYLE Are articles written in a style that is easy to read? (Check readability on a subject with which you are familiar.) Do articles maintain an objective approach to controversial subjects? (Check subjects such as racial integration or communism.)

⊗ AUTHORITY Are articles written by leaders in their subject fields? (Most publishers usually list the names of contributors and their qualifications in the first volume.) Does the edition bear an endorsement of one or more recognized scholarly groups?

[1] As developed by Leslie Janke for *AV Instructional Technology Manual,* McGraw-Hill, New York, 1973.

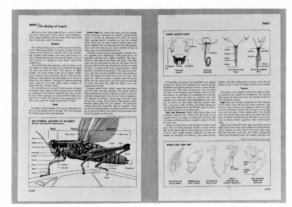

From The World Book Encyclopedia. © 1975 Field Enterprises Educational Corporation. Reprinted by permission of Field Enterprises Educational Corporation.

⊗ FORMAT AND MECHANICAL FEATURES Are pages attractive? Is good-quality paper used? Is the type large and clear? Are pages clearly printed? Are articles clearly and interestingly illustrated?

⊗ ADDED FEATURES Do bibliographies contain materials with up-to-date copyrights? Do they include all types of educational media (not just books)? Do they list items likely to be found in average school libraries? Is the set kept up-to-date through continuous revision? Periodical revision? (Read the introductory section of the first volume to see whether it contains information on how the set is kept up to date.) Are guides or manuals provided to explain how to use it?

teaching with encyclopedias

Students need guidance in working with encyclopedias. Rather than permit or encourage them to memorize or copy information, invite them to obtain information needed to solve some problem or to be used in other appropriate ways. All encyclopedia-related assignments should place a premium upon a student's ability to *interpret* content and, as new facts and ideas are obtained, to reorganize material.

Most encyclopedias contain excellent graphics: charts, maps, diagrams, and pictures—many in

color and some presented as transparent overlays. Be sure your students develop the skills necessary to interpret them.

Finally, encourage your students to be critical of materials they read. Ask them to compare statements and illustrations in two or more encyclopedias that deal with the same topic and to comment on the similarities and differences noted.

Publishers often provide booklets which, if used, enable students to make maximum use of encyclopedias. Also, a number of games have been developed to stimulate student interest in locating and interpreting data given under various encyclopedia entries. Write the distributor or publisher of the set you use to request copies of them.

newspapers
and magazines

Insofar as schools are concerned, magazines and newspapers fall into two types: (1) those published for the public at large, and (2) those written for school readers. In many classrooms they not only provide useful current information and material for specialized reading, but also contribute to improved learning in a number of ways. Typical uses of newspapers and magazines are for:

⊗ Current events study and analysis

⊗ Background studies on important local, national, and world problems

⊗ Practice in improving reading and discussion skills

⊗ Propaganda analysis

⊗ Study of writing and editorial styles

⊗ Foreign language training (magazines such as *Le Petit Journal,* and *Las Americas*)

⊗ Bulletin-board displays and scrapbook collections

⊗ Day-to-day factual data (weather reports, for example) from which data-processing assignments can be developed

Two continuing objectives of teaching with news-

J. Roy Barron, Santa Barbara (California) *City Schools*

papers and magazines should be to develop broad reading interests among students and to stimulate them to assess and evaluate all the materials they read for biases and accuracy in the data presented. To motivate the development of critical skills, you may have students participate in the process of nominating magazines and newspapers to be placed in the school media center. Student groups have often carried out this assignment by critically reviewing sample copies obtained from nearby jobbers and dealers.

Mobile display/storage units such as the one shown above can be used to stimulate interest in newspapers, magazines, and books. These units can be loaded in the media center and wheeled to classrooms where they are used as resource centers for reference or for reading activities.

Individuals wishing to improve and extend the use of newspapers and periodicals in classroom instruction, or through the school media center, may obtain helpful information by writing the American Newspaper Publishers Association (ANPA) Foundation, P. O. Box 17407, Dulles International Airport, Washington, D.C. 20041, and the Educational Press Association of America, Newhouse Center, Syracuse University, Syracuse, N.Y. 13210.

Herbert Breuer

educational comics

Classroom teachers who compete with or sometimes confiscate comic books readily admit that these magazines have a hold on student audiences. Survey figures vary, but we may be fairly certain that as many as half the adults of this country read comics too, and many of them are college graduates.

Most comic-book series have long since ceased to be funny; they have turned into straight picture stories in strip or pamphlet form. Several hundred million comic books are printed each year, and producing them continues as big business. The industry has established its own code of ethics, which some publishers follow scrupulously and others ignore.

Teachers cannot disregard comic books, nor should they wish to, because they occupy so much of the time, attention, and energy of students. Some picture stories presented in comic-book form supplement and contribute positively to regular school studies; others are lurid and objectionable.

The interest of young students and adults in picture stories emphasizes the potential of visual materials for teaching. Many historical, scientific, and literary subjects can be presented well and accurately in comic-book form. *Superman* has even been used to teach grammar. Teachers who use educational comic books such as *True Comics* or *Classics Illustrated* have found that they sometimes assist vocabulary building, stimulate a liking for reading, and help students with limited reading abilities to experience the stories of significant literary works largely in visual form.

Much of the improvement in quality of comics distributed in the United States can be attributed to the influence of the Comics Magazine Association of America. This organization enforces among its members a Comics Code which is intended to raise industry standards. Principal requirements of the Code are that comics presentations shall:

⊗ Not promote distrust of forces of law and justice nor a desire to imitate criminals.

⊗ Create respect for authority.

⊗ Emphasize good as triumphing over evil.

⊗ Never portray the crime of kidnapping.

⊗ Never present lurid, unsavory, or gruesome illustrations.

⊗ Avoid use of profanity, smut, or vulgarities.

CLASSICS IN COMIC FORM

Romeo and Juliet—comics version. Do you consider such materials useful in introducing the classics? If so, under what circumstances would you use them? What steps would you take to encourage students to select comic books of this type? *From Classics Illustrated, Gilbertson and Company.*

CLASSICS *Illustrated*

⊗ Never ridicule religious or social groups.

⊗ Deal with portrayals of family units in ways that show support for the protection of children and family life generally.

The Science Fair Story of Electronics, from which the page below was excerpted, suggests the instructional value of educational comics. Several million copies of this particular publication have been sent free (by Radio Shack, 2617 West 7th Street, Fort Worth, Texas 76107) to schools throughout the United States.

RADIO SHACK, *The Science Fair Story of Electronics, p. 5.*

"SAMUEL F. B. MORSE BUILT HIS FIRST TELEGRAPH SOUNDER IN 1836 FROM MANY MISCELLANEOUS PARTS INCLUDING AN OLD PICTURE FRAME." MORSE WAS ALSO A TALENTED PORTRAIT ARTIST."

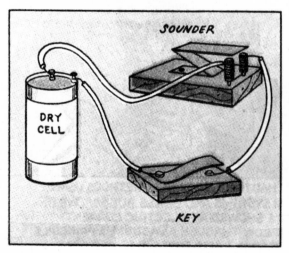

"THIS PROJECT SHOWS HOW A TELEGRAPH SYSTEM WORKS"

SOUNDER

DRY CELL

KEY

"HIS WIRE TELEGRAPH WAS THE FIRST PRACTICAL LONG-RANGE COMMUNICATIONS SYSTEM. IN 1861, STEPHEN FIELD SENT THE FIRST TRANSCONTINENTAL TELEGRAPH MESSAGE TO PRESIDENT LINCOLN."

SAN FRANCISCO

WASHINGTON, D.C.

microforms

The several microform media with which teachers and students should be familiar include: (1) microfilm, (2) microfiche, and (3) microcards. Each type contributes to the capability for storing and retrieving information needed in education.

⊗ MICROFILM Microfilm is a narrow photographic film (usually 35mm or 16mm width) on which various types of images are photographed and stored. The typical procedure with microfilming involves photographing only one or two pages of a book or docu-

Gunn High School, Palo Alto (California)

Dallas (Texas) County Community College

ment on a single frame. The contents of a fairly long book can be placed on only a few feet of film, which can be stored in a small container and read on any of several types of microfilm readers. It is possible to view images directly (from projections on the reader screen) or to have copies produced as paper prints or projectable transparencies.

⊗ MICROFICHE The term *microfiche* is French, translated as "miniature index card." Here the word is used to describe a 4- by 6-inch sheet of fairly rigid film containing space for a large number of frames (60 to 1,000 miniaturized pages).

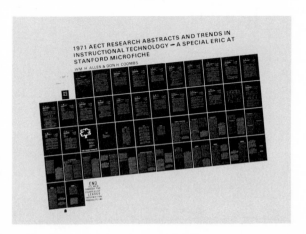

Microfiche transparencies permit the concentration of large amounts of textual and visual data in remarkably little space. Reading the transparencies may be accomplished in several ways: by screen projection, by projection in a portable viewer, by direct viewing in a hand-held enlarger-viewer, and by enlarged print-outs on paper. Low-cost microfiche transparencies for individual use can be produced on duplicator units.

Perhaps the greatest value of the microfiche system is that it provides readers with publications at low cost. The ERIC (Educational Resources In-

formation Center) system, started in the 1960s to ensure wide dissemination of fugitive educational research documents, was one of the first large-scale users of this medium. (ERIC is discussed later in this chapter.) The future publishing of more books in microfiche form appears certain, especially in view of the capability of the medium to reproduce their contents in full color.

⊗ MICROCARDS Microcards are similar in appearance to the microfiche film just described. They are produced as either opaque paper or transparent film reproductions of copy and can be read by means of special hand or desk viewers.

XEDIA Program, Xerox Educational Publications

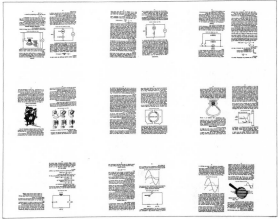

Readex Microprint

Microcard Corporation

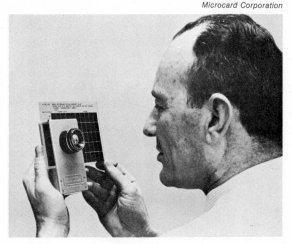

teaching with microforms

As educational materials in microform increase in scope and availability, the schools readily put them to use. Typical applications are:

⊗ To enable media centers to obtain back issues of periodicals or copies of rare books especially needed for projects.

⊗ To save on costs of subscriptions, book or reference purchases, or storage space.

⊗ To assure proper maintenance of periodicals (no coupon tear-outs, no missing articles, files always in correct sequence).

⊗ To speed up the process of skimming large volumes of printed materials.

⊗ To obtain items in a form which permits unusual and functional uses of them (as, for example, projection of a microfilmed drawing for an entire group to study).

⊗ To facilitate independent studies involving microfilms or microfiche, including, for the latter, an increasing number in full color and often accompanied by explanatory or tutorial audio tapes.

Assignments to study materials in microform are almost sure to become more common in schools of the future. This will result from the availability of portable microform readers for the personal use of almost every student and the continually expanding number of resources in microforms. Thus, microform media will offer advantageous means of extending and enriching learning experiences.

XEDIA Program, Xerox Educational Publications

San Diego County (California) Schools

ERIC: an information-sharing system

The Educational Resources Information Center (ERIC) is supported by the National Institute of Education (NIE), an agency of the federal government. Through ERIC's network of sixteen clearinghouses in different parts of the United States, educational personnel receive assistance in locating research reports, articles, monographs, and related information needed to make many different kinds of educational decisions.

Specializations of the various clearinghouses include: (1) counseling and personnel services, (2) early childhood education, (3) educational management, (4) handicapped and gifted children, (4) higher education, (6) information resources (media, telecommunications, information science, librarianship), (7) junior colleges, (8) language and linguistics, (9) reading and communication skills, (10) rural education and small schools, (11) science, mathematics, and environmental education, (12) social studies and social science, (13) teacher education, (14) tests, measurement, and evaluation, (15) the disadvantaged, and (16) career education.

microform equipment

Recent developments in equipment designed to assist in reading microfiche materials seem likely to stimulate widespread use of this medium. Microform industry officials anticipate a wide sale of low-cost viewers that are simple to use, weigh only a few pounds, and enable users to carry thousands of pages of reading material as easily as they now carry briefcases.

Library Resources, Inc.

Three ERIC-related publications provide assistance to users in locating needed information in the system:

⊗ *Resources in Education* (*RIE*).   A monthly journal that abstracts all items entered into the system; indexed by subject, author or investigator, and institution.

⊗ *Current Index to Journals in Education* (*CIJE*). A monthly index to educational literature contained in more than 700 periodicals; indexed by subject and author.

⊗ *Thesaurus of ERIC Descriptors.*   A structured compilation of terms used to index ERIC documents, including *RIE* and *CIJE* articles.

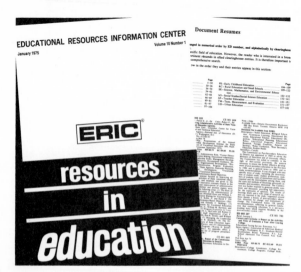

Further information about ERIC and the current titles and addresses of each of its clearinghouses may be obtained from ERIC Central, National Institute of Education, 1200 19th St. N.W., Washington, D.C. 20208. Your school media center may have such information on hand.

Also, more than 600 libraries (mainly in institutions of higher learning) throughout the United States maintain collections of ERIC microfiche which may be consulted. Write the Central ERIC office for locations of those near you.

information sources

There is a great range of source materials that present technical (and sometimes evaluative) data about print-related media and microforms discussed in this chapter. Several listed here are recommended for school media-center collections where they can be consulted by teachers and students. Other items, some more specialized in coverage than those which follow, will be found in Reference Section 6.

⊗ *American Reference Books Annual.* Bohdan S.

Wynar, ed. Libraries, Unlimited, Littleton, Colo. 80120.

⊗ *Books in Print.* R. R. Bowker Co., 1180 Ave. of the Americas, New York, N.Y. 10036. (See also *Books in Print Supplement: Authors, Titles, Subject.*)

⊗ *CEDaR Catalog of Selected Educational Development and Research Programs and Products.* Council for Educational Development, 1518 K St. N.W., Washington, D.C. 20005.

⊗ *A Core Media Collection for Secondary Schools.* Lucille G. Brown, R. R. Bowker, 1180 Ave. of the Americas, New York, N.Y. 10036.

⊗ *Educational Media Yearbook* (annual, since 1973). James W. Brown, ed. R. R. Bowker, 1180 Ave. of the Americas, New York, N.Y. 10036.

⊗ *The Elementary School Library Collection: A Guide to Books and Other Media.* Mary V. Gaver, Ed. Bro-Dart, Inc., 1609 Memorial Ave., Williamsport, Penn. 17701.

⊗ *El-Hi Textbooks in Print.* R. R. Bowker, 1180 Ave. of the Americas, New York, N.Y. 10036.

⊗ *Guide to Microforms in Print.* Albert J. Diaz, National Cash Register (Microcard Editions), Main and K Sts., Dayton, Ohio 45479.

⊗ *Guide to Reference Books (and Supplements).* American Library Association, 50 E. Huron St., Chicago, Ill. 60611.

⊗ *Guide to Reference Books for School Media Centers.* Christina L. Wynar. Libraries, Unlimited, Littleton, Colo. 80120.

San Diego County (California) Schools

⊗ *The Junior High School Library Catalog.* Estelle A. Fidell and Gary L. Bogart, eds. H. W. Wilson Co., 950 University Avenue, New York, N.Y. 10452.

⊗ *Magazines for Libraries: For the General Reader, and School, Junior College, and Public Libraries.* R. R. Bowker, 1180 Ave. of the Americas, New York, N.Y. 10036.

⊗ *Paperbound Books in Print.* R. R. Bowker, 1180 Ave. of the Americas, New York, N.Y. 10036.

⊗ *Periodicals for School Libraries.* Marian H. Scott. American Library Association, 50 E. Huron St., Chicago, Ill. 60611.

⊗ *Subject Guide to Books in Print.* R. R. Bowker, 1180 Ave. of the Americas, New York, N.Y. 10036.

⊗ *Subject Guide to Children's Books in Print: A Subject Index to Children's Books in 7,000 Categories.* R. R. Bowker, 1180 Ave. of the Americas, New York, N.Y. 10036.

summary

It seems assured that, for the predictable future, school personnel will continue to require large amounts of printed text and reference materials. But it is also likely that these resources will evolve and change and that they will reflect improvements that can be traced to continued applications of the principles of systematic teaching and learning discussed throughout this book.

Technological developments in the fields of publishing and information storage and retrieval also seem likely to influence the nature and extent of use of printed instructional materials of the future. The most significant of these developments appear to be: improved color printing equipment and techniques; increased use of empirical and scientific testing and validation of media prior to, during, and after production; and enlargement and improvement of information storage and retrieval capabilities growing out of microform and hard-print reproduction technologies.

Teachers are responsible for applying professional standards and for enlisting the assistance of their students in selecting media of all types. In the case of printed-text and reference materials, for example, they must know and be able to apply criteria related to learner verification, validation, content, treatment and arrangement of content, and certain technical-mechanical standards.

Those who use printed-text and reference materials should regard them as teaching assistants rather than as inflexible tools that are simply read to be remembered. They must make print presentations come alive by also showing films or filmstrips, playing recordings, developing independent and individualized study assignments, conducting field trips, using resource visitors, or introducing other correlated uses of media.

Microforms and computerized data-bank services such as those offered through the Educational Resources Information Center (ERIC) are expected to assume added significance as educational tools of the future. Through them, students and teachers will have ready access to data contained in information systems spread over a broad geographical area. Learning to use these systems, and to profit from the information they provide, constitute learning tasks with which students and teachers must become increasingly involved.

... AND IN THE FUTURE?

chapter purposes

⊗ To review briefly several trends in media and education.

⊗ To make some predictions about the future of media and technology in education based upon these trends.

⊗ To show some examples of instructional modes and audiovisual equipment that suggest possible directions for media utilization in the future.

⊗ To assess teacher responsibilities likely to result from changes in instructional patterns, instructional methods, and instructional technology.

Philips / MCA Disco-Vision

Predicting changes that may occur in education involves risks. A few innovations intended to effect changes seem to explode on the horizon and to spread quickly, some are accepted slowly and painfully, others die almost as soon as they are put forward. Yet despite the difficulties and disappointments, the process of innovation is important.

Many innovative devices and procedures have been developed to help people to communicate and learn. Today, in laboratories and classrooms all over the world, revised modes of instruction based on new applications of technology are being developed. Conceivably, several of these innovations could exert as significant an impact upon education and society as did the printing press, film, television, and computers.

As you read this chapter, consider your own possible roles with some of the innovations and educational developments we think are likely to appear or to grow significantly in importance. How do you think each will influence teaching and learning? Consider, too, what you yourself might wish to do to facilitate using them.

In considering your proposed action, remember that some people cause events to happen, some watch events happen, and still others will be unaware that anything has happened at all. Which kind of person are you? Which kind of teacher will you be?

media and educational trends

During the past twenty years or so, many efforts have been made to improve instruction and learning. Projects supported by federal funds made extensive investigations into instructional procedures and materials. Some have had lasting effects; others have not. The list below is derived from some of the areas in which research was conducted, and many of these areas continue to have implications for the future of media in instruction.

⊗ Increasing emphasis upon innovation

⊗ Changing instructional modes

⊗ Increasing individualization of instruction

⊗ Greater use of new instructional media

⊗ Changing curricular emphases

⊗ Increased attention to teacher education and reeducation

⊗ Increasing access to resources that improve the effectiveness of instruction

⊗ Increasing efforts to reduce the lag between the results of research and practice

⊗ Changing patterns of personnel utilization

⊗ Increasing participation of nonschool agencies in instruction and training

Let us consider some additional areas which clearly involve media.

⊗ SYSTEMATIC EDUCATION There is a major trend to place increased importance upon systematic planning and conducting of educational programs, particularly those which emphasize individualized learning. The systematic approach to instruction recommended throughout this book need not conflict with nor in any way run counter to efforts to provide the open and informal learning environments highly regarded by many people today. The systematic approach may undergird instruction without dominating it, giving direction to teaching and learning and establishing bases for evaluating student achievement.

⊗ INCREASED INDIVIDUALIZATION OF INSTRUCTION A concern for individual differences among students has been expressed for a long time, but it has more often been reflected on in theory or in talk than in action or practice. Within the past ten years, however, efforts to provide more individualization of instruction have been rewarded. Numerous approaches have evolved to facilitate this change, in many of which the use of a variety of media has been the implementing factor that produced the desired results.

⊗ MORE INDEPENDENT LEARNING Schools will almost surely continue to extend opportunities for independent learning. Learning resource centers will contain all types of media; they will be focal points of instructional planning and programming. Places for independent study will be located in learning resource centers as well as in school corridors, in residence halls, and in the corners of ordinary classrooms. School learning centers will provide equipment and materials for students to take home. Increasing numbers of students will have their own technical equipment—portable microform readers, calculators or computers, or other audiovisual equipment—and will continue to use it long after completing their initial schooling.

Typical of the new technology is the pocket-sized computer with programs on tiny magnetic cards (shown at the right, above). The manufacturer sup-

Hewlett Packard Company

plies program strips for various types of problems and for special fields, such as engineering, medicine, surveying, statistics, and mathematics. In addition to the stock programs, users can write and edit their own programs which can be used whenever needed.

⊗ ACCOUNTABILITY AND MEDIA The cost of formal education seems likely to continue to press heavily upon the public. This pressure increases the demand for proof of value received from public investment in school support. Thus, accountability for results is a reality faced by educators at all levels. In addition to the pressure of demands for evidence of the effectiveness of instruction, there is the simultaneous attempt to reduce expenditures. Often the reductions in school budgets tend to eliminate those programs and activities that are most likely to produce desired results of instruction —special services, teacher aides, and school media programs, for example. Proof of results, therefore, is an essential consideration in preserving the necessary elements in school programs.

⊗ MODULARIZED INSTRUCTION Modularizing instruction facilitates the development of a competency based program. Thus, for some learning objectives, units of instruction and the media resources provided for them will be increasingly modularized and self-contained. Pretesting will permit students to compare their own entering levels of competence with the stated performance criteria. Then, they may skip modules for which they have

demonstrated mastery and may choose only the modules or minicourses that their pretests indicate they need. The modular approach thus aids in meeting the goals of accountability and economy for instruction as discussed above.

⊗ CHANGED TEACHER ROLES Increased use of instructional media in systematically designed, modularized learning packages is changing the roles of teachers. Many of the traditional functions of teachers are being replaced by a number of new functions identified as *competencies* to be mastered through preservice and in-service training activities. As packaged resources are adopted, teachers are called upon to (1) make adaptations of the resources to meet the needs of individual students; (2) recommend and guide remedial departures from prescribed instructional programs; (3) lead or participate in deliberations of small groups of students or student teams; (4) work in specialized, sometimes unfamiliar assignments—for example, as nonteaching members of instructional development teams or as evaluators of student work; and (5) direct and help upgrade the performance of coworkers, including paraprofessionals or clerical and technical personnel serving as members of teaching teams.

But, along with these developments, small numbers of uniquely qualified lecturers who communicate well and who are able to motivate students to learn will continue to fill important teaching roles. Their contributions will not be lost in the new modes of instruction. They will often meet with large groups of students in face-to-face situations, and, through television, radio, audio cassettes, and other means, they will spread their talents widely.

With changing roles, teachers will be more concerned with the selection and utilization of instructional media, or their adaptation or creation, for student use.

Modifications in instructional modes, and increased uses of various media, require flexibility of surroundings. The contrast between traditional classrooms of fifty years ago and classrooms today is sharp and significant. The environment provided for learning reflects the educational objectives of the school, and the outcomes of learning are markedly influenced by the spaces and resources characteristic of that environment. On the next two pages we suggest alternative environments.

AN OPTIMUM PLACE
FOR EACH ACTIVITY

Various modes of instruction have various requirements—space, seating, lighting, acoustics, climate, color, and capability for media presentation. In order that large groups can be taught effectively, auditoriums and lecture halls of the future must ensure that people can see and hear, whatever media are used for presentations. Of the infinite possibilities for large-group facilities design, only two are shown here. Brodie Auditorium, below, at the University of Cincinnati, permits flexibility in presentations, and control from the stage of multi-media and lighting. Agua Fria Union Senior High School, Avondale, Arizona, right, contains a turntable divisible auditorium which permits rapid changes of seating capacity and adaptability for teaching requirements.

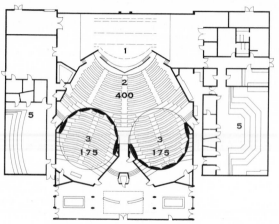

Rossman and Associates, Phoenix, Arizona

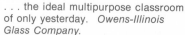

. . . the ideal multipurpose classroom of only yesterday.   *Owens-Illinois Glass Company.*

University of Cincinnati

The contemporary multipurpose classroom can be arranged as required
and permits a variety of study modes
for large groups, small groups, and for
independent study. *San Diego (California) Public Schools.*

Aloha Project

Comfortable isolation is one function
of facilities design and arrangements.
Horace Hartsell.

These viewing chairs are one
approach to student comfort
and isolation in semipublic
areas. *Hewlett Packard
Company.*

the increasing potential of telecommunications

In the last quarter of the twentieth century the most extensive technological changes in education may be in the uses of telecommunication systems. Cable television and extended public broadcasting services, in numerous instructional settings, are examples of telecommunication that most schools, colleges, and universities will be using more and more and in new ways. Other systems may expand in educational use, such as data transmission, from simple facsimile machines to interconnections of complete computer services. Another potential is the use of satellites for direct transmission of information and instruction especially to isolated and impacted areas throughout the world.

data transmission

With facsimile transmission, which requires little technical skill, pictures, printed matter, and graphics can be sent from point to point without actual forwarding of printed paper.

Since facsimile transmission technology is well developed, it is likely that educators will use it for

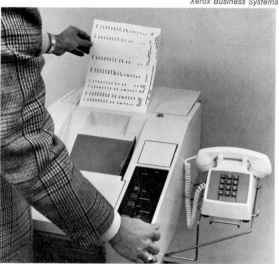

communications, especially for administrative matters, among schools, campuses, districts, states, or provinces, whenever the need is perceived or justified. In addition, it will probably be used in instructional programs whenever a quick transfer of information is needed—for example, the exchange of specialized reference collections among libraries and the provision of data to individual students and their teachers about learning progress. Using data and facsimile transmission in communications courses, students can develop skills in the exchange of information to be shared among school newspapers much as they would use teletype services. And, of course, interconnections and networks provide faster and greater information transfer in sharp contrast with the time-consuming methods of the past.

the potential of satellites

The development of satellites may be attributed to the demand for economical message transmission for commercial services. Nevertheless, the vision of what satellites might accomplish in widespread diffusion of instruction captures the imagination. In the mid-1970s, a brief experiment provided an opportunity to test the concept in a practical way. Coordinated effort by the United States Department of Health, Education, and Welfare, the National Aeronautics and Space Administration, and numerous regional, state, and local institutions made the project possible.

Based upon then current technology, a satellite and low-cost ground stations provided color television and multiple-audio channels, telephone, data, telegraph, and facsimile capability, with coverage over much of North America. Tests were focused in the Appalachian region, the Alaska-Northwest region, and the Rocky Mountain region. By rebroadcasts from Public Broadcasting Stations and television cable systems, programs from the ground were directed from the satellite for simultaneous reception by schools and institutions. A substantial organization of many people supported the program with production work, instructional materials, and miscellaneous field services. The programming included school curriculum and in-service education instruction, and experiments in medical education, communication, and diagnostic services.

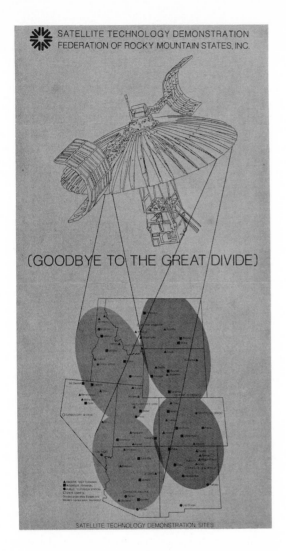

(GOODBYE TO THE GREAT DIVIDE)

SATELLITE TECHNOLOGY DEMONSTRATION, SITES

cable television

Though the potential advantages of cable television for both in-school and general public education were outlined in Chapter 12, Television, a few comments seem warranted here. Even though cable television services are considered desirable for educational purposes, that fact has had only a negligible influence upon the development of cable. However, the interest of educators and the public in the possibilities for using cable for education did encourage recommendations and even the adoption of local and national policies to encourage the establishment of education channels and cable systems. But only limited use has been made of them. Questions of administrative and financial control, regulation, costs, copyrights, conflicts of interest, inequality of services to minorities, types of services that can be provided and by whom, access to channels and by whom, control of content and by whom, and potential impact on existing and potential communication systems have instigated endless discussions, conferences, official studies, and tentative actions. What the outcome of the turbulent evolution of cable television services will be, and what benefits may accrue to people who want to use cable to advance education, is a very open question. The promise of cable television, nevertheless, is great, as indicated by these possibilities:[1]

⊗ Video correspondence courses

⊗ Special education programs for unskilled workers, housewives, senior citizens, and handicapped persons; adult education programs

⊗ Interconnection of school systems to permit teacher and student conferences, seminars, in-service education, and exchange of administrative information

⊗ Extension of home study programs for the especially talented or deprived

In addition to educational topics, consider also the potential of cable services for programs relating to health, legal and consumer problems, and matters of public safety, community welfare, and local current events.

The much discussed potential of cable television

Much too short in duration, but sufficient to demonstrate its feasibility and to reveal the problems involved, the satellite experiment suggested that perhaps by the 1980s satellites will be devoting at least a portion of their transmitting capacity for general education, instruction, health science services, and public information. Further tests of these potentials were to be conducted in India and perhaps other countries.

The technology seems ready; the major developments required are essentially in the areas of the will and abilities of people.

[1] Adapted from *Issues in Broadcasting: Radio, Television, and Cable.* (Ted C. Smythe and George A. Mastroianni, Editors) Mayfield Publishing Co., Palo Alto, CA 94301. 1975.

has not been realized. But limited and experimental programs continue. The voluminous literature about it will keep you informed.

video discs: a revolution?

The advent of video discs is a powerful reminder that change continues to come; as Samuel Hoffenstein, the poet, remarked: "No matter what the morrow brings, inventors are inventing things." The video disc may be the most significant technological development in communications in the last half of this century. The rate of changes in video sciences suggests, however, that any detailed technical discussion here would be untimely.

Briefly, a video disc system is somewhat comparable to that of an audio record player. Through a conventional television receiver, video discs produce sound motion pictures in either color or black and white. Programs may be transferred to video discs from sources such as films or video tapes; the discs, as with audio disc recordings, are then mass-produced by some method of stamping which results in a relatively low cost per disc to consumers.

At this time, there are three video disc systems: Teldec, being sold only in Europe (*Telefunken* of Germany and *Decca* of Great Britain); VideoDisc by RCA of the United States; and Disco-Vision, a development of Philips of Europe and MCA in the United States. The latter uses a relatively new technology—optical lasers—for pickup of sound and picture from a 12-inch disc with a claimed potential of two hours' playing time. VideoDisc uses a mechanical-capacitive pickup system with a one-hour playing time.

Differences between the RCA and Philips/MCA systems are the source of intense competition, and the advantages of each are strongly defended: initial costs, operating costs, program costs, programs available, and so on and on. RCA is stressing economy, simplicity, and reliability, and the use of less complex, less sophisticated technology than Philips/MCA. Disco-Vision advantages claimed are: new technology and capabilities of special interest to educators—programs can be stopped or shown single-frame or can be reversed, skipped ahead, or changed in speed. The system, as announced, can store and retrieve hundreds of thousands of bits of information; for example, as many as 40,000 individual microform frames could be available on one disc. There are clearly implied possibilities for library reference materials, of both print and nonprint resources, in home and school. This is a picture of a prototype search/retrieval remote control shown by Disco-Vision.

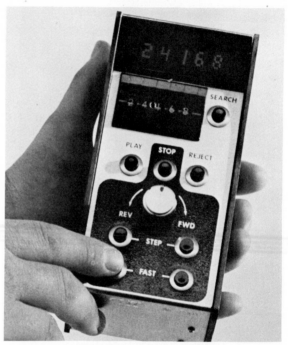

Philips/MCA Disco-Vision

Equipment shown on these pages is to be considered only in prototype form in the process of development; the technical data for this and other systems have been revealed to the public. Current information available to you, the reader, can give you the contemporary facts, but at this time it is possible only to speculate about the future of the video disc. If the reasonable prices promised for disc systems and for programs on discs are realized, and if the systems are perfected, produced, and distributed, you may have a video disc player in your classroom or in your home. If you do, you have a new source of entertainment and education.

ONE VIDEO DISC SYSTEM

Convenience of use and low cost of program material, plus durability of records and light shipping weight, are among the advantages of video disc systems for education. These pictures are prototype publicity pictures provided by MCA Disco-Vision.

extending opportunities to learn

Media are at the heart of programs that are designed to extend opportunities for students to learn what they need to learn, and when and wherever they need to learn.

comprehensive media programs

Comprehensive media programs will increase in number and in the scope and quality of services they provide. The major service unit will usually be a central learning and resource center, but subcenters will be established as needed in the buildings or classrooms of a large school. Specialized study stations may be provided: science laboratory stations, math study stations, and audio and audiovisual carrels as study stations for using the materials in the center collections. A rich variety of printed materials of all kinds, bound and in microforms, will be readily accessible, as well as many different kinds of audio, audiovisual, and visual resources, appropriate for the subjects and objectives of the instructional program. An adequate staff of professional, paraprofessional, and aide/clerical /technical personnel will serve the program; also subject specialists, counselors, and, often, teachers will be available to students as they work in media centers. Necessary types of equipment will be provided for reproducing, creating, and studying media of all kinds, and for communication among students and teachers.

We commend to you the document: *Media Programs—District and School,* revised, prepared by the American Association of School Librarians, ALA, and the Association for Educational Communications and Technology, for a complete review of levels and kinds of media-support services considered appropriate and needed for modern instructional programs. The success of all educational programs is heavily dependent upon the quality and kinds of media resources and facilities provided. Improvement of existing media programs is an essential goal for the future.

To some people in education, this generous position toward media programs may seem unrealistic. Full realization of comprehensive programs may not

yet be within reach. Yet, throughout this book alternative means of providing learning resources have been suggested, many of them depending far more upon ingenuity and resourcefulness than upon money; emphasis has been upon creative construction, on using resources from local agencies and from free and inexpensive sources, and upon continual consideration of the simplest, most economical way to achieve objectives.

But beyond the search for economy, there is a need to seek the optimum way to engage students in learning activities, which may indicate a need for special resources and equipment. One professional responsibility of teachers is to demonstrate the values of adequate and sufficient resources, and to continue to work toward getting them. New priorities must be developed. New adjustments must be made in staffing and use of facilities. Instructional modes must be accepted for the benefit of students, and explained fully to school patrons at all levels in the community. Active involvement of parents and others in the community in the school program, as suggested in numerous chapters in this book, is a desirable and effective way to get public support for media programs. And the final, persuasive evidence is a group of students actively enjoying education, clearly benefiting from it, and giving their enthusiastic support to the program of their school.

preschool and in-school education by television

The use of television to provide learning opportunities both at home and in school has proved successful, and programs correlated with school curricula are made available through both public and school broadcasting services. For special groups, programs such as "Villa Alegre" and "Carrascolendas" are widely used. Instructional programs presented by local educational stations of all types for in-school and home viewing continue to grow in number and quality.

alternative schools and postsecondary education

Schools making special efforts to meet the real and immediate needs of youth are providing alternatives to traditional instruction. Such programs are characterized by markedly revised curricula, close communication and work relationships with the community, both in planning and in giving students work experience and career guidance. Students use media resources in activity assignments to improve their communication skills in speaking, writing, photographing, and recording. More and more, students may leave school—without penalty—and return when they recognize the need; their programs are sufficiently flexible to permit such nontraditional behavior, and media in modular units or minicourses based upon multi-media resources implement study.

Community colleges across the country have been active and successful in developing open enrollment courses, with media basic to instruction. Also, course work is provided by correspondence, by television programs plus other modes of study, and by open laboratory and media-center study stations. As one example of this trend, in Chicago the Community College works in cooperation with the public library to provide alternative modes of study: the libraries provide video cassettes for viewing lectures and also provide related print and nonprint materials for courses, giving students an alternative way by which they may work for college credit or for personal satisfaction.

public library
multi-media services

Many public libraries are extending their activities into nonprint, multi-media fields. In addition to the usual resources in print, loan collections include audio tapes and disc recordings, sound filmstrips, and motion pictures—in 8mm and 16mm films and in video cassette forms.

In some cities, granting of franchises for cable television has been dependent upon requirements that cable operators provide community access channels for public programming. And some libraries have provided training activities in simple video tape techniques for people wishing to produce programs for distribution by cable to the public.

The American Library Association has supported with current policy statements the broadening of library services to include multi-media programs. Check the nonprint services of your own public library system; your students may be able to use them to advantage.

Thus, opportunities to learn continue to be extended by availability of multicultural resources, multilevel study units and modules, and multi-media kits to meet the broad and differentiated requirements of students of different ages in every area of society.

improved printed
learning resources

Although learning resources in printed form have been available in great variety, they continue to change in scope and type. In the future, basic textbooks will continue to change in form; there will be more use of minitexts, for example, as modular instruction increases in popularity. And texts may be replaced by other types of print and nonprint materials. There will also be an increasing use of references in book or in reprint form.

Pamphlets and periodicals have increasingly been used to supplement texts and basic readings; this trend will continue, as will the use of microforms and standard-sized forms of reprints. When instructional programs are developed by systematic methods, the provision of packages and kits of related commercially published materials, supplemented by other materials produced or acquired locally, will become a widely adopted practice. Improved printing technology will permit markedly increased production of attractive and functional publications in full color, with meaningful typography, interesting layouts, and fully integrated illustrations that will stimulate, interest, and guide the reader.

Many books will tend to be less didactic than in the past; they will increasingly incorporate problem-oriented inquiry and study methods, and they will include branching techniques to direct readers to other resources. Attention will be given to preparing materials in print that are for the experience and ability levels typically represented in any group of students.

Print and other media will work together by planned design in most programs of instruction.

trends in equipment for instruction

Much of the equipment available today for instruction was originally developed for commercial entertainment and advertising; television and film equipment are in this category. Later use of such equipment by schools, often in modified or simplified forms, was essentially a spin-off benefit. However, in recent years, equipment developed especially for instruction has markedly increased because of demands by schools and colleges and by business, industry, and government training agencies.

examples of equipment innovation

Illustrated here are several types of equipment that have been developed especially for educational purposes. First, below, is the Caramate by Singer

reader, the Tutorette, by Audiotronics. With magnetic stripe which permits both play and record modes, as well as area for pictorial and verbal material, this type of device is used extensively in language, reading, and speech instruction.

Another device, especially designed for instruction is this Digitor, by Centurion Industries. (A sim-

Education Division; this slide-sound machine uses a slide carrier widely used for 2- by 2-inch slides in projectors and viewers; the sound is carried on a standard audio cassette. Such machines as this, which are available in a number of brands and configurations, are especially useful for independent study.

The unit above, to the right, is an audio-card

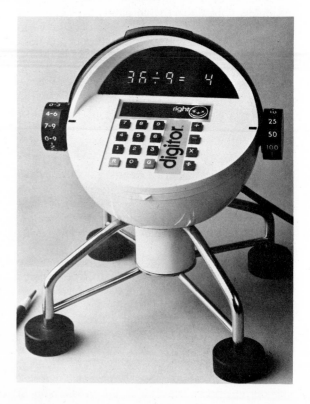

ilar unit is the Mathiputer by Cybernetic Systems, Inc.) Based on computer-type capacities, the machine is effective for practice in a great variety of basic arithmetic problems, and provides instant feedback information as well as total performance scores.

These samples suggest that there are levels of versatility in machines—some are designed for specific teaching purposes; others may be used for many subjects and in many situations. Examples of the latter may be found in machines such as audio cassette recorder/players, and motion-picture equipment—both video and film.

Since there is ample evidence of the success of multi-media kits, education, no doubt, will have extended requirements for appropriate equipment, some of highly specialized types, others more flexible in applications.

equipment for local production

While media resources will continue to be provided in increasing varieties and quantities by commercial producers and publishers, steadily increasing activity is expected in the local production of materials to meet special needs.

All types of equipment for graphic production, including lettering machines, mounting presses, and laminating machines, as well as duplicating machines of several types, are found in many schools. Such equipment speeds preparation of materials, increases product durability, and gives professional finish to prepared materials.

Photographic equipment and television production equipment are continually improved and expanded in variety. Equipment decreases in size, and the quality of work that it permits continues to improve. Work in color film or color video is becoming increasingly common, particularly as costs are reduced in proportion to an increase in the quality and capability of performance.

In selecting production equipment, the purposes for which the equipment is to be used are indicative of the quality and type required. For example, if it is to be used solely for productions made for local use, the result need not have the high technical quality required for a production made for duplica-

tion and widespread distribution. What about the compatibility of the equipment being considered with similar equipment on hand or available from central sources? These concerns will affect selection. As an example, here is a Mini-Studio TV System developed by Philips Audio Video Systems of New York. Not designed for a major production enterprise, but rather to meet school needs, the system can be stored in two luggage cases.

Philips Video Systems Corp.

The Mini-Studio has appeal for students doing production work for class activities and permits an appropriate level of television experience. With accessories, such as a film chain, microscope adapter, and video tape recorder, such a system is entirely satisfactory for making instructional units to be used within the school.

However, if extensive production work for regular class utilization is to be undertaken, more sophisticated equipment, with full editing capability, would be desirable.

This, then, indicates the problem of alternative choices when selecting instructional equipment.

SEEKING APPROPRIATE EQUIPMENT

Media equipment may be simple or complex, inexpensive or costly. It may facilitate instructional procedures for some specific objectives and locations. Or it may be a versatile instrument that enables you to offer many and varied learning activities. The innovative items of media equipment shown here may meet important needs of the present and the future. Some are new in concept and design; and others are welcome improvements on old inventions.

Though only an experimental project, this Eastman Kodak Research Laboratory Interactive Learning Terminal explores the potential of minicomputer-based, multi-media instruction. The student brings lessons (above) to the terminal on microfiche, audiofiche, and a digital tape cassette; the instrument provides multi-media and programming functions needed to study on an individualized basis.

Small Talk is the name of this micro phonograph that plays microrecords about the size of a 25¢ piece that are affixed to printed materials. With its little speaker, or with an earphone, *Small Talk* provides sound to accompany print. *Imperial International Learning Corporation.*

A very inexpensive viewer to present silent Super 8 motion pictures in convenient cartridges is this Montron Corporation development. Hand operated for forward, backward, or hold-frame modes, the viewer images appear on the rear-screen or may be projected on a carrel wall. A hand viewer, also available, uses the same cartridge as the larger model.

Miniaturization provides opportunities for such new equipment designs as these I. A. V.-Standard sound-slide projectors and audio cartridge units. The projector, a modified Kodak Ektagraphic, records and plays synchronized sound. The 2- by 2-inch continuous loop cartridge system also includes self-contained headsets, record/play units, duplicators, and sound synchronizers.

This hand-held individualized instruction device uses interchangeable teaching lessons in inexpensive continuous loop cartridge form. A multiple-choice decision device, each of the forty frames presented requires a response and will advance to the next question only if a correct answer is given. Telor *by ENRICH.*

This picture of the Digi-Log Systems Terminal in use indicates the simplicity of transmitting alphanumeric data from a computer or other inputs using briefcase-size portable instruments. Hard copy can be produced by connecting the instrument to a Teletype machine.

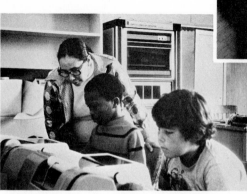

Programs for computer-based instruction are published as are other media. Computer Curriculum Corporation sells or leases courses that can be adapted for use with an available computer, or will provide courses and a complete minicomputer system.

teachers of tomorrow: opportunities

Change may be inevitable, but not all change leads to improvement. Sometimes changes lead only to different ways of doing the same things. If changes in instruction do not lead to improved results, to lowered unit costs, or to other essential educational benefits, perhaps they should be reworked or abandoned. But don't abandon a promising idea until you have fully tested its possibilities; program results should be assessed in the light of adequate evaluation and identification of causes for failure or success. And, based upon the ideas you have found in this book and in related experiences, seek opportunities to:

⊗ Extend your efforts to plan your teaching by a systematic procedure.

⊗ Seek ways to involve students in planning the objectives, procedures, and means of evaluating instruction and learning.

⊗ Select a variety of appropriate student learning activities, including many based solely on uses of instructional media.

⊗ Become expert in preparing various appropriate resources for instruction, such as audio tapes for directing independent study, and slides or transparencies and displays; become especially skilled in motivating and guiding your students in such production work.

⊗ Conduct discussions with your students; assist them in their interaction experiences; guide them in the use of audio and visual techniques in discussion situations.

⊗ Present fewer lectures to transmit information. Prepare the lectures you do present with the assistance of media specialists. Remember that many of your presentations may be made entirely on video tape, film, slide/sound sets, or in other media forms and may be used by students in independent study or for group viewings.

⊗ Provide many opportunities that allow students to develop competencies in creating and using media of many types, audio and visual, printed and projected.

⊗ Each term or year, select from this book some techniques or media that you haven't used before. Incorporate them into your program, and evaluate the results, both in the achievement and in the attitudes of your students.

When you plan to devise or to revise units within a course, squarely face the practical problems: Do not try to do an entire course at once, unless you have been given time, resources, and encouragement to do so. Select one or a few important units which students have found difficult or uninteresting, or choose topics that only some students need to study in depth, and develop an alternate version or several versions. Develop a rationale for what you do and undertake a project of reasonable scope, one that you can complete—including testing and evaluation of results.

We urge you to seek those in the community with whom you have much in common: instructors in such fields as nursing, community services, and health, and business, industry, and government. Seek, too, persons with instruction and communication responsibilities in public information agencies, including newspapers, radio and television stations, libraries, and community information centers. You can learn from them, and they can learn from you; consider attending the workshops they attend and joining their training organizations. Work with them to improve learning and communication with the use of instructional technology.

We hope that your present use of this book and its continued use will help you to improve and professionalize your work and to increase your pleasure and your rewards as you work with students and move toward becoming a master teacher.

Far West Laboratory for Educational Research & Development

San Diego County (California) Schools

Doubleday Multimedia

Montron, Inc.

Centurian Industries

IN CONCLUSION . . .

We are at the end. At the start of this book is the sentence "This is a book about media." But by now you know it is much more a book about people. Emphasis is upon what teachers and students do with media, why, and for what purposes. It is what people *do* that is important. Media, well used, bring people closer together; media aid people in mutually productive enterprises. Media are implements of communication that are valuable and effective according to how wisely they are chosen and how well they are used. And no medium in the world can replace the warm, sympathetic relationships between teachers and students that motivate mutual trust and nurture enthusiasm for learning.

Virginia State College

San Diego County (California) Schools

Montron, Inc.

REFERENCE
SECTIONS

REFERENCE SECTION

1 OPERATING AUDIOVISUAL EQUIPMENT

contents

introduction

Anyone can learn to operate basic audiovisual equipment. Young people do—rapidly. Adults take only a bit longer. Teachers are expected to have satisfactory operating skills for record players, tape recorders, radios, various still-picture projectors, motion-picture projectors, and television equipment.

This section presents step-by-step instructions for operating typical equipment found in most schools and colleges. Insofar as possible, sufficient information is provided here to serve as a guide for self-instruction. In many cases, of course, reference to manufacturers' operating manuals will be desirable.

For most efficient learning, study the text and illustrations; then, following the step-by-step procedures, practice immediately with the equipment. Most institutions provide equipment, materials, and space for self-instruction and practice. Some will also furnish 8mm film loops and other aids for use during your practice. Additional readings and audiovisual materials are listed in Reference Section 6.

Finally, this section can serve you when you need to brush up on your equipment-operating skills or when you are assigned some unfamiliar model of basic equipment.

audio equipment principles

Sound reproduced by audio equipment is most satisfactory when it closely resembles the original performance. The quality of sound reproduction is affected by the type and quality of equipment selected and by the operator's skill in using the equipment. Some important factors are:

☐ Characteristics of the original sound

☐ Environmental conditions in recording and playback

☐ The capability of the recording medium to capture all the frequencies of the original sound

☐ The capability of the playback system to reproduce program material in full frequency and without distortion

☐ The ability of the operator to manipulate the equipment

Another factor affecting sound quality is derived from the principle of *equivalent components*. The saying that a chain is as strong as its weakest link exemplifies the concept. Each component in an audio system should be of equal quality and should have similar performance characteristics—and preferably its quality should be good. It is folly to use a worn stylus to play a disc recording or to connect a fine, expensive amplifier that will reproduce full-frequency sound to a very cheap speaker incapable of reproducing most frequencies without distortion. Even an expert operator can do little to produce fine sound quality with shoddy equipment.

The frequencies that people hear are a practical guide to the desirable frequency response range for sound equipment. Although hearing capacity varies widely with individuals, the normal human ear can respond to and identify tones as low as 30 and as high as 15,000 or more cycles (vibrations) per second. An orchestra will produce tones in this range. Speech tones have a more limited range: understandable speech may be recorded in a frequency range of 1,000 to 3,000 cycles per second. Optimum recording of speech will require a greater frequency range, however, to permit identification of the nuances of individual speech. Undistorted tones from 60 to 8,000 cycles per second will please a majority of listeners, and a top frequency of 12,000 cycles per second is considered high fidelity by most people.

Both monaural and stereo sound equipment may be found in educational institutions. Often musical events and dramatic performances and occasionally panel discussions will be recorded in stereo to preserve the full richness and dimensional quality of the original performance. Much portable recording equipment is monaural, and for most instructional applications this is entirely satisfactory.

controls on sound equipment

Controls for sound reproduction on record players, tape recorders, and motion-picture projectors are similar. Those usually provided are as follows:

☐ AMPLIFIER OFF-ON SWITCH This may be a separate switch, or it may be combined with the volume or tone control.

☐ VOLUME CONTROL(S) Stereo equipment has a volume control for each channel or a balancing control to adjust volume between left and right speaker channels.

☐ TONE CONTROL(S) If only one tone control is provided, it usually serves to eliminate high frequencies. If two controls are provided, one regulates the bass, the other the treble tones.

☐ SPEED SELECTOR This control sets the equipment at the speed required for the recording.

speaker placement

The best placement of a speaker or speakers must finally be determined in each situation by trials, but some general suggestions will be helpful:

☐ Place speakers at about the ear level of listeners.

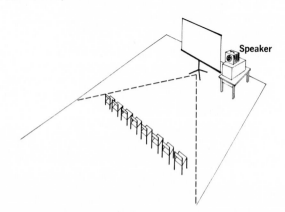

Speaker

☐ When using one speaker in a long room, set the speaker off center at the front of the room and direct it diagonally across the listening area.

□ If a speaker is enclosed in the case of a record player or tape recorder, the same rules for speaker placement still apply.

□ If two speakers are used, move them apart, facing the audience, until the sound seems to come from between them. Check from several positions in the listening area for dead spots or interference and readjust speaker positions if necessary.

microphones

There are many different kinds of microphones. They vary in appearance, sensitivity, directional characteristics, and in other more technical ways.

□ *Unidirectional* microphones, generally designed for use by one person, pick up sound from only one direction.

□ *Bidirectional* microphones pick up sounds from both front and back sides. They are especially useful when recording two or more persons face-to-face or when recording presentations that must include remarks of both speakers and audience.

□ *Omnidirectional* microphones are sensitive to sounds from all directions around the horizontal plane of the unit. These are especially useful when recording round-table discussions or groups of performers.

□ *Microphones for cassette recorders* are basically omnidirectional and will pick up sounds from considerable distances. Often they have built-in remote-control switches to turn the recorder off and on. Some cassette machines are equipped with built-in microphones as well as jacks for external pickups.

□ *Lavalier and lapel microphones* are designed to be worn by the speaker, permitting him more freedom of movement than he would have with a fixed microphone.

□ *Wireless microphones,* which are miniature FM transmitters, provide maximum freedom of move-

3M Company

ment since no cords are required. The microphone broadcasts to a receiver, which feeds the signal into a nearby recorder, public address system, or other sound unit. Because they are actually small broadcasting stations, these microphones must be authorized for use by the Federal Communications Commission; determine the legal status of any wireless microphone you consider purchasing.

MICROPHONE PLACEMENT A microphone placed close to the performer will not pick up many sounds reflected from surfaces in a room. However, while an intimate feeling can be created by recording close to the microphone, the effect of "room presence" may be lost and an unnatural voice quality may result. Always test and retest to determine the best placement of the microphone with reference to performers. During recording sessions involving a number of people, be prepared to readjust microphone positions to accommodate individual behavior of speakers. For example, move the microphone away from the speaker who wants to speak too closely into it and from those who turn from side to side as they speak.

For recording with stereo equipment, two types of microphone arrangements may be available. In one type a single unit contains both left- and right-channel microphones. In other situations, the two microphones are separate units, each feeding a channel in the amplifier. In either case, trial and retrial to achieve proper location of the microphone unit(s) are essential to for good balance.

The occasional need for a long microphone cord raises a special problem. The microphones usually supplied with school tape recorders are not suitable for operation with cords more than 6 to 8 feet long. Occasionally an extension cord can be used, but the results may not be satisfactory. If a long microphone cord is necessary, investigate special *low-impedance equipment:* microphones, transformers, and preamplifier mixers. Local audiovisual dealers or electronics suppliers can advise you, as can your audiovisual engineer or service shop personnel.

When several school recorder microphones are required in one setup, it is generally not practical simply to hook several together through connectors. Investigate inexpensive *high-impedance microphone mixers;* they may solve this problem.

record players

There is a wide range of record players from which to choose for school use, and a variety of record types can be played on most of them. Each record type in the list following is identified by diameter, playing speed, and stylus required:

☐ STANDARD: 10 and 12 inch, 78 rpm, 0.003 (3 mil) stylus. (Obsolete and no longer "standard.")

☐ MICROGROOVE: 7 inch, 45 rpm, 0.001 (1 mil) stylus.

☐ MICROGROOVE (monophonic): 7, 10, and 12 inch; 33⅓ rpm; 0.001 (1 mil) stylus.

☐ STEREOPHONIC: 10 and 12 inch, 33⅓ rpm, 0.0005 (0.5 mil) or 0.0007 (0.7 mil) stylus or elliptical (biradial) stylus (0.2/0.7 mil).

a generalized record player

In the accompanying drawing, principal parts of a typical school record player are identified. Basic controls and features permit the machine to be used for a variety of types of records. With this drawing as your guide, locate the controls provided on the machine available to you.

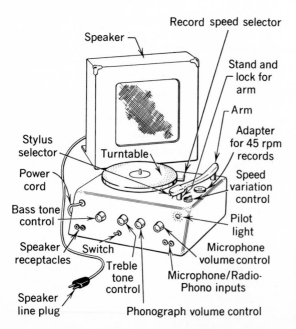

Speaker — Record speed selector — Stand and lock for arm — Arm — Adapter for 45 rpm records — Speed variation control — Pilot light — Microphone volume control — Microphone/Radio-Phono inputs — Phonograph volume control — Treble tone control — Switch — Speaker receptacles — Bass tone control — Power cord — Stylus selector — Turntable — Speaker line plug

SETTING UP AND OPERATING

1. Set the machine gently on a table and open the case.

2. If the speaker is not built in, insert the speaker plug in the speaker receptacle on the phonograph and place the speaker at a location suitable for listening.

3. Plug in the power cord and turn on the power switch.

4. Set the correct turntable speed for the record to be played.

5. Select the proper stylus for the record.

6. Adjust the pickup arm weight control, if provided, according to the type of record to be played.

7. Start the turntable.

8. Unlock and lift the pickup arm and place the stylus gently in the run-in groove near the record edge.

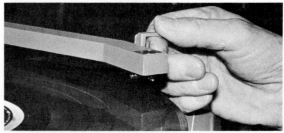

9. Adjust volume and tone controls for optimum listening satisfaction.

10. When the record selection is finished, lift the pickup arm vertically and replace it on its rest.

11. Stop the turntable before removing the record.

PUTTING AWAY

1. Turn off all switches.

2. Set the speed control lever at OFF or in a position that will disengage the rubber drive roller from the turntable. Failure to do this may cause the roller to become deformed, resulting in variation in playing speed and sound distortion.

3. Lock the pickup arm on its stand to prevent damage to the stylus and cartridge during transportation of the machine.

4. Store all cords on brackets or in storage compartments in the case. Never coil cables under the turntable; the drive mechanism could be damaged.

5. Secure lid latches carefully.

care and use of styluses

Give attention to the care of styluses and their replacement when worn. The use of worn or damaged styluses is a major cause of damaged recordings. The illustrations below show the variations in the grooves that cause the stylus to vibrate to produce sound. The fine point of the stylus passes along these grooves with speed and force, creating great friction. A faulty stylus can plough away the delicate variations and destroy the recording's fidelity.

The enlargements below show the grooves in a 78-rpm record (top) and a microgroove record (bottom).

The shadowgraph shows a microgroove stylus in a 78-rpm record groove. The tip rides the groove bottom instead of fitting the shape of the groove as a standard stylus for 78-rpm records would. The correct stylus must be used with each type of record; otherwise severe damage is caused to both record and stylus.

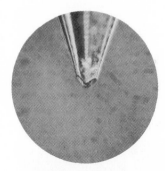

After only a few hours of playing, the tip of a metal stylus is worn to the shape of the groove and has sharp edges. Inexpensive styluses are costly if they destroy records.

Styluses on much-used equipment should be checked frequently with a microscope and worn styluses replaced before they cause record damage.

The cartridge that holds the stylus and creates electrical impulses from stylus vibration is generally quite sturdy but subject to shock. Avoid dropping the pickup arm or allowing it to swing during transportation. Protect both the stylus and cartridge by using care in handling them and by locking the pickup arm on its stand when not in use.

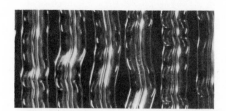

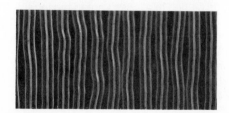

practical pointers

☐ Do *not* play stereophonic records on a monaural record player. (However, long-playing monaural records may be played on a stereophonic machine without damage to record or stylus.)

☐ Keep fingers off record grooves; hold records so that fingers touch only the edge and the label.

☐ If cleaning is necessary, wipe the record grooves gently with a soft cloth dipped in cool water and wrung dry. Never wipe records with fingers, stiff brushes, or harsh cloths.

☐ Store records on edge, in dust jackets, away from direct sunlight or other heat.

☐ Never allow a stylus to sweep across the face of a record, and never add weight or push down on the pickup arm to hold a stylus in the record groove.

☐ Some record players include a built-in stroboscope, a device to help the operator to set turntable speed precisely. Learn to use this feature if it is provided.

☐ Never bump or jar record players when transporting them. Be gentle with equipment.

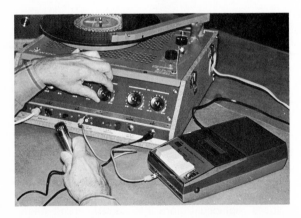

☐ A record player can be used as a public address system if an input and a volume control for a microphone are provided. Practice using a record player as a public address system, for amplifying sound from small tape recorders and from radios, and for making tape recordings with blended music and speech.

power cords

Power cords are permanently attached to most items of audiovisual equipment, and desirably so, as otherwise the cords often become separated from the machine and are not available when needed. When cords are separate, a major responsibility of the operator is to ensure that the cord remains with the equipment. Often, separate power cords have special plugs on them, making them inconvenient to replace.

When power cords are designed for safety grounding and have three prongs on the cord plugs, projectionists should be provided with three/two adapters in the event equipment is used in buildings where obsolete electrical receptacles still exist. The adapter should always be used; do not attempt to defeat the safety grounding by breaking off the grounding tip on a cord plug. Some units are provided with a highly desirable plug that will fit either two- or three-wire receptacles; a squeeze on the button on the plug swings the grounding tip out of the way, enabling it to make safety contact with the surface of a metal receptacle plate.

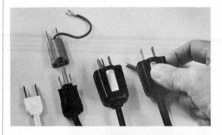

Cords should be tied in a loose knot to the legs or supports of projection stands, as near the floor as possible. A knot is also desirable at the point where the equipment cord and an extension cord are joined. These precautions will help to avoid equipment damage (through accidental falls) and unnecessary power difficulties.

tape recorders

Fortunately, all tape recorders function similarly; only a few fundamental skills are required to operate any of the recorders in common use today.

Following the general discussion of tapes and tape recorders below, more specific details about operation and use will be given, as well as other information about tape speeds, playing time, and track configurations.

magnetic tapes

Examine a piece of audio recording tape. Feel it; break it. One side of the tape is glossy; this is the "backing," or "base." One side is dull; this is the "working" side—a thin coating of metallic oxide.

The tape base is a plastic. Acetate is used widely for low-cost tape. Mylar, a plastic of great strength, is used when a very thin and strong tape base is required.

When a tape is recorded, magnetic fields are set up in the tape. When the tape is played, these magnetic fields generate electrical impulses in the machine which are amplified and reproduced as sound. Though the magnetic fields cannot normally be seen, by a special process a picture of the fields is revealed as shown below.

Tape may be replayed many times without audible changes in sound quality. A program on tape may be removed completely (erased) and replaced with a new program. Moreover, this process can be repeated an almost unlimited number of times.

Tape is supplied on reels of several sizes and in cassettes and cartridges. The oxide is also coated in stripes on cards for card-reading machines.

audio tape machines

While the cassette recorder has increased greatly in popularity in recent years, reel-to-reel machines are still performing heavy service in schools. Cartridge machines, which operate similarly to cassette recorders, are used only occasionally in schools for special applications. Some cartridge machines permit continuous operation without rewinding; these are often used with sound-slide or filmstrip presentations in displays. Although most tape machines in use can record as well as play, schools are also buying play-only machines for individual study or carrel use. Related to the audio tape machines are the audio card machines designed especially for individual instruction and practice.

Regardless of their configuration, all tape recorders have common features necessary for their function—many resemble those of any sound equipment, including sound controls and jacks for input and output. Features unique to tape recorders control tape motion, record sound, and play recorded sound back from the tape. In the drawing of a generalized tape recorder on the next page, these common components are labeled. Examine actual tape recorders and identify the items.

Though functions are the same, the location of parts varies widely among machines; look at the back, sides, and top on each model to find items. Also notice differences in the forms of controls (piano keys, pushbuttons, levers, or knobs) which perform the same functions in different machines. Look also for special instructions which indicate how a machine may differ in some respect from typical recorders.

A GENERALIZED TAPE RECORDER

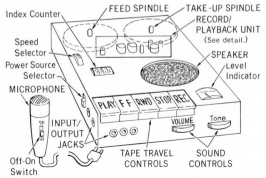

Index Counter
FEED SPINDLE
TAKE-UP SPINDLE
RECORD/
PLAYBACK UNIT
(See detail.)
Speed
Selector
Power Source
Selector
SPEAKER
Level
Indicator
MICROPHONE
INPUT/
OUTPUT
JACKS
PLAY F F RWD STOP REC
VOLUME Tone
Off-On
Switch
TAPE TRAVEL
CONTROLS
SOUND
CONTROLS

RECORD/PLAYBACK UNIT

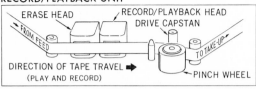

ERASE HEAD
RECORD/PLAYBACK HEAD
DRIVE CAPSTAN
FROM FEED
TO TAKE-UP
DIRECTION OF TAPE TRAVEL
(PLAY AND RECORD)
PINCH WHEEL

In the drawing above, features found on all tape recorders are designated with capital letters (ERASE HEAD); items found only on some machines are identified with small letters (Counter). The same coding applies in the list below except that italic letters are used (*ERASE HEAD, Counter*).

SOUND AND POWER CONTROLS

☐ *SOUND CONTROLS* for adjusting tone normally control playback sound only; volume controls may function on both recording and playback, while some machines provide automatic volume control when recording.

☐ A *Power Off-On Switch,* when provided, may be incorporated in the tone or volume control.

☐ A *Recording Level Indicator,* when provided, may also double as a battery level indicator for machines operating on batteries.

☐ A *Power Source Control Switch* may be used to select the source of power on AC/battery recorders; on some machines, connection of the AC power cord disconnects the batteries.

RECORD/PLAYBACK UNIT

☐ The *RECORD/PLAYBACK HEAD* magnetizes the tape to record program material or reacts to the magnetism on the tape to play the program.

☐ The *ERASE HEAD,* functioning only when the machine is recording, clears previously recorded program material from the tape.

☐ The *DRIVE CAPSTAN* imparts constant forward motion to the tape for playing and recording when the tape is pushed against it by the *PINCH WHEEL.*

TAPE TRAVEL CONTROLS AND INDICATORS

☐ Controls for tape travel on all machines are: *FORWARD* (or *PLAY*), *REWIND, RECORD,* and *STOP.*

☐ A *Fast Forward Control,* when provided, permits skipping program material at a faster-than-play speed.

☐ A *Pause Control,* when pressed, stops tape travel and, when released, allows the tape to continue as before.

☐ An *Index Counter* aids in locating program segments on the tape.

☐ A *Tape Speed Control* (on reel-to-reel machines only) selects the speed of tape travel for playing and recording.

INPUT AND OUTPUT JACKS

☐ A *MICROPHONE JACK* is for microphone input. (Two are provided on stereo recorders.)

☐ An *AUXILIARY JACK* permits recording from other signal sources, such as a radio, a phonograph, or another tape recorder.

☐ An *OUTPUT JACK* connects the recorder to an external speaker or earphone or to another recorder for duplicating tapes.

operating tape recorders

The procedures that follow apply to most tape recorders. Perform operations described whenever your equipment has provision for them. If possible, refer to instructions for the specific machine you are using.

SETTING UP

1. Locate the recording unit where its operation can be closely monitored. For playback, place speakers (if separate) for optimum listening. For recording, place the microphone where it will pick up the desired sounds with a minimum of background noise. (See sections on speaker placement and microphones.)

2. Connect the power cord to a wall receptacle. Turn on the power switch. If the recorder is to be operated on batteries, make certain they have sufficient power to drive the machine at the proper speed; check the battery level by reading the meter—if provided—with the recorder switched to PLAY.

3. To record, insert the microphone cord plug into the microphone input jack. Note whether the microphone has a switch; this must be set at ON for playback as well as for recording.

THREADING REEL-TO-REEL MACHINES

1. Set the tape speed control to the speed required. Activate the PLAY control momentarily; put the empty reel on the spindle that turns.
2. A new reel of tape may have an adhesive or gummed tape seal. Before using the tape, remove and discard one full turn of tape, including the gummed portion. Then put the reel of tape on the spindle that did *not* turn when the machine was on PLAY.
3. Thread the tape through the Record/Playback Unit slot. The dull side of the tape must face the recording heads. Engage the tape end in the slot of the take-up reel. Rotate both reels by hand for about 1½ turns. Then cinch the tape on the take-up reel hub.

INSERTING CASSETTES

1. Before inserting the cassette, make certain that the machine is in the STOP mode. Press the EJECT button to open the cassette compartment.
2. Insert the cassette so that the open side fits against the recorder heads and the full spool is on

the feed spindle. (Remember, the tape plays or records in a counterclockwise direction.)
3. After inserting the cassette and before switching to PLAY or FORWARD, switch to REWIND and rewind the tape until it stops (watch the spindle). The tape should be taut from spindle to spindle.

RECORDING Follow instructions below that apply to your equipment. Whenever possible, make a test recording to check equipment operation and microphone placement, as described below. You will find additional tips on recording techniques in the Practical Pointers at the end of this section.
1. Set the index counter at 000 by turning the wheel or pressing the button adjacent to the counter.
2. To begin recording, engage both the RECORD control and the PLAY control. If the microphone has an OFF-ON switch, this must be set to ON also.

3. Record sound similar to that which you wish ultimately to record, adjusting volume level as necessary to obtain an optimum reading on the level indicator.
4. Stop the machine. Check the tape, which should be taut between the two reels. Then engage the REWIND control. Stop the machine before the tape runs off the take-up reel. Cassette tapes will stop automatically; however, operate the STOP control immediately.
5. Play back the test recording by switching to PLAY. If the microphone has a switch, this must be set to ON. Adjust tone and volume levels to evaluate sound reproduction. You may wish to reposition the microphone or change the recording level before making the final recording.
6. Record the program as described in steps 2 and 3. If you wish to stop the machine briefly, use the PAUSE switch or the microphone switch. The ma-

chine will continue to record when these are released or returned to ON. If the STOP control is used, make certain to engage both the RECORD control and the PLAY control when you wish to continue recording. When the recording is complete, rewind the tape as directed in step 4.

PLAYING RECORDINGS

1. Switch the recorder to PLAY or FORWARD. Adjust VOLUME and TONE controls.

2. Use the FAST FORWARD control to skip program material on the tape and the REWIND control to go back to segments for replay. To prevent the tape from breaking or snarling, be sure the reels are stopped and the tape is taut before restarting, especially when changing direction of tape travel.

3. Use the index counter to locate program segments and to make note of their length. The counter is not a measure of inches, feet, or minutes, but it provides numbers for easy reference.

PUTTING AWAY

Make certain all operating controls are in their OFF position. Store accessories and cords in spaces provided.

combination uses of tape recorders

The value of a tape recorder can be realized only if its capabilities are fully explored. There are numerous methods of making recordings other than recording live material directly with a microphone.

You can record material from another tape recorder, a radio, record player, or television receiver by holding a microphone in front of the speaker. However, the microphone will also pick up room noises. Wired pickup, which eliminates such noises, is preferable.

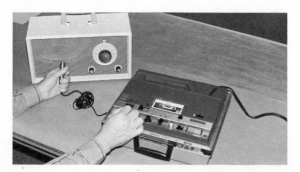

WIRED PICKUP So that recorders can be connected to other sound equipment, machines are provided with input and output jacks. There may be separate jacks for high-level and low-level *input,* or one jack may be used for both. In the first case, it is essential to use the right input connection: use the low-level jack for a microphone or other unamplified or weak signal; use the high-level jack for input of amplified radio or phonograph signals.

High-level signals from the tape recorder normally are *output* at the extension speaker jack; low-level signals, at the earphone or preamplifier jack; however, both output levels may be combined in a single jack. (Check each machine or the manufacturer's manual.)

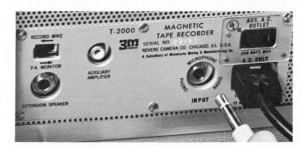

On stereo equipment a set of input and output jacks is provided for each channel, left and right. A variety of jumper cords can be made or procured to connect sound equipment for combination uses.

Since different machines often have different types of input and output jacks or receptacles, proper cords and connectors must be used. Those shown are commonly available or can be assembled.

A special cord with built-in resistance may be required for dubbing from high-level outputs (such as a radio or phonograph) with cassette recorders.

To eliminate room noises when recording from another audio machine, two methods of wired pickup are described.

speaker clip-on method

Attach a clip-on cord to the speaker of the sound source and insert the plug end of the cord in the phono-radio input of the tape recorder. Carefully

balance the volume controls on both the sound source and the tape recorder. Tone must be adjusted at the sound source, since the tape-recorder tone control functions only on playback.

When recording from a record player which has a separate microphone circuit, you can mix voice with sound from a disk recording. Plug a microphone into the record player and use its controls to mix the sound sources. With the clip-on method,

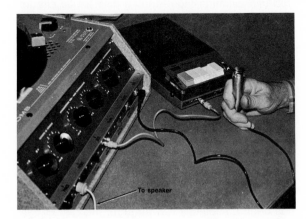
To speaker

the speaker of the record player will carry sound being recorded to the room; when a microphone is used, feedback (squeal) may occur. When this happens, the second method, direct hookup, is recommended.

direct hookup method

The speaker is eliminated from the circuit by connecting a jumper cord directly from the output jack of the sound source to the phono-radio input of the tape recorder. Adjust level by test and by watching the volume-level indicator on the tape recorder. Recording can be monitored with earphones plugged into the sound source or the recorder.

The tape recorder shown below is connected directly to a typical school radio. A cord entering the high-level (phono-radio) input of the tape recorder is connected to a special output on the radio provided for this purpose.

OTHER USES Tape recorders can also be used in combination with projection equipment, both to provide sound and to control presentations of slide series and filmstrips. These uses are described in detail later in this reference section.

Tape recorders are frequently used as sound sources for listening posts. Shown below is a simple jack box for a cassette recorder. The box was made from a tuna can and parts purchased from an electronics parts supplier.

TAPE PLAYING TIMES—MINUTES
REEL-TO-REEL*

reel size, inches	tape length, feet	PLAYING TIME, MINUTES AT SPEED		
		1⅞ ips	3¾ ips	7½ ips
3	150	15	7½	3¾
4	300	30	15	7½
5	600	60	30	15
5	900†	90	45	22½
7	1,200	120	60	30
7	1,800†	180	90	45
7	2,400††	240	120	60
7	3,600†	360	180	90

*Times are for a one-way trip through the machine.
†Extra-thin tape (1 mil).
††½ mil tape.

CASSETTE

designation	per side	total
C-15	7½	15
C-30	15	30
C-60	30	60
C-90	45	90
C-120	60	120

additional facts about tapes

Here are a number of points about tape playing times, speed, and track patterns. These points, and the accompanying chart, will help you plan uses of tapes for instruction.

TAPE PLAYING TIMES Playing time is a principal concern in the selection of tape. Uninterrupted playing times for reel-to-reel tapes depend upon the speed at which the tape is played and upon the length of the tape, which is controlled, in turn, by reel size and tape thickness. As shown in the chart, reel-to-reel tapes are available with uninterrupted playing times of up to six hours (extra-thin tapes on 7-inch reels played at 1⅞ ips). Maximum playing time also depends upon the number of tracks that can be recorded on the tape by the recorder in use. Refer to the chart and to the discussions of tape track formats and tape speeds for help in selecting the tape most suitable for your needs.

Cassette tapes are available with uninterrupted playing times of 5, 7½, 10, 15, 30, 45, and 60 minutes in one direction. Cassettes are designated by their total playing time in two directions: a C-30 cassette, for example, will play a total of 30 minutes, 15 minutes on each side. The C-120 tapes are extremely thin and should be used with care and only when uninterrupted playing time of 60 minutes is imperative.

TAPE SPEED The operating speed of any tape recorder is rated in inches per second (ips) of tape travel across the machine heads. All cassette machines record and play at 1⅞ ips. Reel-to-reel machines may run at one or several speeds; most have two. Standard speeds are exact multiples: 1⅞, 3¾, 7½, 15, and 30 ips or more.

Generally speaking, the faster the tape travels, the better the quality of sound reproduction. For music or high-quality speech recording, the higher speeds are used. When quality is less critical or when economy of tape is important, the slower speeds are preferable.

School machines usually run at a maximum tape speed of 7½ ips. Currently, manufactured recorders operating at this speed will produce undistorted vibrations from about 50 to 12,000 Hz, a satisfactory range for most school requirements.

Refer to the table of tape playing times and speeds. If you learn the principle of tape timing, you can estimate tape running time for all circumstances. Form a basis for estimating running time by starting with a 7-inch reel containing 1,200 feet of standard ¼-inch tape. If it is run at 7½ ips, it will play for thirty minutes in one direction.

TAPE TRACKS Playing times for reel-to-reel tapes in the table are based upon recording single-track. Machines that record the full width of the tape are *single-* or *full-track* machines. But most machines will record two tracks on the same tape. These are called *dual-* or *half-track* machines. Many machines are *quarter-track* (providing two sets of dual track for recording and playing stereo programs).

On a half-track monaural machine, when the tape has been completely run through, the reel or cassette is turned over like a disk record and the second track of the tape is in position for recording. In some machines the tape is automatically reversed. In either case, the full-track playing times of the chart should be doubled for half-track machines.

Tape Recorder Track Patterns

REEL-TO-REEL

Full track

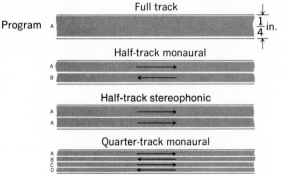

Program

$\frac{1}{4}$ in.

Half-track monaural

Half-track stereophonic

Quarter-track monaural

Quarter-track stereophonic

CASSETTE

Monaural

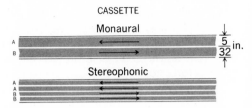

$\frac{5}{32}$ in.

Stereophonic

On quarter-track machines the same turn-over or reversing process occurs. (Track selection may be controlled by a switch.) Maximum playing times in the chart should be multiplied by 4 for quarter-track monaural machines.

In stereophonic tape recording, two tracks are in use simultaneously to carry program material. The stereophonic tape player reproduces sounds from two tracks, playing them through two separate amplifier/speaker systems. (For further discussion, see Audio Equipment Principles at the beginning of this reference section.) Hence, the playing times in the chart are correct for half-track stereo and should be doubled for quarter-track stereo.

Presently monaural cassette recorders are made only in the half-track configuration. Stereo cassette machines are quarter-track, recording two parallel tracks in each direction to produce the stereo sound. Playing times are the same for monaural and stereo cassettes.

TAPE ERASURE Accidental erasure of tape in a recorder is nearly impossible, since all recorders require the manipulation of two controls to place them in RECORD, where erasure occurs. Accidental erasure *can* result if tapes are exposed to a strong magnetic field from any source, such as a speaker magnet.

Cassette tapes can be permanently safeguarded from erasure on the recorder by breaking out the small tabs on the edge opposite the open end. There is a tab to protect each track on the tape; remove the one that is on the left as you read the label for the side you want to protect (side A in the picture). If you want to record on that side later, cover the hole with a piece of thin tape.

Used tapes can be erased before rerecording by running them through a recorder set on RECORD with the volume turned down. Dual-track tape must be run from end to end on both sides. A time-saving bulk tape eraser, such as the one shown below, will completely erase all tracks in a few moments as the tape is manipulated within its magnetic field.

duplicating tapes

Recording from tape to tape requires only two tape recorders and a jumper cord. The machine playing the original material is connected at the output jack; the machine that will make the duplicate tape is connected at the input jack of matching level. For example, if a speaker output jack emitting a high-level signal suitable for driving a large speaker is used, the signal must be fed through the high-level input of the receiving machine.

Special copying machines are available to speed up duplication of tapes. These run at many times the recording speed and may make several copies at once.

In the picture below, several standard cassette recorders are connected to make duplicates from a tape played on the machine in the foreground. The junction box meter measures the playing level.

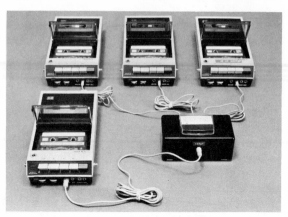

Joel Benedict, Arizona State University

Some pointers on tape duplication are as follows:

☐ Because tone controls operate on a tape recorder only when it is playing, remember that any adjustment to improve tone must be made on the machine that is playing, *not* the one recording.

☐ Test recording and playing levels before making a duplicate tape.

☐ To save tape-duplicating time on reel-to-reel machines, a 3¾-ips tape may be played at 7½ ips and the copying recorder also run at 7½, thus cutting duplicating time in half. The duplicate produced will play only at 3¾ ips.

☐ Speed of a tape may be changed when duplicating: two 1,200-foot tapes running at 7½ ips may be transferred to one 1,200-foot tape running at 3¾ ips. Or a master tape recorded at 3¾ ips may be duplicated on a cassette machine at 1⅞ ips. Single-track tapes may be duplicated on dual track, or vice versa.

editing tapes

There are two simple methods of editing tapes: (1) using two tape recorders and (2) tape splicing.

TWO-MACHINE METHOD After planning editing requirements (a marked script will be helpful), connect two tape recorders as for tape duplication. Play the original tape on one machine and feed the output to the second machine, which will record the edited version. Start the second machine recording, then start the other machine playing. Stop the recording machine to eliminate unwanted portions of the playing tape. (Use the PAUSE control if provided.) Dexterity is required here, and practice is valuable. To avoid thumps and pops in the edited tape, turn down the volume on the recording machine before stopping it; start the machine before turning up the volume.

With this method portions of several tapes can be combined in one recording. Tapes recorded at different speeds can be combined by playing them at their various speeds (1⅞, 3¾, 7½ ips) and re-recording all of them at only one of these speeds. Stereophonic and monophonic recordings can be combined, as well as tapes recorded in different track configurations.

TAPE SPLICING METHOD To the edit tape by splicing, cut and reassemble the tape to eliminate, add, or combine sections. Segments of tape can be identified by writing on the glossy side with a china-marking (grease) pencil. Splice the sections as described below. After a sequence of tape clips has been assembled, duplicate this tape on a new tape, preserving the spliced tape as a master. Note that tapes edited by splicing must have been recorded at the same speed.

splicing tape

To edit tape by cutting and assembling sections or to repair tape that has been accidentally torn, splice the ends as follows:

1. Lap two ends of tape, then cut at a 45-degree angle by holding scissors diagonally in relation to the overlapped tape.

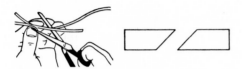

2. Butt the ends together smoothly, then place a short length of splicing tape (such as Scotch Brand No. 41) over the joint. Do not use ordinary cellophane tape.

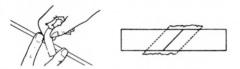

3. Trim off excess splicing tape, cutting into the recording tape very slightly. This eliminates the possibility of a sticky splice.

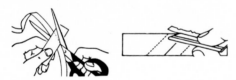

practical pointers

☐ Unless the recorder is equipped with automatic volume control, it is a good practice to make tests before attempting a final recording to ascertain proper recording levels. If the levels are too high, program material will be distorted and the recording unsatisfactory. Proper level can be maintained by observing the recording level indicator and by adjusting the volume control. There are two types of level indicators in common use:

V. U. METER This dial-type indicator shows recording level by needle swing. Dials are usually marked to indicate overrecording levels. The V.U. meter is also used as a battery level indicator on some portable tape recorders.

NEON GLOW LAMPS These are supplied in several types. With a single-glow unit, the lamp should glow on the strongest tones; but a steady glow shows overrecording. With a double-glow unit, one lamp portion glows when the recording level is correct; a second portion glows only on overrecording. With a two-lamp unit, one bulb glows on correct level; the second lights on overrecording.

☐ Keep the machine, and particularly the heads, perfectly clean. After repeated use, heads accu-

mulate dirt and gum, which cause faulty recording. A safe cleaning agent for heads is denatured alcohol. Do *not* use other liquids. As pictured, moisten a swab stick with the liquid; do not use excessive fluid. Never employ metal for cleaning tape-recording heads. Also clean parts other than heads (including tape-driving rollers) with denatured alcohol. Avoid excessive moisture.

☐ After extended use, the record-play head(s) become magnetized in a way that can impose noise on a tape during recording. Use an inexpensive head demagnetizer to demagnetize recorder heads.

☐ Learn to use an earphone or headset. Nearly all recorders are equipped with a jack where earphones can be connected to monitor recording or to listen to prerecorded material. Usually, connecting the earphone shuts off the speaker; thus sound is played through the earphone only and one can listen without disturbing others.

☐ Accessories of many kinds are available to implement tape recording. Investigate bulk tape erasures, head demagnetizers, tape cue devices, and tape-splicing machines. Some recorders accommodate a remote-control mechanism (operated by hand or by foot pedal) to permit intermittent recording and playing. With such a control, recorders may be used for dictation or for intermittent recording and transcribing of discussions and speeches. Attachments for recording from telephones are also available, but there are legal restrictions on their use.

projection equipment principles

When the general principles of how projectors work are understood, learning to operate equipment may seem an easier task. Here is a nontechnical overview of some principles that apply to the use of projection equipment.

projection systems

All projection systems have these common elements:

- ☐ A light source
- ☐ A lens system
- ☐ A surface on which images are projected
- ☐ A means for holding images to be projected

Below is a simple schematic of a projection system. Note that the image to be projected is in the machine upside down; the projection lens inverts the image to its right-side-up position on the screen.

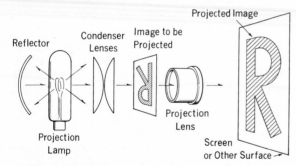

With one exception, all projectors transmit light through a transparent image to the screen; only the opaque projector operates with light reflected from the surface of the material projected. The elements required are the same as for other projectors.

projection lenses

Lenses for projectors are provided in different focal lengths. Their selection is determined by the size of the projected material, the distance from the projector to the screen, and the size of the image required on the screen. Most school projectors are supplied with a lens suitable for traditional class-room projection (e.g., a 2-inch lens on a 16mm motion-picture projector). The image size on the screen can be adjusted by changing the projector-to-screen distance. When the projector is moved *toward the screen,* the image will become *smaller and brighter;* moving the projector *away from the screen* produces a *larger but less bright* image.

Some projectors are equipped with a *zoom* lens which permits adjusting the picture size to the screen size without changing the projector position. This feature is highly desirable under many conditions and increases the convenience of making projector setups. Most zoom lenses are focused in the same manner as standard lenses, and the zoom feature is operated by rotating a portion of the lens barrel.

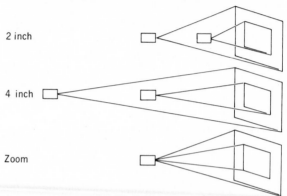

For special conditions, lenses other than the standard ones provided with equipment may be needed. The chart below illustrates the relationship between type of medium, projector-to-screen distance, screen width, and lens focal length.

TABLE OF PROJECTION DISTANCES

type of medium	lens focal length	SCREEN WIDTH, IN. DISTANCE—PROJECTOR TO SCREEN, FT			
		40	50	70	96
35mm slide	5 in.	13	16	22	30
	7 in.	18	22	30	42
Single-frame filmstrip	3 in.	11	14	20	27
	5 in.	19	24	33	46
8 mm	22 mm	17	22	30	Not
Super 8	22 mm	14	17	24	practical
16 mm	1½ in.	13	16	23	32
	2 in.	18	22	31	42

projection screens

As shown below, projection can be from in front of the screen (front projection) or from behind it (rear projection). Front projection is onto a *reflective* surface; rear projection, onto a *translucent* plastic or glass material. Mirrors are frequently incorporated in rear-projection systems used in carrels and in projection devices for individual viewing.

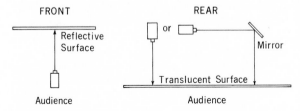

Several types of screens are used for front projection:

☐ *Wall-mounted screens* that are pulled down for use and retracted into a case when not needed.

☐ *Portable stand screens* mounted on tripods or detachable legs. These can be located where desired.

☐ *Projection wall areas,* matte-white wall surfaces painted to serve as screens. These are sometimes covered by sliding chalkboards or display panels when not in use.

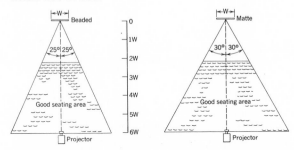

Screen surfaces are often coated with fine glass beads because the highly reflective surface increases brilliance of the screen image. However, the brilliant reflection covers a narrow angle, and viewers seated at the sides of a room receive a subdued and somewhat grayed image from a beaded screen. In contrast, a matte-surface screen, while reflecting a less brilliant image, gives each viewer in the audience a uniform picture brightness. Thus in wide, shallow rooms, matte screens are considered better than beaded-glass screens. Improvements in lens systems and projection lamps have improved image brightness on matte-surface screens. The two charts suggest optimum viewing angles for both matte- and beaded-surface screens. Audience seating arrangements should be planned accordingly.

Screen size, shape, and location are determined by factors such as room size, audience size, distance from projector, and type of projector. For additional information on room arrangements for projection, see Reference Section 4.

practical pointers

☐ Remember that for an image to be rectangular and uniformly clear, the center of the beam of light from the projector must make a 90-degree angle with the screen surface. If this angle is not achieved, a keystone effect (distortion of the picture shape) will result.

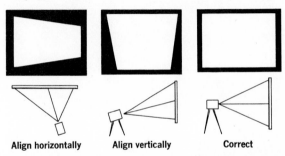

Align horizontally   **Align vertically**   **Correct**

☐ Many types of projection stands, rolling tables, cabinets, and special supports are available for projectors. For example, low stands for overhead projectors permit the operator to sit comfortably, facing the audience and out of their line of view of the screen. Tall stands support classroom television receivers. See manufacturers' catalogs for other examples of stands designed to meet special conditions of projection.

☐ When you obtain equipment, make sure all necessary accessories are provided: cords, reels, spare lamps, and stand and screen if required. When you set up, always check equipment operation.

☐ If preshowing steps in setting up for projection have been completed properly, it is highly unlikely that any adjustments will be necessary during the presentation. If adjustments are essential—as in the placement of the image on the screen or in volume or tone—make them very slowly. The audience, if it is absorbed in the program, is not likely to notice skillful adjustments.

opaque projectors

Flat, printed, or drawn pictures or other materials, as well as some three-dimensional objects, may be projected with an opaque projector. This machine operates with reflected light: the lamp illuminates the material, and the image is reflected by a mirror through the lens to the screen. Since projection by reflected light is much less efficient than by

transmitted light (as with slides), successful projection with opaque projectors can be accomplished only in almost total darkness.

operating opaque projectors

SETTING UP

1. Place the projector on a stand and connect the power cord.
2. Place material to be shown on the platen, face up, with bottom of the material toward the screen.
3. Turn on the motor and lamp. Adjust tilt and level by adjusting the front legs of the machine.
4. Focus the image on the screen. Move the projector toward or away from the screen to obtain a satisfactory picture size. Adjust the focus for a sharp image.
5. For maximum illumination on the screen, locate the projector as close to the screen as possible; rearrange seating if necessary.

OPERATING

This involves only proper placement of materials in the machine in the desired sequence.

PUTTING AWAY

1. When ending a showing with the opaque projector, turn off the lamp and let the fan continue to run a few moments to cool the lamp.
2. Detach the power cord and cover the machine with a dust cover.

practical pointers

☐ Different machines have various means for holding materials in place on the platen: magnets, metal masks, or an endless carrier with a crank to carry materials into the machine and out.

☐ Mount pictures in sequence on long strips of paper to facilitate presentations. Exercise care when inserting and removing books. Pages must lie flat. Avoid damage to pages and binding.

☐ On some machines projected materials are protected from heat and held flat by a heat-resistant glass plate. Remove the glass when projecting of small objects which might scratch it.

☐ Do not show objects which may be damaged by heat. Any objects, especially those made of metal, may become too hot to handle—use care!

☐ The opaque projector can be used as an enlarger: trace the picture projected onto a chalkboard or paper mounted on the wall.

overhead projectors

The overhead projector is one of the simplest visual communication devices available, easy to maintain and to use. Since a bright lamp source transmits light through the translucent material to a screen quite close to the machine, the projected image is bright enough for viewing in a lighted room.

operating overhead projectors

SETTING UP

1. Set the projector on a table or stand at a convenient height for the operator. Attach the head support and head assembly if these are not permanently installed.

2. Adjust the top mirror to locate the light on the screen. Move the machine forward or away from the screen to obtain an image of satisfactory size. Adjust the focus knob until the light on the screen has sharp edges. Time may be saved by projecting a mounted transparency when making these adjustments.

OPERATING

1. Place material on the projector table. Check focus again. Practice pointing to items on the transparency and manipulating overlays. Work without continually looking back toward the screen; keep eye contact with your audience.

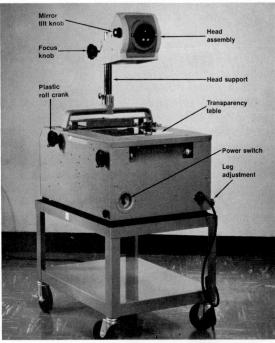

3M Company

2. Practice drawing and lettering on clear plastic sheets or the plastic roll provided with some machines.

practical pointers

☐ Experiment with vertical and tilted screens. Determine how much keystone effect can be tolerated with different types of transparency content.

☐ If room lights near or directly above the screen are extremely bright, they should be adjusted to avoid a washed-out image.

Charles Beseler Company

☐ Overhead projectors are available in a wide variety of types to serve different instructional situations. The machine at the top, above, is designed for large-group instruction, the machine below it for conventional classrooms and smaller groups. Several portable machines are also manufactured and are popular with extension program teachers and traveling lecturers.

2- by 2-inch slide projectors

Slide projectors are available in a variety of types differentiated by the sizes of slides they project and by the way in which slides are carried and changed in the machine. Slides in most common use in schools today are those in 2- by 2-inch mounts; 2¼- by 2¼-inch and 3¼- by 4¼-inch slides are used rarely. Most projectors found in schools now are tray-loading; projectors using single slide carriers are still useful for special applications, such as tachistoscopic projection or when a small number of slides is to be projected.

Several important improvements in 2- by 2-inch slide projection equipment have encouraged the use of slides for instruction. These improvements include the following:

☐ TRAY-LOADING PROJECTORS In these machines the tray not only carries series of slides in proper order but also provides a convenient storage system for slide sets.

☐ REMOTE SLIDE CHANGE Remotely controlled slide projectors offer many conveniences: Slides can be changed from the front of a room by means of a pushbutton on an extension cord. On some machines slides can either be shown in the normal (forward) sequence or they can be repeated, since a reverse button is included with the remote control.

☐ FOCUS CONTROL Some machines permit the operator to adjust focus of the projected picture by a remote-control button; other machines focus the image automatically.

☐ AUTOMATIC SLIDE ADVANCE A built-in timer on some projectors changes slides automatically at predetermined intervals. Also, most machines with a remote control can be adapted to change slides on a signal pulse from a tape recorder for synchronized sound-slide presentations.

a generalized slide projector

Regardless of their apparent differences, all slide projectors operate similarly in principle. Components of a typical projector are shown in the drawings and described in the lists, with variations noted. Items listed in capital letters are found on all projectors. Locate these components and determine which of the variations are provided on your machine.

PROJECTION SYSTEM

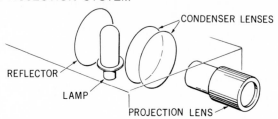

☐ A *PROJECTION LAMP* of the type specified for the projector provides the light source.

☐ A *REFLECTOR* and *CONDENSER LENSES* direct lamp light to the projection lens.

☐ A *PROJECTION LENS,* either a zoom type or a fixed-focal-length lens of appropriate size for the viewing situation, produces the focused image on the screen.

☐ *FOCUS ADJUSTMENT* may be made by turning the projection lens, by turning a focusing knob on the projector body, by pressing a remote-control focus button, or it may be automatic.

SLIDE CARRIERS

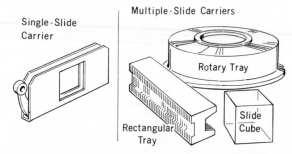

☐ The *Single-slide Carrier* may be permanently installed on the machine, or it may be stored off the projector and inserted in the carrier slot for use.

☐ *Multiple-slide Carriers* can be used for storing as well as projecting slides. *Rectangular Trays* vary in capacity. *Rotary Trays,* mounted horizontally or vertically, according to projector design, may hold up to 140 slides. Slots are numbered for easy location of slides.

Projector trays and cubes, as well as slide file boxes and notebook-size plastic sheets, provide convenient storage for slides. Establish and maintain a simple filing system so that you can easily look through your collection and make selections for a particular use.

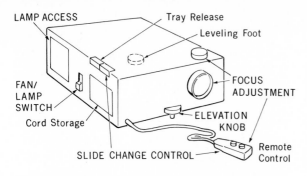

LAMP ACCESS — Tray Release — Leveling Foot — FOCUS ADJUSTMENT — ELEVATION KNOB — Remote Control — SLIDE CHANGE CONTROL — Cord Storage — FAN/LAMP SWITCH

□ A *FAN/LAMP SWITCH* turns on the fan motor and the projection lamp. *High* and *Low* lamp switch positions allow selection of light intensity on some machines.

□ *SLIDE CHANGE CONTROLS* may be located on the projector body or on a *Remote Control Cord*. On single-slide and manual-feed tray projectors, the slide carrier or tray is moved manually to change slides for viewing. On recent models of multiple-loading machines, slide changes are controlled with pushbuttons. Usually both *Forward* and *Reverse* tray motion is provided.

□ A *Tray Release* button unlocks the tray for removal or to change its position for showing slides out of sequence.

□ An *ELEVATION KNOB* raises or lowers the machine front to center the light beam vertically on the screen. A *Leveling Foot* may also be provided to raise one side of the machine for leveling the image horizontally.

□ Compartments for *Cord Storage* may be located on the bottom or sides of the projector body. Removable panels or covers provide LAMP ACCESS for projection system maintenance. Examine

the bottom and sides of the projector for locations of access panels and storage compartments.

slides

SLIDE FORMATS

Slides in 2- by 2-inch mounts may be in one of several formats (35mm, super slide, half frame: see Reference Section 3, ''Photographic Equipment and Techniques''). Remember that when you use the smaller formats, you must increase the projection distance or use a lens with a shorter focal length to produce an image of the same size as you would obtain from a larger slide format.

LOADING SLIDES

To load slides into a slide carrier or tray for projection from in front of the screen, position the slides as follows: As you face the screen, hold the slide so that it reads normally. Invert the slide, so that the image is upside down, and insert the slide into the carrier or tray.

For direct rear-screen projection, slides should be turned so that the image is reversed as well as inverted, since the projection will be from the ''wrong side'' of the screen. In some rear-screen installations, a mirror will reverse the image; loading for these projection systems should be the same as for front-screen projection.

Thumb spots help orient slides for projection. Traditionally, the spot is placed in the top right corner of the slide on the side that is away from the screen when the slide is in position for loading.

HANDLING AND CLEANING SLIDES

Handle slides by their mounts; avoid touching or scratching the film surfaces. Dirty slides can be dusted with a soft brush or wiped with lens tissue or cotton. Cotton dampened with water or film cleaner and applied carefully to avoid scratching will remove grease and other dirt from the film surface.

slide projector operation

Perform the following steps, adapting the procedure as appropriate for your equipment.

SETTING UP

1. Place the projector on a stand and connect the power cord. Connect the remote control cord if used. Insert the lens if it is stored in the projector case.

2. Install the slide carrier or tray. Trays for some machines require alignment of an index number on the tray with a location indicated on the projector.

3. Turn on the lamp, project a slide, adjust the elevation of the front of the projector to center the light on the screen by using the elevation knob and/or other support; then adjust the picture size to fit the screen by moving the projector or by adjusting the zoom lens if provided.

4. Focus the image by turning the lens or focusing knob or by pressing the focus button on the remote control until the image is clear.

OPERATION

☐ SINGLE-SLIDE MACHINES For a smooth presentation, stack slides in proper sequence and in position for convenient loading in the slide carrier. Some carriers are loaded on the right and unloaded on the left; others are loaded alternately on the right and left sides. As you withdraw slides from the machine, stack them neatly, in order, face down. At the end of the showing, be sure to *remove the last slide from the carrier.*

☐ MULTIPLE-LOADING MACHINES Show slides in the desired sequence, using the FORWARD or REVERSE controls to move ahead to or to repeat adjacent slides. Use the RELEASE control button to move the slide tray to slides stored at remote points in the tray. A preview screen on some machines shows the operator the slide next in line to be projected.

PUTTING AWAY

1. At the end of the showing, turn off the projector lamp. Remove the tray or carrier.

2. Level the machine by retracting the elevating knob. Retract the lens or remove it for storage in the projector case.

3. Disconnect, coil and store power and control cords. Place the machine and accessories in the case.

Kodak Carousel

The popular Kodak Carousel projector described below is typical of rotary-tray projectors that are in wide use in schools today. Look for similar features on other machines.

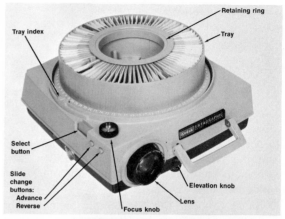

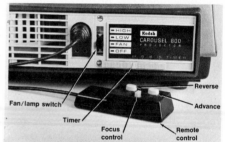

Eastman Kodak Company

☐ REMOTE CONTROL The cord-attached control changes slides, moving the tray forward or backward, and adjusts focus on some models.

☐ CAROUSEL TRAY Black trays carry up to 80 or 140 slides mounted in cardboard; gray trays, 80 cardboard or glass-mounted slides. A retaining ring inserted in the top of the tray prevents slides from falling out when the tray is off the projector. Before placing an 80-slide tray on the machine, check to be sure that the slot in the metal bottom of the tray is aligned with the zero (0) position on the rim.

☐ FAN/LAMP SWITCH This switch has four positions: OFF, FAN, LOW, and HIGH. Use the LOW setting when this produces an adequately bright picture; this setting extends lamp life and may improve the picture quality of thin or overexposed slides.

☐ SELECT BUTTON Push this button to pop slides up for inspection or rearrangement and to release the tray for free rotation to any desired slide or to zero (0) for removal.

☐ SLIDE-CHANGE BUTTONS These are provided on the machine body of some models for changing slides without operating the SELECT button or installing the remote control cord.

☐ AUTOTIMER This control automatically changes slides at intervals for use of the projector unattended, as in exhibits or displays.

☐ AUTOFOCUS Automatic focus is provided on some models: once the first slide is correctly focused on the screen, all slides shown thereafter will be focused automatically.

Accessories available for use with the Kodak Carousel include a stack loader for showing a series of slides without loading them into a tray and a filmstrip adapter which is substituted for the projector lens to show filmstrips.

other slide projectors

The Viewlex projector below is used extensively for filmstrip projection; however, with a single-slide carrier installed in the projector, slides can be loaded manually and shown in sequence.

For independent study, manually operated slide viewers are useful in carrels or at study desks. Remotely controlled tray-loading viewers require less handling of slides by users; these machines may also be programmed by synchronized tapes.

For slide lectures, random-access devices prove helpful to some teachers. The dial of the Kodak Ektagraphic RA 960 provides quick access to any slide in the eighty-slide tray. With the second machine below, a Spindler & Sauppe, buttons are used to access selected slides.

Eastman Kodak Company

Spindler and Sauppe

393

filmstrip projectors

In their general operating characteristics, filmstrip projectors are similar to manual slide projectors. The principal differences are in the format of the projected materials and in the carrier used with them.

Some machines permit projection of double-frame filmstrips which have a picture area of the same size as 2- by 2-inch slides. Most filmstrips used in schools, produced by commercial publishers, are single-frame filmstrips. Many projectors for filmstrips have carriers with built-in masks that permit using either the single- or double-frame format.

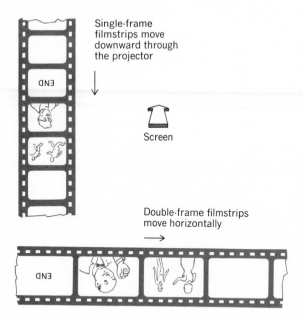

Single-frame filmstrips move downward through the projector

Screen

Double-frame filmstrips move horizontally

operating manual filmstrip projectors

The machine illustrated is quite typical of combination filmstrip and 2- by 2-inch slide projectors used in schools. Other machines differ in appearance and often in the details of mechanical arrangement of parts. With the principles demonstrated here and a manufacturer's operating manual, other machines should pose few problems.

SETTING UP

1. Always remove the projector, including the base cover if not attached, from its case. Set the projector on a firm stand and connect the power cord. Insert the filmstrip carrier.

2. Turn on the lamp and elevate the front of the projector as required. Move the projector toward or away from the screen to obtain the appropriate image size. Focus the beam on the screen to obtain sharp edges of the lighted area. Focus may be adjusted by turning the lens barrel or by using a focusing knob at the side of the lens.

THREADING AND OPERATING

1. Insert the filmstrip in the carrier. For correct orientation of the film, face the screen, read the filmstrip title, turn the filmstrip—left or right—head down, and push the strip gently into the film channel until it stops. Continue to push gently and at the same time turn the film advance operating knob (A).

2. When a focus frame, title, or picture appears on the screen, focus until the image is sharp.

3. If the image is split between two frames or is not precisely centered in the aperture, adjust with the

framing control. Framing controls differ on different machines. The one shown is a lever that is moved up or down to center or "frame" the image.
4. Rotate the operating knob to change the picture on the screen. In most projectors, the filmstrip can be run either forward or backward. Move the strip briskly from frame to frame with a snap-turn; do not move the strip slowly.

PUTTING AWAY

1. Coil the power cord and store neatly.
2. Retract the lens and level the machine.
3. Check to be sure no filmstrips have been left in the case or carrier.
4. Be sure all accessories and spare parts are put in the case before latching the lid.

projector variations

Several variations in filmstrip projectors are used in instruction. Some examples are:

☐ *Cartridge-loading machines* such as the Bell & Howell Autoload model shown below.

Bell & Howell   *Singer*

☐ *Filmstrip viewers* for individual and small-group instruction. These economical and easy to operate machines are especially convenient in individualizing instruction, and filmstrips are available at reasonable cost on a vast variety of subjects.

DuKane Corporation

☐ *Filmstrip viewers with synchronized sound.* Such units provide loud speaker or earphone listening, and some provide various degrees of student control to start, stop, and restart during instruction.

handling filmstrips

☐ After a filmstrip is used, it will be coiled loosely and be too large to put in the can. Roll the strip into a coil small enough to fit the can. *Never* cinch a filmstrip to reduce the size of the roll; cinching produces scratches on the film.

☐ Keep your fingers off the picture surfaces; handle strips only by the edges, the blank leader, or the blank trailer.

☐ Handle filmstrips gently; never force the operating mechanism of the projector.

practical pointers

☐ It is not always necessary to fill the entire screen when projecting filmstrips. Often a smaller but brighter picture will provide improved viewing.

☐ When preparing to project a filmstrip with synchronized sound, it is always advisable to practice the entire showing in advance to become thoroughly familiar with the equipment and the procedures to be used. Three basic techniques are used for sound synchronization with pictures: following a script marked for slide changes, audible tones on the sound recording indicating frame changes, and automatic signal pulses in the recording that cause the projector to change frames automatically. The latter system, of course, requires special equipment.

motion-picture film projection principles

Sound motion-picture film projectors operate on principles similar to those of several less complex equipment items described in preceding sections. For example:

☐ Sound controls on film projectors are identical with those of simple record players.

☐ Film moves through a motion-picture projector in much the same way as through a filmstrip projector.

A motion-picture projector is more complicated than a combined record player and filmstrip projector because

☐ Film moves through the projector at relatively high speed; 16mm pictures with sound are projected one by one at the rate of twenty-four frames per second.

☐ Sound is carried on a track at the edge of the film. The film must move smoothly past the sound drum, where sound is picked up from the track.

☐ The length of 16mm films may be from 400 feet (or less) to 2,000 feet long; 8mm films may be from 50 to 1,200 feet long.

Motion-picture projectors in many configurations are in use today for projecting 8mm and 16mm film, either silent or with optical or magnetic sound. 8mm and 16mm machines may be reel-to-reel, with manual or automatic threading, or cartridge loading. Some projectors are equipped with built-in screens or other special features. On the following pages components and operations common to all

motion-picture film projectors are discussed, then details are given on the specific models most frequently found in schools today.

film for motion pictures

Film is called 8mm or 16mm because of its physical width. The chart compares characteristics of 16mm and Super 8mm films; standard 8mm film is also included since some films of the older format may be in use. It can be seen that these films vary in size of frame, size and location of the sprocket holes, and location and appearance of sound tracks. They also vary in number of frames per foot of length and in the speed at which they are run.

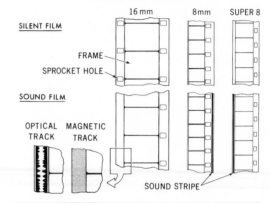

MOTION-PICTURE FILM

FRAMES PER SECOND		FEET PER MINUTE	
		SILENT	SOUND
All Silent	18		
All Sound	24		
	16mm	27	36
	8mm	13.5	18
	Super 8	15	20

Eastman Kodak Company

Technicolor Corporation

Eastman Kodak Company

Bell & Howell

Fairchild Camera and Instrument Corp

Projectors are designed to handle specific film types—the chart will help users to recognize the types of film suitable for a particular projector.

the A-B-C of film projector mechanism

All sound motion-picture projectors must have three basic systems: one for reproducing sound, another for projecting the image carried on the film, and one for transporting the film through the projector. Each is illustrated and described below.

SOUND SYSTEM COMPONENTS (A)

☐ A *Power Cord* supplies AC power.

☐ *Amplifier Controls* turn the amplifier on and off and adjust volume and tone. In some projectors functions may be combined, e.g., the tone control knob may also control the amplifier off-on switch.

☐ A *Speaker,* which may be mounted in the projector case or may be a separate unit.

☐ For optical sound, light from the *Exciter Lamp* passes through the film sound track to the *Sound Drum,* where a photoelectric cell converts the pulsing light into electrical impulses.

☐ For magnetic sound, *Magnetic Heads,* like those on tape recorders, play back or record sound on a magnetic stripe on the film.

PROJECTION SYSTEM COMPONENTS (B)

☐ A *Projection Lamp* is controlled by a *Lamp Switch,* which may be incorporated in the motor switch.

☐ A *Lens,* which is adjusted to focus the image.

☐ An *Elevator Knob,* which raises the projector to direct the light beam at the center of the screen.

☐ A *Framer,* which adjusts the projector aperture so that only one complete picture is seen on the screen.

FILM TRANSPORT SYSTEM COMPONENTS (C)

☐ A *Feed Reel* and a *Take-up Reel* supported on *Reel Arms.* On some machines, drive belts to rotate the reels must be attached in setting up the projector.

☐ A *Motor Switch* controls power for the film transport system motor.

☐ *Toothed Sprocket Wheels* (usually two or more) mesh with sprocket holes in the film to drive it smoothly through the machine.

☐ *Guide Rollers* prevent the film from rubbing against the projector case and *Snubbers* cushion the film against strain.

SOUND SYSTEM (A)

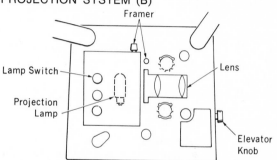

PROJECTION SYSTEM (B)

FILM TRANSPORT SYSTEM (C)

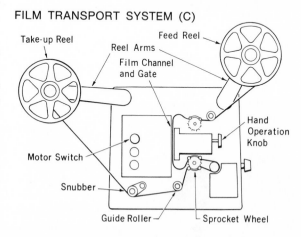

☐ A *Film Channel* incorporates the aperture in the lamp housing, where the light passes through the film, and the Film Gate on the lens housing.

☐ A *Hand Operation Knob,* when turned manually, moves the film through the machine to check threading.

operating 16mm film projectors

The following sequence of steps of motion-picture film projector operation is applicable to all typical classroom projectors, except as noted. The machine used here to illustrate the typical steps is the RCA 416; though obsolete, it is an especially good projector with which to illustrate motion-picture film projection techniques. Instructions for operating other projectors follow this generalized procedure.

SETTING UP

1. Place the projector on a stand or table and open the case.

2. If the speaker is mounted in the cover or is in a separate case, attach the speaker cable at the projector. Place a movable speaker where all will be able to hear.

3. Plug the power cord into a wall receptacle. Secure power and speaker cords to the projector stand legs.

4. Turn on the amplifier.

5. If reel arms or drive belts are not permanently installed, attach these.

6. If there is a film speed selector, set it for the film used. If there is a reverse control, set it at FORWARD.

7. Turn on the motor and lamp. Elevate the projector to center the light on the screen and move the projector toward or away from the screen until the light fills the desired area.

8. Focus the light beam by turning the lens barrel (or the focus knob) until the edges of the lighted area are sharp. Examine the edges for evidence of dirt in the film-channel aperture.

9. Turn off the projector. Open the film channel and clean all surfaces over which film passes with a gate brush. *Note:* Do this every time you thread a film.

10. Check the sound system. Is the amplifier turned on? On some machines the exciter lamp comes on only when the projector is started; on others it lights when the amplifier is turned on.

11. Push the reel of film firmly on the spindle. *Note:* The reel lock on some machines operates simply by pushing the reel on the spindle as far as it will go; others use a snap-down catch.

12. Check to make sure the film is properly wound on the supply reel. As the film comes off the front of the reel, the image should be head down and the sound track should be on the side nearest the projector.

Right   Wrong

13. Attach an empty reel, as large or larger than the supply reel, on the take-up arm spindle.

MANUALLY THREADED PROJECTORS

Refer to threading instructions on the following pages, on the projector, or in the manufacturer's manual for specific types and models of machines. On each machine threaded manually careful attention should be given to the following important points:

1. Teeth on each sprocket wheel must fit through holes in the film.

2. The film must lie smoothly in the film channel.
3. Establish loops above and below the film channel according to lines or embossings on the projector housing (or by a special roller guide on some machines). The lower loop must be made accurately to ensure proper synchronization of sound and picture. If a loop is lost during a showing, the picture will vibrate on the screen; stop the projector and re-form the loop.

4. The film must be under tension as it passes around the sound drum. If film is threaded too loosely here, the sound will be distorted. Some projectors feature special devices to ensure proper tension; on others extra care must be taken in threading around the sound drum to prevent distorted sound.
5. Complete threading by inserting the film end in the take-up reel. Wrap a full turn or more around the reel, turning it clockwise.

AUTOTHREADING AND CARTRIDGE FILM PROJECTORS

Autothreading projectors greatly simplify film-loading procedures. With projector controls properly set, the film is simply inserted in the film slot and the machine is turned on. The projector mechanisms automatically perform all remaining threading operations except insertion of the film end in the take-up reel. The film leader must be in good condition: not warped or buckled, free of torn sprocket holes and misaligned splices, and trimmed smoothly at the end. On most autothreading machines the film can be removed at any point by disengaging the autothreading mechanism and manually unthreading the projector. Refer to instructions for specific machines which follow.

SHOWING THE FILM

The following procedures apply generally to all typical classroom projectors:
1. Set sound-volume control at about ⅓ or ½ of full volume. Set tone at the midpoint. Turn off room lights.
2. Start the showing by the following steps:
    a. Turn on the motor.
    b. Turn on the lamp as the title nears the aperture.
    c. Adjust focus the instant an image appears on the screen.
    d. Adjust volume, then tone.
    e. Correct the framing if necessary.
3. During the entire showing, stay with the projector to:
    a. Correct focus and adjust sound as necessary.
    b. Shut off the machine immediately in case of faulty operation.
4. To close the showing:
    a. As the end title faces, turn off the lamp.
    b. As the sound fades or ends, turn the volume down to zero.
    c. When all the film is on the take-up reel, turn off the motor.
    d. Turn off the amplifier if no more films are to be shown.

REWINDING THE FILM

To rewind the film, reels must turn counterclockwise (opposite to the direction they turned when projecting). Refer to instructions for specific machines on the pages following.

When rewinding, touch the spinning reel with care. Fingers should touch only the sides of the reel rims and should point in the direction of reel rotation to prevent injury to hands.

PUTTING AWAY THE PROJECTOR

1. Coil all cords and place them in the spaces provided.
2. Dismantle the machine and put the reel arms, cables, and spare reel in the positions provided for storage.
3. Retract the elevating mechanism.
4. Be certain all switches are off and all levers are in their forward operating positions (not in reverse or rewind).
5. Close and lock all lids firmly.

practical pointers

☐ Whenever possible, before your audience arrives, run a few feet of film to check both picture and sound quality; then reverse the projector and run the film back to the starting point, ready for the showing.

☐ To judge sound quality and volume best, step a few feet away from the machine; projector noise may cause you to misjudge the sound.

☐ If there is a microphone input on your projector, practice using the machine as a public-address system. Try providing your own commentary in place of the film sound track.

☐ If a film breaks, wind it overlapped on the take-up reel, mark the break with a slip of paper, and continue the showing. *Never,* under any circumstances, make temporary repairs with gummed tape, paper clips, or other materials. Film must be spliced only with special cement in a splicing machine.

Write the type and location of film damage on a slip of paper, return the film to the can without rewinding, and leave the slip in the can with the edge extending; this will signal the film inspector that damage is being reported. *Note:* it is not a crime to break a film, but damage must be reported as a courtesy to the owner and the next user.

☐ Always select a take-up reel *at least* as large as the feed reel carrying the supply of film.

16MM REEL SIZE—RUNNING TIME

Motion-picture film of 16mm width is mounted on reels of several sizes. Conventionally, the 400-foot reel is considered "one reel," other sizes are increments of this size. The table shows the relationship of film footage, reel size, and showing time for sound film.

feet of film	reel	minutes
400	1	11
800	2	22
1,200	3	33
1,600	4	44
2,000	5	55

16mm motion-picture projectors

Instructions for operating five 16mm projectors follow. Threading charts are included, as well as steps for setting up, threading, showing the film, and rewinding film and putting the machine away.

Kodak Pageant projector

SETTING UP

Follow the general procedure outlined on page 398. On the Pageant projector, remove the case front and set up the speaker. Raise the reel arms; install the take-up arm belt. Be sure the rewind tab is in the vertical position before turning the machine on to aim the lens, adjust picture size, and set the speed selector for the film, silent or sound. Turn the projector off.

THREADING

1. Open the feed and take-up sprocket clamps and the gate; push forward and latch gate tab.

2. Turn the hand-threading knob until the white line on the knob is toward you. Engage the film in the upper feed sprocket and set the top loop by aligning the film with the red dot on the rewind tab. Thread film between gate and gate tab. Close the gate.

Eastman Kodak Company

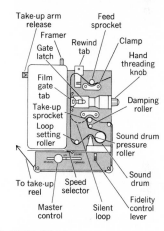

3. Thread the film under the loop-setting roller; the film should just touch the roller. Thread over the sound-drum pressure roller and under the sound drum; then thread behind the damping roller, over the sprocket, and out. Be sure the film sprocket holes engage the sprocket teeth.

4. Press down on the loop-setting roller as far as it will go, then release it; this action forms both top and bottom loops correctly.

5. Feed around the snubber roller and under the two rollers on the bottom of the master control cover. Attach the film end to the take-up rool.

SHOWING THE FILM

Refer to instructions on page 399. A fidelity control lever adjusts sound for optimum quality. To operate the projector in reverse, move the master control lever to REVERSE-MOTOR and, if desired, to REVERSE-LAMP.

AFTER THE SHOWING

☐ *To rewind:* Attach the film from the take-up reel to the empty feed reel. Move the master control to REWIND; set the speed selector at SOUND. Lower the rewind tab to its horizontal position.

☐ *To put away the projector:* Remove the take-up arm drive belt, raise the arm slightly, push the arm release button, and lower the arm to its storage position. Swing the supply arm down, lower the projector, stow speaker and power cords, and replace the cover.

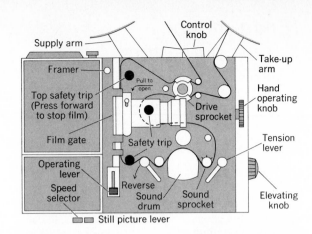

Victor projectors

Kalart-Victor

SETTING UP

Follow the general procedure outlined on page 398. Unlatch and lift off the case top. The speaker is in the detachable door of the case. Pull the two reel arms up; swing the supply arm to the rear and the take-up arm to the front.

THREADING

1. Open the film gate by pulling on the release at the top of the gate.

2. Remove about 3 feet of film from the supply reel and attach the end to the take-up reel. Turn the reel at least two turns to completely engage the film.

3. Push down on the tension levers on either side of the sound drum. Place the film in the groove in the drum. Thread the film around the sound sprocket roller to the right of the drum, engaging the teeth in the sprocket holes. Thread the film tightly around the roller to the left of the drum. Raise both tension levers.

4. Thread the film up to the take-up reel following the path shown. Allow ample clearance around the safety trip behind the lens. Engage sprocket holes with teeth at the bottom of the drive sprocket.

5. Allow a loop of one finger below the lower safety trip. Place the film in the film channel and close the gate. Threadup to the supply reel, engaging the film in the top of the drive sprocket.

6. Raise the operating lever. Turn the hand operating knob to check threading.

SHOWING THE FILM

Refer to instructions on page 399. To operate the projector in reverse, press down on the rear tension lever with the projector running. To project a single frame, push either the upper or lower safety trip to the right. With the projector stopped, hold down the still picture lever.

AFTER THE SHOWING

☐ *To rewind:* The reverse control must be in the forward position. Release the top safety trip. Thread film back to the supply reel. Set the control knob at REWIND. Turn on the motor and raise the operating lever. After the film is rewound, return the control knob to OPERATING POSITION.

☐ *To put away the projector:* Release the reel arms by rotating the control knob to the ARM RELEASE position. Swing the arms upright and push them down. Return the control knob to OPERATING POSITION. Replace speaker door, cord, and case top.

Singer Education Systems projectors

Singer

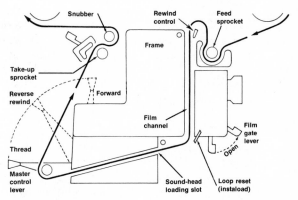

SETTING UP

Follow the general procedure described on page 398. Swing the reel arms into their operating positions. Aim and focus the projector.

MANUAL THREADING

1. The film gate lever is at the bottom of the lens housing. Swing this lever to the right to open the feed sprocket guard and the film channel.

2. Place the master control lever at THREAD.

3. Insert the film under the feed sprocket, engaging sprocket holes in the sprocket teeth. Place the film in the film channel; leave an upper loop approximately 1½ inches high. Close the film gate lever.

4. Insert the film in the sound-head loading slot and bring the film under the idler roller on the master control lever.

5. Open the take-up sprocket (swing the lock lever to the left). Pass the film between the shoe and the sprocket and around the snubber; then attach the head of the film leader in the take-up reel.

6. Set the master control lever at FORWARD. Test threading by rotating the manual advance knob clockwise.

SHOWING THE FILM

Refer to page 399 for general procedures. To start the projector, push the RUN button, then the LAMP button. Adjust focus and sound.

AFTER THE SHOWING

☐ *To rewind:* Set the master control lever at REVERSE/REWIND. Thread the film straight across from the take-up reel (left) to the feed reel (right). Push the RUN button, and pull the REWIND control out.

☐ *To put away the projector:* Follow the procedure outlined on page 400.

INSTANT LOADING MODELS

Set up the projector as described above for manual threading machines. With the master control lever and the REWIND control set at LOAD, draw the film along the film path; it will be guided automatically into position.

To show the film, raise the control lever to PROJECT; run the projector as described above. If the film is stopped during the showing, push the loop reset lever before starting again.

Rewind as for manual threading models.

Special features available on some instant loading models are:

☐ High speed forward, which allows skipping ahead to film segments.

☐ A film counter, which facilitates location of film segments selected for presentation.

Singer

Viewlex

Viewlex and RCA projectors

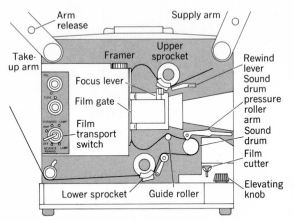

Arm release · Supply arm · Take-up arm · Framer · Upper sprocket · Focus lever · Film gate · Film transport switch · Lower sprocket · Guide roller · Rewind lever · Sound drum pressure roller arm · Sound drum · Film cutter · Elevating knob

SETTING UP

Refer to the general procedure outlined on page 398. Press arm release buttons to raise reel arms.

MANUAL THREADING

1. Unwind about 5 feet of film.

2. Swing open the film gate. Place the film in the film channel and close the film gate.

3. Form a loop above the film gate. Depress the upper sprocket shoe and insert the film.

4. Take the free end of the film over the guide roller, forming a loop as indicated by the guide marker.

5. Depress the sound-drum pressure-roller arm and thread the film around the sound drum and stabilizer roller. Continue around the lower sprocket.

6. Pass the film under the snubber roller and up to the take-up reel. Wrap a full turn or more of film on the reel.

SHOWING THE FILM

Follow the procedure outlined on page 399. Note that a loop-restorer lever and a lens lock are provided.

AFTER THE SHOWING

☐ *To rewind:* wrap the end of the film around the supply reel, counterclockwise. Raise the rewind lever and turn the film transport switch to REVERSE/REWIND. When the film is rewound, turn the projector off and lower the rewind lever.

☐ *To put away the projector:* follow the procedure on page 400.

AUTOMATIC THREADING MODELS

Set up the projector as described for manual threading models. Thread as follows:

1. Be sure the film gate and exciter lamp cover are closed and the rewind lever is down.

2. Push the safe threader in toward the machine.

3. Insert about 2 inches of film into the cutting slot (CS). Engage a sprocket hole in the locating pin, depress the cutter lever, and release.

4. Turn the film transport switch to FORWARD.

5. Insert the film end in the threading slot (TS).

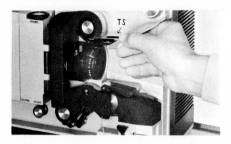

6. When 2 feet or so of leader have passed out of the rear of the machine, turn the film transport switch OFF. Complete threading on to the take-up reel.

7. Disengage the safe threader by pulling it out to the first stop.

Procedures for operating, rewinding, and putting away are like those for manual threading models. To unthread the film, remove the safe threader by pulling it away from the projector.

Bell & Howell projectors

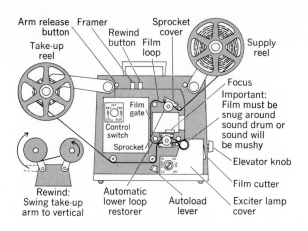

Arm release button · Framer · Rewind button · Sprocket cover · Take-up reel · Film loop · Supply reel · Focus · Film gate · Control switch · Sprocket · **Important: Film must be snug around sound drum or sound will be mushy** · Elevator knob · Film cutter · Exciter lamp cover · Rewind: Swing take-up arm to vertical · Automatic lower loop restorer · Autoload lever

Bell & Howell

SETTING UP

Follow the general procedures outlined on page 398. Press the arm release buttons to swing the reel arms into position. Set the sound-silent switch with the projector running.

MANUAL THREADING

1. Unroll about 5 feet of film from the supply reel. Swing the lens out to open the film gate. Open the three sprocket covers.

2. Thread the film over the top sprocket and close the cover. Thread the film channel, adjusting the upper loop to the case marking. Close the gate.

3. Thread the film around the automatic loop restorer and over the lower sprocket. Close the top sprocket cover.

4. Use special care to draw the film taut around the sound drum before engaging sprocket holes in the lower side of the bottom sprocket wheel; then close the sprocket cover.

5. Continue threading to the take-up reel.

SHOWING THE FILM

Refer to procedures outlined on page 399. To project film in reverse, turn the control switch past REVERSE to LAMP.

AFTER THE SHOWING

☐ *To rewind:* Swing the take-up reel arm to vertical. Thread the film across to the front reel and put two or three turns of film on it (counterclockwise). Set the control switch on REVERSE; press and hold down the rewind button for a moment. Turn off the projector when the film is rewound.

☐ *To put away the projector:* Refer to page 400.

AUTOLOAD MODELS

Set up autoloading models in the same way as manual threading machines. Thread as follows:

1. Inspect the first several feet of film; it must be undamaged and free of tape or obstructions. Trim the end of the leader in the film cutter on the front of the machine.

2. Turn the control switch to FORWARD.

3. Push the autoload lever forward until it locks in position.

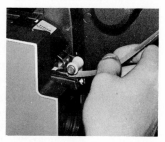

4. Insert the film in the threading slot until it engages the upper sprocket.

5. When about 2 feet of film have come out the back of the machine, pull lightly on the loose end of the film until a click is heard, then stop the projector and attach the film to the take-up reel.

Procedures for operating, rewinding, and putting away the projector are like those for manual threading models.

To unthread film manually, open the threading mechanism door and remove the exciter lamp cover. Turn the supply reel to provide slack in the film. Starting at the upper sprocket, ease the film out of the film path.

8mm film projectors

New developments in projectors have stimulated interest in 8mm sound and silent film. The economy, ease of operation, and convenient size and weight of these projectors, together with improved methods of packaging film, have made these machines attractive devices for use in instruction.

Since 8mm projectors operate on the same principle as 16mm machines, only minimal instructions are provided here; it is recommended that you obtain manufacturers' manuals for the models you use and refer to these during practice. Several models are described here to illustrate the variety of 8mm projector configurations and means of threading or loading.

International Audio Visual Freeway 16mm sound projector

Although at press time a designer's drawing, below, was the only available illustration of the Freeway 16mm projector, the prototype unit was near completion and the several components have been demonstrated. Because of the features promised for it, the inclusion of this projector seems appropriate, for the specifications create an image of a projector often wished for by teachers and students. By the time you read this book, you will know whether, or to what extent, the Freeway 16 has lived up to its advance publicity. And the published specifications, below, provided by the manufacturer are a useful basis for an exercise in projector evaluation. Among the promised features are:

☐ Drop-in-the-slot threading. Quiet operation.

☐ No sprocket wheels, claws, intermittent action, or mechanical shutter mechanism. Extended film life.

☐ Film speed selection: one to fifty frames per second, flicker-free, forward or reverse.

☐ Flying optical scanner and window; no gate.

☐ Solid state sound reader; no sound drum.

☐ No reel arms or external reel drives.

☐ Low silhouette. Estimated weight: 22 pounds.

☐ Internal (or external) speaker; microphone input.

☐ Modular components; long periods between servicing; unconditional two-year guarantee.

☐ Long lamp life; superior screen illumination.

Kodak Supermatic 70 sound projector

This projector is designed for viewing and hearing magnetic sound Super 8 film loaded in Kodak Supermatic cassettes. The image may be projected on a small Ektalite screen, suitable for carrel or small group use; or, with the screen lid dropped, the image can be projected on a wall screen. A companion model—the Supermatic 60—has no sound recording capability, but is otherwise similar to the model 70. A feature of

Eastman Kodak Company

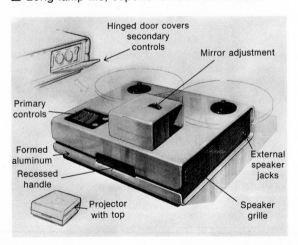

Hinged door covers secondary controls

Mirror adjustment

Primary controls

Formed aluminum

Recessed handle

External speaker jacks

Projector with top

Speaker grille

this machine is the Kodak Supermatic cassette which contains the film and facilitates loading the projector. Film is automatically drawn from the cassette, threaded, and attached to an internal take-up spindle. At the end of a showing, the film returns into its cassette.

To thread the Supermatic 70, raise the lid/screen cover, and move the spindle right in its slot to the position corresponding to the cassette size to be used: to AB for 50- to 100-foot cassettes, to C for 200-foot, and to D for 400-foot cassettes.

Place the cassette in the well and on the spindle in the center, and position with the notch of the cassette over the guide in the projector well. Move the master control lever from OFF to STILL, pause a moment, then move the control to FORWARD. If an image does not appear on the screen in a few seconds, return the control to THREAD and then, again, to FORWARD.

Technicolor Super 8 Optical Sound projector

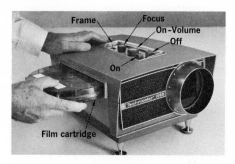

Frame   Focus   On-Volume   Off   On   Film cartridge

Unlike the projector just described, this machine reproduces sound from optical tracks. The cartridge (marked clearly with the word TOP and an arrow to indicate direction of insertion) is pushed firmly into the slot until a loud click is heard. The projector will not start unless the cartridge is properly inserted.

The usual controls for projector operation are on the top of the machine. To stop the projector before the end of a film, push the red OFF button firmly. *Do not* remove the cartridge while the machine is running. An outlet is provided at the rear of the projector for connecting an extension speaker or headphones.

other 8mm projectors

Among the cartridge-loading projectors for silent Super 8 film is the familiar Technicolor unit shown here in a viewer for independent study. The Technicolor cartridge carries a 3-minute film in an endless loop.

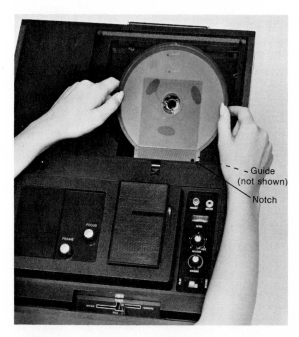

Guide (not shown)
Notch

THREAD
REVIEW
FORWARD
STILL OFF

The Kodak Ektagraphic 120 also provides cassette loading and automatic threading. This machine, as well as the Supermatic 170 sound projector, pulls film from a Kodak Supermatic cassette into the projector where it is threaded and attached to an internal take-up reel. When the film ends, it is rewound into the cartridge automatically. This silent projector has a film capacity limited to 50- and 100-foot Kodak Supermatic cassettes.

*Eastman
Kodak
Company*

Bell & Howell manufactures a number of 8mm projectors, among them this Filmosound 8 which reproduces sound from magnetic tracks. This machine and other models can project *both* Super 8 and regular 8mm film. Other models are cartridge-loading.

Bell & Howell

Fairchild Camera and Instrument Corporation offers Super 8 sound projectors in two distinct configurations. The Cart-Reel 70 series for individual or small-group viewing, upper right, is equipped with an earphone jack, plays sound from magnetic tracks, and is loaded from the front with Fairchild film cartridges.

This Fairchild-Eumig Model 711 R threads automatically from a Kodak film cartridge.

Several untraditional sound motion-picture film machines are now available. The Beseler Cue/See, below, uses standard audio cassettes to carry both sound and pulses to advance the film at any rate from that of still-picture projection up and beyond sound speed. Such machines can be programmed for independent study or for group presentations and permit interesting and creative productions.

Beseler Cue/See

A recent Eastman Kodak Company development, the Supermatic Film Videoplayer, below, plays Super 8mm color sound film through a television receiver, or for transmission over ITFS or closed circuit television distribution systems.

Eastman Kodak Company

television equipment

television receivers

Typical classroom television receivers are operated similarly to home receivers. In the classroom, place receivers so that light from windows and lighting fixtures does not reflect from the receiver screen. The lower edge of the screen should be at least 4 feet above the floor. Rearrange seating if viewing can be improved.

Controls on classroom receivers, as on home sets, allow adjustment for fine tuning, brightness, contrast, and vertical framing. Obtain optimum picture and sound quality by these steps:

1. Select station and adjust fine tuning.

2. Adjust brightness and contrast.

3. Adjust sound volume; adjust tone control if provided.

Color television receivers have only two principal adjustments beyond those for black-and-white:

☐ TINT Adjust tint to provide the most natural tone for flesh (neither too green nor too blue).

☐ COLOR Adjust color to provide an intensity of color to suit personal taste, always seeking the most natural effect. When the color adjustment is too high, colors are unnaturally bright and the picture may appear grainy (noise in the picture); when the color adjustment is too low, colors tend to be weak or to disappear.

simple television cameras

For several reasons, general instructions on how to operate simple television cameras are not practical:

Hewlett Packard

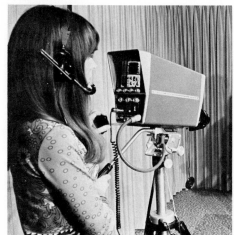

cameras differ markedly according to brand, type, design, and cost. For example, some cameras are fully automatic in electronic adjustments; others have adjustments such as beam, target, and electronic focus mounted on the camera case. For simple guidelines, here are a few points applicable to one type of camera which has a zoom lens for image magnification and which can be used with a simple video tape recorder:

1. Mount the camera on a firm table stand or tripod and attach the lens to the camera.

2. Connect the camera output to the monitor or receiver.

3. Turn on power to the camera and monitor.

4. Uncap the lens. *Do not point the camera at lights or the sun!*

5. Though most cameras have automatic adjustments of these functions, if controls are provided, turn up the target control halfway and turn up the beam until you can see details in the highlights.

6. Zoom the lens all the way out (wide viewing angle) and adjust the vidicon mechanical control at the back of the camera for best focus. Then zoom the lens all the way in (closeup view) and adjust focus for maximum sharpness. The picture should now stay in focus throughout the zoom range from wide shot to closeup *for that subject.*

7. Adjust the monitor for the best possible picture, a compromise of adjustments of both camera and monitor.

video tape recording systems

As with television cameras, video tape recorders differ in complexity, in operating procedures, and in cost according to brand, type, and designed purpose. Machines are not universally standardized in features such as compatibility with other machines and in tape width used (tape may be ¼, ½, ¾, 1, or 2 inches wide). Idiosyncrasies of cameras and recorders are learned by studying the manuals accompanying them and by practicing with them under the guidance of an experienced operator. More than any other audiovisual equipment, video tape recording systems require full training.

However, television technology is developing rapidly, and equipment is increasingly simple, standardized, and dependable. Previous experience with audio equipment and with photographic techniques will be of help to the operators of television recording equipment. In their basic functions, video tape recorders operate much like audio tape recorders, and television cameras function much like still or motion-picture cameras. Each video tape recording system, however, has individual characteristics; one can become familiar with a particular system by referring to the instructions in the operator's manual. The following checklist is typical of lists supplied with equipment.

Sony Corporation of America

SETTING UP

1. Set up and interconnect the system components:
   a. Camera output to recorder input
   b. Microphone to microphone input
   c. Recorder output to monitor input
2. Thread the tape (heads should not be turning).
3. Turn on power to all units.
4. Uncap the camera lens and aim the camera at a subject. Press the RECORD button on the video tape recorder to obtain a picture on the monitor. If a video level control is provided on the recorder, adjust it to the recommended meter reading.
5. Focus the camera, adjust the aperture, and adjust monitor brightness and contrast controls.
6. Have someone speak into the microphone; adjust the audio recording level on the recorder.
7. Set the index counter to 000.

OPERATING

1. Put the recorder into the record mode (usually FORWARD and RECORD). Record a short sequence with both picture and sound. Stop the machine; then rewind the tape.
2. Set control to PLAY.
3. If tracking and tension controls are provided:
   a. Set tracking to provide a clear, noise-free picture.
   b. Set tension control for minimum hooking at the top of the picture.
4. Be sure no more adjustments are necessary before making a complete recording.

practical pointers

☐ Always make and play back a short test recording before making a complete recording to ensure that all components are operating properly. When doing a series of short recordings, replay the end of the last recording to be sure all is working well.

☐ Failure to get a satisfactory picture may indicate that the recording heads need cleaning. Follow procedures recommended by the equipment manufacturer.

☐ When difficulties are encountered, start by checking operating procedures you have followed. Recheck for proper and complete connections and for properly adjusted knobs, levers, and switches.

☐ All moving components in the machine should come to a complete stop before another mode is activated. Do not halt the tape by pressing on the reels; avoid cinching or otherwise causing tape slippage.

video tape care

Video tape on reels should be handled gently and kept perfectly clean. Video tape in cassettes is adequately protected from most damage. Here are some recommended practices for handling and storing tapes.

☐ Tape should be kept entirely dust-free; any dust or dirt on it will cause white or black flashes on the receiver screen. Use tape in a clean environment; avoid smoking in the recording area.

☐ *Do not* remove tape from guides, capstan, or head drum when you have played partially through a program. Wind all tape onto either the supply or the take-up reel first to avoid harm to the tape.

☐ Handle tape by reaching under the reel from two sides and lifting. Do not lift by the top flange or squeeze the flanges together. Cut off ends of the tape that have become creased with use.

☐ Label reels of video tape and cassettes with care. Use pregummed labels on both reels and cassettes and on the boxes. Include date recorded, title of program, performers, recordist, and other essential information. Do not write on the tape end tab when it is on the reel; the pressure dents many layers of tape, thus causing poor recording and playback in the dented areas.

☐ Return reels of tape and cassettes to their boxes immediately after taking them off the recorder. Store tape and cassette boxes vertically on shelves.

☐ When storing tape, avoid extremes of temperature and humidity such as closed car trunks in summer and damp basements in winter.

Hewlett Packard

Sony Corporation of America

equipment maintenance

Although equipment maintenance and repair services are usually provided by your institution, users of the equipment have responsibilities for its care. This section treats information useful to those with equipment operating assignments and includes preventive maintenance and emergency servicing.

general suggestions

☐ Use equipment only for the purposes for which it is designed.

☐ Don't force equipment. If machines do not perform properly and your operational procedures have been correct, report the failure immediately to the appropriate office.

☐ If screws or knobs become loose or if equipment parts seem to be improperly fitted, make sure they are promptly and correctly adjusted.

☐ Always keep equipment covered when not in use, and keep the external portions clean.

☐ Make regular checks of all cords and connectors for loose plug tips, worn or broken insulation, or other indicators of possible problems.

☐ Whenever equipment needing repair or adjustment is returned for service, attach a note to the unit describing the faults as fully and accurately as possible.

☐ In emergencies you may need to make simple adjustments or minor repairs to equipment. In addition to special skills, equipment servicing requires special tools and instruments, spare parts, technical manuals, and convenient bench space where competent technicians can do necessary electrical, electronic, and mechanical work. Unless you are expert in equipment repairing, do not attempt to make major adjustments in components or to disassemble them. Leave such work to experts.

☐ Always remember that many apparent equipment failures may be caused by errors in operation: cords are not fully plugged in; the speaker is not connected; controls are not in the proper position; a threading path has not been followed accurately. Recheck procedures before calling for help.

audio equipment

RECORD PLAYERS

Though generally quite sturdy, school record players are often subject to heavy use and sometimes abuse. Observe these points:

☐ Assist maintenance personnel by observing and reporting any malfunctions or indications of probable faults, such as unreliable turntable speed, friction or resistance in tone-arm movement, apparent excessive noise from the stylus, noises in the sound during play or when controls are operated.

☐ Arrange to have the styluses examined under a microscope at intervals, at least twice a year if a machine is used often. Have styluses and cartridges checked immediately if a tone arm is dropped heavily or banged about during transportation.

TAPE RECORDERS

Recommendations for general care of equipment apply, of course, to any tape recorder, though here are several special pointers about their care:

☐ Clean the heads in your recorder at regular intervals, at least every twenty-five hours. Check them for accumulations of dirt and tape coating every ten hours, and clean if necessary. At the same time, clean the tape drive roller, capstan, and tape

guides. Rub lightly with denatured alcohol and a cotton-tipped swab. *Never* use metal or any hard material (such as fingernails) to clean heads.

☐ To maintain noise-free recording, demagnetize the recorder heads after every 25 to 30 hours of operation. A special electrical device, a "demagnetizer" or "degausser," is made for this purpose.

projection equipment

Though most servicing of projectors will be done by maintenance personnel, there are a number of routine and some emergency services that may be done by operators.

LAMP REPLACEMENTS

Projector lamps are extremely efficient devices, but quite expensive. When replacing lamps, consider the following pointers:

☐ Use only the lamp or lamps specified by the manufacturer. If lamps of more than one wattage are approved for use in a machine, select the one with the lowest wattage that provides a satisfactory picture under expected viewing conditions.

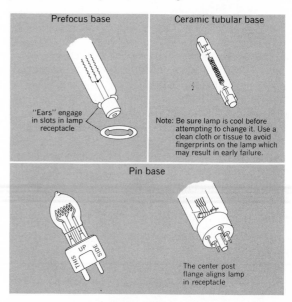

☐ Never, in an attempt to get a brighter picture, install a lamp of higher wattage than that specified as maximum by the manufacturer. Note the different shapes, voltages, and orientation of lamp filaments in the samples shown.

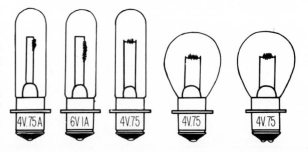

☐ Exciter lamps for motion-picture projectors must be of the *exact* type specified by the manufacturer.

☐ Always disconnect the power cord from its receptacle before attempting any change of lamps.

☐ Always handle lamps by the base when possible, or use a piece of cloth or a special lamp holder. Quartz-type lamps should always be handled by the ends and preferably with a cloth.

☐ Worn-out projection lamps should be replaced to obtain maximum light and to prevent excessive swelling and malformation of the glass envelope.

☐ Lamps must be set firmly and positioned correctly in their bases to obtain maximum brilliance and evenness of illumination on the screen. When a lamp is not fully turned to lock in its socket or is incompletely seated, shadows of the filament may be projected on the screen. A correctly positioned lamp produces an evenly lit picture area if all internal components are also properly fitted.

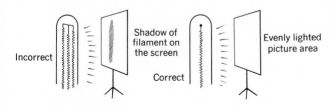

Incorrect Shadow of filament on the screen Correct Evenly lighted picture area

LENS CARE

The optical glass in projector lenses and condensers is easily scratched. Clean lenses only when absolutely necessary, and then only with non-silicone-treated projection lens tissue (available at camera and audiovisual dealers). Moisten the lens surface by breathing on it; then wipe carefully, with a circular motion. Never use handkerchiefs, coarse tissues, or cloth. Keep fingers off lenses; oil deposited will collect grit and impair projection quality.

☐ Never attempt to take a lens apart to clean it. Leave cleaning of the condenser lenses and reflecting mirrors inside the machine to technicians.

television equipment

As with other equipment—optical, electrical, electronic, and mechanical—care should be exercised in the use of television equipment. Any attempts to maintain or adjust the equipment should be restricted to those external controls or parts that are for operating the equipment.

☐ Never attempt to work inside a television receiver—to clean the inside of the safety glass, for example. Even though the receiver is turned off, there may be an electrical shock hazard.

☐ Video tape recorders, like audio recorders, require regular service by maintenance personnel.

☐ Recording heads will need occasional cleaning; need for cleaning *may* be indicated by a faulty picture. Clean the heads *only* with materials and methods indicated by the manufacturer. Sony Corporation of America, for example, provides special cleaning tips for reel-to-reel recorders; note the direction of motion. For U-matic cassette machines, a special cleaning cassette is provided; instructions must be followed exactly.

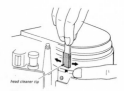

head cleaner tip

lubrication of equipment

Many machines, especially video recorders, do not require lubrication, but for those that do, a few rules will be of help:

☐ Use *only* the grade and type of lubricant that is specified by the manufacturer. *Never* use household or similar oils or synthetic lubricants unless specifically named by the manufacturer.

☐ Follow lubrication instructions in the equipment manual. When oil is required, it is best to oil frequently, but *very little* at a time.

☐ Always wipe excess oil or other lubricant from any part of the equipment. Keep it clean!

REMEMBER . . .

Never *force* controls on any audiovisual equipment. Determine what should turn, when it should turn, how much it should turn, and how much effort is normally required. The strong-arm method of making equipment operate usually compounds trouble.

REFERENCE SECTION

2 DUPLICATING PROCESSES

contents

introduction

Efficient management of the processes of duplicating verbal and visual instructional materials remains an important teacher responsibility. Teachers and students find numerous uses for duplicated materials—for fill-in workbook assignments, contract assignment direction sheets, study guides, points-to-look-for lists, newsletters, examinations, and literally hundreds of similar purposes.

Because such uses of duplicated materials are so common, teachers are customarily expected to (1) know the characteristics of various widely used duplicating processes, (2) know the special advantages and disadvantages of these processes and which to recommend or choose for specific classroom duplicated-materials requirements, (3) be able to prepare master copy for reproducing materials by such means, and (4) operate various types of duplicating equipment.

forms of reproduction

The four principal forms of reproducing multiple copies of various printed, typed, hand-lettered, or hand-drawn items for the classroom are:

☐ SPIRIT DUPLICATING, HECTOGRAPHING This is the well-known process employing aniline-dye carbon papers and a wood- and grain-alcohol fluid to produce masters from which up to 300 or so satisfactory transfer copies can be made.

☐ MIMEOGRAPHING Involves use of the wax-coated stencil through which permanent ink passes, permitting runs as high as 5,000 copies from one master. Modern variations in the preparation of the mimeograph stencil (to be discussed briefly in a later section) have greatly expanded applications of this process in education.

☐ PHOTOCOPYING An especially rapidly expanding field of duplication involving several different process systems: diazo, infrared, dye-transfer, diffusion-transfer, and electrostatic. Details of two of these—electrostatic and infrared—are given here.

☐ PRINTING Also a rapidly changing field of reproduction (particularly offset printing), with numerous applications to the mounting communication needs of classroom teachers and students.

Although other duplicating forms may be mentioned, these four are most commonly encountered in schools.

general pointers in preparing materials

Several general pointers, if observed, will facilitate preparation of good masters for spirit duplicating, mimeograph, photocopy, or offset printing:

☐ ORGANIZE MATERIALS Edit carefully. Insert heads where needed; be consistent in your wording. Be sure your presentation scans—that its statements have parallel construction and that, if you use the outline form, it is correct and consist-ent. Perhaps it will be possible to use contrasting type to emphasize section heads.

☐ CLEAN YOUR TYPEWRITER A clean typewriter is essential for producing good copy. Brush keys carefully or use plastic gum-cleaning materials. Test for smudging or filled-in type faces by typing a few sample lines.

☐ USE A GOOD TYPEWRITER—CORRECTLY An electric typewriter gives best results. If you use a manual typewriter, employ an even, staccato touch and type more slowly than your normal speed.

☐ DEVELOP AN ATTRACTIVE LAYOUT To plan spacing, it is a good practice to type a preliminary rough copy first. Examine this draft for spacing adjustments you can make as you prepare the finished version. Leave enough margin so that copy is not cropped in the duplication process. Use space between blocks of copy to separate ideas or items. Where appropriate, use capital letters, underline, vary type styles (as is possible with the IBM ball-type typewriter), make guide-lettered headings, or rule boxes to set off headings or copy. Use a ruler to produce straight lines.

☐ JUSTIFY LINES FOR BLOCKED EFFECTS You may sometimes wish to have your typed lines appear "in line" on the right-hand margins, as in printing. To achieve this effect on an ordinary manual or electric typewriter, set margins to desired width—and type, but not beyond this width. Fill out extra spaces with X or *:

```
Typed lines which have an**
even right-hand margin give
your copy a look which is**
```

Then go back over your copy and indicate by pencil checks where you can skip spaces. Spread these skips so they do not occur directly under one another, making rivers of white space:

```
Typed˅lines which˅have an**
even right-hand ma˅rgin give
your copy˅a look˅which is**
```

Then retype the copy, skipping the spaces you have marked with checks:

```
Typed  lines which  have an
even right-hand margin give
your copy  a look  which is
```

☐ PROOFREAD AS YOU GO Or at least make necessary corrections before removing the copy from your typewriter.

☐ CHOOSE THE MOST APPROPRIATE REPRODUCTION METHOD It is uneconomical, for example, to prepare a paper offset master to reproduce only ten or fifteen copies of a simple typewritten statement. Photocopies usually cost more per sheet than those produced in other forms, but they are quite economical for short runs.

spirit duplicating

To duplicate copies by the spirit hectographic process, you will use a rotating-drum machine, a special fluid, and a spirit master. If proper grades of master paper, carbon, and fluid are combined, approximately three hundred legible copies may be run from one master.

Ditto, Inc.

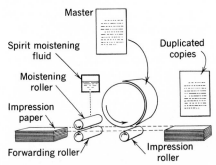

THE SPIRIT (FLUID HECTOGRAPH) PROCESS

preparing the master

A first requirement for satisfactory reproduction with spirit duplicating equipment is a good master. Although preprinted masters (those printed by reproducible carbon) are available for many areas of the curriculum, the majority of masters in use in schools will be teacher- or student-prepared.

To produce a spirit master from which copies may be made, follow these steps:

1. PREPARE THE MASTER PACK The master pack consists of a sheet of spirit master paper and a sheet of spirit carbon paper (with the carbon side up). A backing sheet will ensure clear, sharp copy.

2. INSERT MASTER PACK INTO TYPEWRITER As you see it in the typewriter, the master paper will be nearest to you (on top), the carbon will be next (carbon side up), and the backing sheet, if used, will be against the typewriter platen. Check alignment to be certain that the typing on the master pack will be straight across (parallel with the top edge).

3. TYPE THE MATERIAL Type with the ribbon in normal position (*not* on "stencil"). Leave at least ½ inch of margin at top and bottom and adequate side margins—at least ¾ inch. Don't crowd copy; leave enough white space. Experiment for best key pressure.

4. MAKING CORRECTIONS It is quite easy to correct mistakes on spirit masters. When you do make an error, you simply have spirit carbon where it is not wanted. To make necessary corrections:

☐ IF YOU DO NOT NEED TO REPLACE LETTERS OR WORDS OVER THE MISTAKE (1) cut them out with a razor blade, a sharp knife, or a small pair of scissors; (2) cover them with cellophane tape, a piece of gummed label, or with a special pencil supplied for this purpose by the manufacturer; or (3) block them out. Also use any one of these methods to eliminate extra letters or other unwanted items.

Ditto, Inc.

and t they	and t they	and they
CUT IT OUT or	COVER IT UP or	BLOCK IT OUT

1. To lay out the material, fold back the carbon sheet from the master sheet. Sketch your drawing

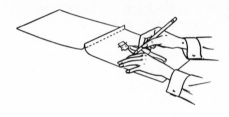

☐ IF YOU MUST MAKE A CORRECTION AT THE POINT OF ERROR First eliminate the error. This you do by (1) lightly scraping off the carbon with a razor blade or a curved scalpel, (2) coating the error with the special pencil referred to above, or (3) erasing the error thoroughly.

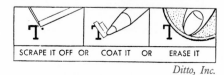

SCRAPE IT OFF OR   COAT IT   OR   ERASE IT

Ditto, Inc.

☐ AFTER ELIMINATING THE ERROR, TYPE THE CORRECTION Remember that you have already used up the carbon at the point of error, so new carbon will be needed there. Experts often tear a small piece of the carbon from a corner of the master for making such corrections in the place where retyping is needed. Remove this slip before going on.

☐ USE SPECIAL SHORTCUTS WHEN APPROPRIATE Scrape off parts of letters with a razor blade, for example, to produce other letters (to change E to F, or e to c, or others). Or cut masters in two, excising entire sections, and then splice them together with cellophane tape. Working over fresh carbon, use a ball-point pen or hard lead pencil to touch up letters (after removing the master from the typewriter), to change commas to semicolons or periods to colons, to strengthen letters, to insert periods, or to make other similar changes. With some letters you may simply strike over them (changing c to e, h to b, and others).

ADDING HAND-DRAWN MATERIALS Frequently teachers and students prepare spirit masters completely by hand or add hand-drawn materials to typed masters using these techniques:

FREEHAND DRAWING AND LETTERING You can add interesting visual materials and increase the communicative power of your spirit duplicated materials by inserting freehand drawings and lettering. To do this, simply follow the directions that are given in the adjoining column.

or lettering lightly in pencil on the face of the master sheet. (Allow about ½ inch at the top for the duplicator clamp.)

2. Fold the carbon back under the master sheet, leaving the slip sheet out. Go over the drawing or lettering on the master, using a sharp, hard pencil or a ball-point pen on a firm smooth surface.

☐ TRACING When transferring illustrations (from magazines, clip art, previously prepared drawings, etc.) or lettering onto masters:
1. Remove the slip sheet from the master.
2. Attach the material to the master sheet with tape.
3. Trace over the lines to be reproduced with a pencil or ball-point pen, using firm, even pressure. The image will appear (reversed, like a mirror image) on the back of the master only.

Remember that color brightens duplicated materials. Spirit carbons are commonly purple, but special carbons are also available in red, blue, green, or black. You may obtain helpful special effects by using one of these carbons for the entire copy or by changing carbons for different copy sections. Color is usually reserved for highlighting important items, such as headings, and for specialized materials in certain fields of study (for instance, indicating losses in materials used in teaching bookkeeping). To have part of a master typed or drawn in red, blue, green, or black, prepare the master in the regular way; but when you come to the part you wish to be in another color, substitute a colored carbon for the regular carbon.

making thermal spirit masters

Clean, sharp-looking copies can be produced at low cost by using thermal spirit masters. This process is frequently used to make duplicate copies of charts, line drawings, newspaper and magazine articles, and complex illustrations as well as ordinary typed copy.

To transfer such copy to a thermal master:

1. Place material to be copied under the thermal spirit master, as shown. Remove the slip sheet.

is fluid in the tank. Place paper in the feed tray and adjust paper grippers and guides to the proper width. It may help to run some blank paper through the machine before attaching the master, thus checking positioning and adjustment of grippers.

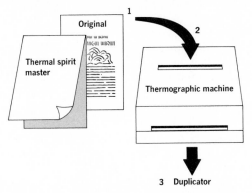

2. Insert this "sandwich" in a thermographic machine (Thermofax, A. B. Dick, others). In only a very few seconds, the heat will cause the carbon dyes to transfer to the master.

3. Carefully peel the master away from the carbon.

4. Attach the master in the usual manner to an ordinary spirit duplicator. Run up to a hundred or more copies, as desired.

operating the spirit duplicator

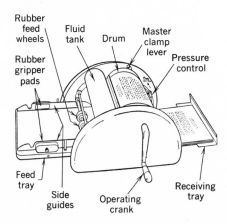

1. READY THE DUPLICATOR It may be a good idea to ask an experienced operator for help the first few times you use the machine. Be sure there

2. ADJUST THE MACHINE Adjust the pressure lever. Be sure the wick is moist. If the machine has been idle for some time, the wick may be dry and need priming. If the wick is very wet, you will produce darker copies—but fewer of them—before the master must be discarded or recharged.

3. ATTACH THE MASTER Turn the handle clockwise to the 6 o'clock position (straight down). Depress the clamp lever to open the clamp that holds the master on the drum. Insert the top edge of the master into the opening (about ¼ inch) so that the side with the carbon faces *up* when the drum is turned. Then return the clamp lever to its normal locked position.

stances should you use them to cleanse the skin. Continued breathing of methanol vapors may cause illness; do not use spirit duplicating equipment in small, poorly ventilated rooms.

4. RUN THE COPIES With some machines it will be necessary to turn the handle first counterclockwise to the 4 o'clock position and then clockwise. Each time you turn it all the way around, you will make a copy from the master. Check flow of spirit to avoid smudgy copies.

A. B. Dick Company

5. REMOVE THE MASTER Stop the handle at the 6 o'clock position. Open the clamp lever, lift out the master without touching the carbon, and close the clamp lever (unless another master is to be run immediately). File the master after stapling to it a specimen piece of the run to protect the carbon side. Set both the copy-control lever and the pressure-control lever at zero. Replace dust cover.

One safety pointer should be mentioned. Most spirit fluids contain methyl alcohol. Avoid undue exposure of the skin to them. Under no circum-

mimeographing

The mimeograph, or stencil, duplicating process has four elements: stencil, ink, paper, and mimeograph machine. The stencil is typed, handwritten, drawn, or reproduced electronically. It is placed on the outside of an inked cylinder. As the paper goes through the mimeograph between cylinder and impression roller, ink flows from inside the cylinder *through* the inking pad and the stencil to produce a printed impression. It is possible to reproduce as many as 5,000 copies from a single well-prepared stencil.

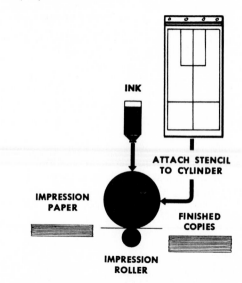

The procedures described here are intended to assist you in learning how to prepare mimeograph stencils and operate the machine.

preparing stencils mechanically

To prepare copy for mimeographing, you need only a mimeograph stencil assembly and some means of imaging it. General suggestions for preparing master copy for reproduction apply to the preparation of mimeograph stencils. Additional pointers follow:

1. SET YOUR TYPEWRITER RIBBON Shift your typewriter ribbon so that letters will strike directly on the stencil rather than on the ribbon. With a typewriter without a "stencil" setting, remove the ribbon entirely.

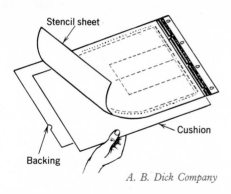

Stencil sheet

Cushion

Backing

A. B. Dick Company

2. INSERT THE STENCIL Place a cushion sheet (supplied in the stencil package) between the stencil sheet and the backing sheet. Leave the backing sheet attached. Insert this assembly into your typewriter, wax stencil side *up* (toward you).

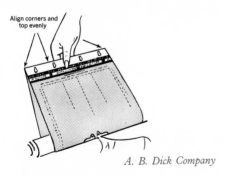

Align corners and top evenly

A. B. Dick Company

3. ALIGN STENCIL AND SET TYPEWRITER MARGINS Align stencil corners and top evenly. Set typewriter side margins within markings printed in light ink on the stencil face. Also observe top and bottom margins, being careful to keep within the 8½- by 11-inch area shown unless you intend to make the run on legal-size (8½- by 14-inch) paper.
4. TYPE THE STENCIL Type with a firm staccato touch. Make necessary adjustments on the typewriter to provide proper key pressure for cutting the stencil. Be sure to watch the hollow letters (o, b, a, others), which tend to fall apart if stroked too heavily.
5. MAKING CORRECTIONS If you make mistakes while typing or drawing on stencils, it is quite easy to correct them.

☐ IF YOU ARE USING A STENCIL WITH A COATED WHITE CUSHION, apply correction fluid over the error, allow it to dry, and retype (with a lighter than normal touch). Coat each character separately, using vertical brush strokes.

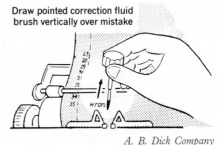

Draw pointed correction fluid brush vertically over mistake

A. B. Dick Company

☐ IF YOU ARE USING A STENCIL WITH A TISSUE (NONCOATED) CUSHION, burnish mistakes by gently rubbing them with the glass rod packed with each bottle of correction fluid. Then apply fluid, allow it to dry thoroughly, and retype, as above.

If you find the number of repairs needed for your stencil has seriously weakened it or splotched its appearance, you may salvage part of it by cutting it apart and cementing a new portion to it. Small mistakes may also be covered with correction fluid (from the top side as you read the stencil).

ADDING HAND-DRAWN MATERIALS Lines, simple drawings, and lettering can be added to the stencil with a stencil stylus. Place the stencil over your layout against a window or on a glass drawing surface illuminated from beneath. Trace lines with the stylus.

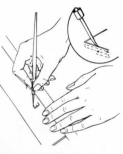

A. B. Dick Company

preparing stencils electronically

An electronic mimeograph stencil is produced by a scanning process involving two cylinders side-by-side. The original copy is laid around one cylinder and a blank stencil around the other. The electronic scanner automatically copies lines of dots in the original, transferring them as microscopic perforations to the stencil. These perforations later permit ink to pass through, re-creating the original image on duplicating paper. The process is capable of producing quite faithful facsimiles of lettering, typing, or illustrations.

A. B. Dick Company

One electronic process widely used by schools permits the preparation of an overhead transparency and a duplicating stencil in one operation. In this case, a clear transparency sheet is placed between the blank stencil and the backing sheet. The remainder of the stencil-making process is the same as that described above. The resulting transparency can be projected immediately. Duplicated paper copies of the same materials can also be given out for student notebooks.

Hagerman (Idaho) Schools

operating the mimeograph

As with the spirit duplicator discussed earlier, several basic steps should be taken in operating the mimeograph machine, as follows:

1. CHECK THE INK SUPPLY See that the drum contains enough ink for the length of run anticipated. Use the ink-measuring rod to determine ink level. Add ink if necessary.

2. ATTACH STENCIL Place the stencil face down on the revolving drum, with the stiff backing sheet *up,* toward you. Clamp the stencil top into place,

A. B. Dick Company

tear off the backing sheet, and clamp the lower portion of the stencil to the cylinder. This stencil "tail" need not be clamped if the blockout is used. Place the blockout strip (about 6 to 9 inches long and 8½ inches wide) over the lower section of the ink pad. This will keep the stencil tail free from ink

A. B. Dick Company

for easy, clean removal. The lower 2 or 3 inches of the blockout must be clamped *under* the stencil tail clamp. It can stay in that position—on the machine—from one run to the next.

3. LOAD THE PAPER SUPPLY Adjust the paper feeder, being sure that wider paper is on top of the pile if there are differences in the paper lots being run at one time.

4. ADJUST THE MACHINE Run a few trial copies and adjust margins up, down, or sideways as needed. Set the counter to ring automatically when the desired number of copies has been run.

photocopying

Recent improvements in photocopy processes and equipment have done much to aid teachers and students in gaining access to valuable verbal and visual materials. Machines described here represent two widely used photocopy processes: electrostatic (xerography) and infrared.

suggestions regarding photocopying

The following suggestions regarding the handling of copy will facilitate uses of either of the two processes:

☐ Be sure to remove all paper clips and staples from copy before commencing the photocopy operation.

☐ Check copy for dog-eared, tissue-thin, or off-size paper or for other irregular items that will require special handling. Patented copy holders (a woven sheath cover, for example) may be used to protect such items against snagging or fouling as they pass through the machine.

5. MAKE THE RUN From time to time, check the quality of impressions. Stand by the machine to stop it if paper jams or if other difficulties arise. During the run, adjust the ink flow as needed.

6. LEAVE THE MACHINE READY FOR THE NEXT USER If the stencil is not to be used again, throw it away. If it is, file it for future use. Place a protective cover directly over the ink pad, sealing it carefully. Check all adjustments to see that they are left in the normal position. Place cylinder in the stop position and replace the dust cover.

☐ Prepare typed copy with a good black carbon or cloth ribbon. As with all other typewritten preparations, be sure keys are clean.

☐ Check copy to be sure it is as free as possible of smudges and marks or poorly made corrections.

☐ Leave a suitable margin around the edges of the copy to avoid crowded effects or the possibility of having some of it cropped off in the process.

☐ Remember that unscreened photographs (glossy black-and-white or color prints, for example) generally reproduce poorly. But screened photographs, such as those appearing in newspapers or magazines, will reproduce quite well.

☐ Set the special quantity dial in advance. Save money (and time) by avoiding a careless double or triple printing before you discover the machine is set for more than the number of copies you need.

Additional information concerning the features of various types of equipment, duplicating supplies, and operating techniques may be obtained from major producers of equipment in this field (addresses of which will be found in Reference Section 6).

Copying books (bound documents) by the photocopy method is relatively simple. Saddle-stitched publications often can be taken apart and reproduced in nearly any photocopier. The copying of bound books intact, however, requires a machine having such capability.

423

Xerox 4000

The Xerox 4000 photocopy machine reproduces up to forty-five copies per minute of nearly any kind of document—from newspaper clippings to pages from books. A dial may be set to produce from one to ninety-nine copies automatically; a lighted counter records the number of copies run. To operate:

1. Lift the document cover and place the original face down, as shown.

2. Close the cover and dial the number of copies desired.

3. Press the START PRINT button.

The Xerox 4000 permits interchanging paper measuring 8 by 10 inches to 8¼ by 14 inches without removing the paper stacks, printing of light originals, and printing automatically on two sides of the paper. A special counter may also be attached to register numbers of copies reproduced for different individuals or departments. This machine will print on paper ranging from 16-pound bond to 32-pound card stock and can also be used to print transparencies.

3M 209 (infrared)

The 3M 209 typifies infrared copying machines for same-size copying of single sheets and bound materials. The step processes below apply to its operations:

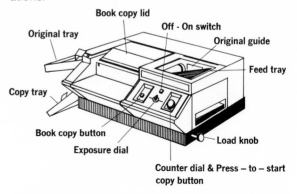

1. Turn on the OFF-ON switch at the start of the day; turn off at end. Push in LOAD KNOB at start of day; pull out at end.

2. Adjust copy registration to keep the image centered up and down on the copy (eliminating a darkened top and bottom margin). One complete turn of the knob moves the image about ⅜ inch.

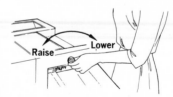

3. For lighter copy, turn exposure dial toward LIGHTER; for a darker copy, turn toward DARKER. It is a good idea to check the first copy and readjust as necessary for lightness or darkness. The dial may be readjusted during the run to affect the quality of the next copy.

4. To make from one to twenty-five copies of any normal original, size 8½ by 11 inches to 8½ by 14 inches:

a. Insert original face down in feed tray. Place paper edge flush against original guide. Original must be flat and free of staples or clips.

b. Adjust counter dial for number of copies. Dial setting can be increased during the run.

c. Press PUSH-TO-START copy button. Copies are automatically made and exited to bottom tray; originals exit into upper tray just before last copy is made.

5. To copy from books one page at a time, follow this procedure:

    a. Push BOOK-COPY button.

    b. Follow lighted instruction. Raise lid, insert only one page (as shown), close lid. Push button again.

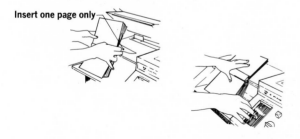

Insert one page only

    c. Wait for and then follow the next lighted instruction. Raise lid, remove bound document, close lid. Push button again. One copy is automatically made and exited to tray.

offset printing

The offset printing process (more accurately, "offset lithography") originated in Germany in about 1800 with the discovery that an image drawn with greasy pigment on a certain kind of stone would retain printing ink while the ink was repelled by remaining moistened areas. In this century the process has been refined; today offset lithography is a *primary* process for printing.

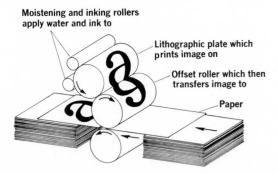

Moistening and inking rollers apply water and ink to

Lithographic plate which prints image on

Offset roller which then transfers image to

Paper

Teachers and students will usually be involved with offset printing only in the matter of preparing copy for the masters from which multiple copies are printed. The masters may be developed as any of the following:

☐ PAPER MASTERS A paper offset master is simply a sheet of rather heavy paper with a special coating that renders its surface receptive to both grease-base images and moisture. An offset master can be made by the *direct image* method—using typewriter, ball-point pen, special pencil, crayon, rubber stamp, or carbon paper. Each of these media contains an oil-based substance which attracts ink on the offset machine. Masters can also be made by the *electrostatic* method (transfer or direct). The Xerox photocopy machine is often used for this purpose. In addition, *transfer* methods (gelatin or photo) will image masters from original copy. In such cases, a light-sensitive sheet receives the image through a camera.

☐ METAL PLATES Metal plates, which offer the advantage of longer runs (more copies) and higher-quality printing, are also popular. With this medium, copy is usually pasted up in a layout that is the same as or directly proportional to the desired size of the finished run. With the *presensitized-plate method,* a film negative is made of the original copy through a special camera. At this stage the size of the original can be enlarged or reduced as desired. The negative image is then transferred (through exposure under an intense light) to a sensitized metal plate. This plate, when developed, becomes the finished offset master.

preparing typed paper masters

There are several important pointers to be observed in producing paper masters for offset printing:

1. Do all your imaging on the white side of the paper master.

2. Use a good typewriter. Although an electric typewriter will produce the best image, any typewriter in good condition can be used.

3. Be sure the typewriter platen, feed rolls, and bail rolls are clean and free of grease.

4. Adjust the paper bail rolls so that they ride outside the area to be covered by the image but be sure they are on the master, thus keeping the master assembly flat when the typewriter keys strike it.

5. Clean the typewriter carefully.

6. Take care not to crease or crack the master when inserting it into the typewriter. (Cracks or creases cannot be removed and will print as fine lines.)

7. Be sure your draft copy is complete before commencing to type.

With a manual typewriter, use a uniform touch—just enough to deposit a clear, unbroken image.

THIS TOUCH IS TOO HEAVY.

THIS TOUCH IS CORRECT.

With an electric typewriter, first make a test master by using different key pressures for a few lines of copy. Select the touch that gives best results. When the copy is printed, letters will usually tend to appear slightly darker.

For a maximum run, allow the typed image to set for thirty minutes. Whenever possible, do all typing first—before adding drawing, handwriting.

Making corrections on typed paper offset plates is relatively simple:

☐ WHEN TYPING WITH A FABRIC RIBBON, erase lightly with a picking action. You need not remove the stain or ghost left from the first image. This will not reproduce.

☐ WHEN TYPING WITH A CARBON RIBBON, use a "gritty" eraser. Remove the image as completely as possible, including the clear waxy substance.

If in the process of making corrections you have removed the master from your typewriter, place a sheet of paper over it before reinserting it. This will protect the image from damage by the feed rolls. Remember that any greasy substance which is allowed to contact the master may print.

drawing and writing on paper masters

Special pens and pencils may be used to write or draw on paper offset masters. Nonreproducing pencils may be used to make guide marks or layout schema of various kinds on copy without fear they will show up in reproductions. Reproducing pencils (easy to erase) can be used to add handwriting, cross-hatching, rules, or underscored lines. The printed copy reproduces with the same pencillike effect. Ball-point pens (containing special reproducing inks), carbon paper, crayons, and drawing fluids can also be used.

When using pencils or ball-point pens, place the master on a hard, smooth surface. This reduces chances of embossing, which can result in poor printing.

preparing copy for metal plates

In preparing copy to be reproduced, eventually, through the medium of a metal offset plate, several procedures are recommended:

☐ TEXT The text of the copy for metal plates may be prepared as "hot" or "cold" type. With hot type, characters are set in metal; cold type involves, for example, photo typesetting, special typewriters, and press-on letters. A large variety of letter styles and sizes are available in both hot and cold types. The cold-type media are usually more economical and convenient for school use.

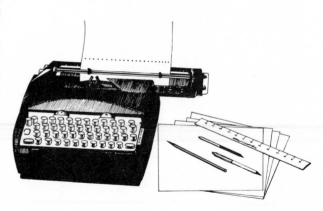

LETTERING TECHNIQUES AND STYLES

TYPEWRITER ─────────────────────
TECHNOLOGICAL INNOVATIONS IN EDUCATION - 12 LETTERS PER INCH (ELITE)
TECHNOLOGICAL INNOVATIONS IN EDUCATION - 10 LETTERS PER INCH (PICA)
TECHNOLOGICAL INNOVATIONS IN EDUCATION - MID-CENTURY IBM EXECUTIVE EXPANDED
TECHNOLOGICAL INNOVATIONS IN EDUCATION - ORATOR IBM SELECTRIC
TECHNOLOGICAL INNOVATIONS IN E-PRIMARY TYPEWRITER

MECHANICAL ──────────────────────
TECHNOLOGICAL INNOVATIONS IN EDUCATION — LEROY 175C
TECHNOLOGICAL INNOVATIONS IN EDUC — LEROY 200C
TECHNOLOGICAL INNOVATIONS IN — LEROY 240C

TRANSFER ─────────────────────────
TECHNOLOGICAL INNOVATIONS IN EDUCATION – NEWS GOTHIC 14 PT
TECHNOLOGICAL INNOVATIONS IN EDUCA - NEWS GOTHIC 18 PT
TECHNOLOGICAL INNOVATION - NEWS GOTHIC 24 PT

TECHNOLOGICAL INNOVATIONS IN EDUCATION - 24 PT CONDENSED
TECHNOLOGICAL INNOVATIONS IN EDUCA - 24 ITALIC
TECHNOLOGICAL INNOVATIONS IN EDU - 24 BOLD

□ FINISHED PASTE-UP To prepare a finished paste-up, begin by drawing the exact shape of the paper on which the item will be printed, using a nonphotographing (light blue) pencil on heavy white paper. Use rubber cement to paste down headlines and text exactly where they are to appear. Line them up with a T square or straight

□ ENLARGING AND REDUCING There may be occasions when it is desirable to have all or part of the printed copy larger or smaller than the original. In such cases, full directions to the cameraman should accompany the paste-up, clearly indicating the size of finished copy you wish. It is recommended that tissues be placed over text and illustrations to show the exact measurements, proportions, or shapes desired. Blocks of typeset and typewritten material can be enlarged or reduced, but care should be exercised to avoid a finished text that is (1) too small to read or (2) so enlarged that unevenness of type or drawings is exaggerated. The best rule, generally, is to set type copy in finished size and to enlarge or reduce only when necessary.

edge. Paste on in their proper locations any line illustrations to be used. If a photograph (*not* a line drawing) is to be printed, block out the area it will occupy with red masking film or black paper. Key the area to the glossy photograph which is to accompany the paste-up. Protect the finished paste-up with a tissue overlay. Be sure all instructions and separate photographs and illustrations accompany the paste-up when it is given to the printer.

copyrights and copying

Increased use of copying and offset printing processes of the types discussed here has intensified problems relating to copyright infringements. Teachers and others who plan to use copies of printed materials developed and copyrighted by others should be familiar with basic copyright regulations. Reference to current literature and to federal law about the subject is strongly recommended.

handling duplicated materials

Teachers intending to plan, prepare, and instruct with duplicated materials of various sorts are reminded that their usefulness will be increased through attention to such simple details as the following:

☐ USE THREE-HOLE PUNCHED PAPER This permits easy notebook filing by students.

☐ USE DIFFERENT COLORS OF PAPER STOCK It facilitates recognition of items and adds interesting variety in appearance.

☐ STAPLE MATERIALS IN ORDER A single staple in the upper-left-hand corner is usually all that is required.

☐ DEVELOP A SYSTEM FOR HANDING OUT DUPLICATED MATERIALS It is often better to do this one unit at a time (one page covering a complete topic or assignment or several pages in a section, for example) rather than all at once. This procedure maintains a proper focus of attention on the items to be studied.

☐ READ THROUGH THE MATERIAL WITH THE CLASS When you read it, at least in overview fashion, you can show its organization and content and can correct orally any minor errors which may have slipped by.

☐ WHEN USING DUPLICATED TESTS, GIVE STUDENTS SEPARATE ANSWER SHEETS This makes it unnecessary to have them mark original copies.

REFERENCE SECTION

3 PHOTOGRAPHIC EQUIPMENT AND TECHNIQUES

contents

introduction

This reference section provides information that teachers and others will find useful in applying basic photography techniques to their work. It is not a substitute for other more complete presentations of the subject. Rather, it is a technical extension of our earlier discussion (Chapter 8) of the ways to use photography in teaching and learning.

Information presented here includes: (1) what to look for in a camera to ensure that it meets your own personal-professional requirements, (2) comparative characteristics of common types of film and alternative ways of processing them, (3) techniques for composing and shooting useful pictures, and (4) several simple but important recommendations regarding copying and close-up work, titling, and film splicing.

Reference Section 6 lists other publications in which you will find additional valuable information on these and related photographic topics.

selecting a camera

In selecting a camera, the most important questions to ask are: Is it easy and convenient to use? Will it provide pictures of the sizes and in the formats you require (as prints or as 2- by 2-inch transparencies, for example)? Are costs of the camera, of film, and of processing pictures for it reasonable?

Cameras range in complexity from those that are fully automatic, with a selection of interchangeable lenses and accessories, to simple aim-and-shoot models. Generally, though, the more complex and expensive cameras are the more versatile; they also require more photographic skill of their users.

You will probably prefer to start with a simple camera—one with which you can get experience and discover your personal requirements.

format choices

☐ STILL CAMERAS The size of the picture or negative a still camera takes affects costs for film and processing and sizes to which pictures may be enlarged satisfactorily. Cameras which produce 2- by 2-inch slides will satisfy most requirements of teachers and students. Slides may be made in any one of the four still camera formats shown here:

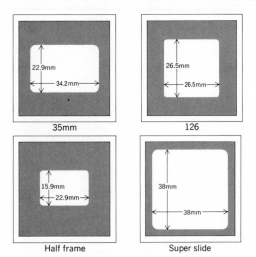

35mm   126

Half frame   Super slide

You will have the best selection of film, and the lowest cost per exposure, if you select a camera

employing either the 35mm or the 126 (cartridge) format.

☐ MOTION-PICTURE CAMERAS Of the two motion-picture formats in common use today, the Super 8 size offers the best economy for most school applications. 8mm equipment is cheaper than 16mm equipment; film cost per unit of running time for 8mm film is about half the cost of film in the 16mm size.

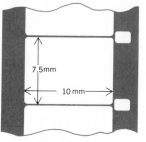

16mm format   Super 8 format

lens speed

The light-transmitting qualities of your camera lens will determine to some extent where and how you will be able to take pictures. Lens speed is determined by the size of the *maximum opening* (aperture) of the lens. A large aperture, like a large window, will allow more light to pass to the film, so that exposures can be shorter (faster) than when the opening is small.

Lens speed is expressed in f/ numbers. Common speeds run from f/1.4 (fast) to f/11 or slower. (The smaller the number, the faster the lens.) Lens speed is embossed on the front ring of the lens housing, as shown in the photograph below. Sometimes lens speed is stated as a ratio (1:2, 1:1.4); the number after the colon can be read as the f/ number indicating the speed.

Fast lenses are more costly than slower lenses,

but fast lenses enable you to make exposures under adverse lighting conditions and with slow films. With a slow lens you will often have to rely on flash or other artificial light, perhaps even out-of-doors. To avoid this problem, select a camera with an f/4 or faster lens.

shutters

The subjects you wish to photograph will influence the characteristics of the shutter you will require for your camera. The single shutter speed of the typical inexpensive camera—usually 1/50 to 1/100 of a second—is just fast enough to permit you to hand hold the camera and to stop the action of subjects moving slowly toward or away from you. But to stop subjects in fast motion, or those passing in front of the camera at 90 degrees, will require shutter speeds up to 1/500 of a second. Most medium-priced and a few lower-priced cameras provide such speeds.

If you expect to make long exposures in dim light (such as pictures in moonlight or other night scenes) or to use slow films, a *Bulb* or *Time* setting on your shutter is likely to be required. With these settings, the shutter can be kept open for any desired length of time.

The shutter may be a *focal plane* type, a sliding curtain located in the camera body; or it may be a *leaf* shutter, part of the camera lens unit. A focal plane shutter is controlled by a knob on the camera, as on the type shown below. The lens shown on the preceding page has a leaf shutter; its speeds are controlled by turning a ring marked with shutter speeds for bulb (B) to 1/500 second (500).

Motion-picture cameras may be equipped with shutters having variable speed controls. Some will expose the film in a range of frames per second, producing slow-motion or time-lapse effects when it is projected. But if you wish to produce animated

motion pictures, you must use a camera which exposes single frames. Check for this feature on any camera you consider purchasing or using.

After considerable use, camera shutters may get out of adjustment, causing incorrect exposures. A reputable camera repairman will check shutter action at little or no cost to determine whether the shutter is functioning properly.

viewfinders and focusing

The camera you select will have a viewfinder in which you will frame and focus the image. Test the finder with your eye (especially if you wear glasses) to see that it is appropriate for you. Two types of viewfinder systems are common:

☐ THROUGH-THE-LENS VIEWING With this system, the scene is viewed through the camera lens. You see exactly what the lens sees and that is what will be recorded on the film. To focus, you adjust the lens-to-film distance until the viewfinder image is sharp.

☐ OPTICAL VIEWFINDERS With the optical viewfinder system, you view the scene through a window near (but separate from) the lens. Focusing may be by (1) *range finder* (which requires aligning a split or superimposed image in the finder) or (2) by *zone focusing* (which requires estimating subject distance and setting the lens at the appropriate range or zone). Optical viewfinders are found on most cameras which do not take interchangeable lenses.

If you expect to use supplementary lenses or extension tubes or rings with your camera, through-the-lens viewing is practically a must. Through-the-lens viewing also helps you to avoid such pitfalls as forgetting to remove the lens cap or failing to compensate for parallax before shooting close-up pictures.

other features

Other features may influence your selection: provision for automatic or semiautomatic exposure control, exposure with flash, and interchangeable lenses, for example. For help in your choice, consult your local photo supply dealer. Types of cameras most likely to fulfill needs of teachers and students are described on the following pages.

A

camera types

still cameras

The variety and number of cameras on the market often seem to present an impossible maze to aspiring photographers. However, the following analysis of camera types may help to simplify the selection problem. Essentially, there are four basic types of camera of particular importance to the teacher-photographer:

B

A. INSTANT-LOADING CAMERAS These feature cartridge loading (126 format); require minimum if any adjustment for taking pictures; and on some even the film advances automatically. Most models have fully automatic exposure control; some have "no go" exposure prevention when light is insufficient or built-in flash that fires automatically when additional light is needed. Most have optical viewfinders with zone focusing. Such features are available on models costing less than $50.

B. COMPACT 35mm CAMERAS To obtain full-frame 35mm slides from a pocket-size camera, you may select one of the 35mm compacts. Most of these have automatic exposure controls with override options for shutter speed or aperture adjustment. They usually do not take interchangeable lenses and are equipped with optical viewfinders. Lenses are slightly shorter (wider angle) than the normal lens for other 35mm cameras. Because of their fast, wide-angle lenses, inconspicuous size, and semiautomatic operation, compacts are especially suited for "shooting from the hip" to capture candid and documentary photographs. Compacts are available for under $100.

C. 35mm REFLEX CAMERAS 35mm reflex camera systems offer maximum flexibility in small-format cameras. Through-the-lens viewing permits accurate framing with a variety of lenses and is especially helpful, if coupled with built-in light metering, for close-up photography. Besides the basic camera, 35mm systems include a wide selection of lenses and special-purpose attachments that permit each photographer to acquire, piece by piece, the items he needs. Costs of basic 35mm cameras range widely; however, a teacher-photographer can start

C

a camera system of good quality with an investment of around $200.

D. POLAROID CAMERAS The Polaroid process features instantaneous picture production—in black-and-white or color prints. The print is processed immediately after exposure. While no negative is produced by this arrangement, Polaroid prints can be copied with good results. Cameras are not expensive, but cost per exposure is higher than is usual for conventional films.

E. OTHER STILL CAMERAS Larger-size negative cameras include 2¼- by 2¼-inch single- and twin-lens reflex cameras, view cameras, and press cameras. At the other end of the negative size spectrum are half-frame 35mm and subminiature cameras. Each type of large- and small-format camera has its own special characteristics suiting it for particular uses. Students and teachers may

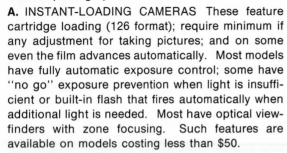

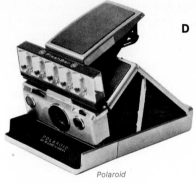

D

Polaroid

E

Bell & Howell

GAF

within the school budget—the cost of 50 feet of Super 8 color reversal film is approximately the same as for twenty 35mm color slides. Films with speeds up to ASA 160 are now available—with these, indoor shooting with available light is possible.

Shooting in the 16mm format may be desirable when films are to be shown to large groups or on television. Although more expensive than 8mm film, the larger 16mm frame permits satisfactory projection on a larger screen. With 16mm cameras, such as the one shown, several high-quality lenses can be mounted on the turret, or the camera can be equipped with a single zoom lens. Automatic exposure systems are also available on 16mm motion-picture cameras.

With some motion-picture cameras, the operator may focus on subjects as close as an inch or so from the lens. Variable speed adjustments also permit him to prolong actions and to resume normal speed without stopping. Fades, dissolves, single-frame exposures, and other special effects are also possible with some models. Versatile titling devices are available which permit the camera operator to make titles and other motion-picture footage from still pictures such as postcards, travel folders, and 2- by 2-inch slides.

Eastman Kodak Company

Paillard, Inc. (Bolex)

have occasion to supplement their own cameras by borrowing one of these, if available, from their school media center. For more information about these and the still cameras mentioned above, consult appropriate items in Reference Section 6.

motion-picture cameras

Of the two motion-picture formats likely to be used for film production by students and teachers, 8mm is the most popular today. Compact models of 8mm motion-picture cameras often come equipped with zoom lenses, built-in exposure controls, and through-the-lens viewing systems. They make motion-picture photography a fairly simple aim-and-shoot operation. The relatively low cost of 8mm film and equipment places this size format

selecting film

Black-and-white and color films come packaged in paper rolls, cartridges, cassettes, sheets, and film packs as well as on reels. Films also differ in characteristics such as speed, color sensitivity, contrast, and grain. Some of these variations are discussed below.

film types

Several distinct film types are available. Those that teachers and students will most often use for *still photography* are:

☐ BLACK-AND-WHITE NEGATIVE FILM for making black-and-white positive prints on paper, for direct viewing, and on film, for projection.

☐ COLOR NEGATIVE FILM for making color prints. Color transparencies and black-and-white prints can also be made from color negatives. Color negative films always have *color* in their names (Koda*color,* Agfa*color*).

☐ COLOR REVERSAL FILM for making color transparencies. Color prints can also be made from transparencies or transparencies can be copied on black-and-white negative film for printing in black and white. Color reversal films always have *chrome* in their names (Ansco*chrome,* Ekta*chrome*).

Films most often used in school *motion-picture photography* are reversal films, in black and white or color, which are processed directly into positives for projection. Negative-type films are available in 16mm only for use when work prints are to be made for editing.

The many special-purpose films (Polaroid, infrared, and copy films, for example) will not be discussed here; information about them is available in photographic stores.

film speeds

Different films require different amounts of light for proper exposure. This quality is expressed as *film speed,* given as an ASA (American Standards Association) rating. The rating usually appears on the film container. Common film speeds vary from ASA 25 (slow) to ASA 500 or higher (fast).

Your choice of a film of a certain speed will depend in part on the light available for exposure. For example, a film speed of ASA 200 or higher will be required to make photographs with a hand-held camera in an average artificially lighted classroom. In bright sun, fast films require extremely short exposures at small apertures. Because of the finer grain of the slower films, it is generally a good rule to use the slowest film which has adequate speed for the light in which it is to be exposed.

When poor lighting conditions or rapid subject motion require fast film speeds, you may choose a film with a high ASA rating (faster film); or you can expose some films at ASA ratings higher than the rating indicated (usually about double that number). Remember that all shots on a roll of film must be exposed at the same ASA rating. Also, if you increase the ASA, special laboratory processing will be required. Inquire at your photo store for further information on this special service and how to order and pay for it.

color characteristics of film

The several different kinds of black-and-white and color films react differently to the colors in light. Panchromatic black-and-white film, for example, is sensitive to all colors; it must therefore be processed in total darkness. Color films also have color sensitivity characteristics to consider when selecting film. Color emulsions are balanced—some for use in daylight, others for use in artificial light. If a color film is to be exposed in light different from that for which it was designed, special filters must be used. (See the data sheet packaged with the film.)

The most economical and reliable source of light for color shooting is the *midday sun.* Daylight color films can be used for copy stand work out-of-doors. Electronic flash units produce light having the same color characteristics as sunlight, thus making it possible to expose daylight-type color films with them without special filtering.

Even when color films are exposed in their specified light, color renditions may be "off." For example, early morning and late afternoon sunlight records on color film as redder than midday sun, for which daylight film is balanced. Photofloods blacken with age and emit redder light. When accurate color reproduction is important, avoid these conditions.

exposing film

The main problem in film exposure is to allow enough, but not too much, light to reach the film to create the desired image in the emulsion. If your camera does not have built-in metering, use a light meter or consult a film data sheet for exposure settings. Once you understand the exposure variables, you can calculate exposures for many natural lighting conditions without these aids. If you wish, you can also manipulate the shutter and aperture controls to expose correctly for special effects, such as decreasing the depth of field (thus making background details fuzzy), or deliberately underexposing.

exposure variables

Four variables affect exposure in your camera: light (reflected from the subject), film speed, camera aperture, and camera shutter speed.

☐ LIGHT The exposure of film is caused by light reflected from the subject to the camera (not by the amount of light falling directly at the location of the camera).

☐ FILM SPEED As discussed earlier, different film types have different ASA ratings, reflecting their sensitivity to light. An ASA 100 film, for example, is four times as fast as an ASA 25 film and thus requires only one-fourth the exposure (length of exposure or size of lens opening).

☐ APERTURE The size of the lens diaphragm opening, or aperture, controls the amount of subject-reflected light that passes through the lens during exposure. Aperture settings are in a series of f/ numbers (f/22, f/16, f/11, f/8, f/5.6, and so on). These openings are standardized size increments called stops, each progressively halving in

area: the f/11 opening is half as large as the next stop, f/8. Therefore, half as much light reaches the film when the aperture is set at f/11 as when it is set at f/8.

☐ SHUTTER SPEED The length of time of exposure is controlled by setting the camera shutter speed. Shutter speeds are stated as fractions of seconds (1/15, 1/30, 1/60, and so on). The different shutter speeds are in intervals of time that approximately double; the exposure is twice as long (twice as much light reaches the film) when the shutter is set at 1/30 as when it is set at 1/60. With T (Time) and B (Bulb) settings, found on most cameras, the shutter can be opened indefinitely for exposures longer than the slowest shutter speed allows.

equivalent exposures

Note that the standardized increments for measuring the three exposure variables—film speed, aperture, and shutter speed—are approximately two to one. This allows easy juggling in the combination of the three to provide many equivalent exposures for a given light-reflecting subject. If you wish to use a smaller lens opening for greater depth of field, you need only to adjust shutter speed an equal number of increments to obtain an equivalent exposure. Some combinations producing equivalent exposures are shown in the table, which is based on an average subject photographed in bright sun.

TABLE OF EQUIVALENT EXPOSURES

FILM SPEEDS	APERTURE SETTINGS*							
	f/2	f/2.8	f/4	f/5.6	f/8	f/11	f/16	f/22
ASA 25	$\frac{1}{2,000}$	$\frac{1}{1,000}$	$\frac{1}{500}$	$\frac{1}{250}$	$\frac{1}{125}$	$\frac{1}{60}$	$\frac{1}{30}$	$\frac{1}{15}$
ASA 50			$\frac{1}{1,000}$	$\frac{1}{500}$	$\frac{1}{250}$	$\frac{1}{125}$	$\frac{1}{60}$	$\frac{1}{30}$
ASA 100	SHUTTER			$\frac{1}{1,000}$	$\frac{1}{500}$	$\frac{1}{250}$	$\frac{1}{125}$	$\frac{1}{60}$
ASA 200	SPEEDS				$\frac{1}{1,000}$	$\frac{1}{500}$	$\frac{1}{250}$	$\frac{1}{125}$
ASA 400						$\frac{1}{1,000}$	$\frac{1}{500}$	$\frac{1}{250}$

*All exposures on a line are equivalent for any film with that ASA rating.

shooting pictures

Basic steps recommended for taking still or motion pictures are outlined below. These procedures will apply to most shooting situations; interpret them to suit your equipment and needs.

1. *Get acquainted with the camera.* Before taking pictures for the first time with any camera, consult the instruction manual or someone experienced with its operation. Locate camera controls; observe the proper method in loading.

2. *Load the camera.* Adjust the metering system, if necessary, for the ASA rating of the film you are using.

3. *Focus the lens and frame the scene.* Use a tripod, if possible, to hold the camera in place while you focus and frame the subject. Adjust the focusing system (zone, range finder, or through-the-lens) to focus the lens on the center of interest. Remember that depth of field (the range within which the camera is in focus and will provide sharp details of objects being photographed) will be affected by lens aperture and distance—the smaller the lens opening, the greater the depth of field; the closer the object, the less the depth of field (see Camera Lenses).

In framing the scene, be sure important elements in the subject are in clear view. Experiment with camera angles and heights. Consider lighting, backgrounds, and foregrounds. With a motion-picture camera, practice any pans, zooms, or motions of subjects planned for the scene.

4. *Make the necessary exposure adjustments.* With fully automatic cameras, no exposure adjustment is necessary. Other cameras require adjustment of aperture and/or shutter speed until the metering system needle indicates proper exposure has been set. For cameras without built-in metering, exposure must be calculated and shutter and aperture settings made accordingly (see Exposing Film). Remember that, everything else being equal, the larger the lens opening, the faster the shutter speed required. Also, ultra-close-ups and exposures with filters will require increased exposures. (See Copying and Close-up Photography.)

5. *Expose the film.* Cock the shutter (if this has not been done previously by advancing the film). Press the shutter-release or cable-release plunger to make the exposure. If additional exposures are to be made immediately, advance the film or wind the camera drive mechanism.

6. *Record your exposure.* Note the exposure number, subject, type of film, shutter speed, aperture, and lighting conditions. (For example, your notation might read, "#1, boys with model planes, Plus X, 1/50 @ f/11, cloudy sky.") While this step is not essential, only by keeping a record of how each exposure was made will you be able to compare results, diagnose errors, and make corrections in later shooting.

SHOOTING CHECKLIST

Depending upon the type of camera you use, you will need to perform certain specific steps each time you take a picture. By making a brief checklist of the steps required for your camera, you can avoid omitting any. Here is a sample list for a typical still camera:

1. ADJUST METER FOR ASA OF FILM.
2. SET SHUTTER SPEED.
3. ADJUST APERTURE.
4. FOCUS LENS.
5. COCK SHUTTER.

---

recognizing errors

When you shoot pictures, you usually hope that your center of interest will be recorded sharply and rendered naturally in color or corresponding tones of gray. If, instead, the subject is too dark, too light, blurred, or otherwise unsatisfactory, the cause is often easily recognized. Some of the most common causes of disappointing photographs are illustrated on the opposite page, with suggested remedies.

Poor color rendition in prints and slides is usually due to incorrect exposure or to exposing film in the wrong light without proper filters. (See Selecting Film.)

Sometimes these "errors" in shooting can be used intentionally to create expressive pictures. For example, in special situations, subject motion, camera motion, or unnatural color might add impact to your photo image.

film processing services

Teachers and students usually send exposed still or motion-picture film to commercial laboratories to be processed. When processing is included in the cost of the film, it must be sent to laboratories specified on mailers packaged with it. For some color films, mailers can be purchased separately—or exposed rolls may be taken directly to a local photo supply dealer who provides for laboratory services.

Special directions may be written on your film mailing package, such as instructions for not mounting slides, requests for special processing for film exposed at a higher-than-normal ASA rating, and special mailing instructions. If not prepaid, you should include the extra cost of the last two services mentioned.

Customized processing for black and white and color is offered by some laboratories. Of course, you will pay for the individual attention given to special processing requirements. Some laboratories provide motion-picture processing, editing, and recording services; be sure you have obtained cost estimates in advance and that you have made clear your processing needs.

OVEREXPOSURE Subject too light, lack of detail in highlight areas. Consult your records, give less exposure under same conditions. Use a faster shutter speed or smaller aperture.

UNDEREXPOSURE Subject too dark, lack of detail in shadow areas. Give more exposure under same conditions. Use a slower shutter speed or larger aperture.

LIGHT LEAK Areas are light-struck (white). If this occurs throughout a roll of film or on more than one roll, check camera for light leaks.

OUT OF FOCUS Subject blurred, foreground or background sharp. Focus carefully on the center of interest.

CAMERA MOTION Entire picture blurred, no sharp areas. Support the camera on a tripod or steady it by other means. Or use a faster shutter speed.

SUBJECT MOTION Subject blurred, other objects at same distance sharp. Use a faster shutter speed or "follow" subject with camera.

camera lenses

For each camera size there is a lens of "normal" focal length—usually the lens which comes on the camera. On many cameras however, the normal lens is mounted so that it can be removed and lenses of other focal lengths substituted to provide wide-angle or telephoto effects.

The focal length of the lens determines the size at which images are reproduced on the film. Lens focal length is the distance from the approximate center of the lens to the film, measured when the lens is focused on infinity. Usually the focal length is embossed on the front ring of the lens housing, stated in millimeters (50mm, 80mm). Normal, wide-angle, and telephoto lens focal lengths for cameras of different sizes are shown in the table.

The figure below illustrates the effect of different focal lengths of lenses used on 35mm cameras. The same proportional differences in image size would occur with lenses specified for cameras of other sizes.

A *wide-angle lens* is frequently useful in photographing interior scenes where space is limited. As you see in the figure, the wider angle of view takes

TABLE OF LENS FOCAL LENGTHS

CAMERA SIZE	LENS FOCAL LENGTHS (mm)*		
	normal	wide angle	telephoto
8mm Super 8mm	12	6	38
16mm	25	15	50
Compact 35mm 126	38	—	—
35mm	50	35	90
$2\frac{1}{4} \times 2\frac{1}{4}$	80	50	120
120, $2\frac{1}{4} \times 3\frac{1}{4}$	105	65	180
4×5	150	75	200

*Focal length of the normal lens is approximately equal to the diagonal of the camera negative. Wide-angle lenses may be shorter than those specified above; telephoto lenses may be longer.

in a larger portion of the subject image on the negative; the effect is similar to moving away from the subject. The wide-angle lenses are useful out-of-doors where you cannot move freely to increase distance in shooting landscape panoramas.

The image produced by a long-focal-length (or *telephoto*) lens is larger than that made by the normal lens, because the narrower angle of view of the telephoto lens captures a smaller area of the subject image on the negative than does the normal lens. The effect is that of moving closer to the subject.

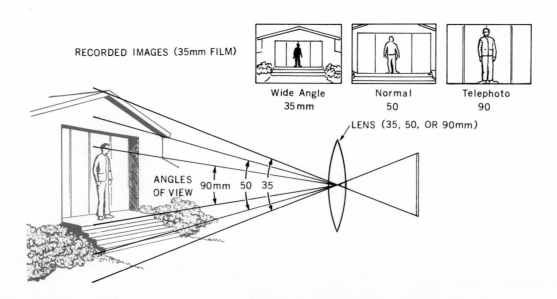

RECORDED IMAGES (35mm FILM)

Wide Angle 35mm · Normal 50 · Telephoto 90

LENS (35, 50, OR 90mm)

ANGLES OF VIEW 90mm 50 35

Wide-angle and telephoto lenses also produce pictorial effects different from those of normal lenses. Distortions, such as exaggeration in relative size and the compression of space, can be accomplished through the selection of lenses and shooting distances.

Robert Pitt

Wide angle lens

Telephoto lens

Zoom lens

Zoom lenses are also popular for use on both still and motion-picture cameras. With them, the camera operator is able to vary the apparent size of subject and width of field without varying the camera-to-subject distance. Nearly all Super 8 motion-picture cameras are equipped with zoom lenses. These can be adjusted for telephoto, normal, or wide-angle effects; some will also take extreme close-ups. Zoom lenses for 16mm motion-picture cameras and for still cameras are generally in the telephoto lens ranges.

depth of field

Picture sharpness often requires a carefully chosen depth of field. Thus, while not an exposure factor in itself, depth of field may influence your selection of shutter speed/aperture combinations.

Depth of field is defined as the "distance between the points nearest to and farthest from the camera

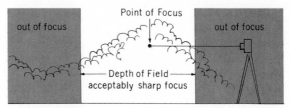

that are still acceptably sharp in the picture." With a given lens, depth of field (or range of acceptable sharpness) varies with the aperture used. With a large lens opening (f/3.5), the depth of field is much less than with a very small opening (f/16). In this way, the adjustable diaphragm on the lens allows some control in the sharpness of focus of objects at different distances from the camera.

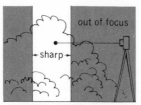

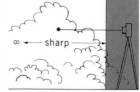

1/1000 sec. at f/2.8   1/30 sec. at f/16

A scale on most lenses indicates the depth of field for each aperture setting. (See the lens pictured in Exposing Film.)

copying and close-up photography

Imaginative teachers and students will have ideas for employing copying and close-up photography. For many copying tasks, you can use your basic camera mounted on a tripod. For copying small materials and for other close-up work, you may need accessory lenses or other attachments. Consult the reference list to supplement the following descriptions of techniques for this.

copying without special equipment

The construction of the camera and its normal lens limits the minimum distance at which the lens can be focused. At this distance, the lens takes in a given area, which is the smallest area that can be used to fill a negative (or slide) without using attachments. Determine the area size by measuring the area seen through the viewfinder. When you are making slides or motion-picture titles, plan or select subjects that are that size or larger—but never smaller. Of course, you can make copies of smaller materials for enlargement into prints, using only a portion of the negative.

close-up attachments

To zero in on small objects, use one of the following devices, alone or in combination: (a) inexpensive supplementary lenses (which fit over the standard camera lens and lengthen its focal length so that it reduces less); (b) extension tubes, or (c) bellows attachments which permit closer focusing of the standard lens; or (d) a special "macro" lens (sub-

A Supplementary lens	B Extension tubes	C Bellows	D Macro lens
Fits over camera lens	Used between camera and lens		Substitutes for camera lens

stituted for the standard lens) which is constructed to focus within inches of the subject.

Tubes, bellows, and macro lenses can be used only on cameras which accommodate interchangeable lenses. These devices require increased exposure. If your camera has through-the-lens metering, this is not a problem—you set it for exposures in the usual way. If it does not have this feature, consult charts supplied with accessories for factors for exposure increase.

framing

Framing of close-up shots is a simple procedure if yours is a single-lens reflex camera. If it is a twin-lens reflex camera or has an optical viewfinder, however, you must check effects of parallax (the difference between what you see through the viewing lens or viewfinder and what the taking lens sees from its different position). Take experimental photographs of a grid with marked center lines to determine how to aim your camera to compensate for parallax when making close-ups.

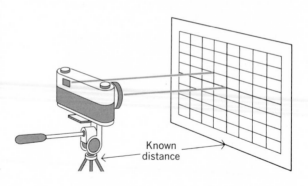

Known distance

copying procedures

Satisfactory copying or close-up photography can be done using the camera on a tripod. A pan-and-tilt head to support the camera is a convenience, as is a flat surface to support materials to be copied. To avoid unwanted perspective effects, adjust the camera position carefully so the lens axis is perpendicular to the plane of the copy surface.

Several types of copy stands are available for close-up photographic work. Some are designed for use with specific cameras only; most can be used with a number of different camera makes. With some stands, cameras are set at a fixed dis-

tance from objects to be photographed; others allow camera adjustment vertically; still others permit both horizontal and vertical adjustment to center the camera lens over the subject materials.

When copying, hold the subject materials flat to avoid distracting shadows and distortions. A sheet of glass, preferably the nonreflecting type, can be placed over paper materials to be copied to hold them tight against the easel.

When using photo-flood lights to illuminate copy, place them at each side of the copy so that the light shines at a 45 degree angle and evenly covers the entire area to be photographed.

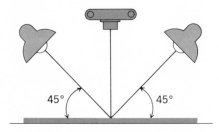

45°   45°

titling

Some of the titles you need for your slide sets, motion-picture films, and photo print series may be made by photographing street signs, billboards, or other items containing the information required. But if ready-made titles such as these are not available or are inconvenient to obtain, you and your students may produce your own titles and photograph them with simple copy equipment.

Lettering for titles may be produced in several ways discussed in Chapter 5, Displaying. A few are mentioned here as reminders.

☐ HAND LETTERING Hand letter the copy on construction paper or on a chalkboard. Colored crayons or chalks produce attractive effects. Frame and focus your camera on such titles so as to avoid showing edges of paper or chalkboard; leave enough space on all sides between the copy and the frame edges to avoid a crowded effect.

☐ READY-MADE LETTERS Dry-transfer stick-on letters, plastic or ceramic letters, and cardboard letters (all of which may be purchased in art supply stores) may also be used to produce effective titles. When properly lighted, the plastic and cardboard letters produce a pleasing three-dimensional effect. Letters or words can also be cut from magazines of newspapers and arranged to spell out titles in interesting combinations of typefaces, sizes, and colors.

```
TYPEWRITER LETTERING

USE CAPITAL LETTERS

DOUBLE SPACE

NOT MORE THAN SIX LINES OF COPY

MAXIMUM LINE LENGTH

32 CHARACTERS
```

☐ TYPEWRITER LETTERING You may also produce titles on the typewriter: use CAPITAL letters, a maximum of 32 characters per line, and no more than six lines of *double spaced* copy per title.

Combinations of real things, sketches, and lettering can also be used to produce effective graphic illustrations for your motion pictures, slide sets, or photo print series. Hand letter colored tags or use cutout arrows to identify parts of a layout (which may show a piece of equipment that has been disassembled or pieces of porcelain of contrasting ages and designs, for example).

Title slides can be made from "sandwiches" in which two or more transparent layers are mounted together. Copy the title, executed on a white background, and then mount it in a mask with another transparency to provide a background. Background frames may be photographed transparencies of relevant scenes, objects, or designs, or they may be designs hand-drawn directly with ink on clear or tinted acetate.

splicing motion-picture film

A good film splice can be made by either of two common methods: tape splicing, using a specially designed adhesive tape; or cement splicing, in which film ends are overlapped and fused together with cement.

To splice film with tape, cut the two film ends so they fit together, and adhere a piece of splicing tape on each side of the film joint. Position the tape so that sprocket holes in it align with holes in the film.

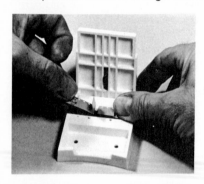

Cement splicing requires a film splicer in which to clamp and trim film ends. Prepare the overlapping surfaces, as shown, for application of a special film cement. When coated, the surfaces must be pressed quickly together and held under pressure for 15 to 20 seconds. "Hot" splices are made in the same way on a splicer with a heated platen which dries the cement sufficiently in 8 to 10 seconds.

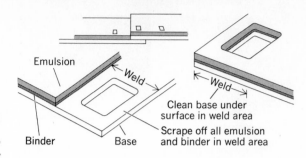

Emulsion

Weld

Weld

Clean base under surface in weld area

Scrape off all emulsion and binder in weld area

Binder

Base

Techniques and equipment for making cement splices differ slightly in detail depending upon characteristics of the film, the type of sound track, and the uses for which the film is intended. A few pointers are given below; consult splicer instruction sheets and references on editing and film splicing for additional suggestions.

☐ Practice making splices with scrap film; this is an inexpensive way to learn the feel of the splicer and to recognize the pressure necessary for proper scraping.

☐ Always use fresh film cement. As cement is exposed to air, it becomes thick and gummy and will not make satisfactory splices. Keep a small quantity sufficient for immediate use in a well-stoppered working bottle.

☐ When splicing film that has a magnetic sound track, be sure the butt of the splice is toward the tail end of the film so the head will drop off the splice, not run into it. This precaution minimizes magnetic head bounce when splices pass the head.

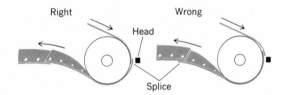

Right

Wrong

Head

Splice

PHYSICAL FACILITIES

contents

introduction

There may have been a time when a classroom needed to be little more than a boxlike area equipped with a few chalkboards, a tack board, a cloak closet, a teacher's desk, and thirty or so desks for students. But that time is past.

Now classrooms need to be flexible, adaptable, and convenient, and many are. However, teaching areas to which you are assigned may be far from ideal, and the learning activities you carry on there may be hampered by inadequate physical facilities. Nevertheless, even with limited funds available, there are many things you and your students can do to make your environment colorful, interesting, and convenient. Facilities over which you may have some control include: working areas, accessory furniture, space dividers, display areas, and work surfaces. The important thing is for you to know the effect of classroom environment on learning and to have some good ideas about how to make changes that will improve conditions for you and your students.

Later in this section, a checklist of *special facilities* problems is provided to remind you that there are some areas over which you may have little or no direct control but you need to identify them and be able to bring them to the attention of those in charge of instructional facilities. But principally, this reference section provides you with a start on how to improve physical facilities on your own initiative.

planning improvements for classroom facilities

A productive approach to improving classroom facilities is to ask the right questions:

☐ What are the sizes and behavior patterns of students? Where will they work? At tables? On the floor? At desks? At display or vertical work surfaces?

☐ Will more than one teacher use the same area at different times? At the same time? Will teacher aides work in the area? What will each user do, when, and for what purposes?

☐ If rooms are to be shared, what rearrangement of furniture or other resources will be required? How often must items be changed, and by whom? Where will they be put when not in use?

☐ What needs to be stored in the room for almost continuous use? For intermittent use? What will be brought to the room from some outside source; how long and for what purposes will it be kept there? Where will it be stored when it is not in use?

☐ What utilities are provided? A sink with hot and cold water? Sufficient electrical outlets and lighting fixtures?

☐ Can the room be darkened easily and adequately for various types of media projection? Are suitable projection screens or projection surfaces provided for individual or small-group use while the classroom is illuminated?

☐ Are furniture and storage units movable? Can they be sent to a warehouse when not needed? Are there movable units to house such varied items as books and pamphlets, maps and globes, filmstrips and films, audio equipment (record players, tape recorders, and accessories), tapes and records?

☐ What facilities do you *wish* you had? Are they within reason and practical possibility? Can your inventiveness and imagination suggest ideas to improve classroom facilities with funds available?

Here follow a number of suggestions, all of which have been tested and found practical where they are in use.

seating arrangements and working surfaces

Seats, desks, and tables should be movable to permit various groupings of students. Furniture should be sturdy but not too heavy to move easily. Chairs should provide good posture support and be of appropriate heights, so that students may write and work in comfort. Trapezoidal, half-circle, or rectangular tables permit arrangements suitable for various learning activities.

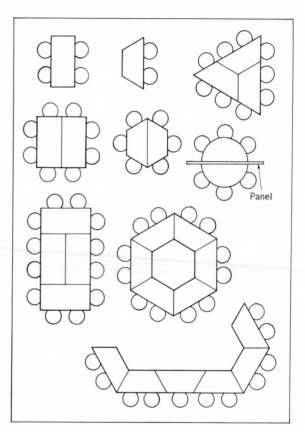

Panel

All elementary and many secondary classrooms should contain areas to serve as library or reference centers, for group or committee work, and for displays. Here are a few suggestions:

☐ Use folding tables to extend working areas. Card tables will work for this purpose if they are sturdy.

☐ Use large plywood sheets (⅜ or ½ inch thick) to make larger working surfaces. Set them up on seats or backs of chairs or on card tables. Be sure

Work surface

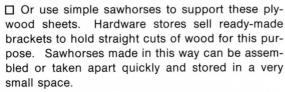

Portland (Oregon) Community College

the height is correct for the students who will work at them. Use paper or small pieces of foam plastic to prevent marring of chairs and slippage of the plywood sheet. If the sheets are to be used for wet work, first cover them with sheet plastic or seal them with lacquer or varnish. Be sure to cover all raw edges; heavy tape serves well for this purpose.

☐ Or use simple sawhorses to support these plywood sheets. Hardware stores sell ready-made brackets to hold straight cuts of wood for this purpose. Sawhorses made in this way can be assembled or taken apart quickly and stored in a very small space.

☐ If your classroom is carpeted, you will soon discover that students often prefer to study or have discussions while sitting on the floor. Develop other informal seating arrangements by making up plywood boxes (rough-grade wood is satisfactory)

to be covered with carpeting. For variety, choose colors that contrast with the floor covering. Students will find it convenient to lean against and sit on these boxes and even to use them as work surfaces. They may also be used to support the plywood sheets mentioned above. Add further practical variety by making some boxes in hexagonal shape, some rectangular, and some square.

☐ If you wish to use the floor as a seating or working area and it is not carpeted, explore sources of inexpensive rug remnants or "throw rugs" to cover only certain areas. Use these rug sections to form comfortable lounging and study corners, for example, as well as sit-on-the-floor areas. Cushions, in various sizes, shapes, and colors, are desirable accessories in lounging and reading areas. Plastic-covered, foam-filled cushions may be most practical.

San Diego (California) City Schools

Portland (Oregon) Community College

solving storage problems

Finding enough storage space in the classroom is frequently a problem. Even when storage space is provided, accumulations of materials are likely to overrun it. The first step in attempting to solve your storage problem is to resolve to avoid storing materials unnecessarily.

Immediate access to materials that are needed is a prime requirement that should influence classroom storage plans. But not everything needs to be immediately accessible all the time. So your second job with respect to storage is to determine which resources should be kept within the classroom and which can be held ready elsewhere (in a more remote location in the building or at a school system storage facility, for example).

For items that are to be stored even temporarily *within the classroom,* a number of arrangements and procedures are suggested:

☐ Use movable cabinets, which may also serve as room dividers. Equip some cabinets with brackets on which to anchor panels to hold tools or materials for the relevant activity (in the art supplies cabinet, for example). Some movable cabinets may be built on casters; others may be small enough to be hand-carried. Consider whether handles will help two persons carry a cabinet.

☐ Cut down on storage problems by setting up a recycling area within the classroom or elsewhere in the building. Use this facility to refurbish instructional materials that can be salvaged; discard the rest.

☐ Find convenient ways to package and move items in and out of remote storage. Inexpensive cardboard boxes (with covers), appropriately labeled, are good for this purpose, as are the smaller hand-carried cabinets just mentioned. Use rolling tables to transport them to the storage room or loading area.

☐ Door surfaces and unused but limited wall areas can be used to store games, flat pictures, charts, and other difficult-to-file materials. A canvas arrangement with rods and cords is shown, but plywood or pressed-wood panels also can be fabricated in the form of oversized, vertical magazine display racks.

Independent Learning Associates, Santa Barbara, California

Charles Beseler Company

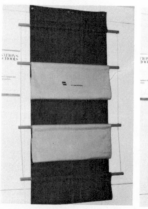

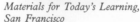

Materials for Today's Learning, San Francisco

space dividers and display areas

The advent of open classroom procedures has stimulated use of commercially and locally prepared area dividers to provide needed arrangements of space. To investigate their possibilities:

☐ Examine catalogs and visit exhibits of commercial producers of such units, seeking good ideas on what could be purchased or made locally to fit your needs.

☐ Think about freestanding panels. These may be made to stand by hinging the panels so they are held erect by their own angular footing. Others are made to stand with a firm base of boxlike construction or with adequately wide and well-braced footings as suggested in the accompanying sketches.

☐ Consider using freestanding racks for materials such as books, magazines, recordings, films; these racks may be semifixed or on casters. Remember, there are casters that perform best on carpets and others serve adequately on both carpet and hard flooring. One rule to observe is this: Be sure the casters are sufficiently large to roll easily over any doorsill or carpet divider strips. Casters with soft rubber tires are recommended on cabinets and tables for AV and TV equipment.

☐ Make room-divider panels serve also as display surfaces. Either one or two sides may be covered with panel board (Celotex type) to receive pins and thumbtacks. Masonite panels may be painted with chalkboard paint, available in several colors from paint stores. Cover one or more panels with flannel or other cloth-board covering, backed, if desired, with Celotex-type wallboard for pinning materials to their surfaces or for attaching items through use of flannel-board techniques, described in Chapter 3. Pegboard in two thicknesses (⅛ and ¼ inch), and with many types of brackets for each, is justifiably popular for room or area dividers. Attach pegboard to wall surfaces or to the backs of cabinets. Use mounting clips or screws with spacers to attach pegboard to supporting surfaces. Paint it with a roller to keep it clean in appearance or to change color schemes.

☐ Under some circumstances, use simple easels to support panels of tack-board, magnetic-board, or cloth-board material. Use these same easels for student art and design activities. Cover unneeded areas of chalkboard with corrugated paper or other tack-board backing (Celotex, for example). Decide whether to make a temporary or a semipermanent installation. Cover large areas of a room wall with taped-up sections of seamless paper (a wide, lightweight cardboard material used in store window displays and for backgrounds for photography). The continuous long length and up to 12-foot width of seamless paper make it a useful base for murals or large displays.

□ Consider hanging from the ceiling lightweight, mobilelike panels of railroad board, seamless paper reinforced with lightweight strips of wood (smooth laths), or cloth. Request permanently installed hooks in appropriate ceiling locations for hanging the display panels. Consider Molly-Bolts, screw eyes, or other appropriate attachments; if ceilings are of the suspended type, request installation of supports from the beams or cement overhead to which to attach wires. Hang display panels from loops in these wires.

San Diego (*California*) *City Schools*

□ For classrooms having ceilings open to solid beams (wood or steel) or to cement overhead areas, consider requesting one of several types of inexpensive, temporary panel supports. Two popular types are: *Pole-Cats* and *Timber-Toppers* (Brewster Corporation).

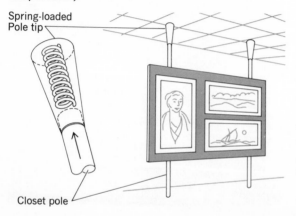

Spring-loaded Pole tip

Closet pole

arrangements for projection

Examples of space arrangements and student groupings for viewing projected visuals are shown on the opposite page. Below are some suggestions for improving classroom projection conditions. When projecting for *an entire class* of, say, thirty students:

□ For general use, choose a screen not smaller than 70 by 70 inches. Remember that an overhead transparency projector with its special projection requirements may be used, as well as slides, filmstrips, and motion pictures; sometimes the overhead may be used simultaneously with one or more of the latter.

□ Remember that screens with matte-white surfaces permit comfortable, undistorted viewing up to 30 degrees from either side of the center line. (See Reference Section 1.)

□ Do not hesitate to rearrange seating to provide comfortable viewing. Arrange a clear path from the projector to the screen.

□ If possible, allow some ambient light into the room during viewing, but not enough to degrade picture quality. As a rule of thumb, the light allowed to enter should permit a student to read a newspaper with some difficulty (representing about one-tenth footcandle).

□ Always place the projection screen with its back toward light coming from windows, doors, or vents. Under some conditions, as in open classroom areas, restrict light on the projection screen surface by using some arrangement of plywood or cardboard to produce a shadow box. Or place the screen between two cabinets, and place a sheet of plywood on top of the cabinets to shadow the screen surface. One screen made by Eastman Kodak Company—the Ektolite—provides a high level of reflection and excellent picture quality in a lighted room, but its screen size was limited in 1972 to a 40- by 40-inch dimension. Because this screen is fragile and will not roll or fold, secure storage is essential. In schools equipped with carpets or rug-covered areas, students may sometimes project visuals on a matte-white sheet of paper supported with tape under the teacher's desk, which provides an effective shadow box. A similarly satisfactory shadow box can often be arranged in a study carrel.

When projecting for *individual or small-group viewing,* consider the following suggestions:

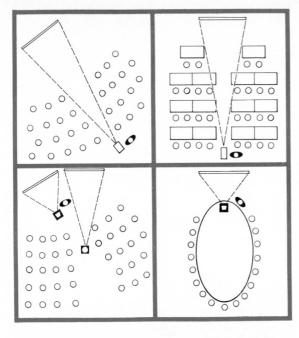

Cleveland (Ohio) Public Schools

Eastman Kodak Company

☐ Place a sheet of white matte-finished paper on a wall, the side of a carrel, or on a freestanding easel or monopod. Some carrels are already equipped with an area painted matte white for use as a permanent screen.

☐ Use a shadow-box hood to eliminate troublesome light.

☐ Consider using simple rear-projection, tabletop screen units, some of which—with tilting mirrors—may be used for both rear-screen and wall projection.

independent study spaces

In almost every school, increasing emphasis on independent study by students has created still other special facilities problems. Study carrels are available from many manufacturers, in a multitude of sizes, styles, and arrangements, and with various built-in technical resources. The terms *wet carrel* and *dry carrel* are widely used to identify those provided with electrical or electronic services and those with furniture only. Minimally, wet carrels are equipped with an electrical outlet, usually a duplex receptacle which will permit connecting at least two units (a tape player and slide viewer, for example) at a time. Some also may have terminals for television or audio program reception, connections for lines to computer or other data services, special reading lights, and built-in facilities for rear-screen projection. The illustrations below suggest the great variety of such arrangements. Of special interest are simple panels that can create carrel or other types of independent study spaces on standard folding or fixed tables or in more difficult situations such as old science laboratories.

Usually, carrels will be clustered in groups in a learning center or library; sometimes corridors are sufficiently large to permit safe use of carrels or tables for independent study. With increasing frequency, carrel-type stations are being provided in classrooms. If you are interested in developing stations for independent study:

☐ Look about your classrooms to locate areas that may be used or adapted for carrel stations; consider how flexibility and variety in the arrangement of these stations may be effected.

☐ Remember that a study station can be the right

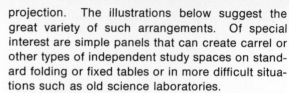

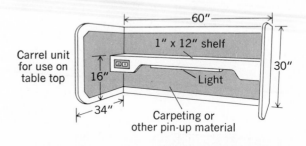

Pacific Telephone Company, Marketing Training, Los Angeles, California

System Development Corporation

Carrel unit for use on table top

60"

1" x 12" shelf

30"

16"

Light

34"

Carpeting or other pin-up material

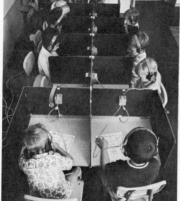

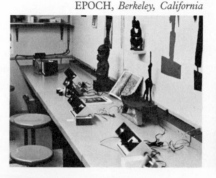

EPOCH, *Berkeley, California*

Santa Barbara City Schools

size, in the right location, and equipped with the proper facilities without being elaborate or costly.

☐ Review the many ways to use space dividers to arrange areas for various types of study activities.

☐ Recognize that, because of safety requirements, electrical facilities must always be arranged with full approval of those responsible for school plant modifications. There are several economical and practical systems for providing separate channels for both power lines and receptacles as well as for low-energy lines such as for audio, television, or computer services (*Wiremold* is one).

☐ When arranging study spaces in a classroom area (or elsewhere), consider (1) your own convenience in assisting students when they need help, (2) provisions for student teams to conduct tutoring sessions, and (3) related arrangements such as convenient spaces for small-group work.

seminar and small-group areas

Although you may find there are ample seminar and group work spaces available to your students in learning resource centers, classroom facilities are also needed to provide comfortable semi-isolated study spaces. The following suggestions may help you to obtain them in your situation:

☐ Refer again to the use of room and area dividers, whereby areas can be provided for small-group work.

☐ Rooms with carpeting and with other desirable acoustical treatment will permit group activities and tutoring without generating objectional noise. General noise, at a reasonable level, often masks and controls specific sounds which might otherwise annoy.

☐ Again, make dividers from bookcases or cabinets or use hanging or freestanding panels to arrange spaces for small-group work.

☐ For group listening or sound film projection, use earphones with multiconnector boxes to control sound and improve student concentration.

☐ Remember, when electrical and electronic equipment is to be used by groups, study areas should be adjacent to wall or floor power receptacles. *Never string electric or other wires across classroom floors or passageways.*

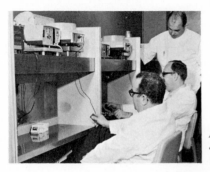

Michigan State University

Portland (Oregon) Community College

Ampex Corporation; Oak Park High School (Ill.)

Cleveland (Ohio) Public Schools

451

special problems

In addition to the recommended ways by which you may improve your classroom environment, there are a number of other problem areas in school facilities that affect instruction. These are areas over which you may not have any large measure of personal control. You may, however, be involved in curriculum revision or other programs that will require modification of the school plant, or you may be asked to help plan a new school. In these projects, you will find references at the end of this section helpful. Also use the checklists below to guide your investigations. A quick review of these lists may help you identify limitations in your present school facility.

In *auditoriums* and other facilities for large-group instruction, the following features require special consideration:

☐ Entries and exits, traffic flow

☐ Aisles, seating plan, sight lines

☐ Light control—from external sources

☐ Lighting control—auditorium and presentation platform areas

☐ Projection facilities

☐ Projection walls and/or screens

☐ Electrical services

☐ Public address/sound systems

☐ Intercommunication devices

☐ Television and audio services

☐ Acoustical treatments (sound control)

☐ Mechanical facilities (plumbing, heating, air conditioning)

☐ Student stations (tablet arms, tables, response stations, counters)

☐ Presentation area (access, storage, problems of changing setups)

☐ Divisible auditoriums (turntable, folding partitions)

☐ Multipurpose rooms (multifunctions in fixed or modifiable spaces)

Though directed toward the analysis of large-group instruction areas, the above checklist has many items that apply to large classrooms, special-purpose rooms, and laboratories. In each type of

outdoor facilities

Where favorable climates permit, modern school designers seek ways to extend classrooms into adjacent outdoor areas. Such designs for elementary schools often provide close relationships between the garden and play areas. To do this, they may employ sliding tempered-glass doors, tinted to avoid glare.

As you consider instructional functions that appropriately and advantageously may be moved to outdoor areas, consider these accessories:

☐ Chairs and tables and other portable furniture that may be moved easily and conveniently in and out of doors.

☐ Planters, pots, and cans in which many types of plants and flowers may be grown and cared for.

☐ Convenient and portable tool and water carriers, with handles or on wheels, to facilitate care of garden areas.

☐ Posts with panels on them or well-braced stands for outdoor exhibits. Sandboxes may be used to stabilize many types of stands.

☐ Tote boxes and trays, in special carriers, for student projects and work materials.

☐ Appropriate sheltered cages to provide temporary housing for visiting pets.

Atlanta (Georgia) Public Schools

room, the items may increase or diminish in importance or in their contribution to the achievement of satisfactory circumstances for instruction.

Electrical and acoustical problems occur in every teaching facility, and a few special items follow for your checklist:

☐ Receptacles (location and number)

☐ General lighting (levels for types of activity to be specified)

☐ Lighting control (for projection of different media, TV viewing)

☐ Location of lighting units (switches, dimmers)

☐ Sound (acoustical) controls (carpets, wall and ceiling treatments, room shapes)

☐ Speakers and speaker placement

☐ Acoustical treatment (floors, doors, corridors, stairwells)

Problems of outdoor areas include traffic and transportation problems, view and weather control, and electrical and other needed services. Here are a few to start a checklist:

☐ Wind and view control (fixed panels, fences, screen planting)

☐ Traffic provisions (parking for mobile laboratories, equipment trucks, vehicles for handicapped persons)

Los Angeles (California) City Schools

☐ Electrical (connections for media production, language, or driver training laboratories)

☐ Sun shelters

☐ Paved and unpaved areas

☐ Drainage

a word about color

Probably the most immediate means of changing your classroom environment is to consider using colors. Here are a few indications of how you can use color easily and economically:

☐ Choose a variety of colors when you select seamless and other paper for display surfaces, both vertical and tabletop.

☐ Use panels of colored paper on the backs of cabinets. Change display-board color backgrounds with each display.

☐ Make recommendations for color combinations when planning the purchase of cabinets and furniture.

☐ Get approval to use fast-drying, no-odor paints to change and refurbish the surfaces of cabinets and panels.

☐ Keep samples of cloth and carpet materials, color charts for paints, and a picture file of room displays showing color combinations; use these for reference when planning new arrangements in teaching areas.

☐ Remember that it costs no more to use interesting colors than it does to use institutional gray or green.

And that last point is one that should be kept in mind! Having an attractive and functional environment for learning is more often the result of imagination and clear thinking than of the expenditure of money. Here, again, the objectives of the school program and the philosophy of those who control the program are fundamental. An often repeated expression is important too: We build our buildings, and our buildings shape our lives. Therefore, the optimum environment for education is a matter of decisions by people who clearly know the outcomes desired and the characteristics of surroundings that stimulate students to achieve these outcomes. Let your own experience contribute to the decisions that determine the environment where learning can best be achieved and enjoyed.

Keep in mind that in comparatively recent times a school media resource center has looked like this:

an exercise in judgments

To exercise your ability to evaluate the need for changes in facilities, try this activity:

First, select a classroom or other instructional area and make a list of the aspects of instruction that do or should take place there: curriculum, methods, and activities. From your observations make two lists:

(1) specific, actual conditions that you believe *facilitate effective instruction* and (2) specific, actual conditions that *hinder effective instruction* and student learning.

Then, make specific recommendations for improving the environment, starting with the simplest and most economical, and putting last the major, more costly changes you recommend. Finally, can you justify each of your proposals?

This type of atmosphere, which served for many years, has been changed in most schools to look more like this:

Look about you in your schools and colleges and identify the facilities that are used today; then consider what facilities may be needed in the future.

5

CLASSIFIED DIRECTORY OF SOURCES

This classified directory of sources provides addresses of companies, institutions, and associations mentioned in the text proper and of other representative major suppliers of media resources and equipment. Write these manufacturers, distributors, or organizations to obtain information you need about their products or services.

The list is organized around each chapter of the text. As of the date we completed this list, the company names and addresses were carefully checked for accuracy. Of course, in time, some of these companies will have merged, changed names or locations, or gone out of business. But because the list is composed of the leading companies or organizations in each field, it should continue to be useful for a considerable period of time.

A compilation such as this should never be considered complete, despite the effort that we made to include major representative sources that would meet the needs of persons for whom it is intended. Many valuable and useful sources had to be omitted because of space limitations. Likewise, the diversification of companies into multi-media fields sometimes made it difficult to classify their materials in every subarea of production.

general source materials

Probably the most complete current set of general source materials in the media field comes from the National Information Center For Educational Media, University of Southern California, University Park, Los Angeles, California 90007. It publishes, and regularly updates, the following indexes in either books or microfiche forms:

1. *Index to Educational Films* (3 vols.)
2. *Index to 35mm Filmstrips* (2 vols.)
3. *Index to Educational Audio Tapes*
4. *Index to Educational Video Tapes*
5. *Index to Educational Records*
6. *Index to 8mm Motion Cartridges*
7. *Index to Educational Overhead Transparencies* (2 vols.)
8. *Index to Educational Slidesets*
9. *Index to Psychology* (multi-media)
10. *Index to Health and Safety Education* (multi-media)
11. *Index to Vocational and Technical Education* (multi-media)
12. *Index to Black History and Studies* (multi-media)
13. *Index to Ecology* (multi-media)
14. *Index to Producers and Distributors*

NICEM also provides an update subscription service of all the specialized areas simultaneously. Monthly issues from September through June cover additional entries to the eight media areas, the five multi-media subject areas, and the directory of producers and distributors.

A second omnibus-type directory of educational materials is published by the Westinghouse Learning Corporation, 100 Park Ave., New York, N.Y. 10017. The original six-volume *Learning Directory* appeared in 1971, with an annual supplement which updates on a yearly basis the extensive original listings. The 1972–1973 supplement indexed another 24,000 items and classified them in 75,000 entries under more than 28,000 different topics. Forty-seven different media classifications were used, from audio cartridge to video tape and workbook. The *Learning Directory* was originally developed to provide instructional materials for 10,000 detailed learning objectives broken down by major fields of knowledge and grade level from kindergarten through twelfth grade.

other basic sources

Another basic source of information about instructional materials and equipment is found in a group of publications of the R. R. Bowker Co., 1180 Avenue of the Americas, New York, N.Y. 10036. Among major publications of this group are: James W. Brown (ed.), *Educational Media Yearbook,* revised annually since 1973; *Previews: News and Reviews of Non-Print Media* (periodical); Roderick McDaniel, *Resources For Learning,* 1971; and Lucy G. Brown, *Core Media Collection for Secondary Schools,* 1975. Bowker also publishes an annual guidebook, *The Audio Visual Market Place* (AVMP), containing general information about the audiovisual industry, names and addresses and other data about producers and distributors of media and media equipment, new publications, and other related items.

The National Audio-Visual Association, 3150 Spring St., Fairfax, Va. 22030, publishes an annual edition of *The Audio-Visual Equipment Directory* which is exceptionally complete in its coverage (except for photographic equipment).

A number of other basic sources contain specialized media and media equipment information. The Educational Products Information Exchange Institute (EPIE), 463 West St. New York, N.Y. 10014, for example, provides results of the laboratory testing of instructional materials and equipment. EPIE members receive in-depth reports on competing products and newsletters (*EPIE Reports* and *EPIE-grams*) on a regular basis; they can also use a hot line for particular questions or problems. Its *Improving Materials Selection Procedures: A Basic How-to Handbook* is an especially useful publication.

The National Audiovisual Center, The National Archives and Records Service, Information Branch, Washington, D.C. 20409, is a key source of information about the offerings and purchase availabilities of audiovisual materials produced by fifty-seven major federal agencies, including the U.S. Civil Service Commission, Department of Agriculture, Department of Defense, Department of State, Foreign Services Institute, Department of Commerce, United States Information Agency, presidential libraries, National Science Foundation, Office of Education, and the Veterans Administration.

Several catalogs and guides present data on the sources of free and inexpensive instructional materials and services. One of the best, revised bi-annually, is Ruth A. Aubrey's *Selected Free Materials for Classroom Teachers* (Fearon Publishers, 9 Davis Drive, Belmont, California 94002). The most extensive and diversified service of this type comes from The Educators Progress Service, Inc., 214 Center St., Randolph, Wis. 53956, which revises most of its guides annually. It offers the following titles: *Educators Guide to Free Films, Educators Guide to Tapes, Educators Index to Free Materials,* and *Elementary Teachers Guide to Free Curriculum Materials.* The Serina Press, 70 Kennedy St., Alexandria, Va. 22305, also provides a number of guides to free or inexpensive materials.

There are several comprehensive sources of data about general manufacturers and distributors, including those involved with audiovisual equipment. Perhaps the most complete of these is the *Thomas Register of American Manufacturers,* issued by the Thomas Publishing Co., Inc., 461 Eighth Ave., New York, N.Y. 10001.

Finally, it should be noted that almost every professional educational magazine includes monthly sections describing new teaching materials and new equipment of interest to those in the special field involved. Coupled with the following reference lists, they will help you to keep up to date about the wealth of materials and equipment which can be used to improve learning.

CHAPTER 1

media and the systematic approach to teaching and learning

general

ERIC Clearinghouse on Information Resources, SCRDT, School of Education, Stanford University, Stanford, Calif. 94305.

Instructional Objectives Exchange, Box 24095, Los Angeles, Calif. 90024.

media and individualized learning

ALOHA Project, 935 Piedmont Road, San Jose, Calif. 95132.

American Institutes for Research in the Behavioral Sciences, 135 N. Bellefield Ave., Pittsburgh, Pa. 15213.

Argyle Publishing Corporation, 200 Madison Ave., New York, N.Y. 10016.

Behavioral Research Laboratories, Box 577, Palo Alto, Calif. 94302.

Encyclopaedia Britannica Educational Corporation, 425 N. Michigan Ave., Chicago, Ill. 60611.

ENRICH (Telor), 760 Kifer Rd., Sunnyvale, Calif. 94086.

Learn-Ease Division, National Blank Book Company, Holyoke, Mass. 01040.

Litton Industries, 9370 Santa Monica Blvd., Beverly Hills, Calif. 90213.

McGraw-Hill Book Company, 1221 Avenue of the Americas, New York, N.Y. 10020.

Modern Teaching Associates, Inc., 1506 W. Pierce St. Milwaukee, Wis. 53246.

Grolier Educational Corporation, 845 Third Ave., New York, N.Y. 10022.

The Welch Scientific Company, 7300 N. Linder Ave., Skokie, Ill. 60076.

John Wiley and Sons, Inc., 605 Third Ave., New York, N.Y. 10016.

CHAPTER 3

community as a learning center

museums (selected examples)

American Museum of Natural History, Central Park West at 79 St., New York, N.Y. 10024.

Anacostia Neighborhood Museum (a branch of the Smithsonian Institution), 2405 Martin Luther King, Jr., Ave. SE, Washington, D.C. 20020.

Brooklyn Children's Museum, 1530 Bedford Ave., Brooklyn, N.Y. 11216.

The Children's Museum, Jamaica Way, Boston, Mass. 02130.

Detroit Institute of Arts, 5200 Woodward Ave., Detroit, Mich. 48202.

Fort Worth Museum of Science and History, 1501 Montgomery St., Fort Worth, Tex. 76107.

Los Angeles County Museum, Los Angeles, Calif.

National Aero-Space Museum, Smithsonian Institution, Washington, D.C. 20560.

National Gallery of Art, Washington, D.C. 20560.

Smithsonian Institution, Washington, D.C. 20560.

zoos, aquariums, zoological gardens (selected examples)

Atlanta Zoo, 518 Atlanta Ave. S.E., Atlanta, Ga. 31315.

Chicago Zoological Society, Brookfield, Ill. 60513.

Cleveland Aquarium (Gordon Park), 601 East 72 St., Cleveland, Ohio 44103.

Kansas City Zoological Gardens, Swope Park, Kansas City, Mo. 64132.

Milwaukee County Zoo, 10001 West Blue Mound Road, Milwaukee, Wis. 53226.

National Fisheries Center and Aquarium, Department of Interior, Fish and Wildlife Service, Washington, D.C. 21240.

New York Zoological Society, The Zoological Park, Bronx, N.Y. 10460.

San Antonio Zoo, Breckenridge Park at Broadway, San Antonio, Tex. 78212.

San Diego Zoo, Balboa Park, San Diego, Calif. 92112.

San Francisco Zoological Society, Zoo Rd. and Skyline Blvd., San Francisco, Calif. 94132.

Steinhart Aquarium, San Francisco, Calif. 94132.

Vancouver Public Aquarium, Vancouver, British Columbia, Canada.

science centers and groups

Franklin Institute, Philadelphia, Pa. 19103.

Greenfield Village, Dearborn, Mich. 41821.

Group for Environmental Education, 1214 Arch St., Philadelphia, Pa. 19107.

Lawrence Hall of Science, University of California, Berkeley, Calif. 94720.

CHAPTER 4

choosing, using, and producing media

Catalog of Audiovisual Materials for Rent and Sale by the National Audiovisual Center, National Audiovisual Center, National Archives and Records Service, Washington, D.C. 20409.

Children's Books in Print, R. R. Bowker, 1180 Ave. of the Americas, New York, N.Y. 10036.

Children's Catalog, with annual supplements, H. W. Wilson Co., 950 University Ave., New York, N.Y. 10452.

Cochran, Lida, and John Johnson, "A Resource List of Information About Media Production," *Audiovisual Instruction,* vol. 19, April 1974, pp. 37–45; May 1974, pp. 53–65; June/July 1974, pp. 8–101.

Doak, Wesley A., and William J. Speed, *International Index to Multi-Media Information, 1970–1972,* Vols. 1–3 (1970, 1971, 1972) from R. R. Bowker Co., 1180 Ave. of the Americas, New York, N.Y. 10036; later volumes from AV Associates, 180 E. California Blvd., Pasadena, Calif. 91105.

EPIE Institute, Educational Products Information Exchange, *EPIE Reports* and *EPIEgrams,* 463 West Street, New York, N.Y. 10014.

Index to Instructional Media Catalogs, R. R. Bowker Co., 1974, 1180 Ave. of the Americas, New York, N.Y. 10036.

Johnson, Harry A., *Multi-Media Materials For Afro-American Studies,* R. R. Bowker Co., New York, N.Y. 1971.

Paperbound Books in Print 1975, R. R. Bowker Co., 1974, 1180 Ave. of the Americas, New York, N.Y. 10036.

Wynar, Christine L., *Guide to Reference Books for School Media Centers,* revised biannually, Libraries Unlimited, Box 263, Littleton, Colo. 80120.

CHAPTER 5

displaying—and some fundamentals of visual communication

art materials and construction supplies

American Art Clay Company, 4717 W. 16 St., Indianapolis, Ind. 46222.

American Crayon Company, 1706 Hayes Ave., Sandusky, Ohio. 44870.

Dick Blick, P.O. Box 1267, Galesburg, Ill. 61401

Milton Bradley Company, 74 Park St., Springfield, Mass. 01101.

Arthur Brown and Bros., 2 W. 46 St., New York, N.Y. 10036.

Eberhard-Faber Company, Inc., Crestwood, Wilkes-Barre, Pa. 18703.

3M Company, 3M Center, St. Paul, Minn. 55101.

Weber Costello Company, 1900 N. Narragansett Ave., Chicago, Ill. 60639.

precut letters

Artype, Inc., 345 E. Terra Cotta Ave., Crystal Lake, Ill. 60014.

Demp-Nock Company, 21433 Mound Rd., Warren, Mich. 48090.

Mittens Designer Letters, 345 5th St., Redlands, Calif. 92373.

Prestype Inc., 194 Veterans Bldg., Carlstadt, N.J., 07072.

lettering devices

Carter's Ink Company, 239 First, Cambridge, Mass. 02141.

Chart-Pak, Rotex Division, Avery Products Corporation, 2620 Susan St., Santa Ana, Calif. 92704.

E. Dietzgen Company, 2425 N. Sheffield Ave., Chicago, Ill. 60614.

Embosograf Corporation of America, 38 W. 21 St., New York, N.Y. 10010.

Keuffel and Esser Company, Education/AV Products, 20 Whippany Rd., Morristown, N.J. 07960.

Morgan Sign Machine Company, 4510 N. Ravenswood Ave., Chicago, Ill. 60640.

Uniline Corp., Union City, Calif. 94587.

Varigraph, Inc., 1480 Martin St., P.O. Box 690, Madison, Wis. 53701.

Wood-Regan Instrument Company, 184 Franklin Ave., Nutley, N.J. 07110.

chalkboards and magnetic boards

Alliance Wall Corporation, Wyncote House, Wyncote, Pa. 19095.

Johns Manville, 22 E. 40 St., New York, N.Y. 10016.

Charles Mayer Studios, Inc., 140 E. Market St., Akron, Ohio, 44308.

Oravisual Company, Inc., P.O. Box 11150, St. Petersburg, Fla. 33733.

Scott-Engineering Sciences, Div. of A-T-O Inc., 1400 S.W. 8th St., Pompano Beach, Fla. 33060.

Weber-Costello Company, 1900 N. Narragansett Ave., Chicago, Ill. 60639.

cloth board materials

Admaster, Inc., 425 Park Ave. S., New York, N.Y. 10016.

American Felt Company, Department KT, Glenville, Conn. 06830.

Bertile Products, Inc., Delta Division, P.O. Box 42, 475 Guy St., Drummondville, Quebec, Canada.

Creative Visuals, Gamco Plaza, Box 1911, Big Spring, Tex. 79720.

Charles Mayer Studios, Inc., 140 E. Market St., Akron Ohio 14308. (Hook-N'-Loop)

Demco Educational Corporation, 2120 Fordem Ave., Madison, Wis. 53701.

Ideal School Supply Company, 11000 S. Lavergne Ave., Oak Lawn, Ill. 60453.

The Instructo Corporation, Cedar Hollow and Mathews Rd., Paoli, Pa. 10301.

Oravisual Company, Inc., P.O. Box 11150, St. Petersburg, Fla. 33733.

bulletin-board materials

Advance Products Company, P.O. Box 2178, Wichita, Kans. 67201 (easels)

Allied-Radio Shack Stores.

Bertile Products, Inc., Delta Division, 475 Guy St. P.O. Box 42, Drummondville, Quebec, Canada.

Demco Educational Corporation, Box 1488, Madison, Wis. 53701.

Edmund Scientific Company (motors), 555 Edscorp Bldg., Barrington, N.J. 08007.

Educators' Manufacturing Company, Subsidiary of Hauserman, Inc., 3401 Lincoln Ave., P.O. Box 1458, Tacoma, Wash. 98421.

Garcy Company of Canada, Ltd., P.O. Box 488, Don Mills, Toronto, Canada.

Hankscraft Motors, Reedsburg, Wis. 53959.

The Instructo Corporation, Cedar Hollow and Mathews Rd., Paoli, Pa. 19301.

Maggie Magnet Company, 39 W. 32 St., New York, N.Y. 10001.

Miami Magnet Co., 7846 W. 2nd Court, Hialeah, Fla. 33014.

Multiplex Display Fixture Company, 1555 Larkin Williams Rd., Fenton, Mo. 63026.

Ronald Eyrich—Magnets, 560 N.E. 42 St., Ft. Lauderdale, Fla. 33308.

Seal, Incorporated, Derby, Conn. 06418.

CHAPTER 6

graphic materials

map and globe sources

Aero Service Corporation, 210 E. Courtland St., Philadelphia, Pa. 19120.

American Map Company, 1926 Broadway, New York, N.Y. 10023.

Benefic Press, 10300 W. Roosevelt Rd., Westchester, Ill. 60153.

George F. Cram Company, Inc., 301 S. LaSalle, Indianapolis, Ind. 46206.

Denoyer-Geppert Company, 5235 Ravenswood Ave., Chicago, Ill. 60640.

Farquhar Transparent Globes, 5007 Warrington Ave., Philadelphia, Pa. 19143.

Hammond, Inc., 515 Valley Street, Maplewood, N.J. 07040.

Hubbard Scientific Company, P.O. Box 442, Northbrook, Ill. 60062.

Rand McNally and Company, 8255 Central Park Ave., Skokie, Ill. 60076.

Mead Educational Services, 1391 Chattahoochee Ave., NW, Atlanta, Ga. 30318.

National Geographic Society, 17th and M Streets, N.W., Washington, D.C. 20036.

A. J. Nystrom and Company, 3333 Elston Ave., Chicago, Ill. 60618.

Tanzer Multi-Maps, RD 3, Shenandoah Rd., Hopewell Junction, New York 12533.

Trippensee Planetarium Company, 301 Cass St., Saginaw, Mich. 48602.

Wards Natural Science Establishment, Inc., 300 Ridge Road E., Rochester, N.Y. 14622.

CHAPTER 7

transparencies for overhead projection

commercial overhead transparencies and manipulative devices

Allyn and Bacon, Inc., AV Department, 470 Atlantic Ave., Boston, Mass. 02110.

Robert J. Brady Company, a Prentice-Hall Company, Bowie, Md. 20715.

Creative Visuals, Gamco Plaza, P.O. Box 1911, Big Spring, Tex. 79720.

Encyclopaedia Britannica Educational Corporation, 425 N. Michigan Ave., Chicago, Ill. 60611.

Ful-Vu Visuals Company, P.O. Box 187, Blackwood, N.J. 08012.

General Aniline and Film Corporation, Audio/Visual Order Department, 140 W. 51 St., New York, N.Y. 10020.

Ginn and Company, Statler Building, Boston, Mass. 02117.

Hammond, Inc., 515 Valley Street, Maplewood, N.J. 07040.

Instructo Products Company, 1635 N. 55 St., Paoli, Pa. 19130.

Keuffel and Esser Company, 20 Whippany Rd., Morristown, N.J. 07960.

Lansford Publishing Co., P.O. Box 8711, San Jose, Calif. 95155.

McGraw-Hill Film Division, 1221 Ave. of the Americas, New York, N.Y. 10020.

NICEM, "Index To Educational Overhead Transparencies," University of Southern California, University Park, Los Angeles, Calif. 90007.

Rand McNally and Company, 8255 Central Park Ave., Skokie, Ill. 60076.

Science Research Associates, Inc., 259 E. Erie St., Chicago, Ill. 60611.

Tecnifax, Scott Education Div., 195 Appleton St., Holyoke, Mass. 01040.

3M Company, Visual Products, 3M Center, St. Paul, Minn. 55101.

Tweedy Transparencies, 208 Hollywood Blvd., East Orange, N.J. 07018.

United Transparencies, Inc., P.O. Box 688, Binghamton, N.Y. 13902.

Visual Materials, Inc., 2549 Middlefield Rd., Redwood City, Calif. 94063.

Xerox Corporation, Haloid St., Rochester, N.Y. 14603.

materials for making transparencies

American Polarizers, Inc., 1500 Spring Garden St., Philadelphia, Pa. 19101

Charles Beseler Company, 8 Fernwood Rd., Florham Park, N.J. 07932.

General Aniline and Film Corporation, A/V Order Department, 140 W. 51 St., New York, N.Y. 10020.

Instructo Products Company, 1635 N. 55 St., Paoli, Pa. 19130.

Keuffel and Esser Company, 20 Whippany Rd., Morristown, N.J. 07960

Eastman Kodak Co., 343 State St., Rochester. N.Y. 14650.

LABELON Corporation, 10 Chapin Street, Canandaigua, N.Y. 14424.

Seal, Incorporated, Derby, Conn. 06418.

J. S. Staedtler, Inc., P.O. Box 68, Montville, N.J. 07045.

Tecnifax, Scott Education Division, 195 Appleton St., Holyoke, Mass. 01040.

3M Company, Visual Products, 3M Center, St. Paul, Minn. 55101.

Transparex, Agfa-Gevaert Inc., 275 North St., Teterboro, N.J. 07608.

United Transparencies, Inc., P.O. Box 688, Binghamton, N.Y. 11702.

CHAPTER 8

photography

general information (publications)

(See also the listings in Reference Section 3, "Photographic Equipment and Techniques.")

Amphoto, 915 Broadway, New York, N.Y. 10010.

Association for Educational Communications and Technology (AECT), 1201 Sixteenth St. N.W., Washington, D.C. 20036.

Audiovisual Services, Media Reasearch, Arizona State University, Tempe, Ariz. 85281.

Center for Understanding Media, Inc., 73 Horatio St., New York, N.Y. 10014

Eastman Kodak Co., 343 State St., Rochester, N.Y. 14650

Morgan and Morgan, Inc., Publishers, 400 Warburton Ave., Hastings-on-Hudson, N.Y. 10706.

Petersen Books, 6725 Sunset Blvd., Los Angeles, Calif. 90028.

Pflaum/Standard (Publishers), 38 W. Fifth St., Dayton, Ohio 45402.

TAB Books, Blue Ridge Summit, Penn. 17214.

CHAPTER 9

still pictures

selected flat-picture sources

American Museum of Natural History, Central Park West at 79 St., New York, N.Y. 10024.

American Petroleum Institute, Committee on Public Affairs, 1271 Ave. of the Americas, New York, N.Y. 10020.

Association for Educational Communications and Technology (AECT), 1201 Sixteenth St. N.W., Washington, D.C. 20036.

Audio-Visual Enterprises, 911 Laguna Rd., Pasadena, Calif. 91105.

Arthur Barr Productions, P.O. Box 7-C, 1029 N. Allen Ave., Pasadena, Calif. 91104.

Creative Educational Society, P.O. Box 227, Mankato, Minn. 56001.

Hi-Worth Pictures, P.O. Box 6, Altadena, Calif. 91001.

Informative Classroom Picture Publishers, 31 Ottawa Ave. N.W., Grand Rapids, Mich. 49502.

Instructional Aids, Inc., P.O. Box 191, Mankato, Minn. 56001

League of Women Voters of the United States, 1730 M St., N.W., Washington, D.C. 20036.

National Geographic Society, 17th and M Sts., N.W., Washington, D.C. 20036.

The Instructo Corporation, Cedar Hollow and Mathews Rd., Paoli, Pa. 19301.

drawing and laminating materials

Bro-Dart, Inc., 1609 Memorial Ave., Williamsport, Pa. 17701.

Demco Educational Corporation, 2120 Fordem Ave., Madison, Wis. 53704.

Laminex, Inc., Box 577, Mathews, N.C. 28105.

Nasco, 901 Janesville Ave., Fort Atkinson, Wis. 53538.

Starex, Inc. (nonphotographic frosted film), P.O. Box 248, Kearny, N.Y. 07032.

F. Weber Company, Division of Visual Art Industries, Inc., Wayne and Windrim Avenues, Philadelphia, Pa. 19144.

filmstrips (silent)

BFA Educational Media, 2211 Michigan Ave., Santa Monica, Calif. 90404.

Basic Skill Films, 1355 Inverness Dr., Pasadena, Calif. 91103.

Stanley Bowmar Company, 622 Rodier Dr., Glendale, Calif. 91207.

Center for Mass Communications, Columbia University Press, 562 W. 113 St., New York, N.Y. 10025.

Coronet Instructional Media, 65 E. South Water St., Chicago, Ill. 60601.

Encore Visual Education, Inc., 1235 South Victory Blvd., Burbank, Calif. 91502.

Encyclopaedia Britannica Educational Corporation, 425 N. Michigan Ave., Chicago, Ill. 60611.

Eye Gate House, Inc., 14601 Archer Ave., Jamaica, N.Y. 11435.

GAF Corporation, Audio/Visual Dept. (Sawyer's Viewmaster), 140 West 51 St., New York, N.Y. 10020.

Informative Classroom Picture Publishers, 31 Ottawa Ave. N.W., Grand Rapids, Mich. 49502.

Imperial Film Company, Inc., 202 Lake Miriam Dr., Lakeland, Fla. 33803.

International Communications Foundation, 870 Monterey Pass Rd., Monterey Park, Calif. 93941.

Jam Handy Organization, 2781 E. Grand Blvd., Detroit, Mich. 48211.

McGraw-Hill Film Division, 1221 Ave. of the Americas, New York, N.Y. 10020.

Media Systems, Inc., 1220 E. 39th South, Salt Lake City, Utah 84121.

Moody Institute of Science, Educational Film Division, 12000 E. Washington Blvd., Whittier, Calif. 90606.

National Film Board of Canada, 1251 Ave. of the Americas, New York, N.Y. 10020.

New York Times, School Service Department, 229 W. 43 St., New York, N.Y. 10036.

Parents Magazine Films, 52 Vanderbilt Ave., New York, N.Y. 10017.

Society for Visual Education, Division of Singer Co., 1345 W. Diversey Pkwy., Chicago, Ill. 60614.

United World Films, Inc., 221 Park Ave. South, New York, N.Y. 10003.

Visual Sciences, P.O. Box 599, Suffern, N.Y. 10901.

Westinghouse Learning Corporation, 100 Park Ave., New York, N.Y. 10017.

Guidance Associates, Harcourt Brace Jovanovich, Inc., 41 Washington Ave., Pleasantville, N.Y. 10570.

International Film Bureau, Inc., 332 S. Michigan Ave., Chicago, Ill. 60604.

Jam Handy School Service, Inc., 2781 E. Grand Blvd., Detroit, Mich. 48211.

McGraw-Hill Film Division, 1221 Ave. of the Americas, New York, N.Y. 10020.

National Film Board of Canada, 1251 Ave. of the Americas, New York, N.Y. 10020.

Parents Magazine Films, 52 Vanderbilt Ave., New York, N.Y., 10017.

Society for Visual Education, Inc., 1345 Diversey Pkwy., Chicago, Ill. 60614.

Time-Life Films, Dept. Al, 43 West 16 St., New York, N.Y. 10011.

Visualized Instructional Productions, Ltd., 1200 Harger Rd., Oak Brook, Ill. 60621.

Westinghouse Learning Corporation, 100 Park Ave., New York, N.Y. 10017.

Weston Woods, Weston, Conn. 06880.

filmstrips (sound)

BFA Educational Media, 2211 Michigan Ave., Santa Monica, Calif. 90406.

Stephen Bosustow Productions, 1649 Eleventh St., Santa Monica, Calif. 90404.

Bowmar Records, Inc., 622 Rodier Dr., Glendale, Calif. 91201.

Cenco Educational Films, 2600 S. Kostner Ave., Chicago, Ill. 60623.

Coronet Instructional Media, 65 E. South Water St., Chicago, Ill. 60601.

Dukane Corporation, Audio-Visual Division, 103 N. 11 St., St. Charles, Ill. 60714. (Publishes the *Educational Sound Filmstrip Directory,* a valuable source list of sound filmstrips, on both a free-loan and purchase basis.)

Encore Visual Education, Inc., 1235 S. Victory Blvd., Burbank, Calif. 91506.

Encyclopaedia Britannica Educational Corporation, 425 N. Michigan Ave., Chicago, Ill. 60611.

Eye Gate House, 14601 Archer Ave., Jamaica, N.Y. 11435.

Films, Inc., 1144 Wilmette Ave., Wilmette, Ill. 60091.

2- by 2-inch slides

American Museum of Natural History, Central Park West at 79 St., New York, N.Y. 10024.

Robert J. Brady Company, a Prentice-Hall Company, Bowie, Md. 20715.

Clay-Adams, Division of Becton, Dickson and Co., 229 Webro Rd., Parsippany, N.J. 07054.

Metropolitan Museum of Art, Fifth Ave. at 82 St., New York, N.Y. 10028.

Museum of Modern Art, 11 W. 53 St., New York, N.Y. 10019.

National Audubon Society, 1130 Fifth Ave., New York, N.Y. 10028.

Society for Visual Education, Division of Singer Company, 1345 Diversey Parkway, Chicago, Ill. 60614.

Ward's Natural Science Establishment, Inc., 300 Ridge Rd. E., Rochester, N.Y. 14622.

handmade filmstrips

Buhl Projector Company, 60 Spruce St., Paterson, N.J. 07501.

Christy's, 212 W. Magnolia Blvd., Burbank, Calif. 91502

Film Makers, P.O. Box 593, Arcadia, Calif. 91006.

CHAPTER 10

audio materials

radio networks

American Broadcasting Company, 1330 Avenue of the Americas, New York, N.Y. 10019.

Columbia Broadcasting System, 383 Madison Ave., New York, N.Y. 10017.

Corporation For Public Broadcasting, 888 16th St., N.W., Washington, D.C. 20006.

Mutual Broadcasting System, 1440 Broadway, New York, N.Y. 10018.

National Broadcasting Company, 30 Rockefeller Plaza, New York, N.Y. 10020.

disc recordings

Bowmar Records, 622 Rodier Drive, Glendale, Calif. 91201.

Caedmon Records, 110 Tremont St., Boston, Mass. 02107.

Capitol Records, 1750 N. Vine St., Hollywood, Calif. 90028.

Columbia Records, 799 7th Ave., New York, N.Y. 10019.

Decca Records, Subsidiary of Brunswick Radio Division, 445 Park Ave., New York, N.Y. 10022.

Educational Recording Services, 6430 Sherbourne Dr., Los Angeles, Calif. 90056.

Educational Record Sales, 157 Chambers St., New York, N.Y. 10007.

Enrichment Teaching Materials, Inc. 246 Fifth Ave., New York, N.Y. 10001.

Folkways Records, 701 Seventh Ave., New York, N.Y. 10036.

National Council of Teachers of English, 1111 Canyon Rd., Urbana, Ill. 61801.

RCA Educational Services, Camden, N.J. 08108.

Spoken Arts, Inc., 310 North Ave., New Rochelle, N.Y. 10801.

H. Wilson Corporation, 555 W. Taft Dr., South Holland, Ill. 60473.

prerecorded tapes

Bell and Howell Co. (audio cards), 7100 McCormick Rd., Chicago, Ill. 60645.

Califone International (audio cards), 5922 Bowcroft St., Los Angeles, Calif. 90016.

Center for Cassette Studies, Inc., 8110 Webb Ave., North Hollywood, Calif. 91605.

Educational Development Laboratories, Division of McGraw-Hill Book Company, 1221 Avenue of the Americas, New York, N.Y. 10020.

Electronic Futures, Inc., 57 Dodge Ave., North Haven, Conn. 06473.

Heath, de Rochemont Corporation, Spring St., Lexington, Mass. 02116.

Holt Information Systems, 383 Madison Ave., New York, N.Y. 10017.

Imperial International Learning Corporation, Box 548, Kankakee, Ill. 60901.

Jeffrey Norton Publishers, 145 E. 49 St., New York, N.Y. 10017.

McGraw-Hill Book Company, Film Division, 1221 Avenue of the Americas, New York, N.Y. 10020.

National Association of Educational Broadcasters (NAEB), 1346 Connecticut Ave. N.W., Washington, D.C. 20036.

National Center For Audio Tapes (NCAT), 384 Stadium Building, University of Colorado, Boulder, Colo. 80302.

Spoken Arts, Inc., 310 North Ave., New Rochelle, N.Y. 10801.

Westinghouse Learning Corporation, 100 Park Ave., New York, N.Y. 10017.

Xerox University Microfilms, 300 North Zeeb Rd., Ann Arbor, Mich. 48106.

international correspondence sources

English-Speaking Union (Pen Friends Division), 16 E. 69 St., New York, N.Y. 10021.

International Friendship League, 40 Mount Vernon St., Boston, Mass. 02108.

International Tape Exchange Program, 834 Ruddiman Ave., North Muskegon, Mich. 49445.

Letters Abroad, 209 E. 56 St., New York, N.Y. 10022.

CHAPTER 11

motion pictures: films and video

general information sources

Center For Advanced Film Studies, American Film Institute, Kennedy Center for the Performing Arts, Washington, D.C. 20566.

8mm Source Directory—Silent and *8mm Source Directory—Sound,* Technicolor Inc., 299 Kalmas Dr., Costa Mesa, Calif. 92626.

Film Review Index, Audio Visual Associates, P.O. Box 324 Monterey Park, Calif. 91754.

Landers Film Reviews, Landers Associates, Box 69760, Los Angeles, Calif. 95155.

Parlato, S. J., Jr., *Films—Too Good for Words: A Directory of Nonnarrated 16mm Films,* R. R. Bowker, 1180 Avenue of the Americas, New York, N.Y. 10036.

U.S. Government Films: A Catalog of Audiovisual Materials for Rent and Sale, National Audiovisual Center, (GSA) Information Branch, Washington, D.C. 20409.

companies (producers)

Academy Films, 748 N. Seward St., Hollywood, Calif. 90038.

ACI Films, Inc., 35 West 45 St., New York, N.Y. 10036.

Association-Sterling Films, Inc., 600 Madison Ave., New York, N.Y. 10022.

Avis Films, Inc., 2408 W. Olive Ave., Burbank, Calif. 91506.

Arthur Barr Productions, P.O. Box 7-C, Pasadena, Calif. 91104.

BFA Educational Media, 2211 Michigan Ave., Santa Monica, Calif. 90404.

Stephen Bosustow Productions, 1649 Eleventh St., Santa Monica, Calif. 90404.

Robert J. Brady Company, a Prentice-Hall Co., Bowie, Md. 20715.

Campus Film Distributors Corporation, 20 E. 46 St., New York, N.Y. 10017.

Carlocke/Langden, Inc., 4122 Main St., Dallas, Tex. 75226.

Carousel Films, Inc., 1501 Broadway, New York, N.Y. 10036.

Churchill Films, Inc., 662 N. Robertson Blvd., Los Angeles, Calif. 90069.

Communications Films Division, 870 Monterey Pass Rd., Monterey, Calif. 91754.

Communications Group West, 6335 Homewood Ave., Suite 204, Hollywood Calif. 90028.

Contemporary Films/McGraw-Hill, 1221 Ave. of the Americas, New York, N.Y. 10020.

Coronet Instructional Media, 65 E. South Water St., Chicago, Ill. 60601.

CRM Educational Films, Del Mar, Calif. 92014.

Dana Productions, a Division of Saparoff Films, Inc., 6249 Babcock Ave., North Hollywood, Calif. 91606.

Walt Disney Educational Materials Co., 802 Sonora Ave., Glendale, Calif. 91201.

Ealing Corp., 2225 Massachusetts Ave., Cambridge, Mass. 02140.

Eastman Kodak, 343 State St. Rochester, N.Y. 14650.

Educational Media, Inc., 809 Industrial Way, P.O. Box 39, Ellensburg, Wash. 98926.

Encyclopaedia Britannica Educational Corporation, 425 N. Michigan Ave., Chicago, Ill. 60611.

Indiana University Audio-Visual Center, Field Services Department, Bloomington, Ind. 47401.

International Film Bureau, 332 S. Michigan Ave., Chicago, Ill. 60604.

International Film Foundation, 475 Fifth Ave., Suite 916. New York, N.Y. 10017.

Johnson Hunt Productions, Film Center, Coronado, Calif. 92118.

Learning Corporation of America, 711 Fifth Ave., New York, N.Y. 10022.

Macmillan Audio-Visual, 209B Brown Street, Riverside, N.J. 08075.

McGraw-Hill Film Division, 1221 Avenue of the Americas, New York, N.Y. 10020.

Modern Learning Aids, Division of Ward's Natural Science, P.O. Box 302, Rochester, N.Y. 14603.

Moody Institute of Science, Educational Film Division, 12000 E. Washington, Whittier, Calif. 90606.

National Educational Media, Inc., 15250 Ventura Blvd., Sherman Oaks, Calif. 91403.

National Film Board of Canada, 680 Fifth Ave., Suite 819, New York, N.Y. 10019.

Perennial Education, Inc., Box 236, 1825 Willow Rd., Northfield, Ill. 60093.

Perspective Films, 369 W. Erie St., Chicago, Ill. 60610.

Petite Film Company, 708 N. 62 St., Seattle, Wash. 98103.

Pyramid Films, Box 1048, Santa Monica, Calif. 90406.

Robert Saudek Associates, 689 Fifth Ave., New York, N.Y. 10022.

Sound Film Loop Source Directory, Technicolor, 1300 Frawley Dr., Costa Mesa, Calif. 92627.

Teaching Film Custodians, Film Library, Audio-Visual Center, Indiana University, Bloomington, Ind. 47401.

Time-Life Films, Department MB-1, 43 W. 16 St., New York, N.Y. 10011.

Thorne Films, Inc., 1229 University Ave., Boulder, Colo. 80302.

United World Films, Inc., 221 Park Ave. South, New York, N.Y. 10003.

Universal Education and Visual Arts, 221 Park Ave. South, New York, N.Y. 10003.

University of Iowa, AV Center, Division of Extension and University Service, Iowa City, Iowa 52240.

Westinghouse Learning Corporation, 100 Park Ave., New York, N.Y. 10017.

Weston Woods, Weston, Conn. 06880.

Wombat Productions, Inc., 77 Tarrytown Rd., White Plains, N.Y. 10607.

Xerox Films, 245 Long Hill Rd., Middletown, Conn. 06457.

free films (selected sources)

American Cancer Society, 219 42 St., New York, N.Y. 10017.

Aluminum Corporation of America, Motion Picture Section, 794 Alcoa Building, Pittsburgh, Pa. 15219.

American Iron and Steel Institute, Public Relations Dept., 1000 16th St. N.W., Washington, D.C. 20036.

Association—Sterling Educational Films, 866 Third Ave., New York, N.Y. 10022.

Bell Telephone Company (Check with your local telephone office.)

Educators Guide to Free Films, Educators Progress Service, Randolph, Wis. 53956.

Films, Inc., 1144 Wilmette Ave., Wilmette, Ill. 60091.

Free-Loan Educational Films, 16mm Sound, Modern Talking Picture Service, Inc., 2323 New Hyde Park Rd., New Hyde Park, N.Y. 11040.

General Telephone Company (Check with your local telephone office.)

General Mills, 9200 Film Center, P.O. Box 1113, Minneapolis, Minn. 55440.

General Motors Corporation, General Motors Building, 3044 W. Grand Blvd., Detroit, Mich. 48238.

National Foundation—March of Dimes, Box 2000, White Plains, N.Y. 10602.

entertainment films

BNA Communications, Inc., 5615 Fishers Lane, Rockville, Md. 20852.

Brandon Films, Inc., 244 Kearney St., San Francisco, Calif. 94108.

Contemporary Films/McGraw-Hill, 1221 Avenue of the Americas, New York, N.Y. 10020.

Films, Inc., 1144 Wilmette Ave., Wilmette, Ill. 60091.

Limbacher, James (ed.), *Feature Films on 8mm and 16mm: A Directory of Feature Films Available for Rental, Sale, and Lease in the United States* (4th ed.), R. R. Bowker Co., 1180 Ave. of the Americas, New York, N.Y. 10036.

Pyramid Films, Box 1048, Santa Monica, Calif. 90406.

Sutherland Learning Assoc., 8425 W. Third St., Los Angeles, Calif. 90048.

United World Films, Inc., 221 Park Ave. South, New York, N.Y. 10003.

Warner Brothers, Non-Theatrical Division, 4000 Warner Blvd., Burbank, Calif. 91505.

prerecorded video tapes:
information and selected sources

Addison-Wesley Publishing Company, Inc., 2725 Sand Hill Rd., Menlo Park, Calif. 94025.

Academy Films, Inc., 748 N. Seward St., Hollywood, Calif. 90038.

Advanced Systems Inc., 1601 Tonne Rd., Elk Grove, Village, Ill. 60007.

Aims Instructional Media Services, Inc., 626 Justin Ave., Glendale, Calif. 91201

Astral Television Films Ltd., 224 Davenport Rd., Toronto, Ontario M5R 1J7 Canada

AV-ED Films, 7934 Santa Monica Blvd., Los Angeles, Calif. 90046.

BNA Communications Inc., 5615 Fishers Lane, Rockville, Md. 20852.

Brigham Young University, Dept. of Motion Pictures Productions, MPS, Provo, Utah 84602

CBS News—Resources Development, 524 West 57 St., New York, N.Y. 10019.

Communications Group West, 6335 Homewood Ave., Hollywood, Calif. 90028.

Cornell University ETV Center, Martha Van Renssaler Hall, Ithaca, N.Y. 14850.

Creative Media, 820 Keosauqua Way, Des Moines, Iowa 50309.

Educational Resources Foundation, 2712 Millwood Ave., Columbia, S.C. 29205.

Encyclopaedia Britannica Educational Corporation, 425 N. Michigan Ave., Chicago, Ill. 60611.

Films Incorporated, 1144 Wilmette Ave., Wilmette, Ill. 60091.

¾″ U-matic video cassette
program sources

Great Plains National Instructional Television Library (catalog revised annually), Box 80669, Lincoln, Nebr. 68501.

Indiana University, Medical Educational Resources, PGM, 1100 West Michigan St., Indianapolis, Ind. 46202.

Intercollegiate Video Clearing House, P.O. Drawer 33000R, Miami, Fla. 33133.

International Film Bureau Inc., 332 South Michigan Ave., Chicago, Ill. 60604.

Learning Corporation of America, 1350 Ave. of the Americas, New York, N.Y. 10019.

McGraw-Hill Films, Dept. AR, 1221 Ave. of the Americas, New York, N.Y. 10020.

Mediascience Ltd., 742 Bay St., Toronto, Ontario, Canada.

Michigan State University, Instructional Television Services, 116 Linton Hall, East Lansing, Mich. 48824.

Modern Talking Picture Service, Inc., 2323 New Hyde Park Rd., New Hyde Park, N.Y. 11040.

Motorola Teleprograms, Inc., 4825 N. Scott St., Suite 26, Schiller Park, Ill. 60176.

National Audiovisual Center, GSA, Attention: Code S, Washington, D.C. 20409.

National Instructional Television Center, Box A, Bloomington, Ind. 47401.

Oxford Films, 1136 North Las Palmas Ave., Hollywood, Calif. 90038.

Perennial Education, Inc., 1825 Willow Rd., Northfield, Ill. 60093.

Pyramid Films, P.O. Box 1948, Santa Monica, Calif. 90406.

The Public Television Library, Video Program Service, 475 L'Enfant Plaza West, S.W., Washington, D.C. 20024.

Raindance Foundation, 51 Fifth Ave., Suite 11D, New York, N.Y. 10003.

Readers Digest, Television Division, 200 Park Ave., New York, N.Y. 10017.

Satellite Video Exchange Society, 261 Powell St., Vancouver, B.C., Canada V6A 1G3.

Smith-Mattingly Prod., Ltd., 310 South Fairfax St. Alexandria, Va. 22314.

Software Distribution, 705 Progress Ave., Unit 33, Scarborough, Ont., Canada.

Telstar Productions, Inc., 366 North Prior, St. Paul, Minn. 55104.

Time-Life Multi-Media, Time and Life, Rockefeller Center, New York, N.Y. 10020.

Transmedia, Ltd., 7, Grape St., London, WC2H 8DR, England.

TV Ontario, MKT Branch, 2180 Younge St., Toronto, Ontario M4S 2C1, Canada.

Video Bluebook (guide to packaged programs), Knowledge Industry Publications, Inc., 2 Corporate Park Dr., White Plains, N.Y. 10602.

Western Instructional Television, 1549 N. Vine St., Los Angeles, Calif. 90028.

CHAPTER 12

television

general information (educational)

Agency for Industrial Television, Box A, Bloomington, Ind. 47401.

Aspen Program On Communications and Society, Suite 232, 770 Welch Rd., Palo Alto, Calif. 94304.

Diamant, Lincoln (ed.), *The Broadcast Communications Dictionary,* Hastings House, Publishers, 10 East 40 St., New York, N.Y. 10016.

ITV Center, San Jose State University, San Jose, Calif. 95192.

National Association of Broadcasters, Television Information Office, 745 Fifth Ave., New York, N.Y. 10022.

National Association of Educational Broadcasters, 1346 Connecticut Ave., N.W., Washington, D.C. 20036.

National Educational Television, 10 Columbus Circle, New York, N.Y. 10019.

NET Film Service, Indiana University, Bloomington, Ind. 47401.

Prange, Werner, *Index to College Television Courseware,* 1st. ed., 1974, University of Wisconsin—Green Bay, Wis. 54302. (Regular revisions planned.)

The Video Handbook 2d ed., 1974, Media Horizons, 750 Third Ave., New York, N.Y. 10017.

Winslow, Ken (ed.), *Videoplay Program Source Guide* 4th ed., 1975, C. S. Tepfer Publishing Company, Inc., Ridgefield, Conn. 06877. (Updated annually.)

television networks

American Broadcasting Company, 1330 Ave. of the Americas, New York, N.Y. 10019.

Children's Television Workshop, 1 Lincoln Plaza, New York, N.Y. 10023.

Columbia Broadcasting System, 383 Madison Ave., New York, N.Y. 10017.

Corporation for Public Broadcasting, 888 Sixteenth St., N.W., Washington, D.C. 20006.

National Broadcasting Company, 30 Rockefeller Plaza, New York, N.Y. 10020.

National Educational Television, 10 Columbus Circle, New York, N.Y. 10019.

CHAPTER 13

real things, models and demonstrations

general information

Junior museums in your locality.

State, municipal, or private museums (over 5,000 in operation); see also the listings in Chapter 3.

Anthropology, science, and history departments in state universities and colleges.

models

Bico Scientific Co., National Biological Supply Co., Inc., 2325 S. Michigan Ave., Chicago, Ill. 60616.

Central Scientific Company, 2600 South Kostner Ave., Chicago, Ill. 60623.

Denoyer-Geppert Company, 5235 Ravenswood Ave., Chicago, Ill. 60640.

Frey Scientific Co., 465 S. Diamond St., Mansfield, Ohio 44903.

Hubbard Scientific Company, P.O. Box 105, Northbrook, Ill. 60062.

A. J. Nystrom and Company, 3333 Elston Ave., Chicago, Ill. 60618.

Rand McNally and Co., P.O. Box 7600, Chicago, Ill. 60680.

Ward's Natural Science Establishment, Inc., 300 Ridge Rd. E., Rochester, N.Y. 14622.

Welch Scientific Company, 7300 N. Linder Ave., Skokie, Ill. 60076.

aquariums and terrariums:
insect and animal collections

Aquarium and Science Supply Company, Box 11841, Philadelphia, Pa. 19128.

Aquarium Systems Inc., 33208 Lakeland Blvd., East Lake, Ohio 44094.

Denoyer-Geppert Company, 5235 Ravenswood Ave., Chicago, Ill. 60640.

Eduquip-Macalaster Corporation, 1085 Commonwealth Ave., Boston, Mass. 02215.

Fisher Scientific Company, Stansi Educational Materials Division, 125 N. Wood St., Chicago, Ill. 60622.

Frey Scientific Company, 465 S. Diamond St., Manfield, Ohio 44903.

Jewel Aquarium Company, Inc., 5005 West Armitage Ave., Chicago, Ill. 60639.

Lab-Aids Inc., 130 Wilbur Place, Bohemia, N.Y., 11716.

Nasco, 901 Janesville Ave., Fort Atkinson, Wis. 53538.

National Pet Supply Company, 3105 Olive St., St. Louis, Mo. 63103.

Ward's Natural Science Establishment, 300 Ridge Rd. E., Rochester, N.Y. 14622.

calculators

Hewlett-Packard, 1501 Page Mill Rd., Palo Alto, Calif. 94304.

Monroe, The Calculator Co., Division of Litton Industries, 550 Central Ave., Orange, N.J. 07051.

Olympia USA, Inc., Rt. 22, Box 22, Somerville, N.J. 08876.

Sharp Electronics, Paramus, N.J. 07652.

Singer Business Machines, 2350 Washington Ave., San Leandro, Calif. 94577.

Texas Instruments, P.O. Box 22283, Dallas, Tex. 75222.

scientific experimental laboratory
equipment and supplies

American Optical Corporation, Scientific Instrument Division, Buffalo, N.Y. 14215.

Beckman Instruments, Inc., Cedar Grove Operations, 89 Commerce Rd., Cedar Grove, N.J. 07009.

Central Scientific Company, 2600 S. Kostner Ave., Chicago, Ill. 60623.

Edmund Scientific Company, Dept. YS, 300 Edscorp Building, Barrington, N.J. 08007.

Fisher Scientific Company, Stansi Educational Materials Division, 1259 N. Wood St., Chicago, Ill. 60622.

General Biological Supply House, Inc., 8200 S. Hoyne Ave., Chicago, Ill. 60626.

Malicks' Fossils, Inc., 5514 Plymouth Rd., Baltimore, Md. 21214.

Spectra-Physics, 1250 W. Middlefield Rd., Mountain View, Calif. 94040.

C. H. Stoelting Company, 424 N. Homan Ave., Chicago, Ill. 60624.

Swift Instruments, Inc., Technical Instrument Division, P.O. Box 562, San Jose, Calif. 95106.

Ward's Natural Science Establishment Inc., 300 Ridge Rd. E., Rochester, N.Y. 14622.

W. M. Welch Scientific Company, 7300 N. Linder Ave., Skokie, Ill. 60076.

planetariums

Astro Tec Manufacturing, Inc., 231 Locust St., Canal Fulton, Ohio 44614.

Jack C. Coffey, Inc., 104 Lake View Ave., Waukegan, Ill. 60085.

Demco Educational Corporation, 2120 Fordem Ave., Madison, Wis., 53704.

Denoyer-Geppert Company, 5235 Ravenswood Ave., Chicago, Ill. 60640.

Farquhar Transparent Globes, 5007 Warrington Ave., Philadelphia, Pa. 19143.

Trippensee Planetarium Company, 301 Cass St., Saginaw, Mich. 48602.

Viewlex, Inc., 1 Broadway Ave., Holbrook, N.Y. 11741.

CHAPTER 14

games, simulations, and dramatizations

Academic Games Associates, Inc., 430 E. 33 St., Baltimore, Md. 21218.

Avalon Hill Company, 4517 Hartford Rd., Baltimore, Md. 21214.

Changing Times Education Service, Kiplinger Washington Editors, 1729 "H" St., N.W., Washington, D.C. 20006.

Didactic Systems, Inc., 6 N. Union Ave., Cranford, N.J. 07016.

Digital Equipment Corporation, 146 Main St., Maynard, Mass. 01954.

Educational Methods, 500 N. Dearborn St., Chicago, Ill. 60610.

Foreign Policy Association, 345 E. 46 St., New York, N.Y. 10017.

Garrard Publishing Company, Champaign, Ill. 61829.

Herder and Herder, 232 Madison Ave., New York, N.Y. 10020.

Information Resources, Ins., P.O. Box 417, Lexington, Mass. 02173.

Instructional Development Corporation, P.O. Box 805, Salem, Oreg. 97308.

Interact, P.O. Box 262, Lakeside, Calif. 92040.

International Learning Corporation, 245 S.W., 32 St., Ft. Lauderdale, Fla. 33301.

Intext Educational Publishers, Scranton, Pa. 19515.

Joint Council on Economic Education, 1212 Ave. of the Americas, New York, N.Y. 10036.

Listener Educational Enterprises, 6777 Hollywood Blvd., Hollywood, Calif. 90028.

McGraw-Hill Book Co., Inc., 1221 Ave. of the Americas, New York, N.Y. 10020.

Local Telephone Companies, Educational Services Offices (telephone devices).

Macmillan Co., 866 Third Ave., New York, N.Y. 10022.

Chas. E. Merrill Publishing Co., 1300 Alum Creek Drive, Columbus, Ohio 43216.

New Century/Appleton-Century-Crofts, 440 Park Ave. South, New York, N.Y. 10016.

Jeffrey Norton Publishers, Inc., 145 E. 49 St., New York, N.Y. 10017.

Parker Bros., Inc., 190 Bridge St., Salem, Mass. 01970.

Science Research Associates, 259 E. Erie St., Chicago, Ill. 60611.

Scott, Foresman and Company, 1900 E. Lake Ave., Glenview, Ill. 60025.

Selchow and Righter Co., 2215 Union Blvd., Bayshore, N.Y. 11706.

Simile II, P.O. Box 1023, La Jolla, Calif. 92037.

Social Studies School Service, 10000 Culver Blvd. Culver City, Calif. 90230.

Simulated Environments, Inc., University City Science Center, 3401 Market St., Philadelphia, Pa. 19104.

Simulation/Gaming/News Service, Box 3039, University Stn., Moscow, Idaho 83843.

3M Company, 3M Center, St. Paul, Minn. 55101.

Transnational Programs Corp., 54 Main St., Scottsville, N.Y. 14546.

Urban Systems, Inc. 1033 Massachusetts Ave., Cambridge, Mass. 02138.

Urbex Affiliates, Inc., P.O. Box 2198, Ann Arbor, Mich. 48106.

Western Publishing Company, Education Division, 850 Third Ave., New York, N.Y. 10022.

WFF'N PROOF, 1111 Maple Ave., Turtle Creek, Pa. 15145.

Zuckerman, David W., and Robert E. Horn, *The Guide to Simulations/Games for Education and Training,* Information Resources, Inc., P.O. Box 417, Lexington, Mass. 02173. (New editions every 2 to 3 years.)

CHAPTER 15

free and inexpensive supplementary materials

selected associations

Aluminum Association, 750 Third Ave., New York, N.Y. 10017.

American Gas Association, 1515 Wilson Blvd., Arlington, Va. 22209.

American Iron and Steel Institute, 1100 16th St., N.W., Washington, D.C. 20036.

American Petroleum Institute, 1271 Ave. of the Americas, New York, N.Y. 10020.

American Telephone and Telegraph, Information Dept., 195 Broadway, New York, N.Y. 10007.

American National Red Cross, 18th and E. St., N.W., Washington, D.C. 20066.

Farm Film Foundation, 1425 H St., N.W., Washington, D.C. 20005.

National Canners Association, 1133 Twentieth St., N.W., Washington, D.C. 20036.

National Dairy Council, 11 N. Canal St., Chicago, Ill. 60606.

National Education Association, 1201 Sixteenth St. N.W., Washington, D.C. 20036.

The National Foundation—March of Dimes, Box 2000, White Plains, N.Y. 10602.

National Geographic Society, 1146 Sixteenth St. N.W., Washington, D.C. 20036.

National Wildlife Federation, 1412 Sixteenth St. N.W., Washington, D.C. 20036.

The Smithsonian Institution, Washington, D.C. 20025.

selected commercial organizations

Aluminum Company of America, Motion Picture Section, 1501 Alcoa Building, Pittsburgh, Pa. 15219.

American Airlines, 633 Third Ave., New York, N.Y. 10017.

American Telephone and Telegraph Company, 195 Broadway, New York, N.Y. 10007.

Bell Telephone Laboratories, from your local Bell System Business Office.

Binney and Smith, 380 Madison Ave., New York, N.Y. 10017.

Denoyer-Geppert Company, 5235 Ravenswood Ave., Chicago, Ill. 60640.

A. B. Dick Company, 5700 W. Touhy Ave., Chicago, Ill. 60648.

Eastern Airlines, International Airport, Miami, Fla. 33148.

General Motors Corporation, Public Relations Staff, Detroit, Mich. 48202.

Goodyear Tire and Rubber Company, Public Relations Department, 1144 East Market St., Akron, Ohio. 44316.

New York Stock Exchange, Manager, School and College Relations, 11 Wall St., New York, N.Y. 10005.

Sandia Laboratories, P.O. Box 5800, Albuquerque, N.M. 87115.

Standard Oil Company of California, Public Relations Department, 225 Bush St., San Francisco, Calif. 94104.

Summy-Birchard Company, 1834 Ridge Ave., Evanston, Ill. 60204.

Timken Company, 1835 Dueber Ave. S.W., Canton, Ohio 44706.

United Air Lines, P.O. Box 8800, O'Hare International Airport, Chicago, Ill. 60666.

United States Steel Corporation, Public Relations Department, 71 Broadway, New York, N.Y. 10006.

selected professional associations

Teachers and students who want vocational or other information about a special field or profession should write directly to the national headquarters office of the appropriate professional association. The following list suggests a few such associations:

American Dental Association, 211 E. Chicago Ave., Chicago, Ill. 60611.

American Historical Association, 400 A St. S.E., Washington, D.C. 20003.

American Library Association, 50 E. Huron St., Chicago, Ill. 60611.

American Medical Association, 535 N. Dearborn St., Chicago, Ill. 60610.

Modern Language Association of America, 60 Fifth Ave., New York, N.Y. 10003.

National Association of Educational Broadcasters, 1346 Connecticut Ave. N.W., Washington, D.C. 20036.

National Council of Teachers of English, 1111 Kenyon Rd., Urbana, Ill. 61801.

National League for Nursing, 10 Columbus Circle, New York, N.Y. 10019.

selected sources of information
on foreign countries

Australian News and Information Bureau, 636 Fifth Ave., New York, N.Y. 10020.

Belgian Embassy, 3330 Garfield St. N.W., Washington, D.C. 20008.

Consulate General of Ireland, 580 Fifth Ave., New York, N.Y. 10036.

Embassy of the Union of Soviet Socialist Republics, Press Department, 1125 Sixteenth St. N.W., Washington, D.C. 20036.

German Travel Advisors Information Center, 866 Third Ave., New York, N.Y. 10022.

Information Service of India, 3 East 64 St., New York, N.Y. 10021.

Italian Cultural Institute, 686 Park Ave., New York, N.Y. 10021.

Japan National Tourist Organization, 45 Rockefeller Plaza, New York, N.Y. 10020.

Swedish Information Service, 825 Third Ave., New York, N.Y. 10017.

Swiss National Tourist Office, 10 West 49 St., New York, N.Y. 10020.

selected sources on international affairs

Bureau of International Education and Cultural Affairs, U.S. Department of State, Washington, D.C. 20520.

Food and Agricultural Organization of the United Nations, North American Liaison Office, 1325 C St. S.W., Washington, D.C. 20437.

International Labor Office, 7917 15th St. N.W., Washington, D.C. 20005.

Pan American Union, 17th and Constitution, Washington, D.C. 20006.

United States National Commission for UNESCO, Washington, D.C. 20520.

United Nations Office of Public Information, New York, N.Y. 10017.

CHAPTER 16

print, multimedia,
and microforms

general sources of information (See also Reference Section 5, Chapter 4 listings.)

American Library Association, 50 E. Huron St., Chicago, Ill. 60611.

Association For Educational Communications and Technology, 1201 16th St. N.W., Washington, D.C. 20036.

BFA Educational Media, 2211 Michigan Ave., Santa Monica, Calif. 90404.

Bro-Dart, 1609 Memorial Ave., Williamsport, Pa. 17701.

Brown, James W. (ed.), *Educational Media Yearbook,* R. R. Bowker Co. (A Xerox Education Co.), 1180 Ave. of the Americas, New York, N.Y. 10036. (Revised annually since 1973.)

Choice, 100 Riverview Center, Middletown, Conn, 06457.

Comics Magazine Association of America, 300 Park Ave., New York, N.Y. 10010.

ERIC Central, National Institute of Education, 1200 19th St. N.W., Washington, D.C. 20208.

Gaddy, Dale, *A Microform Handbook,* 1974, National Microfilm Association, 8728 Colesville Rd., Silver Springs, Md. 20910.

Libraries, Unlimited, Inc., Box 263, Littleton, Colo. 80120.

Microfilm Programs for Educational Institutions, 1975, Eastman Kodak Co., 343 State St., Rochester, N.Y. 14650.

Wynar, Bohdan S., *American Reference Books Annual,* 1974, Libraries Unlimited, Box 263, Littleton, Colo. 80120.

Xerox University Microfilms, 300 North Zeeb Rd., Ann Arbor, Mich. 48106.

publishers (selected examples)

Appleton-Century-Crofts, Inc., 440 Park Ave. South, New York, N.Y. 10016.

Bantam Books, Inc., 666 Fifth Ave., New York, N.Y. 10019.

R. R. Bowker Company (a Xerox Co.), 1180 Ave. of the Americas, New York, N.Y. 10036.

Thomas Y. Crowell Company, 666 Fifth Ave., New York, N.Y. 10019.

Doubleday and Company, Inc., Garden City, N.Y. 11530.

Educational Technology Publications, 140 Sylvan Ave., Englewood Cliffs, N.J. 07632.

Follett Publishing Company, 1010 W. Washington Blvd., Chicago, Ill. 60607.

Gage Educational Publishing, Ltd., 165 Commander Blvd., Agincourt, Ontario, Canada.

Ginn and Company, Statler Building, Boston, Mass. 02117.

Harcourt Brace Jovanovich, 757 Third Ave., New York, N.Y. 10017.

Harper & Row, Publishers, Inc., 10 E. 53 St., New York, N.Y. 10022.

Holt, Rinehart and Winston, Inc., 383 Madison Ave., New York, N.Y. 10017.

Houghton Mifflin Company, 110 Tremont St., Boston, Mass. 02107.

Learning Corporation of America, 711 Fifth Ave., New York, N.Y. 10022.

Little, Brown and Company, 34 Beacon St., Boston, Mass. 02106.

Macmillan Company, 866 Third Ave., New York, N.Y. 10022.

McGraw-Hill Book Company, 1221 Ave. of the Americas, New York, N.Y. 10020.

Penguin Books, Inc., 7110 Ambassador Rd., Baltimore, Md. 21207.

Pergamon of Canada, Ltd., 207 Queen's Quay West, Toronto, Canada.

Pocket Books, Inc., Division of Simon and Schuster, Inc., 630 Fifth Ave., New York, N.Y. 10020.

Prentice-Hall, Inc., Englewood Cliffs, N.J. 07632.

Random House, Inc., 457 Madison Ave., New York, N.Y. 10022.

Scott, Foresman and Company, 1900 E. Lake Ave., Glenview, Ill. 60025.

Simon and Schuster, Inc., 630 Fifth Ave., Rockefeller Center, New York, N.Y. 10020.

John Wiley and Sons, Inc., 605 Third Ave. New York, N.Y. 10016.

H. W. Wilson Company, 555 W. Taft Dr., South Holland, Ill. 60473.

encyclopedias

Americana Corporation, (*Encyclopedia Americana*), 375 Lexington Ave., New York, N.Y. 10022.

Collier Macmillan School and Library Services, 866 Third Ave., New York, N.Y. 10022.

Columbia University Press (*Columbia Encyclopedia*), Center for Mass Communication, 440 W. 110 St., New York, N.Y. 10025.

F. E. Compton and Company, Division of Encyclopaedia Britannica, 425 N. Michigan Ave., Chicago, Ill. 60611.

Encyclopaedia Britannica Educational Corporation (*Britannica* and *Britannica Junior*), 425 N. Michigan Ave., Chicago, Ill. 60611.

Field Enterprises Educational Corporation (*World Book Encyclopedia*), Merchandise Mart Plaza, Chicago, Ill. 60654.

Grolier Educational Corporation, 845 Third Ave., New York, N.Y. 10022.

multi-media kits

Allyn and Bacon, 470 Atlantic Ave., Boston, Mass. 02210.

Behavioral Research Laboratories, P.O. Box 577, Palo Alto, Calif. 94302.

BFA Educational Media, 2211 Michigan Ave., Santa Monica, Calif. 90404.

Bowmar Publishing Corporation, 622 Rodier Dr., Glendale, Calif. 91201.

Childrens Press Books, 1224 W. Van Buren St., Chicago, Ill. 60607.

Denoyer-Geppert Company, 5235 Ravenswood Ave., Chicago, Ill. 60640.

Educational Media, Inc., P.O. Box 39, Ellensburg, Wash. 98926.

ENRICH Telor, 760 Kifer Rd., Sunnyvale, Calif. 94086.

Ginn and Company, Statler Building, Boston, Mass. 02117.

Imperial International Learning, Rt. 45 S., P.O. Box 548, Kankakee, Ill. 60901.

Listener Educational Enterprises, 6777 Hollywood Blvd., Hollywood, Calif. 90028.

McGraw-Hill Book Company, Film Division and Webster Division, 1221 Ave. of the Americas, New York, N.Y. 10020.

Science Research Associates, Inc., 259 E. Erie St., Chicago, Ill. 60611.

Scott Education Division (Scott Graphics), 104 Lower Westfield Rd., Holyoke, Mass. 01040.

Scott, Foresman and Company, 1900 E. Lake Ave., Glenview, Ill. 60025.

Ward's Natural Science Establishment, 300 Ridge Rd. E., Rochester, N.Y. 14622.

microforms

Arcata Microfilm Corp., 700 S. Main St., Spring Valley, N.Y. 10977.

Greenwood Press, Inc., Educational Film Division, 51 Riverside Ave., Westport, Conn. 06880.

Eastman Kodak Company, 343 State St. Rochester, N.Y. 14650.

Micro Photo Division, Bell and Howell, Old Mansfield Rd., Wooster, Ohio 44691.

Smithsonian Institution, Photographic Services, Washington, D.C. 20560.

The New York Times Company, 229 W. 43 St., New York, N.Y. 10036.

3M-International Microfilm Press, 3M Center, St. Paul, Minn. 55101.

U.S. Historical Documents Institute, Inc., 1647 Wisconsin Ave., Washington, D.C. 20007.

University Microfilms-Xerox, 300 N. Zeeb Rd., Ann Arbor, Mich. 48106.

John Wiley and Sons, Inc., 605 Third Ave., New York, N.Y. 10016.

microform projectors and readers

Dukane Corporation, 2900 Dukane Drive, St. Charles, Ill. 60174.

Ednalite Corporation, 200 N. Water St., Peekskill, N.Y. 10566.

Karl Heitz, Inc., 979 Third Ave., New York, N.Y. 10022.

Micro Information Systems, Inc., 467 Armour Circle N.E., Atlanta, Georgia 30324.

Microfilming Corporation of America, 21 Harristown Rd., Glen Rock, N.J. 07452.

Realist, Inc., N. 93 W. 16288 Megal Drive, Menominee Falls, Wisc. 53051.

3M Company, 3M Center, St. Paul Minn. 55101.

CHAPTER 17

. . . and, in the future?

Audiotronics, 7428 Bellaire Ave., P.O. Box 3997, North Hollywood, Calif. 91609.

Computer Curriculum Corporation, 1032 Elwell Court, Palo Alto, Calif. 94303.

Digi-Log Systems, Inc., Babylon Rd., Horsham, Pa. 19044.

Digitor: Centurion Industries Inc., 2549 Middlefield Rd., Redwood City, Calif. 94063.

Disco-Vision: MCA Disco-Vision, Inc., 100 Universal City Plaza, Universal City, Calif. 91608.

ENRICH Telor, 760 Kifer Road, Sunnyvale, Calif. 94086.

Hewlett-Packard, 1501 Page Mill Rd., Palo Alto, Calif. 94304.

Imperial International Learning Corporation, Box 548, Kankakee, Ill. 60901.

International Audio Visual, Inc., 15818 Arminta St., Van Nuys, Calif. 91406.

Mathiputer: Cybernetic Systems, Inc., 9615 Acoma Southeast, Albuquerque, N.M. 87123.

Montron Corporation, Audio Visual Products Division, 185 East Dana St., Mountain View, Calif. 94041.

Radio Corporation of America (RCA), Consumer Electronics Division, 600 North Sherman Drive, Indianapolis, Ind. 46201.

Sharp Electronics Corp., Paramus, N.J. 07652.

Xerox Corporation, Office Systems Division, 1341 West Mockingbird Lane, Dallas, Tex. 75247.

Xerox Corporation, Xerox Square, Rochester, N.Y. 14644.

REFERENCE SECTION 1

audiovisual equipment

record players

Audiotronics, Inc., 7428 Bellaire Ave., North Hollywood, Calif. 91605.

Califone International, 5922 Bowcroft, Los Angeles, Calif. 90016.

Hamilton Electronics, 2003 W. Fulton St., Chicago, Ill. 60612.

Newcomb Audio Products Company, 12881 Bradley Ave., Sylmar, Calif. 91342.

Radio Matic of America, 760 Ramsey Ave., Hillside, N.J. 07205.

RCA Victor Division, Radio Corporation of America, Camden, N.J. 08102.

V-M Corporation, 304 Territorial Rd., Benton Harbor, Mich. 49022.

tape recorders

Akai Electric Company, Ltd., 2139 E. Del Amo Blvd., Compton, Calif. 90220.

Ampex Corporation, 401 Broadway, Redwood City, Calif. 94063.

Audiotronics Corporation, 7428 Bellaire Ave., North Hollywood, Calif. 91605.

Bell and Howell Company, 7100 McCormick Rd., Chicago, Ill. 60645.

Newcomb Audio Products Co., 12881 Bradley Ave., Sylmar, Calif. 91342.

Norelco, 100 E. 42 St., New York, N.Y. 10017.

RCA Victor Division, Radio Corporation of America, Camden, N.J. 08102.

Califone International, Inc., 5922 Bowcroft St., Los Angeles, Calif. 90016.

Sony Corporation of America, 9 West 57 St., New York, N.Y. 10019.

Telex Communications Division, 9600 Aldrich Ave. S., Minneapolis, Minn. 55420.

V-M Corporation, Box 1247, Benton Harbor, Mich. 49022.

3M Company, Audio Products Div., 3M Center, St. Paul, Minn. 55101.

filmstrip projectors (silent and sound)

American Optical Corporation, 14 Mechanic St., Southbridge, Mass. 01550.

Bell and Howell Company, 7100 McCormick Rd., Chicago, Ill. 60645.

Jack C. Coffey Company, 104 Lake View Ave., Waukegan, Ill. 60085.

Dukane Corporation, St. Charles, Ill. 60174.

Eastman Kodak Co., 343 State St., Rochester, N.Y. 14650.

Kalart Victor Corporation, Plainville, Conn. 06062.

LaBelle Industries, Inc., 510 S. Washington St., Oconomowoc, Wis. 53066.

Singer Education Systems, 3750 Monroe Ave., Rochester, N.Y. 14603. (Graflex).

Standard Projector and Equipment Co., 1919 Pickwick Ave., Glencoe, Ill. 60025.

Viewlex Audiovisual Inc., Holbrook, L.I., N.Y. 11741.

video tape recorders

Akai America, Ltd., 2139 E. Del Amo Blvd., Compton, Calif. 90220.

Ampex, 401 Broadway, Redwood City, Calif. 94063.

Concord Electronics Corporation, 1935 Armacost, Los Angeles, Calif. 90025.

International Video Corporation, 990 Almanor Ave., Sunnyvale, Calif. 94086.

JVC Industries, Inc., 50–35 56th Rd., Maspeth, N.Y. 11378.

Panasonic/Matsushita Electric Corporation, Pan Am Bldg., 200 Park Ave., New York, N.Y. 10017.

Philips Audio Video Systems Corporation, One Philips Parkway, Montvale, N.J. 07645.

Radio Corporation of America, Front and Cooper St., Camden, N.J. 08108.

Sony Corporation, 9 West 57 St., New York, N.Y. 10019.

3M Video Learning Systems, 3M Center, Bldg. 236-IN, St. Paul, Minn. 55101.

programmed instruction equipment

Audiotronics Corporation, 7428 Bellaire Ave., N. Hollywood, Calif. 91605.

Bell and Howell, Audio-Visual Products Division, 7100 McCormick Rd., Chicago, Ill. 60645.

Califone International, Inc., 5922 Bowcroft St., Los Angeles, Calif. 90016.

Educational Development Laboratories, McGraw-Hill Book Co., 1221 Ave. of the Americas, New York, N.Y. 10020.

Grolier Educational Corporation, 845 Third Ave., New York, N.Y. 10022.

North American Phillips Corporation, 100 E. 42 St., New York, N.Y. 10017.

Scott Engineering Services-Omnilab, Division of A-T-O, Inc., 1400 S.W. Eighth St., Pompano Beach, Fla. 33060.

Singer Company, GPL-TV, Link Division, Binghamton, N.Y. 13402.

Viewlex, Inc., Holbrook, N.Y. 11741.

Welch Scientific Company, 7300 N. Linder Ave., Skokie, Ill. 60076.

2- by 2-inch slide projectors

Allied Impex Corporation (Bauer projectors), 168 Glen Cove Rd., Carle Place, N.Y. 11514.

American Optical Corporation, Eggert Rd., Buffalo, N.Y. 14215.

Bausch and Lomb, Inc., 36374 Bausch St., Rochester, N.Y. 14602.

Bell and Howell Company, 7100 McCormick Rd., Chicago, Ill. 60645.

Eastman Kodak Company, 343 State St., Rochester, N.Y. 14650.

Graflex Division-Singer Company, 3750 Monroe Ave., Rochester, N.Y. 14603.

Honeywell, Inc., 4800 E. Dry Creek Rd., Denver, Colo. 80217.

Kalart Victor Corporation, Plainville, Conn. 06062.

LaBelle Industries, Inc., 510 S. Washington St., Oconomowoc, Wis. 63066.

E. Leitz, Inc., 468 Park Ave., South, New York, N.Y. 10016.

Spindler and Sauppe, 1329 Grand Central Ave., Glendale, Calif. 91201.

Standard Projector and Equipment Company, 1911 Pickwick Ave., Glenview, Ill. 60025.

Viewlex Company, Inc., Holbrook, L.I., N.Y. 11741.

16mm sound projectors

Bell and Howell Company, 7100 McCormick Rd., Chicago, Ill. 60645 (Bell and Howell).

Eastman Kodak Company, 343 State St., Rochester, N.Y. 14650 (Pageant).

International Audio-Visual, Inc., 15818 Arminta St., Van Nuys, Calif. 91406. (Freeway 16)

Kalart Victor Corporation, Hultenius St., Plainville, Conn. 06062. (Victor)

Singer Educational Systems, 3750 Monroe Ave., Rochester, N.Y. 14603. (Graflex)

Viewlex, Inc., Holbrook, L.I., N.Y. 11741. (Viewlex/RCA)

8mm sound projectors

Bell and Howell Company, 7100 McCormick Rd., Chicago, Ill. 60645.

Charles Beseler Company, 8 Fernwood Rd., Florham Park, N.J. 07932.

Eastman Kodak Co., 343 State St., Rochester, N.Y. 14650.

Fairchild Camera and Instrument Corporation, Industrial Products Division, 75 Mall Drive, Commack, L.I., N.Y. 11725.

Jayark Instruments Corporation, 420 Madison Ave., New York, N.Y. 10017.

Paillard, Inc., 1900 Lower Road, Linden, N.J. 07036.

Technicolor, Inc., Commercial and Educational Division, 1300 Frawley Dr., Costa Mesa, Calif. 92627.

opaque projectors

American Optical Corporation, 14 Mechanic St., Southbridge, Mass. 01550.

Bausch and Lomb Incorporated, 36374 Bausch St., Rochester, N.Y. 14602.

Charles Beseler Company, 8 Fernwood Rd., Florham Park, N.J. 07932.

Buhl Projector Co., 1009 Beech Ave., Pittsburgh, Pa. 15233.

Projection Optics Co., 271 Eleventh Ave., East Orange, N.J. 07018.

overhead and standard
lantern-slide projectors

American Optical Corporation, Scientific Instrument Division, Buffalo, N.Y. 14215.

Bausch and Lomb, Inc., 36374 Bausch St., Rochester, N.Y. 14602.

Bell and Howell Company, Audio-Visual Products Division, 7100 McCormick Rd., Chicago, Ill. 60645.

Charles Beseler Company, 8 Fernwood Rd., Florham Park, N.J. 07932.

Keystone Division of Berkey Photo, Keystone Pl., Paramus, N.J. 07652.

GAF Corporation, 140 W. 51 St., New York, N.Y. 10020.

Tecnifax, Products of Scott Education Division, Holyoke, Mass. 01040.

3M Company, Visual Products Division, 3M Center, St. Paul, Minn. 55101.

H. Wilson Corporation, 546 W. 119 St., Chicago, Ill. 60473.

television receivers (school models)

Admiral Corporation, 3800 W. Cortland St., Chicago, Ill. 60647.

Conrac Corporation, 600 North Rimsdale Ave., Covina, Calif. 91722.

Electrohome, 9314 W. 122 St., Palos Park, Ill. 60464.

General Electric Company, Visual Communications Products Department, Electronics Park, Syracuse, N.Y. 13201.

GPL Division, General Precision, Inc., Pleasantville, N.Y. 10570.

Jerrold Electronics Corporation, 200 Witmer Rd., Horsham, Pa. 19044.

RCA Victor Division, Radio Corporation of America, Camden, N.J. 08101.

Setchel Carlson S-C Electronics, Inc., 530 Fifth Ave., St. Paul, Minn. 55112.

Sony Corporation, 47–47 Van Dam St., Long Island City, N.Y. 11101.

Westinghouse, 200 Park Ave., New York, N.Y. 10017.

listening centers and laboratories

Advance Products Company, Inc., Central at Wabash, P.O. Box 2178, Wichita, Kan. 67201.

Ampex Corporation, 401 Broadway, Redwood City, Calif. 94040.

Avedex, Inc., 7326 Niles Center Rd., Skokie, Ill. 60076.

Avid, Avid Corporation, 10 Tripps La., East Providence, R.I. 02914.

Califone International, Inc., 5922 Bowcroft St., Los Angeles, Calif. 90016.

Educational Developmental Laboratories, Division of McGraw-Hill Book Company, Huntington, N.Y. 11744.

Hamilton Electronics Corporation, 2726 W. Pratt Ave., Chicago, Ill. 60645.

North American Philips Corporation (Norelco), 100 E. 42 St., New York, N.Y. 10017.

P/H Electronics (Moni-Com), 117 East Helena St., Dayton, Ohio 4504

Radio Corporation of America, Education Service, Camden, N.J. 08108.

Raytheon Learning Systems Co., 455 Sheridan Ave., Michigan City, Ind. 46360.

Tandberg of America, Labriola Ct., Armonk, N.Y. 10504.

Westinghouse Electric Corporation, Gateway Center, 201 Liberty Ave., Pittsburgh, Pa. 15230.

television systems

A-V Systems, Inc., 44 Railroad Ave. Glen Head, N.Y. 11545.

Akai, 2139 E. Del Amo Blvd., Compton, Calif. 90220.

Audiotronics Corporation, 7428 Bellaire Ave., N. Hollywood, Calif. 91605.

Buhl Optical Co., 1009 Beech Ave., Pittsburgh, Pa. 15233.

Concord Electronics Corp., 1935 Armacost Ave., Los Angeles, Calif. 90025.

General Electric Company, Telecommunication Products, Electronics Park, Syracuse, N.Y. 11201.

GTE Sylvania, Burlington Rd., Bedford, Mass. 01730.

International Video Corporation, 990 Almanor Ave., Sunnyvale, Calif. 94086.

JVC Industries, Inc. 30–35 56th Rd., Maspeth, N.Y. 11378.

Panasonic, 200 Park Ave., New York, N.Y. 10017.

Philips Audio Video Systems Corp., 1 Philips Parkway, Montvale, N.J. 07645.

Radio Corporation of America, Closed Circuit Video Equipment, New Holland Pike, Lancaster, Pa. 17604.

Sony Corporation, 9 West 57 St., New York, N.Y. 10019.

Whitehouse Products, Inc. (Beacon), 360 Furman St., Brooklyn, N.Y. 11201.

desk-top multi-media viewers

Chas. Beseler Co., 8 Fernwood Rd., Florham Pk., N.J. 07932.

DuKane Corporation, 2900 Dukane Dr., St. Charles, Ill. 60174.

Fairchild Camera and Instrument Corporation, 75 Mall Rd., Commack, N.Y. 11725.

General Electric Company, P.O. Box 43, 1 River Rd., Schenectady, N.Y. 12305.

General Learning Corporation, Media Division, 250 James St., Morristown, N.J. 07960.

North American Phillips Corporation, 100 E. 42 St., New York, N.Y. 10017.

Retention Communication Systems (RCS), 2 Penn Plaza, New York, N.Y. 10001.

Singer Educational Systems, 3750 Monroe Ave., Rochester, N.Y. 14603.

Viewlex, Inc., Holbrook, L.I., N.Y. 11741.

screens

Da-Lite Screen Company, State Road 15 N., Warsaw, Ind. 46580.

Draper Shade and Screen Company, 411 S. Pearl St., Spiceland, Ind. 47385.

Eastman Kodak Co., 343 State St., Rochester, N.Y. 14650.

Polacoat, Inc., 9750 Conklin Rd., Cincinnati, Ohio 45242.

Radiant Corporation, 8220 N. Austin Ave., Morton Grove, Ill. 60053.

Scott Education Division, 104 Lower Westfield Rd., Holyoke, Mass. 01040.

Singer Education Systems, 3750 Monroe Ave., Rochester, N.Y. 14603.

REFERENCE SECTION 2

duplicating processes

spirit duplicators

A. B. Dick Co., 5700 W. Touhy Ave., Chicago, Ill. 60648.

Bell and Howell, Ditto Division, 6800 McCormick Rd., Chicago, Ill. 60645.

Bohn Rex-Rotary Div. VIN, Inc., 444 Park Ave. South, New York, N.Y. 10016.

Copy-Rite Corp., 2061 N. Southport Ave., Chicago, Ill. 60614.

Deltek Business Machines, 23209 Miles Ave., Warrensville Heights, Ohio 44128.

Heyer, Inc., 1850 S. Kostner Ave., Chicago, Ill. 60623.

Standard Duplicating Machines Corp., 1935 Revere Beach Pkwy., Everett, Mass. 02149.

mimeograph duplicators

A. B. Dick Co., 5700 W. Touhy Ave., Chicago, Ill. 60648.

Bohn Rex-Rotary Div., VIN, Inc., 444 Park Ave. South, New York, N.Y. 10016.

Copy-Rite Corp., 2061 N. Southport Ave., Chicago, Ill. 60614.

Deltek Business Machines, 23209 Miles Ave., Warrensville Heights, Ohio. 44128.

Heyer, Inc., 1850 S. Kostner Ave., Chicago, Ill. 60623.

Roneo Div. of Facit-Addo, Inc., 501 Windsor Dr., Syracuse, N.J. 07894.

Speed-O-Print Business Machines Corp., 1801 W. Larchmont Ave., Chicago, Ill. 60613.

patented reproduction devices

A. B. Dick Co., 5700 W. Touhy Ave., Chicago, Ill. 60648.

APECO Corp., 2100 W. Dempster, Evanston, Ill. 60204.

GAF Corp., 140 W. 51 St., New York, N.Y. 10020.

Royal Typewriter Co. (Royfax), Div. of Litton Industries, 150 New Park Ave., Hartford, Conn. 06106.

3M Company, 3M Center, St. Paul, Minn. 55101.

Viewlex, Inc., 84 Broadway, Holbrook, N.Y. 11741.

Xerox Corp., Midtown Towers, Rochester, N.Y. 14603.

offset reproducers

Addressograph Multigraph, Multigraphic Div., 1800 W. Central Rd., Mt. Prospect, Ill. 60056.

Bohn Rex-Rotary Div., VIN, Inc., 444 Park Ave. South, New York, N.Y. 10016.

Graphic Communications Corp., 25 Graphic Pl., Moonachie, N.J. 07074.

Itek Business Products, 1001 Jefferson Rd., Rochester, N.Y. 14603.

REFERENCE SECTION 3

photographic equipment and film

There is a broad spectrum of specialized photographic equipment. However, major sources will provide most of the various products needed, including animation equipment; blimps; cameras, 8mm and 16mm, silent or sound; magnetic or optical sound; cartridges; changing bags; film; editing equipment; reels; rewinds; synchronizers; timers; viewers; lenses; lighting equipment; tripods and pan heads; and processing equipment. (See also the listings for Chapter 8, Photography.)

general (sample listings)

Berkey Marketing Companies, P.O. Box 1060, Woodside, N.Y. 11377.

International Camera Corp., 852 W. Adams St., Chicago, Ill. 60607.

Ponder and Best, 1630 Stewart St., Santa Monica, Calif. 90406. (also Vivitar still cameras).

Willoughbys, 110 W. 32 St., New York, N.Y. 10001.

still cameras (selected)

Bell and Howell Mamiya Co., (Mamiya), 2201 W. Howard St., Evanston, Ill. 60202.

Canon U.S.A. (Canon), 10 Nevada Dr., Lake Success, N.Y. 11040.

Eastman Kodak Co., (Kodak), 343 State St., Rochester, N.Y. 14650.

Ehrenreich Photo-Optical Industries, Inc. (Bronica, Nikon, Nikkormat), Garden City, N.Y. 11530.

Karl Heitz, Inc., (Alpa, Tessina), 979 Third Ave., New York, N.Y. 10022.

Honeywell, Inc. (Pentax), Photographic Products Div., P.O. Box 20083, Denver, Colo. 80222.

Konica Camera Corp. (Konica), Woodside, N.Y. 11377.

E. Leitz, Inc. (Leica), Rockleigh, N.J. 07647.

Minolta Corp. (Minolta), 101 Williams Dr., Ramsey, N.J. 07446.

Paillard, Inc. (Hasselblad, Bolex, Topcon), 1900 Lower Rd., Linden, N.J. 07036.

Polaroid Corp. (Polaroid), 549 Technology Sq., Cambridge, Mass. 02139.

Yashica (Yashica), 50–17 Queens Blvd., Woodside, N.Y. 11377.

motion picture cameras (selected)

Hervic Corp. (Beaulieu), 14225 Ventura Blvd., Sherman Oaks, Calif. 91403.

Eastman Kodak Co. (Kodak), 343 State St., Rochester, N.Y. 14650.

Bell and Howell Co. (Bell and Howell), 7100 McCormick Blvd., Chicago, Ill. 60645.

Ehrenreich Photo-Optical Industries (Fujica), Garden City, L.I., N.Y. 11530.

Minolta Corp. (Minolta), 101 Williams Dr., Ramsey, N.J. 07446.

E. Leitz, Inc. (Leicina), Rockleigh, N.J. 07647.

Paillard, Inc. (Bolex), 1900 Lower Rd., Linden, N.J. 07036.

film (unexposed)

Agfa Gavaert, Inc. (Agfa), 275 N. St., Teterboro, N.J. 07608.

Eastman Kodak Co. (Kodak), 343 State St., Rochester, N.Y. 14650.

Fuji Photofilm, Inc. (Fuji), 350 Fifth Ave., New York, N.Y. 10001.

GAF Corp. (GAF), 140 W. 51 St., New York, N.Y. 10020.

Ilford, Inc. (Ilford), Paramus, N.J. 07652.

Polaroid Corp. (Polaroid), 549 Technology Sq., Cambridge, Mass. 02139.

3M Company (Dynachrome), 3M Center, St. Paul, Minn. 55101.

REFERENCE SECTION 4

planning improvements for classroom facilities

general information

Association for Educational Communications and Technology, 1201 Sixteenth St. N.W., Washington, D.C. 20036.

American Library Association, 50 E. Huron St., Chicago, Ill. 60611.

Educational Facilities Laboratories, 477 Madison Ave., New York, N.Y. 10022.

Educational Research Information Center (ERIC), U.S. Office of Education, Washington, D.C. 20202.

U.S. Office of Education, Washington, D.C. 20202.

classroom furniture

Advance Products Company, P.O. Box 2178, Wichita, Kan. 67201.

American Seating Company, 901 Broadway, Grand Rapids, Mich. 49502.

Arlington Aluminum Company, 19303 W. Davison, Detroit, Mich. 48223.

Brewster Corporation, 50 River St., Old Saybrook, Conn. 06475.

Brunswick Corporation, 1 Brunswick Plaza, Skokie, Ill. 60076.

Fleetwood Furniture Company, 25 Washington St., Zeeland, Mich. 49464.

Gaylord Brothers, Inc., P.O. Box 61, Syracuse, N.Y. 13201.

General Electric Company, 60 Washington Ave., Schenectady, N.Y. 12309.

Globe-Wernicke Company, 1329 Arlington, Cincinnati, Ohio 45225.

E. F. Hauserman Company, 5711 Grant Ave., Cleveland, Ohio 44105.

Heywood-Wakefield Company, Menominee, Mich. 49858.

MSG-Macalester, Rt. 111 and Everett Turnpike, Nashua, N.H. 03060.

Charles Mayer Studios, Inc., 140 E. Market St., Akron, Ohio 44308.

Media Systems Corporation, an affiliate of Harcourt Brace Jovanovich, Inc., Department E, 250 W. Main St., Moorestown, N.J. 08057.

Milton Bradley Co., 74 Park St., Springfield, Mass. 01101.

Neumade Products, Inc., 720 White Plains Rd., Scarsdale, N.Y. 10583.

Radio Corporation of America, Front and Cooper Sts., Camden, N.J. 08102.

Smith System Manufacturing Company, 1405 Silver Lake Rd., New Brighton, Minn. 55412.

Toledo Metal Furniture Company, 1400 N. Hastings St., Toledo, Ohio 43607.

Virco Manufacturing Corporation, 15134 S. Vermont Ave., Los Angeles, Calif. 90044.

H. Wilson Corporation, 555 W. Taft Dr., South Holland, Ill. 60473.

Worden Company, 199 E. 17 St., Holland, Mich. 49423.

light control and room
darkening equipment

Aeroshade, Inc., P.O. Box 559, Waukesha, Wis. 53186.

Colonial Plastics Corporation, 5107 Glen Alden Dr., Richmond, Va. 23231.

CLASSIFIED
DIRECTORY
OF SOURCES

Draper Shade and Screen Company (drapes), 411 S. Pearl St., Spiceland, Ind. 47385.

E. I. duPont de Nemours and Co., Wilmington, Del. 19898.

Levelor-Lorentzen, 720 Monroe St., Hoboken, N.J. 07030.

Plastic Products, Inc. (Luxout), Box 1118, Richmond, Va. 23208.

acoustical materials

Celotex Corporation, 1500 N. Dale Mabry Hwy., Tampa, Fla. 33607.

Johns-Manville, 22 E. 40 St., New York, N.Y. 10016.

Owens-Corning Fiberglas Corporation, 717 Fifth Ave., New York, N.Y. 10022.

space dividers, in-classroom storage

Alliance Wall Corp., Wyncote House, Wyncote, Pa. 19095.

Jack C. Coffey Co. (Luxor), 104 Lake View Ave., Waukegan, Ill. 60085.

Demco Educational Corporation, 2120 Fordem Ave., Madison, Wis. 53704.

Marsh Chalkboard Co., Dover, Ohio 44622.

Peabody Space Dividers, North Manchester, Ind. 46962.

6

REFERENCES:
PRINT AND AUDIOVISUAL

In supplementing the preceding text material, this reference section lists items suggested for reading, listening, and viewing. Specific citations are organized according to text chapters and reference sections; they are further divided into print and audiovisual categories.

Many students will want to go well beyond this list to obtain additional information from a still larger number of helpful publications about educational media—especially those which appear after this book goes to press. The following entries of general references and of the products and services of several prominent media-related organizations provide the data on which to base such searches.

general references

A few general references are valuable points of departure for this further investigation. First, we suggest that you become acquainted with and use the ERIC system and the resources and services of the ERIC Clearinghouse on Information Resources (ERIC/IR), School of Education, Stanford Center for Research and Development in Teaching, Stanford, Calif. 94305. See especially two publications of the latter: *ERIC: What It Can Do for You / How to Use It,* James W. Brown, et al. (1975), and *How to Prepare for a Computer Search of ERIC: A Non-Technical Approach,* Judith Yarborough (1975). See also the ERIC system-related publications, *Resources in Education* (RIE), which contains abstracts and indexes of recently completed educational research and research-related reports as well as descriptive accounts of current educational practice, including many specifically related to media; and *Current Index to Journals in Education* (CIJE), which indexes articles from more than 700 periodicals that, together, represent the core of educational literature. You may also ask to be placed on the ERIC/IR mailing list to receive free publication lists and copies of occasional news briefs concerning its operations.

Several other general indexes contain reference entries that direct readers to periodical literature pertaining to educational media. The *Media Review Digest,* covering reviews of films, filmstrips, records, tapes, and miscellaneous media, is published regularly by Pierian Press, Ann Arbor, Mich. 48106, with supplements in issues of *Audiovisual Instruction.* The *International Index to Multi-Media Information* is published by Audio-Visual Associates, 180 E. California Blvd., Pasadena, Calif. 91105.

The *Education Index* (published ten times a year, with annual cumulations, by the H. W. Wilson Company, 950 University Ave., New York, N.Y. 10452) provides a basic index to educational literature found in professional periodicals, references, proceedings, yearbooks, and similar publications. The same company's *Readers' Guide to Periodical Literature* is also a useful guide to articles appearing in some 160 periodicals of general interest.

The R. R. Bowker Company, 1180 Avenue of the Americas, New York, N.Y. 10036, publishes several critical and important items on educational media. Prominent among them are three magazines, *Previews, School Library Journal,* and *Library Journal,*

and its *Educational Media Yearbook* (in print and regularly updated since 1973), edited by James W. Brown. The latter contains a wealth of reference material on the state of the art of media, special organizations, national and international developments in media, research, periodicals, articles, and media resources of special current interest to the field.

Two other key references contain useful materials that relate closely to the contents of this book. The first, *AV Instructional Technology Manual for Independent Study* (5th edition), James Brown and Richard Lewis, editors, McGraw-Hill, 1977, offers a variety of "hands-on" media exercises; the second, *Transparency Masters,* McGraw-Hill, 1977, provides drawings, lists, and charts in the format of large transparencies.

media-related organizations

Of the many media-related professional associations, three in particular provide basic information services to the field. They are: (1) the Association for Educational Communications and Technology (AECT), at 1201 16th St., N.W., Washington, D.C. 20036, (2) the American Association of School Librarians (AASL), and (3) the American Library Association (ALA) both at 50 E. Huron St., Chicago, Ill. 60611. Each publishes extensively. Magazines of particular interest are *Audiovisual Instruction* (including *Learning Resources*) and *AV Communication Review* from AECT; and *American Libraries* and *School Media Quarterly* from the ALA and AASL.

Other organizations provide valuable general information about media and the media field. The Educational Product Information Exchange Institute (EPIE), 463 West St., New York, N.Y. 10014, evaluates learning materials and equipment. It publishes *EPIE Reports* and *EPIEgrams,* which are based on these evaluations. The movement toward "learner verification" of learning materials has received much of its impetus from this organization. The National Audio-Visual Association, 3150 Spring St., Fairfax, Va. 20030, publishes directories of equipment and distributors.

CHAPTER 1

media and the systematic approach to teaching and learning

print

Allen, William H., "Intellectual Abilities and Media Design," *AV Communication Review,* vol. 23, Summer 1975, pp. 139–170.

Bruner, Jerome, "The Process of Education Revisited," *Phi Delta Kappan,* vol. 53, September 1971, pp. 18–21.

Curl, David H., and Carlton W. H. Erickson, *Fundamentals of Teaching with Audiovisual Technology,* Macmillan, New York, 1972, pp. 1–80.

Davis, John A., "Instructional Development: Key to Instructional Change," *Audiovisual Instruction,* vol. 17, June–July 1972, p. 43.

Kemp, Jerrold E., *Instructional Design (A Plan for Unit and Course Development)* (2d ed.), Fearon Publishers, Belmont, Calif., 1976.

Mager, Robert F., *Preparing Instructional Objectives* (2d ed.), Fearon Publishers, Belmont, Calif., 1975.

Molstad, John A., "Selective Review of Research Studies Showing Media Effectiveness: A Primer for Media Directors," *AV Communication Review,* vol. 22, Winter 1974, pp. 387–407.

Olson, David R., *Media and Symbols: The Forms of Expression, Communication, and Education.* Twenty-third Yearbook of the National Society for the Study of Education, University of Chicago Press, Chicago, Ill., 1974.

Pipe, Peter, *Objectives: Tool for Change,* Fearon Publishers, Belmont, Calif., 1975.

Popham, W. James, "Objectives '72," *Phi Delta Kappan,* vol. 53, March 1972, pp. 432–435.

Schlack, Lawrence B., and John W. Kofel, "Media in the Classroom," *School Media Quarterly,* Spring 1975, pp. 204–209.

Selecting Media for Learning, AECT, Washington, D.C., 1974.

Taylor, Sandra, et al., *The Effectiveness of CAI,* ERIC Drs., Arlington, Va., 1974.

Teaching Strategies: A Systems Approach, Harper & Row Publishers, Inc., New York, 1974.

Wittich, Walter A., and Charles F. Schuller, *Audiovisual Materials: Their Nature and Use* (6th ed.), Harper & Row, New York, 1976.

audiovisual materials

Learning with Today's Media, 16mm film, 35 min., sound, color, Encyclopaedia Britannica Education Corporation, Chicago, Ill., 1973.

Lectures Are a Drag, 16mm film, 19 min., sound, color, Churchill Films, Los Angeles, Calif., 1974.

Networks for Learning, sound filmstrip, color, with script booklet, AECT, Washington, D.C., 1975.

Teaching Strategies: A Systems Analysis, I and II, two 16mm films, sound, b and w, Harper & Row Publishers, Inc., New York, 1974.

Teaching Styles, audio-tape cassette, Harper & Row Publishers, New York, 1974.

Writing Behavioral Objectives Series: General; set of fifteen transparencies, Applications to Elementary School Students; set of ten transparencies, Application to High School Students; set of ten transparencies, Application to College Students; set of ten transparencies, Lansford Publishing Co., San Jose, Calif., 1974.

CHAPTER 2

media and individualized learning

print

Alexander, Lawrence T., and Stephen L. Yelon, *Learning System Design,* McGraw-Hill, New York, 1974.

Building Independent Learning Skills, Learning Handbooks, Boulder, Colo., 1974.

Burns, Richard, "Methods for Individualizing Instruction," *Educational Technology,* vol. 11, June 1971, pp. 55–56.

Buterbaugh, James G., and Robert G. Fuller, "Personalized System of Instruction (PSI): An Alternative," *Audiovisual Instruction,* vol. 20, March 1975.

Cook, Donald A., *Personalized System of Instruction (PSI): Potential and Problems,* Educational Products Information Exchange (EPIE), New York, 1974.

Creativity, Harper & Row, New York, 1974.

Espinosa, Leonard, and John Morlan, *Easy-To-Make Devices for Learning Centers,* Personalized Learning Associates, San Jose, Calif., 1974.

Frase, Larry E., "The Concept of Instructional Individualization," *Educational Technology,* vol. 12, July 1972, p. 45.

Hendershot, Carl H., *Programmed Learning and Individually Paced Instruction Bibliography,* Hendershot Bibliography, Bay City, Mich. (regularly updated versions).

Kemp, Jerrold E., *Planning and Producing Audiovisual Materials* (3d ed.), Thomas Y. Crowell Co., New York, 1975.

Media and Library Programs. Issue of *Audiovisual Instruction,* vol. 20, September 1975. Seven articles on topics related to this theme.

Media Centers: Readings from Audiovisual Instruction, AECT, Washington, D.C., 1975.

Media Programs: District and School, AECT Publications, Washington, D.C., 1975.

audiovisual materials

Independent Study, 16mm film, 12 min., sound, color, Journal Films, Chicago, Ill., 1970.

Individualized Instruction, five sound filmstrips, color, with audio tapes, forty-eight case study brochures and manual, National Education Association, Washington, D.C., 1970.

Individualizing Teaching and Learning. Set of six cassette tapes, Lansford Publishing Co., San Jose, Calif., 1974.

Learning Styles, audiotape/cassette, Harper & Row, New York, 1974.

Learning Styles: Primary Materials, 16mm film, b and w, Harper & Row, New York, 1974.

Media: Resources For Discovery, 35mm sound filmstrip, color, eight strips with disc or cassette, Encyclopaedia Britannica Educational Corporation, Chicago, Ill., 1975.

Networks For Learning, sound filmstrip, color, AECT, Washington, D.C., 1975.

Open Classroom, 16mm film, sound, b and w, Harper & Row, New York, 1974.

Open School, 16mm film, sound, color, Harper & Row, New York, 1974.

Personalized System of Instruction: An Alternative, 16mm film, 14 min., b and w, University of Nebraska Instructional Media Center, Lincoln, Nebr., 1974.

CHAPTER 3

the community as a learning center

print

Aronstein, Lawrence W., and Edward Olsen, *Action Learning: Student Community Service Projects,* Association for Supervision and Curriculum Development, Washington, D.C., 1974.

Blackman, N., M. Shukyn, and Sister Rita Arthur, "Community as Classroom: Three Experiments," *National Association of Secondary School Principals Bulletin,* vol. 55, May 1971, pp. 147–158.

Burman, Lionel A., "Museums and Visual Education," *Visual Education,* March 1971, pp. 31–33.

Chalmers, John J., "Exploring Local Resources" (Alberta, Canada), *Audiovisual Instruction,* vol. 16, May 1971, pp. 70–74.

Fernhoff, R., "Making the Most of Your Field Trip," *Arithmetic Teacher,* vol. 18, March 1971, pp. 186–189.

Ford, Barbara, "Creature Comforts at the Zoo," *Saturday Review,* vol. 55, no. 32, Aug. 5, 1972, pp. 40–48.

Lynch, Dennis, "YCC: An Experience in the Outdoors," *American Biology Teacher,* vol. 34, no. 3, March 1972, pp. 147–149, 159.

Matters of Choice: A Ford Foundation Report on Alternative Schools, Ford Foundation, Office of Reports, New York, 1974.

Riordan, Robert C., *Alternative Schools In Action,* Phi Delta Kappa, Educational Foundation, Bloomington, Ind., 1972.

Rogers, Lola Eriksen, *Museums and Related Institutions,* U.S. Department of Health, Education, and Welfare, Washington, D.C., 1969.

Smith, Vernon, Daniel J. Burke, and Robert D. Barr, *Optional Alternative Public Schools,* Phi Delta Kappa, Educational Foundation, Bloomington, Ind., 1974.

Wasserman, Paul (ed.), *Museum Media,* Gale Research, Detroit, Mich., 1974.

audiovisual materials

Building Bridges to the Future, filmstrip, 11 min., color, 33⅓ rpm record, script book, AECT, Washington, D.C., 1974.

Community Action Motion Pictures, nine 16mm films, 13–30 min., sound, color, Centron Educational Films, Lawrence, Kans., 1970.

Open School, 16mm film, sound, color, Harper & Row, New York, 1973.

Your Community Is a Classroom, 16mm film, 28 min., sound, color, American Iron and Steel Institute, New York, or Capital Film Services, Lansing, Mich., 1968.

CHAPTER 4

choosing, using, and producing media

print

Brown, James, Kenneth D. Norberg, and Sara Srygley, *Administering Educational Media* (revised), McGraw-Hill, New York, 1972.

Brown, Lucy G., *Core Media Collection for Secondary Schools,* R. R. Bowker Company, New York, 1975.

Bullough, Robert V., *Creating Instructional Materials,* Charles E. Merrill, Columbus, Ohio, 1974.

Curl, David H., and Carlton W. H. Erickson, *Fundamentals of Teaching with Audiovisual Technology* (2d ed.), Macmillan, New York, 1972.

Dale, Edgar, *Audiovisual Methods in Teaching* (3d ed.), Holt, Rinehart, and Winston, New York, 1969.

"Evaluation and Selection of Media," twelve articles, issue of *Audiovisual Instruction,* vol. 20, April 1975.

Gagné, Robert M., and Leslie J. Briggs, *Principles of Instructional Design,* Holt, Rinehart, and Winston, New York, 1974.

Kemp, Jerrold E., *Planning and Producing Audiovisual Materials* (3d ed.), Thomas Y. Crowell Co., New York, 1975.

Komoski, P. Kenneth, "An Imbalance of Product Quantity and Instructional Quality," *AV Communication Review,* vol. 22, Winter 1974, pp. 357–386.

Media Review Digest, Pierian Press, Ann Arbor, Mich. (annual index, successor to *Multi Media Reviews* index).

Raskin, Bruce W., *Great Ideas from "Learning,"* Education Today Co., Inc., Palo Alto, Calif., 1975.

Selecting Media for Learning, AECT Publications, Washington, D.C., 1974.

Student-Produced Media, issue of *Audiovisual Instruction,* vol. 19, October 1974.

audiovisual materials

At the Center, 16mm film, 28 min., sound, color, American Association of School Librarians, 50 East Huron, Chicago, Ill., 1970.

Media Programs: District and School, filmstrip, sound, with 33⅓ rpm record; book also available; AECT, Washington, D.C., 1974.

Multimedia Center, 35mm filmstrip, sound, color, 50 frames, Library Filmstrip Center, Wichita, Kans., 1970.

Producing Effective Audiovisual Presentations, series of six filmstrips, sound, color, Media Research and Development, Arizona State University, Tempe, Ariz., 1974. Includes: *Planning; Graphics; Photography; Photographic Copying; Sound Recording;* and *Presentation Systems.*

Setting the Stage for Learning, 16mm film, 22 min., sound, b and w, Churchill Films, Los Angeles, Calif., 1974.

CHAPTER 5

displaying and some fundamentals of visual communication

print

Dwyer, Francis M., "Color as an Instructional Variable," *AV Communication Review,* vol. 19, Winter 1971, pp. 339–416.

Espinosa, Leonard, and John Morlan, *Easy-To-Make Devices for Learning Centers,* Personalized Learning Associates, San Jose, Calif., 1974.

Frye, Roy, *Graphic Tools For Teachers: Practical Techniques* (4th ed.), Graphic Tools, Mapleville, R.I., 1975.

Gilmour, Frederick T., *A Guide to Making a Display,* RISE, Williamsport Area School District, Williamsport, Pa., 1975.

Hoff, Syd, *Art of Cartooning,* Stravon Educational Press, New York, 1973.

Kemp, Jerrold E., *Planning and Producing Audiovisual Materials* (3d ed.), Thomas Y. Crowell, New York, 1975.

Minor, Ed, and Harvey R. Frye, *Techniques for Producing Visual Instructional Media,* McGraw-Hill, New York, 1970.

Morlan, John E., *Preparation of Inexpensive Teaching Materials* (2d ed.), Thomas Y. Crowell, New York, 1973.

Wiman, Raymond V., *Instructional Materials,* Charles A. Jones, Worthington, Ohio, 1972.

audiovisual materials

Audio-Visual Production Techniques Series, 8mm film, standard 8 or Super 8 Technicolor cartridges, 3–4 min. each, color and b and w, McGraw-Hill Films, New York, 1969. Includes: *Lettering: Leroy 500 and Smaller; Lettering: Leroy 700 and Larger; Lettering: Prepared Letters; Lettering: The Felt Pen (Applications); Lettering: The Felt Pen (Basic Skills); Lettering: Wricoprint; Lettering: Wrico signmaker.*

Basic Educational Graphics (multimedia series), Scott Education Division, Holyoke, Mass., 1967. Ten filmstrips with eight records; twelve color projectuals; instructor's and participant's manuals, plus student materials kit and demonstration kit.

Display and Presentation Boards, 16mm films, 15 min., sound, color, International Film Bureau, Inc., Chicago, Ill., 1971.

CHAPTER 6

graphic materials

print

Bartz, Barbara S., *Designing Maps for Children,* Queens University, Kingston (Ontario), September 1970.

Brown, Jerome C., *Cartoon Bulletin Boards,* Fearon Publishers, Belmont, Calif., 1971.

Dreyfuss, Henry, *Symbol Sourcebook: An Authoritative Guide to International Graphic Symbols,* McGraw-Hill, New York, 1972.

Frye, Roy, *Graphic Tools For Teachers* (4th ed.), Mapleville, R.I., 1975.

Making the Most of Charts: An ABC of Graphic Presentation, U.S. Government Printing Office, Washington, D.C., November 1970.

Rowe, Mack, "Charts: What Makes Them Effective," *AV Guide,* vol. 51, January 1972, pp. 4–7.

audiovisual materials

Editorial Cartoons, 35mm filmstrip, silent, b and w, Visual Education Consultants, Madison, Wisc., 1960.

Exploring the World Of Maps, kit with five sound filmstrips, five discs, teacher's guide, and selection of maps, National Geographic Society, Washington, D.C., 1973.

Map and Atlas Survey, 35mm filmstrip, sound, color, 51 frames, Library Filmstrip Center, Wichita, Kans., 1970.

Map and Globe Skills, series of nine 8mm film loops, Leonard Peck Productions, Wayne, N.J., 1974. Includes *Oceans and Continents; Cardinal Directions; Guidelines of the Earth; Latitude, Longitude; Why the Seasons?; Flat Maps of a Round World; Map Keys and Symbols; and Scale.*

Map Skills Games, A-E, Xerox Education Publications, Columbus, Ohio, 1974.

What Is a Graph?, 35mm filmstrip, 26 frames, b and w, Visual Education Consultants, Madison, Wisc., 1965.

CHAPTER 7

transparencies for overhead projection

print

Allan, David W., "Enlarging and Color Separation: An Easy Way for Producing Top Quality Overhead Transparencies," *Audiovisual Instruction,* vol. 15, April 1970, p. 104.

Bretz, Rudy, *"Will My Visual Be Visible?" or The Bretz Rule of Thumb,* Rand Corporation, Santa Monica, Calif., 1974

Index to Educational Overhead Transparencies, NICEM, University of Southern California, University Park, Los Angeles, Calif. Revised regularly.

Kemp, Jerrold E., *Planning and Producing Audiovisual Materials* (3d ed.), Thomas Y. Crowell, New York, 1975.

Richards, Oscar W., and Patricia Mackin, "Colored Overhead Transparencies: Contrast Gain or Seeing Loss?" *AV Communication Review,* vol. 19, Winter 1971, pp. 432–436.

Teaching Your Overhead Projector Some New Tricks, 3M Company, 3M Center, St. Paul, Minn., 1975.

audiovisual materials

Media Production Series, Transparencies, Media Systems, Inc., Salt Lake City, Utah, 1974. Slide sets or filmstrips on *"Color Lift" Transparencies* and *Thermofax Transparencies.*

(The) *Overhead Projection Series* (Multimedia series), Scott Education Division, Holyoke, Mass., 1971, five filmstrips with five records, eighteen review projectuals, instructor's and participant's manuals.

Overhead Transparencies (series), 8mm standard 8 or Super 8 technicolor cartridge, color and b and w, each running approximately 3–4 min., McGraw-Hill Films, New York, 1968. Includes *Making Overlays; Adding Color; Principle of Diazo Process; Diazo Process; Heat Process; Picture Transfer (Shelf Paper); Mounting and Masking.*

Transparency Preparation, set of ten transparencies, Lansford Publishing Co., San Jose, Calif., 1975.

Use of Overhead Projector, set of eight transparencies, Lansford Publishing Co., San Jose, Calif., 1975.

CHAPTER 8

photography

print

Coynik, David, *Movie Making: A Worktext for Super-8 Film Production,* Loyola University Press, Chicago, Ill., 1974.

Eble, Kenneth E., "Camera Reading," *Phi Delta Kappan,* vol. 52, January 1971, p. 285.

Glenn, George D., and Charles B. Scholz, *Super 8 Handbook,* Howard W. Sams, Indianapolis, Ind., 1974.

Glimcher, Sumner, and Warren Johnson, *Movie Making: A Guide to Film Production,* Pocket Books, New York, 1975.

Kemp, Jerrold E., *Planning and Producing Audiovisual Materials* (3d ed.), Thomas Y. Crowell, New York, 1975.

Kuhns, William, *Movies In America,* Pflaum/Standard, Dayton, Ohio, 1974.

Lerner, Susan, "Elementary School Pupils Make Animated Movies," *Audio Visual Notes,* Eastman Kodak Co., Rochester, N.Y., 1973.

Malkiewicz, J. Kris, *Cinematography: A Guide for Film Makers and Film Teachers,* Van Nostrand Reinhold, New York, 1973.

Oddie, Alan, "Film: A Sprocket-Holed Road to Change," *Media and Methods,* vol. 8, no. 5, January 1972, pp. 19–22.

Photography: How It Works, Eastman Kodak Co., Rochester, New York, 1974.

Rosenbery, Kenneth, and Edward Troy, "Student-produced Film," *Teaching Strategies and Classroom Realities,* Mildred G. McCoskey (ed.), Prentice-Hall, Englewood Cliffs, N.J., 1971.

Ryan, Mack, "Preparing a Slide-Tape Program: A Step-by-Step Approach," Parts I and II, *Audiovisual Instruction,* vol. 20, no. 7 and no. 9, September 1975 and November 1975.

Seeing and Making Films, issue of *Audiovisual Instruction,* vol. 19, February 1974.

Visual Literacy: Audiovisual Instruction, vol. 17, May 1972, entire issue.

audiovisual materials

The American Super-8 Revolution, 16mm film, 31 min., color, International Film Foundation, New York, 1974.

A Movie about Light, 16mm film, sound, color, 8 min., AECT Publication, Washington, D.C., 1972.

Audio-Visual Production Techniques Series, 8mm standard 8 or super 8 technicolor cartridge, 2–4 min. each, b and w, McGraw-Hill, New York, 1969.

Basic Film Editing, 16mm film, 17 min., sound, color, Pyramid Films, Santa Monica, Calif., 1974.

Filmmaking, 16mm film, 27 min., sound, color, John Wiley and Sons, New York, 1973.

Films about Filmmaking Series, 16mm film, sound, color, International Film Bureau, Chicago, Illinois. Includes: *A Film about Filmmaking,* 17 min.; *Making a Sound Film* 13 min.; *A Film about Editing,* 14 min.; and *A Film about Cinematography,* $17\frac{1}{2}$ min.

Film Arts Starter Kit, multi-media kit, audio slides, strips, film making material, Film Makers, Arcadia, Calif., 1974.

Grammar of Film, 16mm film, 30 min., sound, color, Center for Mass Communication, Columbia University, New York.

CHAPTER 9

still pictures

print

Brown, Roland, "The Gut Impact of the Visual Media," *Media and Methods,* vol. 11, May/June 1975, pp. 28–29.

Diffor, John C., and Mary F. Horkheimer, *Educators Guide to Free Filmstrips,* Educators Progress Service, Randolph, Wisc., 1975 (updated yearly).

Educational Sound Filmstrip Directory, DuKane Corporation, St. Charles, Ill. (updated at irregular intervals).

Fransecky, Roger B., and John L. Debes, *Visual Literacy: A Way to Learn—A Way to Teach,* Association for Educational Communications and Technology, Washington, D.C., 1972.

Freudenthal, Juan R., "The Slide as a Communication Tool," *School Media Quarterly,* vol. 2, Winter 1974, pp. 109–115.

How Good Are the "Best" Filmstrips? Educational Products Information Exchange (EPIE), New York, 1973–74.

Image, Object and Illusion, Scientific American, New York.

Index to Educational Slide Sets, and *Index to 35mm Filmstrips,* NICEM, University of Southern California, University Park, Los Angeles, Calif. Regularly updated.

Levie, W. Howard (ed.), *Research on Learning from Pictures: A Review and Bibliography,* Indiana University Publications Office, Bloomington, Inc., 1973.

Minorities in Filmstrips, EPIE Report No. 44, Educational Products Information Exchange Institute, New York, 1972.

Spitzer, Dean P., and Timothy O. McNerny, "Operationally Defining Visual Literacy," *Audiovisual Instruction,* vol. 20, September 1975, pp. 30–31.

Visual Literacy is the theme of the entire issue (eleven articles) of *Audiovisual Instruction,* vol. 17, no. 5, May 1972.

Ward, Nel, and Sue Hardesty, "Filmstrips and Sound-Slide Sets '74: The Great and the Super Great," *Media and Methods,* vol. 11, March 1975, pp. 30–35.

audiovisual materials

Audio-Visual Production Techniques Series, 8mm standard 8 or super 8 technicolor cartridge, 2–4 min, each, b and w, McGraw-Hill, New York, 1969.

How Does a Picture Mean?, filmstrip, 76 frames, b and w, National Education Association, Washington, D.C., 1968.

How to Plan Filmstrip-Tape Multi-Media Lessons, filmstrip (color filmstrip and tape), 36 min., sound, Mt. San Jacinto College, Gilman Hot Springs, Calif., 1969.

Media Production Series, Still Pictures, Media Systems, Inc., Salt Lake City, Utah, 1974. Slide sets and/or filmstrips *Dry Mounting with Heat Press; Cold Mounting; Heat Laminating;* and *WRICO Lettering and Dry Mounting.*

CHAPTER 10

audio materials

print

Bergamini, Edwin S. (ed.), *Sight and Sound Tape Directory,* Drorbaugh Publications, New York, 1974.

Burstein, Herman, *Questions and Answers about Tape Recording,* Tab Books, Blue Ridge Summit, Pa., 1974.

Index to Educational Audio Tapes and *Index to Educational Records,* NICEM, University of Southern California, University Park, Los Angeles, Calif.. (revised annually).

Maleady, Antoinette O., *Record and Tape Reviews Index,* Scarecrow Press, Metuchen, N.J., 1973.

Suhor, Charles, "Talking, the Spoken Word as a Way of Teaching," *Media and Methods,* vol. 11, April 1975, pp. 32–36.

Wittich, Walter A., and Raymond H. Suttles, *Educators Guide to Free Tapes, Scripts, and Transcriptions,* Educators Progress Service, Inc., Randolph, Wis. (revised annually).

audiovisual materials

Listening Is Communicating, set of four sound filmstrips, color, four discs or cassettes, Imperial Film Co., Lakeland, Fla., 1974.

Tips on Tapes for Teachers, Open reel or audio cassette, 27 min., National Center For Audio Tapes, Educational Media Center, Stadium Bldg., University of Colorado, Boulder, Colo.

Some Plain Talk on Tape about Tape Recording, 3¾ IPM tape recording, 9 min., Eastman Kodak, Rochester, N.Y., 1971.

Utilizing the Tape Recording in Teaching, two-color filmstrips, two audio cassettes and narration guide, Media Systems, Inc., Salt Lake City, Utah, 1974.

CHAPTER 11

motion pictures:
films and video

print

The Compleat Video-Cassette User's Guide: Principles and Practice of Programming, Knowledge Industry Publications, White Plains, N.Y., 1973.

Diffor, John C., and Mary F. Horkheimer, *Educators Guide To Free Films,* Educators Progress Service, Randolph, Wis., 1975 (annual revisions).

Eller, M. Linda, "Individualized Learning Using TV," *Educational Broadcasting,* vol. 8, no. 4, July/August 1975.

Film Evaluation Guides, Supplements, and Cards, Educational Film Library Association, New York. Issued regularly.

Gordon, Robert M., "Classrooms without Chalk Boards," *Educational Technology,* vol. 15, April 1975, pp. 38–40.

Hoban, Charles F., Jr., "The State of the Art of Films in Instruction," *Audiovisual Instruction,* vol. 20, April 1975, pp. 30–34.

Index to Educational Video Tapes, Index to 8mm Motion Cartridges, and *Index to 16mm Educational Films,* NICEM, University of Southern California, University Park, Los Angeles, Calif., 1975 (revised annually).

Jones, Emily S., *Manual of Film Evaluation,* revised, Education Film Library Association, New York, 1975.

Videoplay Program Source Guide, Tepfer Publishing Co., Ridgefield, Conn., 1975.

audiovisual materials

Conventional Media (multi-media series), Scott Education Division, Holyoke, Mass., 1971. Includes one film, *Using Motion Film in the Classroom,* (10.5 min., color with guide, 16mm or super 8mm sound cartridge); ten review projectuals; one sound color filmstrip (including 3¾ IPS tape), plus instructor's and participant's manual.

Learning to See and Understand Visual Literacy, sound slide with cassette and disc, 160 slides, Center for Humanities, Inc., White Plains, N.Y. 1973.

Television Series, each 30 min., b and w, Great Plains National Instructional Television Library, Lincoln, Nebr., 1971. Includes *Television: A Broader Look* (UF-110); *Television: Effective Instruction* (UF-103); *Television: Following Up the Lesson* (UF-109); *Television: Implications for Instruction* (UF-101); and *Television: A Potent Medium* (UF-102).

Utilizing Instructional Television, two color filmstrips, two audio cassettes and guide, Media Systems, Inc., Salt Lake City, Utah, 1974.

CHAPTER 12

television

print

Bensinger, Charles, et al., *Petersen's Guide to Video Tape Recording,* Petersen Publishing Co., Los Angeles, Calif., 1973.

Harwood, Don, *Everything You Always Wanted to Know about Tape Recording,* VTR Publishing Co., Bayside, N.Y., 1975.

Murray, Michael, *The Videotape Book,* Bantam Books, New York, 1975.

NAEB Directory of Educational Telecommunications, National Association of Educational Broadcasters, Washington, D.C., 1974.

Ronse, Virginia J., "Community Television: It's Working in Columbus, Indiana," *Educational and Industrial Television,* vol. 7, April 1975, pp. 32–34.

Sohn, David, and Jerry Kosirski, "A Nation of Vidiots," *Media and Methods,* vol. 11, April 1975, pp. 24–31; 52–57.

Witherspoon, John P., *State of the Art (A Study of Current Practices and Trends in Educational Uses of Public Radio and Television),* Advisory Council of National Organizations, Washington, D.C., August 1, 1974.

audiovisual materials

Closed Circuit Television (0875-01), audio-tape, open reel, or cassette, 15 min., National Center for Audio Tapes, Stadium Bldg., University of Colorado, Boulder, Colo., 1974.

Reynolds, Robert, with Glen Pensinger, *Making Effective Video Tapes* (series), available in ½-inch EIAJ, ¾-inch U-Matic, or 2-inch Quad, b and w, Instructional Television Center, San Jose State University, San Jose, Calif. Includes: *Shooting Good Videotapes,* 22 min., 1974. *Mechanics of Dubbing and Editing ½" Video Tapes,* 26 min., 1974. *This Tape Shows How This Tape Was Made,* 13 min., 1974. *How Do Yuh Make This Thing Work?* (Rover Type Camera Systems), 13 min., 1975.

Sight and Sound of Videotape, Production Series, six 20-min. video tapes, ¾ U-matic video cassette or EIAJ ½-inch reel-to-reel, 3M Co., St. Paul, Minn., 1974. Includes *The Show Must Go On, or Setup, Operation, and Care of the Videotape System; A Juggler of Paradoxes or How to Perform On Television; The World's Your Stage or A Practical Guide to Sets and Props for Videotape Production; The Necessary Art or Lighting for Videotape Production; Teaching an Old Camera New Tricks, or Camera Techniques for Videotape; Putting It All Together or How to Produce a Videotape Program.*

Videotapes on Videotape, a series of videotapes, ½-inch EIAJ or ¾ U, Smith-Mattingly Productions, Alexandria, Va., 1973. Includes *Basic Lighting Techniques,* 20 min.; *Basic VTR Sound Techniques,* 30 min.; *Introducing CCTV/VTR,* 7 min.; *Introduction to CATV,* 27 min.; *The Portapak,* 30 min.; *Set-Up, Production Techniques;* and *Preventive Maintenance for the Single Camera VTR System,* 40 min.; *VTR Feedback,* 18 min.; *VTR Production Planning,* 28 min.; *A Video Essay-One Approach to Editing Videotape,* 25 min.; *The Video cassette,* 40 min., and *Welcome to the Wonderful World of VTR,* 10 min.

CHAPTER 13

real things, models, and demonstrations

print

Lukas, Terrence, "An Easy and Inexpensive Technique for Group Viewing of Small Biologicals," *Learning Resources,* vol. 2, April 1975, pp. 23–24.

Morlan, John, and Leonard Espinosa, *Electric Boards You Can Make,* Personalized Learning Associates, San Jose, Calif., 1974.

Roberts, Edward M., *Fingertip Math,* Texas Instruments Learning Center, Dallas, Tex., 1975 (calculator use).

audiovisual materials

Demonstrations, 16mm film, 18 min., sound, b and w, Pennsylvania State University, University Park, Pa., 1959.

Science through Discovery, 16mm film, 28 min., sound, b and w, DuArt, New York, 1966.

We Grew a Frog, 16mm film, 13 min., sound, color, International Film Bureau, Chicago, Ill., 1971.

CHAPTER 14

games, simulations, and informal dramatizations

print

Belch, Jean, *Contemporary Games,* vols. I and II, Gale Research Co., Book Tower, Detroit, Mich., 1973 and 1974.

Bruner, J. S., "Play Is Serious Business," *Psychology Today,* January 1975, pp. 80–83.

Cathorne, Edythe O, "Toys and Games—'The First Reading Tools,'" *School Library Journal,* vol. 21, April 1975, pp. 24–27.

Chapman, Katherine, et al., *Simulation/Games in Social Studies: What Do We Know?* (ED 093 736), ERIC DRS., Arlington, Va., 1974.

Creating and Using Learning Games, Learning Handbooks, Palo Alto, Calif., 1975.

Fluegelman, Andrew (ed.), *New Games Handbook,* New Games Foundation, San Francisco, Calif., 1976.

Golick, Margie, *Deal Me In!: The Use of Playing Cards in Teaching and Learning,* Jeffrey Norton Publishers, New York, 1973.

Harpole, Charles, "ERIC Report: Gaming and Simulation in Speech Communication Education," *Speech Teacher,* vol. 24, January 1975, pp. 59–64.

Heyman, Mark, *Simulation Games for the Classroom,* Fastback No. 54, *Phi Delta Kappa,* Bloomington, Ind., 1975.

Kaplan, Sandra Nina, et al., *A Young Child Experiences,* Goodyear Publishing Co., Palisades, Calif., 1975.

McCord, Cindy, and Shirley Ross, *Animal Rhythms,* Personalized Learning Associates, San Jose, Calif., 1974 (puppetry).

Simulation/Gaming/News, University Station, Box 3039, Moscow, Idaho.

Stadsklev, Ron, *Handbook of Simulation Gaming in Social Education,* I, Institute of Higher Education, University, Ala., 1974.

Zeleny, Leslie D., *How to Use Simulations: How to Do It Series, No. 26,* National Council for the Social Studies, Washington, D.C., 1974.

audiovisual materials

Children Are Creative, 16mm film, 10 min., sound, color, BFA Educational Media, Santa Monica, Calif., 1965.

Communication Games, set of eight transparencies and kit, Lansford Publishing Co., San Jose, Calif., 1975.

Mrs. Ryan's Drama Class, 16mm film, 35 min., b and w, McGraw-Hill, New York, 1971.

Simulation, series of twelve transparencies, Lansford Publishing Co., San Jose, Calif., 1975.

CHAPTER 15

free and inexpensive materials

print

Aubrey, Ruth H. (ed.), *Selected Free Materials for Classroom Teachers,* Fearon Publishers, Belmont, Calif., 1975, revised biennially.

Educators Progress Service, Randolph, Wisc., publishes many guides to various types of free or inexpensive materials, most of them revised annually.

Forte, Imogene, Mary Ann Pangle, and Robbie Tupa, *Center Stuff For Nooks, Crannies and Corners,* Library of Contemporary Education, Riverside, N.J., 1974.

How to Locate Useful Government Publications: No. 11 in *How to Do It* series, National Council for the Social Studies, Washington, D.C., 1968.

CHAPTER 16

print, multimedia, and microforms

print

Broadus, Robert N., *Selecting Materials for Libraries,* H. W. Wilson Company, New York, 1973.

Cheney, Frances Neel, *Fundamental Reference Sources,* American Library Association, Chicago, Ill., 1971.

Eliminating Ethnic Bias in Instructional Materials: Comment and Bibliography, Association for Supervision and Curriculum Development, Washington, D.C., 1974.

Hanna, Paul R., "Printed Materials Will Supplement Videorecordings," *College Management,* vol. 6, May 1971, pp. 30–31.

Introduction to Micrographics, National Microfilm Association, Silver Spring, Md., 1975.

Printed Media and the Reader, Harper & Row, New York, 1974.

Schrank, Jeffrey, *The Seed Catalog: A Guide to Teaching/ Learning Materials,* Beacon Press, Boston, Mass., 1974.

Spigai, Frances G., *Invisible Medium: The State of the Art of Microform and a Guide to the Literature,* ERIC at Stanford, Stanford University, Stanford, Calif., 1973.

audiovisual materials

Encyclopedia Transparencies Series, GAF Corporation Reprographic Products, New York, 1970.

Encyclopedias—Usage, Techniques, and Encyclopedias —Basic Knowledge, two sound filmstrips, forty-seven frames each, sound, color, Library Filmstrip Center, Wichita, Kans., 1970.

Introduction to Micrographics, 35mm sound filmstrip, 30 min., color, with audio cassette, National Microfilm Association, Silver Spring, Md., 1975.

The Library: Our Learning Resource Center, sound filmstrip, color, three sets with disc or cassette, McGraw-Hill Films, New York, 1975.

Using Books, 35mm filmstrip, forty-one frames, color, McGraw-Hill Films, New York, 1971.

CHAPTER 17

. . . and, in the future?

print

Baker, Justine C., *The Computer in the School,* Fastback No. 58, *Phi Delta Kappa,* Bloomington, Ind., 1975.

Computers in Education Resource Handbook (revised) (ED 093 294), Department of Computer Science, University of Oregon, Eugene, Oreg., 1974.

Dickson, Edward M., and Raymond Bowers, *The Video Telephone: Impact of a New Era in Telecommunications,* Praeger Publishers, New York, 1974.

Evans, Art, "Videodisc on the Horizon," *Audiovisual Instruction,* vol. 20, May 1975, pp. 31–33.

Fransecky, Roger B., "The Video Explosion: Choosing a Future," *Audiovisual Instruction,* vol. 20, May 1975, pp. 28–30.

Free, John, "The Coming Age of Fiber Optics: How Wires of Glass Will Revolutionize Communications," *Popular Science,* August 1975, pp. 82–85; Ill.

Toffler, Alvin (ed.), *Learning for Tomorrow: The Role of the Future in Education,* Random House, New York, 1972.

"Videodiscs: The Expensive Race to Be First," *Business Week,* Sept. 15, 1975, pp. 58–61; 64; 66.

audiovisual materials

The Communications Explosion, 16mm film, 25 min., sound, color, McGraw-Hill Films, New York, 1967.

Friesen, Paul, *Innovation and Change in Education* (audiotape cassette), Educational Technology Publications, Engelwood Cliffs, N.J., 1971.

Gattegno, Caleb, *Eye of the Future,* cassette tape, 57 min., Xerox University Microfilms, Ann Arbor, Mich., 1971.

Six Cents a Day, 35mm sound filmstrip, 10 min., with audio cassette, Association of American Publishers, New York, 1974 (six major trends).

Teacher and Technology, 16mm film, 49 min., sound, b and w, Ohio State University, Columbus, Ohio, 1967.

REFERENCE SECTION 1

operating audiovisual equipment

print

Educational Products Information Exchange Institute, selected reports, New York, published since 1967.

How to Select a Microform Reader or Reader-Printer, National Microfilm Association, Silver Spring, Md., 1974.

Locatis, Craig N., "Evaluating Instructional Hardware," *Audiovisual Instruction,* vol. 20, April 1975, pp. 12–14.

The Audio-Visual Equipment Directory, National Audio-Visual Association, Fairfax, Va. (new edition annually).

Wittich, Walter A., and C. F. Schuller, D. W. Hessler, and J. C. Smith, *Student Production Guide to Accompany Instructional Technology* (5th ed.), Harper & Row, New York, 1975.

audiovisual materials

Individualized AV Equipment Operation Service, choice of slide set or filmstrip for nineteen different types of equipment, Media Systems, Inc., Salt Lake City, Utah, 1974.

Using Technology: The Equipment, series of eight video tape cassettes, 20 min. each, Great Plains National Instructional Television Library, Lincoln, Nebr., 1974.

REFERENCE SECTION 2

duplicating processes

audiovisual materials

Copying and Duplicating, transparency series, 3M Company, Visual Products Division, 3M Center, St. Paul, Minn. n.d.

The Spirit Duplicator: Operation, Audio-Visual Production Techniques Series, 8mm film, color, 3 min., technicolor silent cartridge Super 8 or standard 8, McGraw-Hill Films, New York, 1969.

The Spirit Duplicator: Preparing Masters, Audio-Visual Production Techniques Series, 8mm film, color, 3 min., technicolor silent cartridge Super 8 or standard 8, McGraw-Hill Films, New York, 1969.

Reprographics, set of fifteen transparencies, Lansford Publishing Co., San Jose, Calif., 1975.

REFERENCE SECTION 3

photographic equipment and techniques

print (see also chapter 8)

Barry, Les, *Getting Started in Photography,* Amphoto, New York, 1970.

Bruce, Helen Finn, *Your Guide to Photography; A Practical Handbook,* Barnes and Noble, New York, 1971.

Index 1975 to Kodak Information, Eastman Kodak Co., Rochester, N.Y., 1975.

Photography: How It Works, Eastman Kodak Co., Rochester, New York., 1974.

audiovisual materials

Photographic Art Series, each 20 min., b and w, sound, Indiana University Audiovisual Center, Bloomington, Ind. Includes *Photography as an Art, Language of the Camera Eye,* and *Point of View: Professional Photographic Techniques.*

Photography: Black and White Outdoors, 16mm film, 18 min., reel, b and w, optical sound, Petite Film Company, Seattle, Wash., 1974.

REFERENCE SECTION 4

physical facilities

print

Carnahan, David J., "More Questions than Answers: Large Group Learning Spaces," *Audiovisual Instruction,* vol. 15, March 1970, pp. 97–98.

Facilities for Learning, Issues of *Audiovisual Instruction,* vol. 15, October 1970, and vol. 18, December 1973.

Educational Facilities Laboratories, New York, offers many publications, such as *Fewer Pupils/Surplus Space* (1975); *Found Spaces and Equipment for Children's Centers* (1972); *High School: Process and Place* (1972); *Places and Things for Experimental Schools* (1972); *Places for Environmental Education* (1971); *Schools/More Space: Less Money* (1971); McVey, Gerald F., *Sensory Factors in the School* (1974).

Learning Environment (What Research Says to the Teacher Series), Association of Classroom Teachers, National Education Association, Washington, D.C., 1971.

Zifferblatt, Steven M., "Architecture and Human Behavior: Toward Increased Understanding of a Functional Relationship," *Educational Technology,* vol. 12, August 1972, pp. 54–57.

audiovisual materials

Large Group Teaching Auditoriums, 35mm filmstrip, color, with reel or cassette tape, Educational Media, Inc., Washington, D.C., 1969.

Room to Learn, 16mm film, 22 min., color, sound, Association Films, New York, 1970. (Underwritten by Educational Facilities Laboratories.)

Space Is Not Enough: Planning Facilities for Media, filmstrip, 16 min., sound, color, National Audio-Visual Association, Fairfax, Va., 1969.

INDEX

INDEX